Chemistry Now!

Shea Mullally
B.Sc., Ph.D., C.Chem., MRSC

Gill & Macmillan

Gill & Macmillan Ltd

Goldenbridge

Dublin 8

with associated companies throughout the world

www.gillmacmillan.ie

© Shea Mullally 2000

0 7171 2237 9

Design: The Unlimited Design Company, Dublin

Illustrations: Mike Geraghty

Print origination in Ireland by:

The Unlimited Design Company

CONTENTS

PREFACE

Chemistry Now! is designed to reflect the aims of the new syllabus. It provides a relevant and modern approach not only for students who will end their study of chemistry at Leaving Certificate level but for almost all students of chemistry.

Considerable care has been taken in the layout of each page in order to stimulate and sustain the student's enjoyment of chemistry in all its aspects - scientific, social, environmental, economic, technological and historical. The text is written to help students develop skills in scientific method, laboratory procedures and techniques, analysis, problem solving and other important techniques.

Chemistry Now! is written specially for the new syllabus. It follows the order of the syllabus rigidly, with two exceptions. In Unit 3, the 'mole' concept precedes the gas laws because the equation of state for an ideal gas, $PV = nRT$, cannot be understood without a knowledge of the mole.

In Unit 9, 'hardness in water' and 'water treatment' precede the pH scale, which results in a renumbering of the mandatory experiments.

The following are features of the text.

- **Mandatory experiments:** All mandatory experiments are fully integrated with the text in a clear, concise format.
- **Suggested activities:** All suggested activities are included at the relevant point in each unit.
- **Key terms:** Each unit ends with a list of the main points. These provide the student with a quick check-list which helps revision.
- **Worked examples:** The text contains numerous worked examples, which enable the student to understand a particular calculation with minimum help from the teacher.
- **Boxes:** Key definitions and concepts are highlighted throughout the text.
- **Photographs:** The use of photographs enhances the learning process for many chemical processes and emphasises the social and applied aspects involved. The use of colour throughout the book brings the subject of chemistry to life.
- **Exercises:** The text contains hundreds of questions. Each section in each unit finishes with a selection of relevant questions.
- **Organic Chemistry:** Unit 5 and in particular Unit 7 are written to reflect the new style of teaching suggested for the syllabus.

ACKNOWLEDGMENTS

For permission to reproduce photographs, grateful
acknowledgment is made to the following:
Science Photo Library; National Portrait Gallery,
London; Zefa; Mansell Collection; Hulton Picture Library;
Bruce Coleman Ltd; ET Archive; BASF; Bayer;
Denis Baker; Popperfoto; National Gallery of Ireland;
Bettman Archive; National Medical Slide Bank;
Barnaby's Picture Library; NHPA; Image Select;
Bord Gais Eireann; National Galleries of Scotland;
Greenhill Photo Library; Bildarchiv Preussicher
Kulturbesitz; Exxon; BIC; M. C. O'Sullivan & Co.;
Dublin Corporation; Steve Treacy; Premier Periclase;
IFI; Aughinish Alumina.

MANDATORY EXPERIMENTS

H = Honours level only

vii

THE PERIODIC TABLE AND ATOMIC STRUCTURE

1.1 The Periodic Table

1.2 Atomic Structure

1.3 Radioactivity

1.4 Electronic Structure of Atoms

1.5 Oxidation and Reduction

Similar foodstuffs are arranged together on shelves in a supermarket.

1.1 THE PERIODIC TABLE

What is an Element?

Aluminium, carbon, wood and plastic are common materials. The first two, aluminium and carbon, are elements, while the other two are chemical compounds. **Elements are substances that cannot be broken up into any simpler substances**. No matter how you try to break up aluminium, you still end up with the element aluminium.

There are ninety-two elements which occur in nature. Scientists have made another twelve artificially. Most students have seen the periodic table of the elements hanging in a science laboratory. The elements in the periodic table are arranged in an orderly manner, like the goods in a supermarket, which are arranged neatly on shelves. In a supermarket, fruits are grouped together, dairy products are grouped together and other similar products are

grouped with each other. In the periodic table, elements with similar chemical properties are arranged together in groups, like the goods in a supermarket.

• Vertical columns are called **groups**. Elements in the same group have similar properties.

• Horizontal rows are called **periods**. On the left-hand side are reactive metals. On the right-hand side are non-metals.

• The top number of each element is the **atomic number**; the bottom number is the **relative atomic mass** (mass number).

THE MODERN PERIODIC TABLE

Group	I Alkali metals	II Alkaline earth-metals											III	IV	V	VI	VII Halogens	0 Noble Gases
1																		2 He Helium 4.0
2	3 Li Lithium 6.9	4 Be Beryllium 9.0											5 B Boron 10.8	6 C Carbon 12.0	7 N Nitrogen 14.0	8 O Oxygen 16.0	9 F Fluorine 19.0	10 Ne Neon 20.2
3	11 Na Sodium 23.0	12 Mg Magnesium 24.3											13 Al Aluminium 27.0	14 Si Silicon 28.1	15 P Phosphorus 31.0	16 S Sulphur 32.1	17 Cl Chlorine 35.5	18 Ar Argon 39.9
4	19 K Potassium 39.1	20 Ca Calcium 40.1	21 Sc Scandium 45.0	22 Ti Titanium 47.9	23 V Vanadium 50.9	24 Cr Chromium 52.0	25 Mn Manganese 54.9	26 Fe Iron 55.9	27 Co Cobalt 58.9	28 Ni Nickel 58.7	29 Cu Copper 63.5	30 Zn Zinc 65.4	31 Ga Gallium 69.7	32 Ge Germanium 72.6	33 As Arsenic 74.9	34 Se Selenium 79.0	35 Br Bromine 79.9	36 Kr Krypton 83.8
5	37 Rb Rubidium 85.5	38 Sr Strontium 87.6	39 Y Yttrium 88.9	40 Zr Zirconium 91.2	41 Nb Niobium 92.9	42 Mo Molybdenum 95.9	43 Tc Technetium 99	44 Ru Ruthenium 101.1	45 Rh Rhodium 102.9	46 Pd Palladium 106.4	47 Ag Silver 107.9	48 Cd Cadmium 112.4	49 In Indium 114.8	50 Sn Tin 118.7	51 Sb Antimony 121.8	52 Te Tellurium 127.6	53 I Iodine 126.9	54 Xe Xenon 131.3
6	55 Cs Caesium 132.9	56 Ba Barium 137.3	57-71	72 Hf Hafnium 178.5	73 Ta Tantalium 181.0	74 W Tungsten 183.9	75 Re Rhenium 186.2	76 Os Osmium 190.2	77 Ir Iridium 192.2	78 Pt Platinum 195.1	79 Au Gold 197.0	80 Hg Mercury 200.6	81 Tl Thallium 204.4	82 Pb Lead 207.2	83 Bi Bismuth 209.0	84 Po Polonium 210	85 At Astatine 210	86 Rn Radon 222
7	87 Fr Francium (223)	88 Ra Radium (226)	89-103	104 Unq Unnil-quadium (262)	105 Unp Unnil-pentium (262)	106 Unh Unnil-hexium (263)												

Transition elements (columns Sc through Zn and corresponding)

Example box:
1
H
Hydrogen
1.0
Atomic Number
Mass Number

False–colour scanning tunnelling microscope (STM) image of a surface. The technique allows superficial atoms to be identified. The STM has been used to follow the path of the AIDS virus in blood.

Early Ideas of Elements

Between 400 and 450 B.C. the Greek philosophers **Leucippus** and **Democritus** proposed that all matter consisted of very small particles called atoms (Greek 'atomos', meaning indivisible). However, there was no evidence at the time to support this theory and the Ancient Greeks believed that all matter resulted from the combination of four elements: fire, earth, water and air. The idea that atoms existed gained popularity mainly due to the genius of scientists such as **Robert Boyle** and **Humphrey Davy**. Boyle suggested that acids contained particles which were responsible for the action of acids. Davy used electricity to discover many elements.

History of the Periodic Table

In the early nineteenth century, chemists became interested in the physical and chemical similarities which existed between the elements. In 1817, **Johann Dobereiner** noticed that three elements, calcium, strontium and barium, had not only similar chemical behaviour but that the pattern of their atomic weights (relative atomic masses) had a simple relationship. The atomic weight of strontium (88) is midway between the atomic weights of calcium (40) and barium (137). He also noticed similar behaviour between the three elements lithium, sodium and potassium. He called these sets of elements 'triads'.

In the years 1863–66, **John Newlands** developed his 'Law of octaves'. Newlands listed the elements in order of their increasing atomic weights and noticed that every eighth element had similar chemical properties. Initially his theory was ridiculed because it worked only for the first sixteen elements. However, his classification of the elements was important in the overall development of the periodic table because first, he assigned a number to each

element and second, he arranged the elements in a way which showed that they had periodic properties. In 1869, **Lothar Meyer** proposed a periodic classification of the elements based on the arrangements of the atomic volumes of the elements.

The modern classification of the periodic table is based on **Dmitri Mendeleev's** work. Mendeleev constructed a periodic table on the basis of atomic weights (relative atomic masses) and grouped together elements which had similar behaviour. He not only left gaps in the table for elements which had not yet been discovered, but he also predicted the properties of the 'unknown' elements with remarkable clarity. The table shows his predictions for germanium.

Property	Predicted by Mendeleev	Actual
Appearance	Light grey metal	Dark grey metal
Oxide type	GeO_2 high melting point	GeO_2 high melting point
Oxide density	4.7 g cm^{-3}	4.70 g cm^{-3}
B.p. of chloride	Less than 100°C	86°C
Chloride density	1.9 g cm^{-3}	1.887 g cm^{-3}

Mendeleev's periodic table had some anomalies. Tellurium (atomic weight 127.6) seemed to be incorrectly placed before iodine (atomic weight 126.9). He assumed that when the atomic weights (relative atomic masses) were more accurately determined that the atomic weight of tellurium would be less than that of iodine. However, this was not the case.

In 1913, **Henry Mosely** solved the problem when he discovered that the positive charge on the nucleus of an element is a definite amount. This positive charge on the nucleus is called the atomic number of the element.

Today, the periodic table is arranged according to the atomic numbers (Z) of the elements; tellurium ($Z = 52$) and iodine ($Z = 53$) are in the order given by Mendeleev. As we have seen, the periodic table is made up of vertical columns called groups and of horizontal rows called periods.

The first group or Group I contains the elements lithium, sodium, potassium, rubidium,

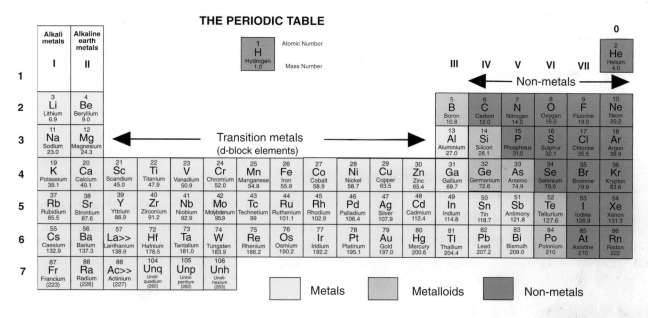

THE PERIODIC TABLE

caesium and francium while the first period contains only two elements, hydrogen and helium. There are eight main groups, which are usually designated using the Roman numerals I, II, III, IV, V, VI, VII, while the remaining group, the noble gases is indicated by a 0. The elements between groups II and III are called the d-block elements. There are sixteen non-metals which are positioned in the right-hand side of the periodic table. There are some elements called metalloids between the metals and non-metals: these have some properties like metals and some like non-metals.

Activity 1.1

Arranging Elements in Order of Atomic Mass

The modern periodic table is arranged in order of the atomic numbers of the elements. Make a periodic table of the first 54 elements arranging the elements in order of their relative atomic masses.
What do you notice?

The Periodic Law and Periodic Properties

The chemical and physical properties of the elements are periodic functions of the atomic number.

In general, the elements in the same group of the periodic table will have similar chemical properties. The similarities are not as strong in some groups as in others. The alkali metals (group I) show marked similarities to one another, as do the halogens (group VII). On the other hand, the elements in group IV range from the non-metal, carbon, to the metal, lead, and have correspondingly different chemical properties. The electronic structure of the elements and the periodic table help us to understand the systematic changes in properties of the elements.

Group I elements, the alkali metals

The elements of group I are called the alkali metals because they all react with water to form alkaline solutions. They all have one electron in their outer shell. They have some unusual properties for metals: they are soft, float on water, have low melting points and must be stored in oil because they are so reactive with water or with oxygen. Some of the properties of the best known alkali metals, lithium, sodium and potassium are given in the table below.

Element	Relative atomic mass	Density /g cm^{-3}	Reaction with water
Lithium	6.9	0.53	Reacts steadily
Sodium	23.0	0.97	Reacts vigorously
Potassium	39.1	0.86	Reacts violently

3

Although the properties of the alkali metals are similar, there is a gradual change in properties as we go down the group. This illustrates an important feature of the periodic table: similar properties within each group.

Activity 1.2

Demonstration of the Reaction of Some Alkali Metals with Water

Introduction

Lithium, sodium and potassium all react with cold water. They increase in reactivity as we go down the group. Lithium reacts steadily with water forming lithium hydroxide and releasing hydrogen gas. The others react more vigorously forming alkaline solutions and releasing hydrogen gas. For safety, your teacher will carry out the procedure described below. You should observe the reactions.

$$2Li + 2H_2O \rightarrow 2LiOH + H_2$$

$$2Na + 2H_2O \rightarrow 2NaOH + H_2$$

$$2K + 2H_2O \rightarrow 2KOH + H_2$$

Requirements

Safety glasses, safety screen, plastic basin, water, lithium, sodium and potassium, scalpel or knife, filter paper, litmus paper.

Procedure

(1) Half fill the plastic basin with water.
(2) Cut a very small piece of lithium with a scalpel. Dry off the paraffin oil it has been stored in with filter paper.
(3) Carefully, place the lithium in the water and observe what happens.
(4) Test the white trail of alkali formed with litmus paper.
(5) Repeat the procedure for sodium and potassium. Great care must be taken with potassium as it reacts violently with water.

Results

Copy and complete the data table below.

Element	Reaction with water	Effect of litmus on solution	Flame colour observed
Lithium			
Sodium			
Potassium			

Group II elements, the alkaline earth metals

The alkaline earth metals each have two electrons in their outer shell. They are also reactive, but not as reactive as the alkali metals. The reactivity increases going down the group: beryllium shows no reaction in water, magnesium reacts slowly, while calcium and the other alkaline earth metals react quite rapidly with water.

Group VII elements, the halogens

The halogens all require one electron in order to attain the electronic configuration of the noble gases. They are reactive non-metals. The smallest, fluorine, is a pale yellow gas, chlorine is a green gas, bromine is a red-brown liquid, while the largest, astatine, is a synthetic element and is radioactive.

Group 0 elements, the noble gases

The noble gases all have complete shells; consequently, they are relatively unreactive. They exist mainly as gases of uncombined atoms. Until the 1960s no compounds of the noble gases were known. Today many compounds of xenon, krypton and radon are known.

Exercise 1.1

•

(1) Draw a simple diagram of the periodic table. Indicate on the diagram:
(a) group I (b) group VII (c) group VIII.

(2) Where would you find the following on the periodic table:
(a) the alkali metals (b) the halogens (c) the noble gases?

(3) Using simple sketches break the periodic table up into the following:
(a) groups and periods (b) metals, non-metals and metalloids.

(4) The elements in the same group of the periodic table have similar chemical properties. Discuss this in relation to each of the following:
(a) the alkali metals (b) the halogens.

(5) Describe, briefly, some characteristic properties of the elements in each of the following:
(a) group II (b) group VIII.

(6) Write down three ways in which lithium, sodium and potassium are similar to each other.

(7) (a) Describe, using equations, how the alkali metals react with water.
(b) In relation to the size of the alkali metal atoms, how would you describe their chemical reactivity?
(c) What safety precautions should you adopt during an experiment to demonstrate the reaction of the alkali metals with water?

(8) Where in the periodic table would you find the following:
(a) metals (b) non-metals (c) metalloids?

(9) Why did Newlands use the word 'octave', when he arranged the elements in order of their atomic mass?

(10) Explain each of the following:
(a) periodic properties, groups, periods
(b) Newlands' Octaves and Dobereiner's Triads.

(11) Discuss, briefly, the role played by each of the following in the development of the periodic table:
Johann Dobereiner, John Newlands, Humphrey Davy, Henry Mosely and Dmitri Mendeleev.

(12) Mendeleev placed tellurium before iodine in the periodic table. Initially, it seemed as if these elements were incorrectly positioned. Discuss how this apparent 'mistake' was explained.

(13) (a) Outline some of Mendeleev's predictions for germanium. How do his predictions compare with the observed properties of germanium?
(b) How does the modern periodic table compare with Mendeleev's version?

(14) Why was Mendeleev's periodic table more successful than Newlands' periodic table?

(15) Identify the group and the period to which each of the following atoms belongs in the periodic table. In each case, state whether the particular atom is a metal, a non-metal or a metalloid:
C, N, Cr, Br, Li, Mg.

(16) Describe, briefly, how the elements within each of the following groups are similar to each other:

(a) alkali metals (b) alkaline earth metals (c) halogens (d) noble gases.

(17) (a) Which elements are the noble gases?
(b) Why were they once called the inert gases and are now called the noble gases?

1.2 ATOMIC STRUCTURE

Composition of Matter

Matter is the substance from which all things are made. Billions of different substances are known, all of which are made up of solids, liquids, gases or liquid crystals. Substances in these states of matter are all composed of very small particles, which may be atoms, molecules or ions.

Dalton's Ideas on the Structure of Matter

In 1803, **John Dalton**, an English teacher and chemist who was interested in the science of weather, speculated on the composition of gases in the atmosphere. From his experiments, he suggested the following:
1 All matter is composed of extremely small indivisible particles, called atoms.
2 An element is a form of matter composed of only one kind of atom; each atom of the same kind has the same properties. For instance, atoms of a given element have the same mass.
3 A compound is a form of matter consisting of atoms of two or more elements chemically combined in fixed proportions.
4 Atoms of two or more elements can combine in a chemical reaction and rearrange themselves to form new chemical combinations.

Today we know that atoms are divisible, and are made up of particles. Nevertheless, Dalton's ideas are essentially correct.

Law of Conservation of Mass
Matter is neither created nor destroyed in the course of a chemical reaction.

Minute Size of Atoms

Today, even the most powerful electron microscope cannot give us a clear view of a single atom. However, it is possible to look at groups of atoms using field ion micrographs. Atoms are so small that approximately 50 million of them would fit side by side through a 1 cm gap.

1 cm

The Electron

Towards the end of the nineteenth century, scientists began to propose that the atom, which was thought to be the smallest particle of matter consisted of even smaller particles.

In 1808, **Humphrey Davy** discovered several elements by decomposing compounds using electricity. This prompted him to propose that elements are held together in compounds by electrical attraction.

Michael Faraday reported that when an electrical current is passed through a molten salt or a salt solution, the salt is decomposed into elementary substances. Faraday's laws of chemical electrolysis prompted **George Johnston Stoney** from University College Galway to suggest that units of electrical charge are associated with atoms and he called the units electrons.

Cathode rays were discovered in 1859 by **Julius Plucker**. Two electrodes are sealed in a glass tube from which the air is nearly evacuated. When a high voltage is imposed across the electrodes, electrical rays called cathode rays emanate from the negative electrode, the cathode, and move towards the other electrode, the anode.

Several others, notably **William Crookes**, performed many clever experiments using cathode rays. By applying electric and magnetic forces to the cathode ray beams it was possible to learn about the masses and charges of the beams. In this way **Joseph Thomson** discovered the electron in 1897. He was able to show that the cathode rays consisted of negative particles called electrons.

Michael Faraday 1791-1867
The English chemist and physicist Michael Faraday was born in Surrey, the son of a poor blacksmith. He was largely self-taught, never thought too highly of himself but was essentially a very happy man. He is best known today for his research into electricity and magnetism. His reputation, in his own time, was based on his remarkable lecturing skills; the public at the time looked on electricity as a useful toy and was more interested in the more useful inventions of Stephenson and Joule. He was apprenticed to a bookbinder in 1805. Having attended a lecture by Humphrey Davy, he presented Davy with a bound version of the lecture. Davy was so impressed with the detailed notes and drawings that he offered him a job as a laboratory assistant. From there on the relationship between the greatest scientists of their day flourished. He also collaborated with the mathematician, James Clerk Maxwell who expressed many of Faraday's ideas in mathematical form.

Sir Joseph John Thomson 1856–1940
J. J. Thomson was born in Manchester in 1856. His father was a bookseller and publisher. Thomson was Cavendish Professor of experimental physics, Cambridge University from 1894–1919. He was described as humble, devout, generous, a good conversationalist and had an uncanny memory. He valued and inspired enthusiasm in his students. Thomson was awarded the Nobel Prize for physics for his investigations of the passage of electricity through gases. In 1897, he discovered the electron through his work on cathode rays. Thomson's son, Sir George Paget, shared the Nobel Prize for physics with C. J. Davisson in 1937. Seven of Thomson's trainees were also awarded Nobel Prizes. J. J. Thomson is buried in Westminster Abbey close to some of the World's greatest scientists, Newton, Kelvin, Darwin, Hercshel and Rutherford.

Thomson, in further work, was able to show that no matter what material the cathode was made of, the value of the ratio of charge to mass was always the same. The value is

$$e/m = -1.7588 \text{ C g}^{-1}$$

where the Coulomb (C) is the unit of electrical charge.

Thomson's Cathode Ray Tube

The negatively charged cathode emits electrons, which are attracted towards the positively charged anode. A small hole bored in the anode allows some electrons to pass through and these are shown on a zinc sulphide screen. Zinc sulphide has the ability to convert the kinetic energy of the moving electrons into visible light. In this way the electrons are 'seen'. The beam of electrons can be bent or deflected by either moving the electrical plates across

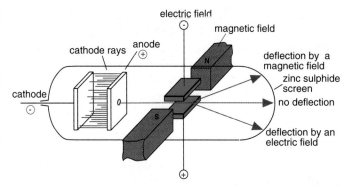

the tube or by a magnet. By measuring the speed of the beam and the amount of deflection Thomson was able to measure the ratio e/m.

Later experiments by **Robert Milliken** in 1909 determined the precise charge of the electron,

$$e = -1.6 \times 10^{-19} \text{ C.}$$

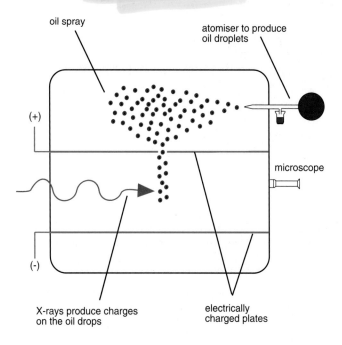

The charged oil drops fall due to gravity. The fall is stopped by adjusting the voltage across the two plates. The voltage used and the mass of an oil drop were used to calculate the charge of the electron.

He then calculated the mass of the electron from the ratio e/m as

$$m = 9.11 \times 10^{-31} \text{ kg.}$$

The Nuclear Atom

Thomson's discovery of the electron posed problems concerning the other part or parts of the atom. If the atom was electrically neutral, then the negative charge or electrons must be balanced by an equal amount of positive charge. These positive particles, now called protons,

are assumed to be a component of all atoms. Thomson pictured the atom as a sphere which is positively charged and has the negatively charged electrons imbedded in it rather like 'pieces of fruit in a plum pudding'.

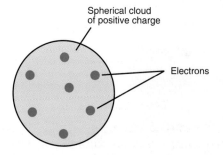

The discovery of radioactivity by **Henri Becquerel** in 1896 showed that atoms spontaneously emit rays. **Ernest Rutherford** explained the nature of these radioactive rays. In 1911, he reported the results of a series of experiments using α-particles performed by himself and his fellow workers, Geiger and Marsden.

A beam of α-particles was directed against a very thin sheet of gold. The majority of the α-particles passed through the gold foil but some were deflected from their straight line path while others bounced back from the foil. Rutherford explained this by proposing that the centre of the atom consisted of a small heavy, highly positively charged body, called the nucleus of the atom, with the electrons spinning around it very much like the planets moving around the sun. The nucleus was very small compared to the atom, the atom being mainly empty space. This explained why the majority of the α-particles passed through the gold foil.

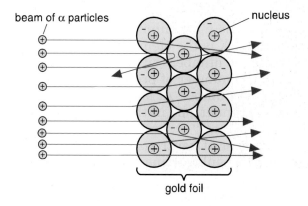

In 1920, Rutherford suggested the name proton for the positive particle in the nucleus and predicted the existence of another particle in the nucleus which, unlike the electron or the proton, would be electrically neutral. The

Lord Ernest Rutherford 1871–1937

Ernest Rutherford was born in Nelson, New Zealand in 1871. He began work in J. J. Thomson's laboratory in 1895. He later moved to McGill University in Montreal where he became one of the leading figures in the field of radioactivity. From 1907 on he was professor at the University of Manchester where he worked with Geiger and Marsden. He was awarded the Nobel Prize for Chemistry in 1908 for his work on radioactivity. In 1910, with his co-workers Geiger and Marsden, he discovered that α–particles could be deflected by thin metal foil. This work enabled him to propose a structure for the atom. Later on he proposed the existence of the proton and predicted the existence of the neutron. He died in 1937 and like J. J. Thomson is buried in Westminster Abbey. He was one of the most distinguished scientists of this century.

unknown particle would explain why the mass of the helium atom is four and not two.

In 1932, **James Chadwick**, one of Rutherford's colleagues discovered that particle which was called the neutron. Further experiments showed that a neutron had the same mass as a proton.

The relative masses and charges of the three fundamental sub-atomic particles are shown at the bottom of the page.

Atomic Numbers, Mass Numbers and Isotopes

An atom is identified by two numbers, the atomic number and the mass number:

$$^{A}_{Z}X$$

where A is the mass number, Z is the atomic number and X is the element.
• The atomic number (or proton number), Z, is equal to the number of protons in the nucleus.
• The mass number, A, is the total number of protons and neutrons in the nucleus of the atom.
• The number of neutrons in the atom can be calculated by subtracting the atomic number from the mass number $(A - Z)$.

$^{39}_{19}K$ tells us that potassium has a mass number, $A = 39$ and an atomic number, $Z = 19$ and has 19 electrons, 19 protons and 20 neutrons $(39 - 19)$.

Isotopes

Dalton's atomic theory stated that all atoms of an element are the same. This statement is incorrect.

The identity of a particular atom is decided by the number of protons in the nucleus. The number of neutrons in the nucleus may vary in an atom, resulting in different atoms of the same element.

For instance, an atom of carbon contains 6 protons in the nucleus. However, it may contain 6, 7, or even 8 neutrons: this results in different atomic masses of the same element, i.e. carbon can be represented as $^{12}_{6}C$ as $^{13}_{6}C$ or as $^{14}_{6}C$.
The different types of carbon are called isotopes of carbon.

The isotopes of hydrogen can be represented as $^{1}_{1}H$, as $^{2}_{1}H$ or as $^{3}_{1}H$.

Isotopes are atoms of an element having the same atomic number but different mass numbers.

Sodium and aluminium each have only one naturally occurring isotope, but most elements are mixtures of isotopes. Naturally occurring oxygen, for example, is a mixture containing 99.759% $^{16}_{8}O$, 0.037% $^{17}_{8}O$ and 0.204% $^{18}_{8}O$.

Most elements have naturally occurring isotopes, while some elements have isotopes formed as a result of radioactivity.

Determination of Relative Atomic Masses

Atomic masses are very small numbers. The mass of the hydrogen atom is only 1.67×10^{-33} kg. It is convenient to compare the atomic masses of elements with a standard: the standard used is the mass of the carbon-12 isotope. The atomic masses of all other atoms are related to the mass of carbon-12.

Particle	Mass (kg)	Charge (C)	Mass (amu)	Charge (e)
Electron	9.10953×10^{-31}	-1.60219×10^{19}	0.00055	-1
Proton	1.67265×10^{-27}	$+1.60219 \times 10^{-19}$	1.00728	$+1$
Neutron	1.67495×10^{-27}	0	1.00866	0

Relative atomic mass, A_r, is defined as the ratio of the average mass per atom of the natural composition of an element to $\frac{1}{12}$ of the mass of an atom of carbon-12.

For instance, the relative atomic mass (also known as the atomic weight) of chlorine, $A_r(Cl)$ is 35.453.

It is possible to measure the charge-to-mass ratio (e/m) for positive ions in the same way that Thomson studied cathode rays. Experiments of this type can be used to measure the actual masses of atoms (or molecules).

The Mass Spectrometer

A mass spectrometer is an instrument which separates ions by mass-to-charge ratio.

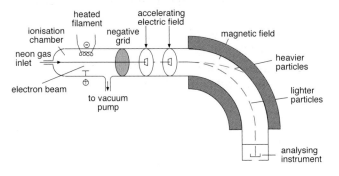

The processes that occur in a mass spectrometer are:
- **Vaporisation**: positive ions of the element under investigation are generated by bombarding the gas, or vapour of the substance, with electrons.

- **Generation of ions**:

$$X \rightarrow X^+ + e^-$$

- **Acceleration**: the stream of positive ions from the ionisation chamber is then accelerated by a negative grid through slits to form a positive beam.
- **Separation and detection**: the beam is then deflected, first by an electrical field and then by a magnetic field and focused on to an analyser. Just as a heavy lorry is buffeted about less than a light car in a strong wind, the larger ions are deflected less than the smaller ions by the magnetic field. Therefore, the different ions focus on the analyser in different positions according to their size. The analyser then gives out a different electrical impulse depending on the number of atoms with a particular mass.

For a gas such as neon three isotopes are separated, Ne-20, Ne-21 and Ne-22. The mass spectrum of neon is shown below.

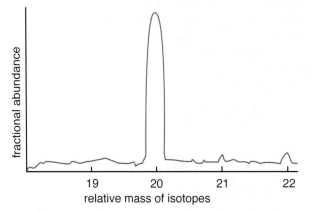

Example 1.1

A sample of chromium, Cr, was analysed using a mass spectrometer. Calculate the approximate relative atomic mass, A_r, of chromium if the sample contained the following: 4.35% Cr-50, 83.80% Cr-52, 9.50% Cr-53 and 2.36% Cr-54.

Solution:
(1) Convert the % abundances into fractional abundances.
(2) Multiply each of the fractional abundances by the isotope mass.
(3) Add up to give the relative atomic mass.
Approximate relative atomic mass, $A_r(Cr)$ is 52.060. (The more correct value of $A_r(Cr)$ is 51.996.)

% abundance	Fractional abundance	Isotope mass	Product
4.35	0.0435	50	$50 \times 0.0435 = 2.175$
83.80	0.8380	52	$52 \times 0.8380 = 43.576$
9.50	0.0950	53	$53 \times 0.0950 = 5.035$
2.36	0.0236	54	$54 \times 0.0236 = 1.274$
			Total = 52.060

Exercise 1.2

●

(1) How would you physically describe the size of the atom?

(2) In 1903, John Dalton proposed some ideas about the nature of matter. Outline each of his ideas.

(3) (a) What is the Law of Conservation of Mass?
(b) Discuss the Law of Conservation of Mass in relation to Dalton's ideas on atomic theory.

(4) (a) What are cathode rays?
(b) Explain how a cathode ray tube can be used to 'see' electrons.

(5) What early evidence showed that electrons are:
(a) negatively charged (b) the same in all substances?

(6) Discuss the role played by each of the scientists named below in the discovery of the electron.
Humphrey Davy, Michael Faraday, Julius Plucker, William Crookes, Joseph Thomson, Robert Milliken, George Johnstone Stoney.

(7) The discovery of radioactivity by Henri Becquerel in 1896 led to the discovery of the proton by Ernest Rutherford in 1920.
(a) Describe Rutherford's experiments using α-particles on gold foil.
(b) What did Rutherford propose as a result of these experiments?
(c) How did Rutherford's model of the atom differ from that of Thomson?

(8) Explain the observations below in relation to the experiments performed by Geiger and Marsden using α-particles directed at a thin gold foil.
(a) Some α-particles passed through undeflected.
(b) Some α-particles rebounded.
(c) Some α-particles were deflected.

(9) Describe, briefly, the properties of the electron, the proton and the neutron. In your answer, indicate:
(a) the relative mass (b) the relative charge
(c) the location within the atom of each species.

(10) (a) Explain the meaning of the following terms: atomic number, mass number and isotope.
(b) What are the atomic number and the mass number of each of the following:

$$^{41}_{19}\text{K} \quad ^{31}_{15}\text{P} \quad ^{26}_{12}\text{Mg} \quad ^{27}_{13}\text{Al} \quad ^{56}_{26}\text{Fe} \quad ^{32}_{16}\text{S}$$

(11) Using a periodic table, determine the number of electrons, protons and neutrons in each of the following atoms:
Ca, Na, F, O, Br, Sc, P, Hg, Cu, H, Li, Se.

(12) Draw simple diagrams illustrating the main isotopes of (a) hydrogen (b) carbon.

(13) Silicon has three main isotopes containing 14, 15 and 16 neutrons, respectively. What are the mass numbers of the isotopes?

(14) (a) Define 'relative atomic mass'.
(b) Describe an experiment which is used to determine the relative atomic masses of elements.
(c) Calculate the relative atomic mass of an element which consists of 60.4% of atoms with a mass of 69 and 39.6% of atoms with a mass of 71. What is that element?

(15) Calculate the relative atomic masses (r.a.m.) of the following elements given the mass number and the % abundance.

Element	Mass number	% abundance
O	16	99.759
	17	0.037
	18	0.204
Si	28	92.20
	29	4.70
	30	3.10
Mg	24	78.70
	25	10.13
	26	11.17

(16) Calculate the relative atomic masses (r.a.m.) of the following elements given the mass number and the fractional abundance.

Element	Mass number	Fractional abundance
Ir	191	0.373
	193	0.627
Pb	204	0.15
	206	0.236
	207	0.226
	208	0.523

(17) What are the fundamental processes which take place in a mass spectrometer?

(18) Describe how a mass spectrometer is used to determine relative atomic mass.

(19) Chlorine has two isotopes, Cl-35 and Cl-37. Explain why the relative atomic mass is not the average of the two masses.

(20) Hydrogen has three isotopes, hydrogen, deuterium and tritium. Explain how this arises.

1.3 RADIOACTIVITY

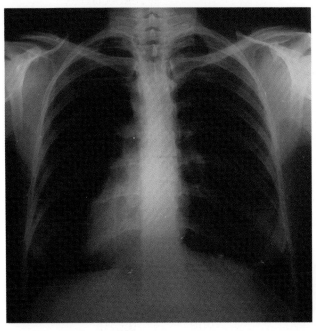

Chest X–ray of a young woman.

Students of Chemistry tend to think of the atomic nucleus as that part of the atom which has no real influence in chemical reactions: the simplest view taken is that its positive charge is there merely to bind the electrons in atoms and molecules. However, since the advent of the atomic bomb, the nucleus and its properties have had an large impact on our lives. Today, we are familiar with nuclear power, nuclear medicine and radiocarbon dating.

Nuclear chemistry is the study of atomic nuclei, the changes in structure of the nuclei and the consequences of those changes. Nuclei which change their structure spontaneously are said to be radioactive. A radioactive nucleus decomposes forming another nucleus and producing one or more particles.

Nucleus → New nucleus + particles

Radioactivity is the break up of the nucleus by the emission of particles and radiant energy.

Antoine Henri Becquerel 1852–1908
In 1896, Becquerel discovered the phenomenon of radioactivity. In 1899 he noted that radiation coming from a uranium salt could be deflected by a magnetic field. He concluded that this radiation was partly composed of electrons.

Radioactivity was first discovered in 1896 by the French scientist, **Henri Becquerel**. He noticed that a photographic plate which was wrapped in black paper became fogged when it was left near a uranium salt. Two of Becquerel's colleagues, **Pierre and Marie Curie** examined this phenomenon in more detail and found that all uranium salts showed radioactivity. They also noticed that the radioactivity caused a discharge on a gold leaf electroscope. They concluded that the radioactivity must carry an electrical charge. The Curies noticed that impure uranium sulphide (pitchblende) was far more radioactive than its uranium content indicated. In 1898, Marie Curie isolated two other radioactive elements from pitchblende. She named the elements polonium (after her native Poland) and radium (after the word radioactivity).

Pierre and Marie Curie, taken from a newspaper in 1903.

The discovery that radioactivity carried an electrical charge prompted **Ernest Rutherford** to investigate the effects of electric fields on the paths of nuclear radiation. He noticed that the rays took three different paths and called the three types alpha (α), beta (β) and gamma (γ) rays.

Types of Radioactive Decay

Alpha (α) Particles

Rutherford found that alpha rays were repelled from the positively charged electrode and concluded that alpha rays were positively charged particles. He later identified them as the nuclei of helium atoms, $^4He^{2+}$. We can think of an alpha particle as a tightly bound cluster of two protons and two neutrons. Alpha particles are formed when heavy radioactive isotopes decay. For example:

$$^{226}_{88}Ra \rightarrow ^{222}_{86}Rn + \alpha$$

$$^{241}_{95}Am \rightarrow ^{237}_{93}Np + \alpha$$

Alpha particles are not very penetrating but are damaging rays. If they disturb the DNA in cells they can cause cancer.

Beta (β) Particles

Rutherford also discovered that beta rays were attracted to a positively charged electrode which suggested that they were negative particles. Beta particles are electrons ejected from the nucleus of an atom (they are not electrons from the outer electron cloud). For example, carbon-14 produces a beta particle:

$$^{14}_{6}C \rightarrow ^{14}_{7}N + \beta$$

Beta particles are moderately penetrating and can penetrate up to 1 cm of flesh.

Gamma (γ) Rays

Gamma rays are neutral rays and are unaffected by electrical fields. They are similar to photons of light but have a higher frequency. They can be thought of as a stream of high energy photons. For example, when cobalt is excited to a higher energy state it emits gamma rays:

$$^{60}_{*}Co \rightarrow ^{60}Co + \gamma$$

Gamma particles are very penetrating and can ionise molecules causing radiation sickness which may result in cancer.

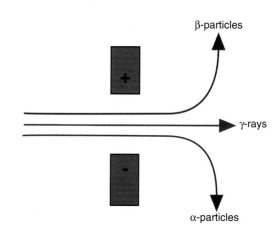

Property	Alpha particles	Beta particles	Gamma rays
Nature	Helium nuclei	Electrons from the nucleus	High frequency electromagnetic radiation
Mass	4	Negligible	None
Effect of electric or magnetic fields	Deflected	Deflected	No deflection
Electric charge	+2	−1	None
Penetration ratio	1	100	10 000
Absorption	By paper	By aluminium sheet	By lead (a few cm)

Activity 1.3

Use of Video to Demonstrate the Properties of Radioactivity

Watch a suitable video on radioactivity. Radioactivity is usually detected using a Geiger-Müller tube, or GM tube. It contains argon gas, which is ionised by the particles and rays emitted by the radioactive substance.

$$Ar\ (g) \rightarrow Ar^+\ (g) + e^-$$

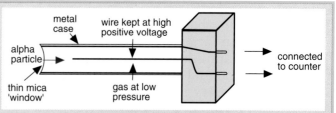

This produces a sudden flow of electrons to a high voltage wire, releasing a short pulse of current which is amplified and counted by an electronic pulse counter. The number of counts produced in a given time indicates the penetrating power of the radiation.

Chemical Reactions and Nuclear Reactions

Chemical reactions occur in the outer parts of atoms: they involve the transfer or sharing of the outer electrons.

In **ionic reactions**, electrons are transferred:

Na (2,8,1) + Cl (2,8,7) → [Na⁺ (2,8)] [Cl⁻(2,8,8)]

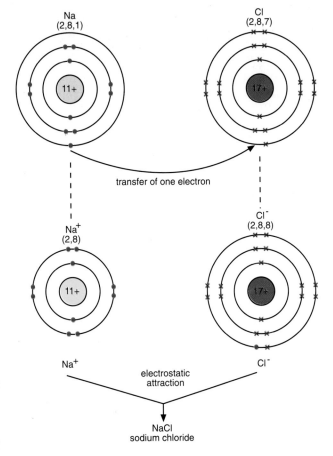

In **covalent reactions**, electrons are shared:

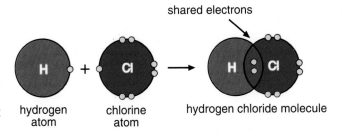

H (1) + Cl (2,8,7) → HCl

Nuclear reactions involve changes in the central part of the atom, the **nucleus**. During a nuclear reactions one element is converted into another by means of the following:

- **Radioactive decay:**

$$^{226}_{88}\text{Ra} \rightarrow {}^{222}_{86}\text{Rn} + \alpha$$

$$^{241}_{95}\text{Am} \rightarrow {}^{237}_{93}\text{Np} + \alpha$$

- **Nuclear fission:** large nuclei break up to form two smaller nuclei of roughly the same size together with the release of large quantities of energy.

- **Nuclear fusion:** light nuclei join together to make heavier nuclei. Energy is released, due to an overall loss of mass.

Law of Conservation of Mass–Energy ($E = mc^2$)

Nuclear reactions confirm the law of conservation of mass–energy. In the nuclear reaction below, carbon-14 decays by losing a beta particle to form nitrogen-14.

$$^{14}_{6}\text{C} \rightarrow {}^{14}_{7}\text{N} + {}^{0}_{-1}e$$

Mass before = 14,
Mass after = 14 + 0 = 14

Charge before = 6,
Charge after = 7 +(−1) = 6

In the reaction, the total mass and the total charge are the same after the beta decay as they were before. Thus, the law of conservation of mass–energy has been confirmed.

Half Life

The rate of decay of a radioactive isotope is shown by its half life. **The half life of a radioactive sample is defined as the time required for half of the atoms of the isotope to decay.** Half lives can vary from fractions of a second to millions of years: the shorter the half life, the more active the isotope. For example, the radioactive isotope iodine-131 has a half life of 8 days. It is used in medicine to measure how much iodine is taken up by the thyroid gland. Uranium-238 is a very stable isotope as it has a half life of 4.5 billion years, while polonium-218 is very unstable as it has a half life of 0.16 milliseconds.

Barium is used in radiography of the oesophagus, stomach and small intestine.

Uses and Social Aspects of Nuclear Radiation

• **Widespread occurrence:** nuclear radiation is all around us in the rocks, in the soil, in animals and in plants. Some rays come from outer space, some from testing atomic weapons and a very small amount from the operation of nuclear power stations. The major source of background radiation comes from a natural build-up of radon and thoron inside our homes. The level of background radiation is too low to have a serious effect on our health but does account for some deaths from cancer.

• **Damage to biological systems:** exposure to radioactive isotopes must be prevented by use of concrete and lead shields.

• **Use of radiotracers:** radioactive isotopes are easy to detect and are used to trace what happens in chemical and biological processes. For example, carbon-14 is used to determine the age of many objects, and is also used with phosphorus-32 to investigate various metabolic pathways. Thallium-201 is used to assess the damage done to heart muscles after a heart attack, while iodine-131 is used to diagnose and treat thyroid illnesses.

• **Treatment of cancer:** gamma rays from cobalt-60 are used to kill cancer cells inside the body. Less penetrating beta rays from phosphorus-32 or from strontium-90 are used to treat skin cancers.

• **Sterilisation:** medical equipment is sterilised by exposure to gamma rays from ^{60}Co which kill the bacteria present. Radiation is also used to kill bacteria present in food, to prolong the shelf life of the food. For example,

radiation slows down the growth of moulds in fruit and sprouting of stored vegetables such as potatoes.

• **Nuclear power:** the heat generated in nuclear reactions is used to produce electricity. The disposal of the radioactive waste produced causes problems.

Exercise 1.3
•

(1) Why were the photographic plates used by Becquerel fogged up when they were left in the dark near some uranium salts?

(2) What did Marie and Pierre Curie discover from their experiments on pitchblende?

(3) The discovery of radioactivity prompted Rutherford to do his famous experiment. Describe that experiment.

(4) Why is a gold leaf electroscope discharged by radioactive materials?

(5) (a) What is radioactivity?
(b) What are the main types of radioactivity?
(c) What is a nuclear reaction?

(6) What is the difference between a nuclear reaction and a chemical reaction?

(7) (a) Discuss alpha, beta and gamma rays in relation to the following:
(i) Mass, (ii) speed and (iii) ionising power.
(b) Why are gamma rays not affected by an electrical field?

(8) Americium-241 emits α-particles and a new atom.
(a) What is the mass number of the new atom?
(b) What is the atomic number of the new atom?
(c) Write down the symbol of the new atom.

(9) Uranium-238 emits α-particles and a new atom.
(a) What is the mass number of the new atom?
(b) What is the atomic number of the new atom?
(c) Write down the symbol of the new atom.

(10) Carbon-14 is a beta emitter.
(a) What is a beta particle?
(b) What is the mass number of the new atom?

(c) What is the atomic number of the new atom?
(d) Write down the symbol of the new atom.

(11) (a) What is the mass and the electrical charge of a gamma particle?
(b) Explain why gamma rays are more damaging than alpha particles or beta particles.

(12) (a) Briefly, discuss the penetrating power of radioactive particles.
(b) How is radioactivity detected?

(13) (a) What is meant by radioactive half-life?
(b) How are half-lives used in age determination?

(14) Radioisotopes have widespread uses. Write a note on their uses in the following:
(a) the food industry
(b) medicine
(c) power stations.

(15) Discuss the main problems associated with radioactivity.

(16) Marie Curie developed mobile radiography units during World War I. What is radiography?

(17) (a) What is background radiation?
(b) What are the main sources of this radiation?
(c) Find out about radon in homes using an ENFO briefing sheet.

1.4 ELECTRONIC STRUCTURE OF ATOMS

Energy Levels in Atoms

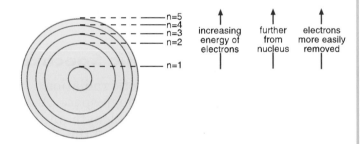

Niels Hendrik David Bohr
1885–1962
Niels Bohr was born in Copenhagen in Denmark in 1885. His father was a professor of physiology at the University of Copenhagen. Niels attended the same university and was a distinguished soccer player as well as a brilliant physics student.
Bohr studied at J. J. Thomson's Cavendish Laboratory and at Rutherford's laboratory. At the young age of 28, while working with Rutherford, he invented the first effective model and theory of the structure of the atom. His work ranks as one of the truly great examples of an imaginative mind at work. He was awarded the 1922 Nobel Prize for physics for his study of the structure of atoms.
During World War 2, Bohr and his family escaped from occupied Denmark to the United States. He and his son, Aage, acted as advisers at the Los Alomos Atomic Laboratories, where the atom bomb was developed. Thereafter, Bohr concerned himself with developing peaceful uses of nuclear energy.
Aage Bohr, Niels' son, shared the Nobel Prize for physics in 1975.

Niels Bohr suggested a model for the way electrons are arranged in atoms. He suggested that electrons move around the nucleus in orbits, much the same way as the planets move around the sun.

The orbitals are called shells: each shell holds a certain number of electrons. The first shell is nearest to the nucleus and is filled up with electrons first. The second shell is next nearest to the nucleus: it starts to fill up with electrons when the first shell is full. When the second shell is full, the third shell starts to fill up, and so on.

Classification of the First Twenty Elements: Arrangement of Electrons in Atoms

The noble gases are very unreactive because they have a stable electronic structure: they have full outer shells. The first shell is full and stable when it contains 2 electrons (like helium), the second shell is full and stable when it contains 8 electrons (like neon), the third shell is full and stable when it contains 18 electrons (like argon). Atoms can be classified on the basis of the arrangement of their outer electrons.

The electronic structures of the first three noble gases can be written simply as

He (2)   Ne (2, 8)   Ar (2, 8, 8)

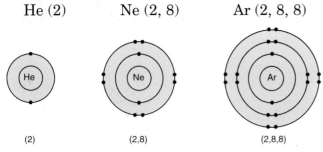

(2)   (2,8)   (2,8,8)

Other elements are arranged on a similar basis. When the first shell is full at helium, electrons go into the second shell. When the second shell is full at neon, electrons go into the third shell and so on. The electronic structure of lithium is 2,1; beryllium is 2,2; boron is 2,3; carbon is 2,4; nitrogen is 2,5; oxygen is 2,6; fluorine is 2,7.

The first 20 elements in the periodic table are arranged as shown at the bottom of the page.

Each horizontal row across the periodic table is called a period, while each vertical column down the table is called a group. Because elements fit into a particular group on the basis of the number of their outer electrons it is possible to explain why elements in the same group have similar properties.

Emission and Absorption Spectra

Neon lights at BASF, Ludwigshafen.

Neon signs and sodium street lights are familiar aspects of everyday life. Neon signs emit light with a reddish orange colour, while sodium lights emit light with a yellow colour. The neon sign is an example of a gas discharge tube or cathode ray tube, similar in some ways to the ones used by Rutherford and Moseley. The many experiments conducted using such tubes not only led to the discovery of the electron, but also told us something about how the electrons are arranged in atoms.

Evidence from Atomic Spectra

Light is a form of electromagnetic radiation composed of waves moving through space. A wave is characterised by its wavelength and frequency. Visible light forms part of the electromagnetic spectrum and extends from the violet end of the spectrum, which has a wavelength of about 400 nm, to the red end, which has a wavelength of about 800 nm. White light can be dispersed or separated by means of a prism into the familiar colours of the rainbow. The resulting pattern consists of a continuous spectrum of colours, a gradual blending of one colour to the next.

Period 1	H							He
Configuration	1							2
Period 2	**Li**	**Be**	**B**	**C**	**N**	**O**	**F**	**Ne**
Configuration	2,1	2,2	2,3	2,4	2,5	2,6	2,7	2,8
Period 3	**Na**	**Mg**	**Al**	**Si**	**P**	**S**	**Cl**	**Ar**
Configuration	2,8,1	2,8,2	2,8,3	2,8,4	2,8,5	2,8,6	2,8,7	2,8,8
Period 4	**K**	**Ca**						
Configuration	2,8,8,1	2,8,8,2						

White light passing through a triangular prism is split into the rainbow colours of the visible spectrum.

The light emitted from a gaseous element in a gas discharge tube gives a different type of spectrum, called a line spectrum. The line spectrum shows only certain colours or specific wavelengths of light. The line spectra of several atoms are shown in the photographs.

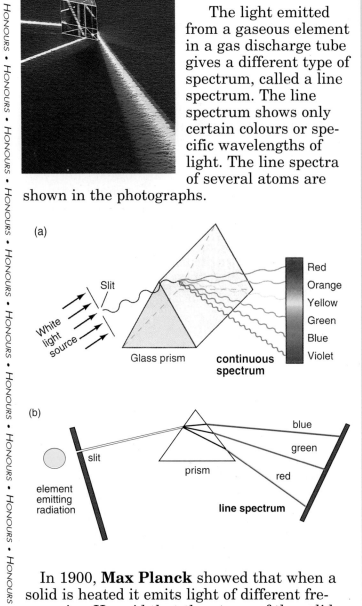

(a)

White light source — Slit — Glass prism — **continuous spectrum**

Red
Orange
Yellow
Green
Blue
Violet

(b)

element emitting radiation — slit — prism — blue, green, red — **line spectrum**

In 1900, **Max Planck** showed that when a solid is heated it emits light of different frequencies. He said that the atoms of the solid vibrated with a definite frequency, f. He related this frequency of the vibrating atom to certain energies using the formula

$$E = hf,$$

where h is called Planck's constant.

The vibration energies of the atoms are said to be quantised; that is the energies are limited to certain discrete values. The values can be 1 or 2 or 3 or any other whole number. This idea that the energy emitted was quantised was explained and extended in 1905 by **Albert Einstein**. In the photoelectric effect experiment, he showed that electrons were emitted when light was shone on the surface of a metal. Electrons are ejected only when the light has a particular frequency. Einstein assumed that an electron was ejected from a metal when it was struck by a photon or bundle of energy. When the photon hits the metal its energy, hf, is absorbed and taken up by the electron.

Fingerprints of elements

When atoms of an element are excited, they return to their state of lowest energy by emitting radiation at specific wavelengths. When this radiation is passed through a spectrometer, the element's characteristic emission lines are seen as a 'fingerprint'.

Emission spectrum of cadmium. Cadmium's predominant lines are red, green and blue. Many fainter lines are not visible in the photograph.

Emission spectrum of copper. The emission spectrum of copper is complex, with blocks of lines appearing in blue, green and red. Many fainter lines are not visible in the photograph.

This photograph of the emission spectrum of sodium shows a single bright yellow line. In fact there are two emission lines but they are so close together that they are indistinguishable on the photograph.

Emission spectrum of hydrogen. Hydrogen's predominant lines are red, cyan and blue. Many fainter lines are not visible in the photograph.

Energy Levels

The emission and absorption of energy is illustrated below.

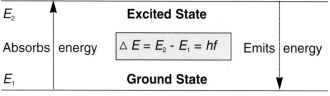

E_2	**Excited State**	
Absorbs energy	$\Delta E = E_2 - E_1 = hf$	Emits energy
E_1	**Ground State**	

When an electron moves from its lowest energy level (ground state), E_1, to the next highest energy level (excited state), E_2, it must gain energy by absorbing a photon of energy, hf.

If the electron moves from the higher energy state, E_2, to the lower state, E_1, it loses energy by emitting light of a particular frequency.

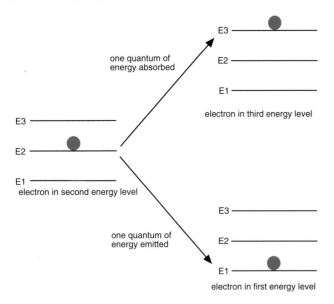

All atoms when sufficiently excited emit light, not only in the visible region but also in the ultraviolet and infrared regions of the electromagnetic spectrum. The line spectrum of the hydrogen atom is relatively simple. In the visible region it consists of four lines (red, blue-green, blue and violet). The lines in the visible region are due to the emission of light by electrons falling from all the higher energy levels back down to the second energy level, E_2. The set of lines in the visible region is called the **Balmer series**, named after the person who discovered it.

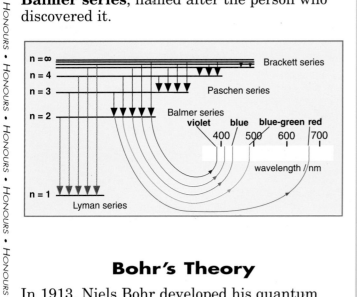

Bohr's Theory

In 1913, Niels Bohr developed his quantum theory based on the hydrogen atom. His theory was extremely successful in explaining both the structure of the hydrogen atom and the line spectrum of the element, but was unsuc-

cessful in predicting the atomic structure of other atoms. His work, nevertheless, forms the basis of modern atomic theory. Bohr's theory includes the following points.

(i) An electron can have only specific energy values in an atom, which are called its energy levels. These energy levels were called orbits or shells. Each level is designated by a value of n (1, 2, 3, 4 …).

(ii) The electrons with the lowest energy level are found nearest the nucleus. An electron in the second energy level (second shell or orbit) has the next lowest energy, while electrons in the other energy levels have higher energies.

(iii) When electrons are in the lowest energy level they are said to be in the ground state. When the electrons are excited they absorb energy and jump into a higher outer energy level, called an excited state.

When an electron falls back to a lower energy level it emits a definite amount of energy, hf. This energy is emitted in the form of a quantum of light and produces a characteristic spectral line.

Activity 1.4

Viewing of Emission Spectra

When excited elements return to their normal state they produce an emission spectrum as they relax. Use a spectroscope to view the spectrum. The spectroscope splits the radiation from a particular element into different frequencies. The radiation from the element is split up by a prism or a diffraction grating into a series of bright lines against a dark background.

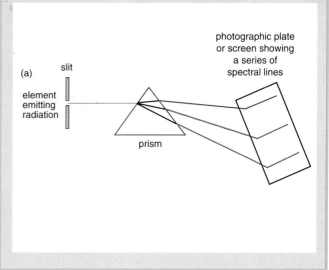

18

Social and Applied Aspects

• **Atomic absorption spectrometry (AAS):** AAS is used to analyse identity and quantity of elements. It is used extensively in quality control in the food industry.

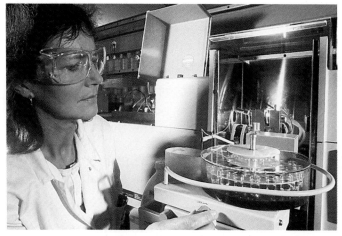

Laboratory assistant checking the graphite tube furnace of an atomic absorption spectrometer. Elements can be determined with the help of the graphite tube which is heated to 2700°C.

• **Street lights:** sodium street lights and neon lights emit light with distinct colours.
• **Fireworks:** fireworks displays are a familiar and fascinating sight to most people. The spectacular colours seen are due to the heating of several different metallic compounds ; strontium gives a red colour while copper gives a blue-green colour.

Fireworks.

Mandatory experiment 1

To identify some elements using flame tests.

Introduction

Many metals emit light of a distinct colour when they are heated; sodium has a distinct yellow colour, potassium has a violet colour, while barium has a green colour.

Some elements seem to emit light with similar colours. For instance, when strontium and lithium are heated in a flame they both emit red light. The light from each can be separated by means of a spectroscope into different distinct colours. These colours correspond to the line spectrum of the particular element. Each element has a characteristic line spectrum which can be used to identify it.

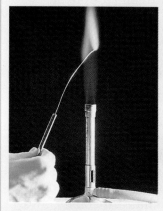

(a) sodium flame test

(b) potassium flame test

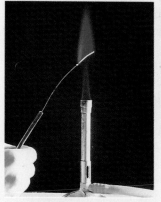

(c) strontium flame test

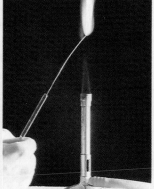

(d) copper flame test

Metal flame tests. Colour is due to the fact that the metal salt is vaporised by the flame and excited to emit radiations that are characteristic of the metal.

Requirements

Safety glasses, Bunsen burner, nichrome wire, clock glasses, concentrated hydrochloric acid, compounds of sodium, lithium, potassium, calcium, barium, copper and lead.

Procedure

(1) Clean the nichrome wire thoroughly by first dipping it in concentrated hydrochloric acid and then heating it to redness in the hot part of a Bunsen flame. Continue heating until no flame colour is seen.
(2) Dip the wire into a clean sample of the hydrochloric acid and then into a sample of the salt to be tested.
(3) Hold the wire in the Bunsen flame. Observe and record the colour.
(4) Repeat the complete procedure for each salt, making sure that the nichrome wire is cleaned each time in a fresh sample of hydrochloric acid.

Atomic Orbitals and Energy Sublevels

Bohr's theory firmly established the concepts of atomic energy levels. The next developments in explaining atomic structure were mainly due to the work of the following.

In 1927, **Werner Heisenberg** showed that it is impossible to know simultaneously, and precisely, the position and the speed of an electron. This means that it is impossible for us to describe how an electron moves in an atom; at best we can describe the probability of finding an electron.

In 1923, **Louis de Broglie** calculated that matter had wave characteristics. Einstein had previously shown that light not only has wave properties, but also has particle properties.

Electrons were shown in 1927 by **Davisson** and **Germer** to have the same dual particle and wave characteristics as light.

In 1926, **Erwin Schrödinger** developed an equation that describes the electron in terms of its wave character. His theory is the basis of quantum mechanics or wave mechanics and describes the wave properties of particles such as the electron. The solutions to his wave equations can describe the probability of finding an electron of a particular energy at various points in space.

An **atomic orbital** is the region in space where the probability of finding an electron of a particular energy is relatively high.

The simple 'orbits', similar to the movement of the planets around the Sun, are now best described as atomic orbitals. There are a number of different atomic orbitals whose boundary surfaces (or outer 'shape') are as follows.

The s-orbitals are spherical in 'shape'.

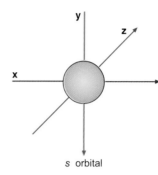

s orbital

They are called:
1s-orbitals if they are in the first energy level ($n = 1$)
2s-orbitals if they are in the second energy level ($n = 2$)
3s-orbitals if they are in the third energy level ($n = 3$)
4s-orbitals if they are in the fourth energy level ($n = 4$)

The p-orbitals are dumb-bell in 'shape'.

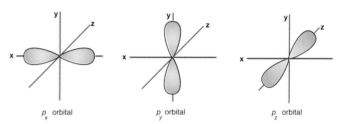

p_x orbital p_y orbital p_z orbital

There are no p-orbitals in the first energy level.
They are called:
2p-orbitals if their electrons are in the second energy level ($n = 2$)
3p-orbitals if their electrons are in the third energy level ($n = 3$)

There are three different types of p-orbital oriented at right angles (in three dimensions) to each other. They are called p_x, p_y and p_z orbitals respectively depending on how they are arranged in space.

The d-orbitals are more complex in 'shape'. There are five different types of d-orbitals depending on how they are arranged in space. The third energy level is the lowest energy level where we can find d-orbitals.

The maximum number of electrons in an energy level is $2n^2$, where $n = 1, 2, 3, 4 \ldots$

n	$2n^2$	Maximum number
1	$2(1)^2$	2
2	$2(2)^2$	8
3	$2(3)^2$	18

This means there is
a maximum of 2 electrons in the first energy level, (both electrons in the 1s-orbital)
a maximum of 8 electrons in the second energy level, (two electrons in the 2s-orbital and two electrons in each of the three 2p-orbitals)
a maximum of 18 electrons in the third energy level, (two electrons in the 3s-orbital, two electrons in each of the three 3p-orbitals, two electrons in each of the five 3d-orbitals).

Electrons in a particular orbital can spin about their own axis. When there are two electrons in an orbital they spin in different directions to each other. This is usually indicated by $\uparrow$ for the electron spinning one way and by $\downarrow$ for the electron spinning the other way.

Pauli's exclusion principle: a particular orbital can only hold a maximum of two electrons if each electron is paired with another electron with the opposite spin $\uparrow\downarrow$.

Electronic Configurations of Atoms

The way that electrons are arranged in an atom is called the electronic configuration of the atom. Most electronic configurations can be explained in terms of the **Aufbau principle** (or build-up principle). The electrons fill into successive sublevels in a specific order.

A **sublevel** is a group of orbitals which have the same energy.

```
1s
2s  2p
3s  3p  3d
4s  4p  4d  4f
5s  5p  5d  5f
6s  6p  6d
7s  7p
```

The first energy level contains a maximum of two electrons in the 1s sublevel. The second main energy level contains eight

electrons, two electrons in the 2s sublevel and six in the 2p sublevel. The electrons fill up sublevels in the following order:

1s, 2s, 2p, 3s, 3p, 4s, 3d, 4p, 5s, 4d, 5p, 6s, 4f, 5d, 6p, 7s, 7f.

For the most part the build-up corresponds to the increase in energy of the sublevels, with some exceptions. The electronic configurations of the first eighteen elements follow the pattern we would expect: 1s before 2s, 2s before 2p and then 3s before 3p.

Each sublevel is filled up in accordance with Hund's rule of maximum multiplicity.

Hund's rule of maximum multiplicity: when electrons fill up a sublevel they do so singly and not in pairs.

Nitrogen ($Z = 7$) has the electronic configuration $1s^2, 2s^2, 2p_x^1, 2p_y^1, 2p_z^1$ where the outer electrons fill singly into the $2p_x$ ($\uparrow$), into the $2p_y$ ($\uparrow$) and then into the $2p_z$ ($\uparrow$) orbital.

The next element, oxygen ($Z = 8$), has the configuration $1s^2, 2s^2, 2p_x^2, 2p_y^1, 2p_z^1$ where the outer electron moves into the $2p_x$ orbital and pairs up ($\uparrow\downarrow$) with the electron already there.

● ● ● ● ● ● **Example 1.2** ● ● ● ● ● ●

Write the electronic configuration for the element silicon.

Solution

Silicon has 14 electrons; the 1s, 2s, 2p and 3s orbitals are full. The remaining two electrons fill the next lowest orbitals, the $3p_x$ and the $3p_y$ singly, i.e. one electron in the $3p_x$ and one electron in the $3p_y$. The electronic configuration is

$1s^2, 2s^2, 2p_x^2, 2p_y^2, 2p_z^2, 3s^2, 3p_x^1, 3p_y^1$.

The first eighteen elements fill their orbitals and sublevels in a similar manner. However, the next element, potassium ($Z = 19$), does not have its outer electron in the 3d sublevel but in the 4s sublevel. The 4s sublevel is filled before the 3d level because the 4s is marginally lower in energy than the 3d; this has been shown experimentally.

3d 4s 4p.

The electronic configurations of the first thirty-six elements are shown in the table:

Element	1s	2s	2p	3s	3p	3d	4s	4p	Electronic configuration
$_1$H	↑								$1s^1$
$_2$He	↑↓								$1s^2$
$_3$Li	↑↓	↑							$1s^2\,2s^1$
$_4$Be	↑↓	↑↓							$1s^2\,2s^2$
$_5$B	↑↓	↑↓	↑						$1s^2\,2s^2\,2p^1$
$_6$C	↑↓	↑↓	↑ ↑						$1s^2\,2s^2\,2p^2$
$_7$N	↑↓	↑↓	↑ ↑ ↑						$1s^2\,2s^2\,2p^3$
$_8$O	↑↓	↑↓	↑↓ ↑ ↑						$1s^2\,2s^2\,2p^4$
$_9$F	↑↓	↑↓	↑↓ ↑↓ ↑						$1s^2\,2s^2\,2p^5$
$_{10}$Ne	↑↓	↑↓	↑↓ ↑↓ ↑↓						$1s^2\,2s^2\,2p^6$
$_{11}$Na	↑↓	↑↓	↑↓ ↑↓ ↑↓	↑					$1s^2\,2s^2\,2p^6\,3s^1$
$_{12}$Mg	↑↓	↑↓	↑↓ ↑↓ ↑↓	↑↓					$1s^2\,2s^2\,2p^6\,3s^2$
$_{13}$Al	↑↓	↑↓	↑↓ ↑↓ ↑↓	↑↓	↑				$1s^2\,2s^2\,2p^6\,3s^2\,3p^1$
$_{14}$Si	↑↓	↑↓	↑↓ ↑↓ ↑↓	↑↓	↑ ↑				$1s^2\,2s^2\,2p^6\,3s^2\,3p^2$
$_{15}$P	↑↓	↑↓	↑↓ ↑↓ ↑↓	↑↓	↑ ↑ ↑				$1s^2\,2s^2\,2p^6\,3s^2\,3p^3$
$_{16}$S	↑↓	↑↓	↑↓ ↑↓ ↑↓	↑↓	↑↓ ↑ ↑				$1s^2\,2s^2\,2p^6\,3s^2\,3p^4$
$_{17}$Cl	↑↓	↑↓	↑↓ ↑↓ ↑↓	↑↓	↑↓ ↑↓ ↑				$1s^2\,2s^2\,2p^6\,3s^2\,3p^5$
$_{18}$Ar	↑↓	↑↓	↑↓ ↑↓ ↑↓	↑↓	↑↓ ↑↓ ↑↓				$1s^2\,2s^2\,2p^6\,3s^2\,3p^6$
$_{19}$K	↑↓	↑↓	↑↓ ↑↓ ↑↓	↑↓	↑↓ ↑↓ ↑↓		↑		$1s^2\,2s^2\,2p^6\,3s^2\,3p^6\,4s^1$
$_{20}$Ca	↑↓	↑↓	↑↓ ↑↓ ↑↓	↑↓	↑↓ ↑↓ ↑↓		↑↓		$1s^2\,2s^2\,2p^6\,3s^2\,3p^6\,4s^2$
$_{21}$Sc	↑↓	↑↓	↑↓ ↑↓ ↑↓	↑↓	↑↓ ↑↓ ↑↓	↑	↑↓		$1s^2\,2s^2\,2p^6\,3s^2\,3p^6\,4s^2\,3d^1$
$_{22}$Ti	↑↓	↑↓	↑↓ ↑↓ ↑↓	↑↓	↑↓ ↑↓ ↑↓	↑ ↑	↑↓		$1s^2\,2s^2\,2p^6\,3s^2\,3p^6\,4s^2\,3d^2$
$_{23}$V	↑↓	↑↓	↑↓ ↑↓ ↑↓	↑↓	↑↓ ↑↓ ↑↓	↑ ↑ ↑	↑↓		$1s^2\,2s^2\,2p^6\,3s^2\,3p^6\,4s^2\,3d^3$
$_{24}$Cr	↑↓	↑↓	↑↓ ↑↓ ↑↓	↑↓	↑↓ ↑↓ ↑↓	↑ ↑ ↑ ↑ ↑	↑		$1s^2\,2s^2\,2p^6\,3s^2\,3p^6\,4s^1\,3d^5$
$_{25}$Mn	↑↓	↑↓	↑↓ ↑↓ ↑↓	↑↓	↑↓ ↑↓ ↑↓	↑ ↑ ↑ ↑ ↑	↑↓		$1s^2\,2s^2\,2p^6\,3s^2\,3p^6\,4s^2\,3d^5$
$_{26}$Fe	↑↓	↑↓	↑↓ ↑↓ ↑↓	↑↓	↑↓ ↑↓ ↑↓	↑↓ ↑ ↑ ↑ ↑	↑↓		$1s^2\,2s^2\,2p^6\,3s^2\,3p^6\,4s^2\,3d^6$
$_{27}$Co	↑↓	↑↓	↑↓ ↑↓ ↑↓	↑↓	↑↓ ↑↓ ↑↓	↑↓ ↑↓ ↑ ↑ ↑	↑↓		$1s^2\,2s^2\,2p^6\,3s^2\,3p^6\,4s^2\,3d^7$
$_{28}$Ni	↑↓	↑↓	↑↓ ↑↓ ↑↓	↑↓	↑↓ ↑↓ ↑↓	↑↓ ↑↓ ↑↓ ↑ ↑	↑↓		$1s^2\,2s^2\,2p^6\,3s^2\,3p^6\,4s^2\,3d^8$
$_{29}$Cu	↑↓	↑↓	↑↓ ↑↓ ↑↓	↑↓	↑↓ ↑↓ ↑↓	↑↓ ↑↓ ↑↓ ↑↓ ↑↓	↑		$1s^2\,2s^2\,2p^6\,3s^2\,3p^6\,4s^1\,3d^{10}$
$_{30}$Zn	↑↓	↑↓	↑↓ ↑↓ ↑↓	↑↓	↑↓ ↑↓ ↑↓	↑↓ ↑↓ ↑↓ ↑↓ ↑↓	↑↓		$1s^2\,2s^2\,2p^6\,3s^2\,3p^6\,4s^2\,3d^{10}$
$_{31}$Ga	↑↓	↑↓	↑↓ ↑↓ ↑↓	↑↓	↑↓ ↑↓ ↑↓	↑↓ ↑↓ ↑↓ ↑↓ ↑↓	↑↓	↑	$1s^2\,2s^2\,2p^6\,3s^2\,3p^6\,4s^2\,3d^{10}\,4p^1$
$_{32}$Ge	↑↓	↑↓	↑↓ ↑↓ ↑↓	↑↓	↑↓ ↑↓ ↑↓	↑↓ ↑↓ ↑↓ ↑↓ ↑↓	↑↓	↑ ↑	$1s^2\,2s^2\,2p^6\,3s^2\,3p^6\,4s^2\,3d^{10}\,4p^2$
$_{33}$As	↑↓	↑↓	↑↓ ↑↓ ↑↓	↑↓	↑↓ ↑↓ ↑↓	↑↓ ↑↓ ↑↓ ↑↓ ↑↓	↑↓	↑ ↑ ↑	$1s^2\,2s^2\,2p^6\,3s^2\,3p^6\,4s^2\,3d^{10}\,4p^3$
$_{34}$Se	↑↓	↑↓	↑↓ ↑↓ ↑↓	↑↓	↑↓ ↑↓ ↑↓	↑↓ ↑↓ ↑↓ ↑↓ ↑↓	↑↓	↑↓ ↑ ↑	$1s^2\,2s^2\,2p^6\,3s^2\,3p^6\,4s^2\,3d^{10}\,4p^4$
$_{35}$Br	↑↓	↑↓	↑↓ ↑↓ ↑↓	↑↓	↑↓ ↑↓ ↑↓	↑↓ ↑↓ ↑↓ ↑↓ ↑↓	↑↓	↑↓ ↑↓ ↑	$1s^2\,2s^2\,2p^6\,3s^2\,3p^6\,4s^2\,3d^{10}\,4p^5$
$_{36}$Kr	↑↓	↑↓	↑↓ ↑↓ ↑↓	↑↓	↑↓ ↑↓ ↑↓	↑↓ ↑↓ ↑↓ ↑↓ ↑↓	↑↓	↑↓ ↑↓ ↑↓	$1s^2\,2s^2\,2p^6\,3s^2\,3p^6\,4s^2\,3d^{10}\,4p^6$

There are two exceptions among the first thirty-six elements to the Aufbau principle: chromium ($Z = 24$) and copper ($Z = 29$).

(i) Chromium has the electronic configuration $1s^2, 2s^2, 2p^6, 3s^2, 3p^6, 3d^5, 4s^1$ and not the expected configuration $1s^2, 2s^2, 2p^6, 3s^2, 3p^6, 3d^4, 4s^2$. Chromium prefers to have a half-filled 3d sublevel and a half-filled 4s sublevel rather than a 3d sublevel with 4 electrons and a full 4s sublevel because it is energetically more feasible.

(ii) In a similar way copper prefers to have a full 3d sublevel and a half-filled 4s sublevel rather than a 3d sublevel with 9 electrons and a full 4s sublevel. The electronic configuration of copper is $1s^2, 2s^2, 2p^6, 3s^2, 3p^6, 3d^{10}, 4s^1$ and not the expected $1s^2, 2s^2, 2p^6, 3s^2, 3p^6, 3d^9, 4s^2$ configuration.

Electronic Configurations of Ions

The electronic configuration of an ion is found in a similar way to that of an atom. However, it must be remembered that negative ions have one or more extra electrons than the particular atom from which they come and that positive ions have one or more electrons fewer than the particular atom from which they come.

• • • • • Example 1.3 • • • • •

Write the electronic configuration for the magnesium ion, Mg^{2+}.

Solution

Take away 2 electrons from the configuration of magnesium.

Magnesium has 12 electrons; the 1s, 2s, 2p and 3s orbitals are full.

The electronic configuration of Mg is $1s^2, 2s^2, 2p_x^2, 2p_y^2, 2p_z^2, 3s^2$.

The magnesium ion has 10 electrons; the 1s, 2s and 2p orbitals are full.

The electronic configuration of Mg^{2+} is $[1s^2, 2s^2, 2p_x^2, 2p_y^2, 2p_z^2]^{2+}$.

• • • • • Example 1.4 • • • • •

Write the electronic configuration for the sulphide ion, S^{2-}.

Solution

Add 2 electrons to the configuration of sulphur.

Sulphur has 16 electrons; the 1s, 2s, 2p and 3s orbitals are full. There are 4 electrons in the 3p energy level. The electronic configuration of S is $1s^2, 2s^2, 2p_x^2, 2p_y^2, 2p_z^2, 3s^2, 3p_x^2, 3p_y^1, 3p_z^1$.

The sulphide ion has 18 electrons; the 1s, 2s and 2p, 3s and 3p orbitals are full. The electronic configuration of S^{2-} is $[1s^2, 2s^2, 2p_x^2, 2p_y^2, 2p_z^2, 3s^2, 3p_x^2, 3p_y^2, 3p_z^2]^{2-}$.

Some Periodic Properties

In accordance with the periodic law, the physical and chemical properties of the elements vary periodically according to the atomic number.

Atomic Radius

An atom does not have a definite size. An isolated atom does not have a clearly defined boundary because the outer electron cloud does not end abruptly but thins out gradually as the distance from the nucleus increases.

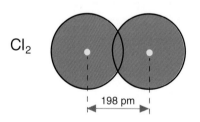

Cl₂   198 pm

Values of atomic radii are obtained from measurements of distances between atoms in chemical bonds. For instance, it is possible to measure the Cl—Cl bond distance in the Cl_2 molecule. The atomic radius of the chlorine atom is assumed to be half of the Cl—Cl bond distance, i.e. 198/2 pm which is 99 pm. In turn the atomic radius of Cl (99 pm) can be subtracted from the C—Cl bond distance (176 pm) to give the atomic radius of C (77 pm). Other atomic radii are calculated from bond distances in a similar manner.

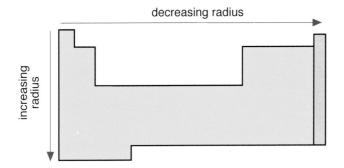

decreasing radius

increasing radius

Values of atomic radii show the following trends:

(a) Within a period, the atomic radius decreases from left to right with increasing atomic number (nuclear charge). For instance, the largest atom in the second period is lithium, while the smallest is neon.

(b) Within each group, the atomic radius increases from top to bottom. For instance, in group VII, the smallest atom is fluorine, while the largest is astatine.

These general trends are explained as follows:

(i) The larger the value of n, the larger the size of the orbital. This means that each successive member in a group of elements gets bigger as another 'shell' of electrons is added on. Each extra 'shell' of electrons also screens or shields the outer electrons from the nuclear charge.

The shielding of the outer electrons reduces the effect of the nuclear charge. As a result, the effective nuclear charge that an outer electron experiences is not the full nuclear charge; thus the outer electron cloud experiences less pull from the nucleus and the atomic size increases.

(ii) However, within each period, as the nuclear charge increases across, the outer electrons are pulled in closer to the nucleus and the radius of the atom decreases.

Ionisation Energy

When an electron is removed from a neutral atom a positive ion is left behind. The process is called ionisation and can be described as

$$A(g) \rightarrow A^+ + e^-$$

Values of this energy are usually quoted for one mole of atoms. For example, the first ionisation energy of the lithium atom is 519 kJ mol^{-1}.

The **first ionisation energy** is the minimum energy required to remove the most loosely bound electron from a neutral atom in the gas state.

Values for the first ionisation energies are given in the table below and display a periodic variation.

First Ionisation Energies of the Elements (in kilojoules per mole)

IA	2A	3B	4B	5B	6B	7B		8		1B	2B	3A	4A	5A	6A	7A	0
H 1310																	He 2370
Li 519	Be 900											B 799	C 1090	N 1400	O 1310	F 1680	Ne 2080
Na 494	Mg 736											Al 577	Si 787	P 1060	S 1000	Cl 1260	Ar 1520
K 418	Ca 590	Sc 632	Ti 661	V 649	Cr 653	Mn 715	Fe 761	Co 757	Ni 736	Cu 745	Zn 908	Ga 577	Ge 761	As 967	Se 941	Br 1140	Kr 1350
Rb 402	Sr 548	Y 636	Zr 669	Nb 653	Mo 695	Tc 699	Ru 724	Rh 745	Pd 803	Ag 732	Cd 866	In 556	Sn 707	Sb 833	Te 870	I 1010	Xe 1170
Cs 377	Ba 502	La 540	Hf 531	Ta 577	W 770	Re 761	Os 841	Ir 887	Pt 866	Au 891	Hg 1010	Tl 590	Pb 715	Bi 774	Po 812	At -	Rn 1040
Fr -	Ra 510	Ac 669															

The rare earth elements and the actinides have been omitted

The first ionisation energy is plotted against atomic number in the graph below.

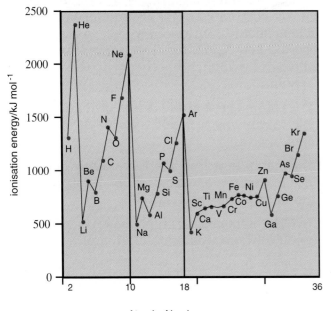

Atomic Numbers

The following features are noticed.

(a) The values increase across a period and decrease down a group.

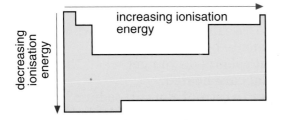

The values increase across a period due to the increased nuclear charge and also due to the decrease in the size of the atomic radius. For instance, it is more difficult to pull an electron off a neon atom than off a lithium atom because the outer electron is more tightly bound and is also nearer to the nucleus.

The values decrease down a group due to the increase in size of the atoms. The outer electrons are screened from the effect of the nuclear charge by the inner 'shells' of electrons. The further away an electron is from the nucleus the easier it is to remove. For example, helium is a small atom; the electron to be removed is close to the nucleus. The electron is more tightly bound than, say, in a bigger atom such as radon. Thus, the ionisation energy for helium is larger than that of radon.

(b) Exceptions to the general trends occur.

Group II elements, such as beryllium or magnesium, have higher ionisation energies than group III elements, such as boron or aluminium. This is because beryllium and the other group II elements have full s-orbitals which make them more stable. Consequently it is more difficult to remove an outer electron from beryllium than from boron.

For a similar reason, the ionisation energies of the group V elements, such as nitrogen and phosphorus, are higher than the group VI elements, such as oxygen and sulphur. The group V elements have half-filled p-orbitals which make them more stable.

The ionisation energies of the transition metals do not increase across a period as rapidly as those of the main group elements. The values are almost constant; the ionisation energies of the elements going from scandium to zinc are similar because in each case the same 4s electron is being removed.

(c) Second and successive ionisation energies are higher.

Greater energy is required to remove a second outer electron than the first outer electron because in this case the electron is being removed not from a neutral atom but from a positively charged ion.

$$A^+(g) \rightarrow A^{2+}(g) + e^-$$

Third and successive electrons are accordingly more difficult to remove; the third and higher ionisation energies are extremely high.

Evidence for Energy Levels

Values of the first and successive ionisation energies provide evidence for the existence of energy levels. For example, the first, second, and third ionisation energies for beryllium are 900, 1757 and 14 848 kJ mol^{-1}, respectively. The extremely large jump in energy from the second to the third ionisation energy corresponds to the removal of an electron from the inner 1s-orbital and not the outer 2s-orbital. Ionisation energies can be determined from the line spectra of atoms.

Chemical Properties and Electronic Structures

The chemical properties of an element depend on its electronic structure. The elements in Group I (alkali metals) have similar chemical properties because they have similar electronic configurations.

Exercise 1.4

•

(1) (a) Why are the noble gases so unreactive?
(b) Draw simple electron structures for the first three noble gases.
(c) How do the electrons occupy the outer parts of atoms?

(2) Write out simple electron structures for the following elements:
Li, B, O, F, Mg, Ar.

(3) Use the electronic structures of the halogens to explain why they are in Group VII of the periodic table.

(4) Why do the elements in the same group have similar properties?

(5) (a) Give a brief description of the wave nature of light.
(b) Describe the electromagnetic spectrum.
(c) Explain what is meant by saying that the light spectrum is continuous.

(6) Sodium street lights and fireworks are examples of atomic absorption spectrometry. Explain how they emit particular colours.

(7) Neon signs emit light with a red-orange colour. Describe, briefly, how discharge tubes similar to those in neon lights can tell us about the electron.

(8) Explain the meaning of the following terms:
(a) continuous spectrum (b) line spectrum.

(9) (a) What is meant by the term atomic emission spectra?
(b) How do emission spectra help us to understand the arrangement of electrons in atoms?

(10) In 1900, Max Planck showed that when a solid is heated it emits light of various frequencies.
(a) What did he notice about the frequencies?
(b) What is the meaning of 'quantised'?
(c) How did Einstein's photoelectric effect explain Planck's findings?

(11) What is the meaning of the term 'orbit', which was proposed by Niels Bohr in 1913?

(12) How does the model of the atom proposed by Bohr differ from that proposed by Rutherford?

(13) The model of the nuclear atom proposed by Rutherford did not account for the fact that an electron rotating around a central nucleus should continuously lose energy. Describe how Niels Bohr overcame this problem when he applied his theory to the hydrogen atom.

(14) Explain the main features of Bohr's theory. Explain why it is not suitable for all atoms.

(15) Explain how line spectra provide evidence for energy levels. Draw a simple diagram of any line spectrum.

(16) What is meant by the terms (i) excited state and (ii) ground state of an atom?

(17) Draw an energy level diagram showing each of the following:
(a) ground state (b) excited state (c) emission of energy (d) absorption of energy.

(18) All atoms emit light when they are sufficiently excited. Discuss this in relation to the line spectra of the hydrogen atom. In your answer refer to spectral transitions in the visible, the infrared and the ultraviolet regions.

(19) (a) Describe an experiment which identifies elements by heating them.
(b) What colours do you get when the following metals are heated:
(i) copper (ii) strontium (iii) sodium?
(c) What is the difference in the light observed when a metal is heated and that formed from the spectrum of white light?

(20) Describe, giving experimental details, how you would:
(a) identify some elements using flame tests
(b) observe the sodium spectrum using a spectroscope.

(21) Neon lights emit light.
(a) Why is it necessary to heat the neon to produce light?
(b) Why does neon gives a different coloured light to sodium street lights?

(22) Fireworks produce a sparkling array of colours. Use your knowledge of spectra to explain why.

(23) Discuss, briefly, the role played by (a) Bohr, (b) Heisenberg, (c) de Broglie and (d) Schrödinger, in the development of atomic theory.

(24) Define each of the following terms:
(a) energy level (b) atomic orbital.

(25) Draw diagrams of: (a) s-orbitals
(b) p-orbitals.

(26) Write the electronic configurations for the first ten elements in the periodic table.

(27) Define each of the following terms:
(a) Aufbau principle (b) Pauli's exclusion principle (c) Hund's rule of maximum multiplicity.

(28) Give the electronic configurations of the following atoms:
B, N, Li, C, Mg, F, Sc, V, Cr, Fe, Ar, Cu, H

(29) Name the atoms with the following electronic configurations:

(a) $1s^2, 2s^2$

(b) $1s^2, 2s^2, 2p_x^1, 2p_y^1, 2p_z^1$

(c) $1s^2, 2s^2, 2p^6, 3s^1$

(d) $1s^2, 2s^2, 2p^6, 3s^2, 3p^6, 4s^2, 3d^3$

(e) $1s^2, 2s^2, 2p^6, 3s^2, 3p^6, 4s^1, 3d^5$

(f) $1s^2, 2s^2, 2p^6, 3s^2, 3p^6, 4s^2, 3d^5$

(g) $1s^2, 2s^2, 2p^6, 3s^2, 3p^6, 4s^1, 3d^{10}$

(30) (a) Explain why the 4s orbital is filled before the 3d orbital.
(b) Explain why chromium and copper do not have the electronic configurations one might expect.

(31) Write the electronic configurations of the following ions:
Br^-, S^{2-}, Al^{3+}, H^-, Cr^{3+}, Cu^{2+}

(32) (a) What is the maximum number of electrons in the level $n = 3$?
(b) How many orbitals are there in the 3d subshell?
(c) What is the maximum number of electrons in the 3s orbital?
(d) Explain why the maximum number of electrons in each orbital is the same.

(33) (a) Explain why an atom does not have a definite size.
(b) Describe how the atomic radius of an atom is determined.
(c) List the main factors which influence the size of atoms.

(34) If the Cl—Cl bond distance in the chlorine molecule is 198 pm and the Br—Cl bond distance is 213 pm, calculate the atomic radius of the bromine atom.

(35) Given the following bond distances:
Br—Cl = 213 pm, I—Br = 247 pm, Cl—Cl = 198 pm, what is the I—Cl bond distance?

(36) In accordance with periodic trends, arrange the following sets of atoms in order of increasing atomic radius:
(a) O, P, S
(b) Br, Cl, Se

(37) Determine which species in each of the following pairs has the bigger radius. In each case explain why.
(a) O^{2-} and O (b) K^+ and K (c) Cl^- and Cl

(38) (a) Define first ionisation energy.
(b) Discuss the general trends in ionisation energies.
(c) Give reasons for the exceptions which occur in the general trends.
(d) Why is the second ionisation energy of an element always larger than the first ionisation energy?

(39) Explain why the difference between the first and second ionisation energies of sodium is much greater than the difference between the first and second ionisation energies of magnesium.

(40) The first, second and third ionisation energies for beryllium are 900, 1757 and 14 848 kJ mol^{-1}, respectively. Explain how this may be taken as evidence for the existence of energy levels.

1.5 OXIDATION AND REDUCTION

False-colour scanning micrograph of a flake of rusty bodywork on a car. Rusting is an electrolytic process; when iron is moist, current flows between different points in its surface. It is an oxidation–reduction reaction.

The properties of the elements in the periodic table depend on which group they belong to. For example, the elements in group I have similar properties because they all can lose 1 electron, while the elements in group VII have similar properties because they can all gain 1 electron.

The slow corrosion of bronze, the rapid rusting of iron, the burning of a fuel, the production of electricity in a cell and the use of electricity to produce elements such as copper or chlorine are all examples of oxidation reactions. Corrosion of metals is a damaging oxidation reaction, while the burning of a fuel is a beneficial oxidation reaction.

Gain or Loss of Oxygen

As the name implies, the term oxidation was used initially when referring to reactions which involved oxygen: oxidation is about gaining oxygen.

$$2Mg(s) + O_2(g) \rightarrow 2MgO(s)$$

Here, the magnesium gains oxygen and is oxidised to a white powder, magnesium oxide, while at the same time, the oxygen is reduced to magnesium oxide (reduction is about losing oxygen).

The dual process of oxidation and reduction is called a **redox reaction**.

Many reactions involve an increase in the oxygen content of a substance:

$$C(s) + O_2(g) \rightarrow CO_2(g)$$

$$CH_4(g) + 2O_2(g) \rightarrow CO_2(g) + 2H_2O(l)$$

$$4Fe(s) + 3O_2(g) \rightarrow 2Fe_2O_3(s)$$

Reduction may be considered the reverse of this process: a decrease in the oxygen content. Many metals are extracted from their ores in this way. For example:

$$2Fe_2O_3(s) + 3C(s) \rightarrow 4Fe(s) + 3CO_2(g)$$

Here, the iron oxide is reduced to iron, while the carbon is oxidised to carbon dioxide.

The definition was later extended to include reactions which did not involve oxygen.

$$2Na(s) + Cl_2(g) \rightarrow 2NaCl(s)$$

(Reduced / Oxidised)

Here, the sodium is oxidised to sodium chloride, while the chlorine is reduced to sodium chloride.

Redox in Terms of Electron Transfer

Recall
• Electrons have a negative charge.
• Loss of electrons from a neutral atom makes a positive ion.

If we examine the reaction between zinc and copper oxide, we see that the zinc is oxidised to zinc oxide while at the same time the copper oxide is reduced to copper.

$$Zn(s) + CuO(s) \rightarrow ZnO(s) + Cu(s)$$

(Oxidised / Reduced)

This reaction can be divided into two 'half reactions':

$$Zn(s) \xrightarrow[\text{oxidised}]{\text{loses } 2e^-} Zn^{2+} + 2e^-$$

$$Cu^{2+} + 2e^- \xrightarrow[\text{reduced}]{\text{gains } 2e^-} Cu(s)$$

Zinc cannot be oxidised to zinc oxide without becoming a zinc ion, while the copper oxide cannot be reduced to copper without becoming a copper atom. In each case electrons are involved: a zinc atom loses electrons to become a zinc ion, while a copper ion gains electrons to become a copper atom.

$$Zn(s) \rightarrow Zn^{2+}(s) + 2e^-$$

$$Cu^{2+}(s) + 2e^- \rightarrow Cu(s)$$

The oxide ion, O^{2-}, does not alter throughout the reaction, but simply exchanges partners from the copper to the zinc.

Redox reactions can now be defined in terms of electron transfer.

Oxidation Is Loss of electrons	**OIL**
Reduction Is Gain of electrons	**RIG**

Oxidising and Reducing Agents

In an oxidation–reduction reaction, the substance which is oxidised is the reducing agent, while the substance which is reduced is the oxidising agent.

$$CuO(s) + H_2(g) \rightarrow Cu(s) + H_2O(l)$$

(Oxidised: $H_2 \rightarrow H_2O$; Reduced: $CuO \rightarrow Cu$)

Hydrogen is oxidised: hydrogen is the reducing agent.
Copper oxide is reduced: copper oxide is the oxidising agent.

The official name for copper oxide, CuO, is copper(II) oxide, while the official name for zinc oxide, ZnO, is zinc(II) oxide.

Social Aspects of Oxidising and Reducing Agents

• **Swimming pools:** chlorine is used to sterilise water by killing bacteria. Chlorine, Cl_2, reacts with water forming two acids, HCl and HOCl.

$$Cl_2(aq) + H_2O(l) \rightleftharpoons HCl(aq) + HOCl(aq)$$

The active ingredient is chloric(I) acid, HOCl, which kills the bacteria by oxidising them.
• **Domestic bleach:** this works in a similar manner. In this case, the active ingredient is NaOCl instead of HOCl.
• **Anti-oxidants:** vitamin C (ascorbic acid) is used to stop the browning of fruit and discoloration of meat. It works by reducing the oxygen in the air.

Introducing Electrochemistry: The Electrochemical Series

Some metals, such as silver and gold, are found free in nature uncombined with other elements, while others, such as sodium and potassium, are only found chemically combined with other elements. Some metals are resistant to corrosion, while others are not. The most useful way of comparing the reactivity of metals is to look at the position of the metal in the electrochemical series.

The electrochemical series is similar to the activity series of metals: the metals are arranged in order of their ability to be oxidised.

Decreasing reactivity $\longrightarrow$
Li—K—Ca—Na—Mg—Al—Zn—Fe—Sn—Pb—H—Cu—Hg—Ag—Au

The position of a metal in the electrochemical series not only tells us how reactive a metal is but also tells us how a compound of a metal reacts with another substance.

If magnesium and iron are exposed to oxygen, the magnesium reacts, forming a white oxide, much more quickly than the iron forms iron oxide. The magnesium reacts faster than the iron because it loses electrons more readily.

The ability of a metal to lose electrons can tell us whether a particular reaction will take place or not. For example, when potassium is placed in a solution of zinc ions the potassium will donate electrons to the zinc ions, thereby changing them into zinc metal, while the potassium metal changes into potassium ions.

$$2K(s) + ZnCl_2(aq) \rightarrow 2KCl(aq) + Zn(s)$$

However, copper does not donate electrons to zinc ions because copper is lower in the electrochemical series than zinc. When copper is placed in aqueous zinc chloride no reaction occurs.

$$Cu(s) + ZnCl_2(aq) \rightarrow No\ reaction$$

Displacement reactions can result when one metal has more ability than another to donate electrons.

The electrochemical series can be useful to us in many ways. Here are two examples.
1 Scrap iron can be used to extract copper from its ores.

$$Cu_2S(s) + Fe(s) \rightarrow FeS(s) + 2Cu(s)$$
copper ore   scrap iron   copper

Here, the iron donates electrons to the copper ions, thereby changing them into copper metal, while at the same time the iron metal changes into iron sulphide.

2 This series can tell us how to predict chemical reactions, such as how we may extract a metal from its ore, the effect of heat on a carbonate or nitrate, or how easily we may oxidise a metal.

Mandatory experiment 2

Part (a) To investigate some redox reactions of the group VII elements

Introduction

As most halogens are harmful substances, this experiment should be done by the teacher using a fume cupboard.

(1) Chlorine can oxidise iron(II) to iron(III) in solution. The green colour of Fe^{2+} in an iron(II) chloride solution turns yellow as it is oxidised to Fe^{3+} when chlorine is bubbled through it.

$$2FeCl_2(aq) + Cl_2(g) \rightarrow 2FeCl_3(aq)$$

$$\text{Fe(II)} \xrightarrow{\text{oxidised}} \text{Fe(III)}$$
[green]   [yellow]

(2) Chlorine can oxidise sulphite ions, SO_3^{2-}, to sulphate ions, SO_4^{2-}. The presence of the sulphate ions formed as a result of the oxidation can be tested in the usual way with barium chloride (see page 45).

$$Na_2SO_3(aq) + Cl_2(g) + H_2O(l) \rightarrow Na_2SO_4(aq) + 2HCl(aq)$$

$$\text{S(IV)} \xrightarrow{\text{oxidised}} \text{S(VI)}$$

The presence of the sulphate ion, SO_4^{2-}, is indicated by the formation of a white precipitate which is insoluble in dilute hydrochloric acid.

$$SO_4^{2-}(aq) + Ba^{2+}(aq) \rightarrow BaSO_4(s)$$
                             white

Bromine and iodine will also oxidise sulphite ions to sulphate ions.

(3) Bromine and iodine are not as strong oxidising agents as chlorine. If bromine water is added to a clear solution of potassium iodide the solution turns black as the iodide ion is oxidised to molecular iodine. The end colour in this test depends on the concentration of the reactants. The presence of iodine, I_2, is confirmed by adding a drop of 1,1,1-trichloroethane to the test-tube: the iodine is seen as a purple colour.

$$I^- \xrightarrow{\text{oxidised}} I_2$$
clear (−1)   black (0)

Requirements

(1) Safety glasses, fume hood, chlorine gas, test-tubes, iron(II) chloride solution.
(2) Safety glasses, fume hood, test-tubes, sodium sulphite, chlorine water, bromine water, iodine dissolved in potassium iodide solution, barium chloride and dilute sulphuric acid.
(3) Safety glasses, fume hood, test-tubes, bromine water and potassium iodide.

Procedure

(1) Oxidation of iron(II) by chlorine gas.
(a) In the fume hood, bubble some chlorine gas through the iron(II) chloride solution.
(b) Observe what happens.

(2) Oxidation of sulphite by chlorine, bromine and iodine.
(a) In the fume hood, add 2 cm³ of chlorine water to 2 cm³ of sodium sulphite solution.
(b) Test the resulting solution with barium chloride solution. Record your observations.
(c) Repeat the procedure above using, in turn, the bromine and iodine solutions. Record your observations.

(3) Oxidation of iodide by bromine.
(a) In the fume hood, place 2 cm³ of potassium iodide into a test-tube. Add 2 cm³ of bromine water. Stopper the tube and mix well.
(b) Record what happens.

Mandatory experiment 2

Part (b) To investigate the displacement reactions of some metals

Introduction

A metal can displace another metal from its salt if it has more ability than the other to donate electrons. The position of the metal in the electrochemical series tells us whether a displacement reaction will occur.

When zinc metal is placed in a solution of copper(II) sulphate, the copper ions are displaced by zinc ions and copper metal is formed.

$$Zn(s) + CuSO_4(aq) \rightarrow ZnSO_4(aq) + Cu(s)$$

Other metals, such as magnesium, which are higher in the electrochemical series than copper will also displace copper from its salts in aqueous solution.

However, if copper is placed in a solution of sodium chloride no displacement of sodium will occur because sodium is higher in the electrochemical series than copper.

Requirements

Safety glasses, 50 ml beakers, copper sulphate solution, sodium chloride solution, zinc granules, and magnesium ribbon.

Procedure

(1) Place some copper sulphate solution into two beakers.
(2) Place some sodium chloride solution into another two beakers.
(3) Add some zinc to some copper sulphate solution. Observe what happens.
(4) Add some magnesium to some copper sulphate solution. Observe what happens.
(5) Add some zinc to some sodium chloride solution. Observe what happens.
(6) Add some magnesium to some sodium chloride solution. Observe what happens.

Results

Copy and complete the data table below.

Metal	Reaction with sodium chloride solution	Reaction with copper sulphate solution
Zinc		
Magnesium		

Electrolysis

Electrolysis is the process where an electric current is used to bring about a chemical reaction.

Electrolysis uses electricity to decompose or split up chemical substances called electrolytes. The electrolyte may be a molten salt or an aqueous solution of a salt. Two electrodes are used to connect a power supply to the electrolyte. The cathode is the electrode attached to the negative pole of the power supply, while the anode is attached to the positive pole of the power supply.

Chemical changes take place at the electrodes as soon as the current begins to flow. The dual process of oxidation and reduction causes the chemical changes; oxidation occurs at the anode and reduction occurs at the cathode. The current is transferred through the electrolytic solution (electrolyte) by means of mobile ions.

Changes occur at the surfaces of the electrodes during the chemical reaction either by decomposition from solution or by evolution of gases, or sometimes both.

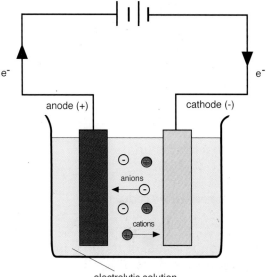

Predicting the Products

Predicting the products formed depends on how easily oxidation and reduction takes place. This depends on:

• The electrolyte used: if the electrolyte is a molten salt only one cation and one anion are present which means that the products formed are easily predicted. An aqueous solution often contains several different ions in solution. This makes it difficult to predict the chemical reactions.

• The type of the electrode: in some cell reactions the electrodes react with the electrolyte (electroplating). Such electrodes are called active electrodes. In most electrolysis reactions the electrodes are inert and do not take part in the cell reaction.

Electrolysis of Aqueous Copper Sulphate using Active Electrodes

If the positive electrode, the anode, is made from copper, the copper electrode changes into copper ions which pass into solution. The copper ions in solution then pass over to the other electrode, the cathode, where they accept electrons and are deposited on the cathode as copper atoms.

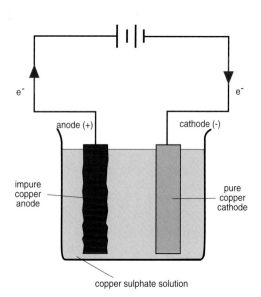

Anode (+) reaction: oxidation

$$Cu(s) \rightarrow Cu^{2+} (aq) + 2e^-$$

Cathode (−) reaction: reduction

$$Cu^{2+} (aq) + 2e^- \rightarrow Cu(s)$$

If the anode is made of impure copper and the cathode of pure copper, the impure copper is refined as the anode dissolves into copper ions which move through the cell solution and are deposited as pure copper atoms on the cathode. This process is an example of electroplating.

Activity 1.5

To demonstrate ionic movement during electrolysis

Introduction

Many common ions are highly coloured. For example, the manganate(VII) ion is an intense purple colour, the chromate(VI) ion is yellow, while the copper(II) ion is green. During electrolysis ions gather at the anode and the cathode. If the ions are highly coloured it is possible to 'see' them migrating towards the anode and cathode.

Requirements

Safety glasses, DC electricity supply, electrical leads, crocodile clips, microscope slide, copper chromate crystals, sodium chloride solution and filter paper.

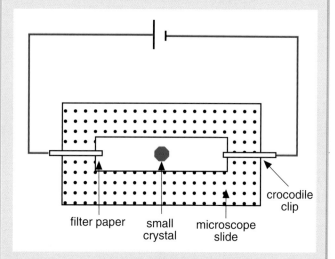

Procedure

(1) Make a conducting solution by dipping filter paper into the sodium chloride solution.

(2) Arrange the circuit as in the diagram. Place some copper chromate crystals in the centre of the filter paper.

(3) Switch on the power and observe what happens.

Electrolysis of Aqueous Electrolytes

The prediction of the products formed at the anode and the cathode in the electrolysis of an aqueous electrolyte is more difficult than when a molten electrolyte is used. An aqueous solution of an electrolyte contains not only the ions from the electrolyte, but also hydrogen ions, hydroxide ions and water molecules. Often, one of the ions from the electrolyte does not like to take part in the cell reaction. For instance, the sulphate ion, SO_4^{2-} (very low down in the electrochemical series) prefers to be a 'spectator' during most cell reactions.

The products obtained during electrolysis can also depend on the type of electrodes being used; the electrodes may be inert like carbon or platinum which do not take part directly in the cell reaction or they may be active electrodes like copper which do take part in the cell reaction.

Activity 1.6

Electrolysis of Sodium Sulphate Solution using Inert Electrodes

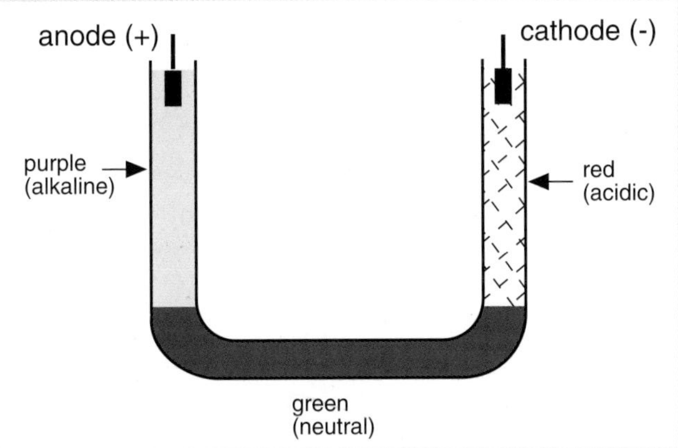

Set up the apparatus as shown in the diagram. When aqueous sodium sulphate undergoes electrolysis the following reactions occur.

Anode (+) reaction: oxidation

Even though the negative sulphate and hydroxyl ions are attracted towards the positive electrode, the anode, neither of them is liberated. Oxygen gas is released at the anode when water molecules are oxidised.

$$2H_2O(l) \rightarrow O_2(g) + 4H^+(aq) + 4e^-$$

Cathode (−) reaction: reduction

Here, again, although the sodium ions and the hydrogen ions migrate towards the negative electrode, the cathode, neither of them is liberated. Hydrogen gas is released at the cathode when water molecules are reduced.

$$4H_2O(l) + 4e^- \rightarrow 2H_2(g) + 4OH^-(aq)$$

The overall reaction is

$$6H_2O(l) \rightarrow 2H_2(g) + O_2(g) + 4H^+(aq) + 4OH^-(aq)$$

Both H^+ and OH^- appear as products: the H^+ ions gather at the anode, while the OH^- ions gather at the cathode. Thus the anode region becomes acidic, while at the same time the cathode region becomes alkaline. Place some universal indicator in the electrolytic cell and observe the colour changes. The anode region should turn purple, while the cathode region should turn red. The neutral region in the middle should remain green.

Electrolysis of Acidified Water using Inert Electrodes

Pure water is a poor conductor of electricity. It contains very few hydrogen ions and hydroxyl ions. Some dilute sulphuric acid is added to increase the concentration of ions present.

Electrolyte: dilute sulphuric acid
Ions/molecules: hydrogen ions, $H^+(aq)$, hydroxyl ions, $OH^-(aq)$, water molecules, H_2O, and sulphate ions, $SO_4^{2-}(aq)$.

Electrode: inert – platinum or carbon

Anode (+) reaction: oxidation

Water loses electrons and oxygen gas is released at the anode.

$$2H_2O(l) \rightarrow O_2(g) + 4H^+(aq) + 4e^-$$

Cathode (−) reaction: reduction

Water accepts electrons and forms hydrogen gas at the cathode.

$$4H_2O(l) + 4e^- \rightarrow 2H_2(g) + 4OH^-(aq)$$

The overall reaction is

$$2H_2O(l) \rightarrow 2H_2(g) + O_2(g)$$

where two moles of hydrogen gas are produced at the cathode, while one mole of oxygen gas is produced at the anode.

To demonstrate the electrolysis of water a Hoffman voltameter is often used.

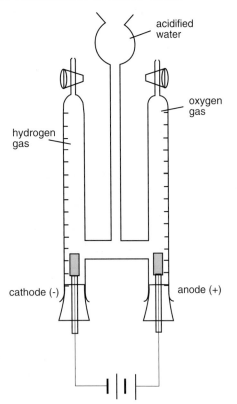

acidified water

oxygen gas

hydrogen gas

cathode (-)

anode (+)

Activity 1.7

Electrolysis of Aqueous Potassium Iodide using Inert Electrodes

e^-

e^-

anode (+)

cathode (-)

I_2 formed

H_2 formed

potassium iodide solution

Set up the apparatus as shown in the diagram. When aqueous potassium iodide undergoes electrolysis the following reactions occur.

Anode (+) reaction: oxidation

Iodide ions, I^-, are easy to discharge at the anode and are converted into iodine molecules, I_2, when they lose an electron.

$$2I^-(aq) \rightarrow I_2(s) + 2e^-$$

Cathode (−) reaction: reduction

Although the sodium ions and the hydrogen ions migrate towards the negative electrode, the cathode, neither of them is liberated. Hydrogen gas is released at the cathode when water molecules are reduced.

$$2H_2O(l) + 2e^- \rightarrow H_2(g) + 2OH^-(aq)$$

The overall reaction is

$$2H_2O(l) + 2I^-(aq) \rightarrow H_2(g) + I_2(g) + 2OH^-(aq)$$

The overall effect of the cell reaction is to change aqueous potassium iodide into aqueous potassium hydroxide. Place a few drops of phenolphthalein in the electrolytic cell and observe the solution turning purple as it becomes more alkaline.

Social and Applied Aspects of Electrolysis

• **Extraction of metals:** metals such as zinc can be extracted from aqueous solutions of their salts by electrolysis. Others like sodium and aluminium can be extracted from their molten salts using electrolysis (see pages 344–46).

•**Purification of metals:** impure copper can be purified by electroplating.

• **Electroplating:** chromium is often plated onto iron to protect it from corrosion, while nickel is often plated with silver to make electroplated nickel silver cutlery (EPNS) (see page 342).

Exercise 1.5

•

(1) Which substance is oxidised and which substance is reduced during each of the following reactions:
(a) fuel + oxygen → carbon dioxide + water
(b) iron + oxygen → iron oxide
(c) food + oxygen → carbon dioxide + water

(2) Explain oxidation and reduction in terms of the following:
(a) gain or loss of oxygen (b) transfer of electrons.
Use a relevant example in each case.

(3) (a) Write an equation for the reaction of sodium, Na, with chlorine, Cl_2.
(b) Explain in terms of electron transfer which substance has been oxidised and which substance has been reduced.

(4) Write an equation for the reaction between zinc and copper(II) oxide.
(a) Show which substance has been oxidised and which substance has been reduced.
(b) Explain your answer in terms of electron transfer.

(5) (a) Write an equation for the reaction of calcium, Ca, with oxygen, O_2.
(b) Use the reaction to explain the meaning of the following terms:
(i) oxidation (ii) reduction (iii) oxidising agent (iv) reducing agent (v) redox reaction.

(6) Define oxidation and reduction in terms of electron transfer. Using the definition, show which substance has been oxidised, which substance has been reduced and name the oxidising and the reducing agent in each of the following examples.

(a) $Mg + Cl_2 \rightarrow MgCl_2$

(b) $2Mg + O_2 \rightarrow 2MgO$

(c) $CuO + H_2 \rightarrow Cu + H_2O$

(d) $2FeCl_2 + Cl_2 \rightarrow 2FeCl_3$

(e) $Cl_2 + KBr \rightarrow 2KCl + Br_2$

(f) $Fe + CuSO_4 \rightarrow Cu + FeSO_4$

(g) $Br_2 + 2KI \rightarrow 2KBr + I_2$

(7) (a) Why does iron corrode?
(b) Describe, using equations, how iron rusts.

(8) Chlorine is used in bleach and in swimming pools to sterilise and disinfect water. Explain how it does this.

(9) (a) What is the electrochemical series?
(b) List the following metals in order of their activity in the electrochemical series: sodium, silver, copper, magnesium and potassium.
(c) Why is potassium the most reactive metal in this series?

(10) (a) How can the ability of a metal to be oxidised be used to determine whether or not a particular chemical reaction can occur ?
(b) Explain your answer using the following:

(i) $2K(s) + ZnCl_2(aq) \rightarrow 2KCl(aq) + Zn(s)$

(ii) $Cu(s) + ZnCl_2(aq) \rightarrow$ No reaction

(11) Use equations to explain how scrap iron can be used to extract copper from its ores.

(12) (a) What is a displacement reaction?
(b) Describe, using equations, what happens when zinc metal, Zn, reacts with copper sulphate solution, $CuSO_4$.

(13) Describe an experiment to demonstrate the movement of ions in solution.

(14) Chlorine can oxidise iron(II) to iron(III) in solution. The green colour of Fe^{2+} in an iron(II) chloride solution turns yellow as it is oxidised to Fe^{3+} when chlorine is bubbled through it. Use half equations to explain what is happening.

(15) (i) Define electrolysis.
(b) Describe the cell reaction which occurs when copper sulphate solution is electrolysed using copper electrodes.

(16) (a) Write equations for (i) the anode reaction and (ii) the cathode reaction, when acidified water is electrolysed.
(b) Why are platinum or graphite electrodes used during this reaction?

(17) Write the anode and cathode reactions for the following reactions:
(a) electrolysis of aqueous sodium sulphate
(b) electrolysis of aqueous potassium iodide.

(18) Describe, briefly, the use of electrolysis in the following:
(a) electroplating (b) purification of copper (c) EPNS cutlery.

Some of the more important key terms are listed below. Others terms not listed may be located by means of the index.

1. The Periodic Law and Periodic Properties: The chemical and physical properties of the elements are periodic functions of the atomic number.

2. Dalton's Ideas on the Structure of Matter:
- All matter is composed of extremely small indivisible particles, called atoms.
- An element is a form of matter composed of only one kind of atom; each atom of the same kind has the same properties. For instance, atoms of a given element each have the same mass.
- A compound is a form of matter consisting of atoms of two or more elements chemically combined in fixed proportions.
- Atoms of two or more elements can combine in a chemical reaction and rearrange themselves to form new chemical combinations.

3. Law of Conservation of Mass: Matter is neither created nor destroyed in the course of a chemical reaction.

4. The electron: The electron was the first sub-atomic particle to be identified. Thomson discovered the electron through his work on cathode rays. Later experiments by Robert Milliken determined the charge and mass of the electron,

$$e = -1.6 \times 10^{-19} \text{ C}$$

$$m = 9.11 \times 10^{-31} \text{ kg}$$

5. The proton and neutron: The proton was the second fundamental particle to be discovered. Rutherford and two of his students, Geiger and Marsden, bombarded gold foil with a beam of α-particles and some bounced back. This suggested that the atom consisted of a very small positively charged nucleus. Rutherford proposed the existence of the proton and predicted the existence of the neutron.

6. Fundamental particles: Until the 1920s only three fundamental particles were thought to exist. Today, quarks and leptons are also thought to be fundamental particles.

7. Atomic number: The atomic number (or proton number), Z, is the number of unit positive charges on the nucleus.

8. Mass number: The mass number, A, is the total number of protons and neutrons in the nucleus of the atom.

9. Isotopes: Isotopes are atoms of an element having the same atomic number but different mass numbers.

10. Relative atomic mass: Relative atomic mass, A_r, is defined as the ratio of the average mass per atom of the natural composition of an element to $\frac{1}{12}$ of the mass of an atom of carbon-12.

11. Mass spectrometer: A mass spectrometer is an instrument which separates ions by mass-to-charge ratio.

12. Radioactivity: Radioactivity is the break-up of the nucleus by the emission of particles and radiant energy.

13. Types of Radioactive Decay:
Alpha particles are positively charged particles which are the nuclei of helium atoms. We can think of an alpha particle as a tightly bound cluster of two protons and two neutrons. Alpha particles are not very penetrating but are damaging rays.
Beta particles are negative particles. Beta particles are moderately penetrating and can penetrate up to 1 cm of flesh.
Gamma rays are neutral rays and are unaffected by electrical fields. They are similar to photons of light but have a higher frequency. Gamma particles are very penetrating and can ionise molecules.

14. Chemical Reactions and Nuclear Reactions:
Chemical reactions occur in the **outer** parts of atoms: they involve the transfer or sharing of the outer electrons.
Nuclear reactions involve changes in the **central** part of the atom, the nucleus.

15. Continuous spectrum: When a beam of white light is dispersed by means of a prism a gradual blending of colours emerges. This is called a continuous spectrum.

16. Line spectrum: The line spectrum shows only certain colours or specific wavelengths of light. Each line consists of electrons jumping from one energy level to another.

17. Planck's constant:
$$E = hf$$

During electron transitions photons are emitted or absorbed. ΔE is related to the frequency f of the photon by Planck's constant, h.

18. The Bohr model of an atom:
(a) An electron can have only specific energy values in an atom, which are called its energy levels. These energy levels were called orbits or shells.
(b) The lowest energy level, $n = 1$, is the level nearest to the nucleus.
(c) When electrons are in the lowest energy level they are said to be in the ground state. When the electrons are excited they absorb energy and jump into a higher outer energy level, called an excited state.

19. Heisenberg uncertainty principle:
It is impossible to know simultaneously, and precisely, the position and the speed of an electron. At best we can describe the probability of finding an electron.

20. Dual nature of the electron:
Electrons were shown to have the same dual particle and wave characteristics as light by Davisson and Germer.

21. Quantum mechanics:
Erwin Schrödinger developed an equation that describes the electron in terms of its wave character. His theory is the basis of quantum mechanics or wave mechanics.

22. Atomic orbitals:
The region in space where the probability of finding an electron is relatively high is called an atomic orbital. The s-orbitals are spherical in shape. The p-orbitals are dumb-bell in shape. The d-orbitals are more complex in shape.

23. Energy levels
The maximum number of electrons in an energy level is $2n^2$, where $n = 1, 2, 3, 4 \ldots$
There is:
a maximum of 2 electrons in the first energy level,
a maximum of 8 electrons in the second energy level,
a maximum of 18 electrons in the third energy level.

24. Pauli's exclusion principle:
A particular orbital can only hold a maximum of two electrons if each electron is paired with another electron with the opposite spin ($\uparrow\downarrow$).

25. Electronic configurations of atoms:
The way that electrons are arranged in an atom is called the electronic configuration of the atom. Most electronic configurations can be explained in terms of the Aufbau principle (or build-up principle). The electrons fill into successive sublevels in a specific order.
A sublevel is a group of orbitals which have the same energy.

26. Hund's rule of maximum multiplicity:
When electrons fill up a sublevel they do so singly and not in pairs.

27. Atomic radius:
An atom does not have a definite size. An isolated atom does not have a clearly defined boundary because the outer electron cloud does not end abruptly but thins out gradually as the distance from the nucleus increases.

28. Ionisation energy:
The first ionisation energy is the minimum energy required to remove the most loosely bound electron from a neutral atom in the gaseous state.

29. Oxidation:
The term oxidation was used initially when referring to reactions which involved oxygen: oxidation is the process of gaining oxygen.

30. Redox reaction:
The dual process of oxidation and reduction is called a redox reaction.

31. Oxidation and reduction:
Oxidation Is Loss of electrons OIL;
Reduction Is Gain of electrons RIG

32. Electrochemical series of metals:

Decreasing reactivity $\longrightarrow$
Li—K—Ca—Na—Mg—Al—Zn—Fe—Sn—Pb—H—Cu—Hg—Ag—Au

33. Electrolysis:
Electrolysis is the process where an electric current is used to bring about a chemical reaction.

The white cliffs of Dover. The cliffs are composed of calcium carbonate, which consists of calcium ions and carbonate ions. The binding force between the ions is an ionic bond.

2.1 CHEMICAL COMPOUNDS

Metal burning in air.

When magnesium burns in oxygen it forms a white solid, magnesium oxide. The compound, magnesium oxide, has very different properties to the elements from which it was composed. The ninety-two naturally occurring elements can form millions and millions of different compounds. When atoms come very close to each other they sometimes form compounds, sometimes they do not. When atoms approach each other closely their outer electron shells repel each other (because like charges repel each other). However, under certain circumstances, atoms can overcome the repulsive forces which keep them apart and combine with each other.

A **compound** is a substance made by combining two or more elements.

Compounds have very different properties to the elements from which they are made.

Some simple common compounds

Compound	Formula
Water	H_2O
Carbon dioxide	CO_2
Sodium chloride	$NaCl$
Aluminium oxide	Al_2O_3
Methane	CH_4

A **chemical bond** is a strong attractive force that exists between certain atoms in a substance.

When atoms combine with each other and form new compounds, they do so by forming chemical bonds. There are three distinct types of chemical bond:
(1) Ionic bonding occurs when oppositely charged ions are attracted to each other by electrostatic attraction, such as in sodium chloride.
(2) Covalent bonding occurs when atoms are held together by sharing of electron pairs, such as in methane.
(3) Metallic bonding occurs when positive metallic ions are held together by a sea of mobile electrons, such as in sodium metal.

Stability of Noble Gases and the Octet Rule

Some elements, such as helium, tend not to form chemical bonds, either with themselves or with other atoms. The other noble gases, neon, argon, krypton and xenon and radon, rarely form chemical bonds. The noble gases dislike forming chemical bonds because they all have full outer shells of electrons. Atoms are extremely stable when they have full outer shells of electrons. With the exception of helium (which has a duet of electrons) all the noble gases have a complete octet of outer electrons.

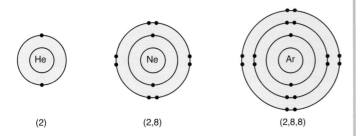

(2)   (2,8)   (2,8,8)

Uses of Noble Gases

The noble gases are extremely stable because they have a full outer shell of electrons (a duet or an octet). The uses of the noble gases reflect that stability.
• Helium gas is used in weather balloons, because of its low density and because it is non-flammable.
• Neon is used in neon street lights.
• Argon is used in electric light bulbs to prevent evaporation of the tungsten element. It is also used in welding to prevent oxidation of the metals with air.

Chemical Bonds and Noble Gas Electron Configurations

Elements that are not noble gases do not have full outer shells of electrons. These atoms try to attain the electronic configuration of the nearest inert gas by forming chemical bonds. However, it should be remembered that not all atoms always attain a noble gas configuration when they bond with other atoms. Nitrogen and sulphur do not always attain an inert gas structure when they combine with other atoms to form chemical bonds. In general, however, atoms combine with each other by forming an octet of electrons in their outer shell. The outer shell is often called the valence shell.

Formulae for Compounds – Valency

The chemical formula of a compound tells us the number and type of atoms in a compound. For example:
The formula of magnesium oxide is MgO. This formula tells us that one magnesium atom combines with one oxygen atom.

The formula of water is H_2O. This formula tells us that two hydrogen atoms combine with one oxygen atom.

Other formulae, such as $CaCO_3$, seem more complicated. This formula tells us that one calcium ion combines with one carbonate ion to form calcium carbonate. It also tells us the order in which the atoms are arranged in the compound, i.e. the calcium is bonded to the carbon, which is in turn bonded to the three oxygens.

When an atom reacts with another atom it can form only a limited number of chemical bonds.

The number of bonds that an atom can form is predicted by the valency of the atom.

The **valency** of an atom tells us the number of chemical bonds that an atom can form, i.e. it is the combining power of an atom.

The valency of an atom is worked out using the following rules.
(1) The **valencies of the s block and p block elements** are determined from the number of electrons lost, gained or shared by the particular atom during the formation of a chemical bond.

This can be easily seen from the group number in the periodic table.

Group number	I	II	III	IV	V	VI	VII	VIII
Valency	1	2	3	4	3	2	1	0

For example, the formula of magnesium oxide is MgO. Magnesium has a valency of 2 and oxygen also has a valency of 2. Because both atoms have the same valency, they combine with each other in the same ratio; in this case one magnesium atom combines with one oxygen atom.

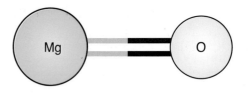

The formula of water is H_2O. Hydrogen has a valency of 1, while oxygen has a valency of 2. In order to balance their combining power (valency), one oxygen atom must combine with two hydrogen atoms.

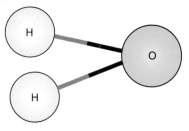

The formula of aluminium oxide is Al_2O_3. Aluminium has a valency of 3, while oxygen has a valency of 2. In order to balance their combining power (valency), two aluminium atoms must combine with three oxygen atoms.

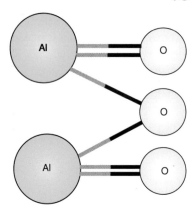

(2) The d-block elements have variable valency.

Iron usually forms two different chlorides.

Iron(II) chloride has the formula $FeCl_2$: here, iron has a valency of 2 (divalent).

Iron(III) chloride has the formula $FeCl_3$: here, iron has a valency of 3 (trivalent).

Copper usually forms two different oxides.

Copper(I) oxide, Cu_2O : here, copper has a valency of 1 (monovalent)

Copper(II) oxide, CuO : here, copper has a valency of 2.

Chromium usually has a valency of 3 or 6 in its common compounds.

Chromium(III) oxide, Cr_2O_3: here, chromium is trivalent.

Sodium chromate(VI), $Na_2Cr_2O_7$: here, chromium has a valency of 6.

Manganese forms compounds with many different valencies.

Manganese(IV) oxide, MnO_2: here, manganese has a valency of 4.

Potassium manganate(VII), $KMnO_4$: here, manganese has a valency of 7.

The roman numerals I, II and III etc. are used to indicate the valency of an element in a particular compound. By convention, roman numerals are placed in round brackets after the particular element to indicate its valency.

• The **transition metals** are strictly defined as those elements having a partially filled d or f subshell in any common oxidation state.
• The **d-block transition elements** are those transition elements with an unfilled d subshell in common oxidation states

(3) The **valency of complex ions** can be worked out using the simple rules which follow. The valency of the hydroxide ion, OH^-, is 1, because it combines with one hydrogen atom to form H_2O. (H has a valency of 1.)

Valencies and some common compounds of some complex ions

Complex ion	Formula	Valency	Compound	Name
Ammonium	NH_4^+	1	NH_4Cl	Ammonium chloride
Nitrate	NO_3^-	1	KNO_3	Potassium nitrate
Hydroxide	OH^-	1	$NaOH$	Sodium hydroxide
Cyanide	CN^-	1	HCN	Hydrogen cyanide
Manganate(VII)	MnO_4^-	1	$KMnO_4$	Potassium manganate (VII)
Ethanoate	CH_3COO^-	1	CH_3COOH	Ethanoic acid
Hydrogencarbonate	HCO_3^-	1	$NaHCO_3$	Sodium hydrogencarbonate
Carbonate	CO_3^{2-}	2	$CaCO_3$	Calcium carbonate
Sulphate	SO_4^{2-}	2	$MgSO_4$	Magnesium sulphate
Sulphite	SO_3^{2-}	2	$NaSO_3$	Sodium sulphite
Thiosulphate	$S_2O_3^{2-}$	2	$Na_2S_2O_3$	Sodium thiosulphite
Dichromate(VI)	$Cr_2O_7^{2-}$	2	$K_2Cr_2O_7$	Potassium chromate(VI)
Phosphate	PO_4^{3-}	3	$(NH_4)_3PO_4$	Ammonium phosphate

The valency of the nitrate ion, NO_3^-, is 1, because it combines with one hydrogen atom to form HNO_3.

The valency of the hydrogencarbonate ion, HCO_3^-, is 1, because it can combine with one hydrogen atom to form H_2CO_3.

The valency of the sulphate ion, SO_4^{2-}, is 2, because it combines with two hydrogen atoms to form H_2SO_4.

The valency of the sulphite ion, SO_3^{2-}, is 2, because it combines with two hydrogen atoms to form H_2SO_3.

The valency of the phosphate ion, PO_4^{3-}, is 3, because it combines with three hydrogen atoms to form H_3PO_4.

The valencies of some complex ions are given in the table on the opposite page.

Limitations of the Octet Rule

• Hydrogen does not complete an octet: it completes a helium-like configuration, a duplet of electrons.

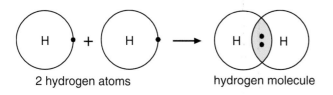

2 hydrogen atoms   hydrogen molecule

• The octet rule breaks down for boron, which forms many compounds with three electron pairs. Boron trichloride, BCl_3, is a molecule with three electron pairs: an **incomplete octet**.

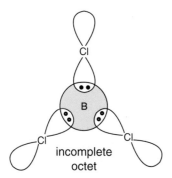

incomplete octet

• The octet rule works well for elements such as C, N, O and F which are in the second period but the rule seems to break down for elements in the third period.

For example, in phosphorus(V) chloride, PCl_5, the phosphorus has to use its d electrons in order to accommodate an extra pair of electrons. The molecule PCl_5 is said to have an **expanded octet**.

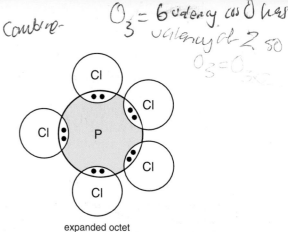

expanded octet

• There are many other exceptions to the octet rule. However, it can be used as a good guide to explain many chemical bonds.

Exercise 2.1

(1) The noble gases are stable gases. Explain, using the electronic configurations of the first three noble gases why they are stable.

(2) The use of the noble gases is related to their stability. Describe some uses of the noble gases.

(3) What is meant by completing an octet? Give an example.

(4) (a) What is a chemical bond?
(b) Describe, briefly, the main types of chemical bonds. Give an example of each type.
(c) Explain why the noble gases do not easily form chemical bonds.

(5) Explain, using selected atoms from the periodic table, why, on the one hand, you might expect atoms to combine with each other, while, on the other hand, you would expect them not to combine with each other.

(6) (a) Explain why an atom can only form a limited number of chemical bonds.
(b) What is valency?
(c) Write down the usual valency of each of the following atoms:
H, Be, Mg, Br, K, N, Si, P, Cl, B, C and F.
(d) How would you determine the valency of the complex ions in the following compounds:
(i) $MgSO_4$ (ii) Na_2CO_3 (iii) $Ca(NO_3)_2$
(iv) H_3PO_4

(7) Give the formulae of the compounds formed when the following pairs of elements combine with each other:

(a) Na and Cl (b) Mg and Cl (c) K and Br
(d) Mg and O (e) N and H (f) C and H
(g) O and H (h) O and O (i) P and Br
(j) C and Cl (k) O and F (l) N and N
(m) Al and O (n) Si and O (o) Mg and N
(p) Li and O

(8) What are the formulae of the following
compounds:
(a) calcium chloride (b) magnesium oxide
(c) sodium nitride (d) phosphorus(III) oxide
(e) phosphorus(V) chloride (f) mercury(I) sulphide (g) iron(II) oxide (h) iron(III) oxide
(i) magnesium nitrate (j) sodium sulphate
(k) potassium carbonate

(9) Transition metals have variable valency.
Predict the formulae of the compounds
formed when the following react:
(a) iron(II) and chlorine, iron(III) and chlorine
(b) iron(II) and oxygen, iron(III) and oxygen
(c) copper(II) and chlorine, copper(III) and
chlorine
(d) copper(II) and oxygen, copper(III) and
oxygen.

(10) The octet rule has some limitations.
Describe the limitations in relation to the
following:
(a) hydrogen chloride, HCl
(b) boron trichloride, BCl_3
(c) phosphorus(V) chloride, PCl_5.

2.2 IONIC BONDING

Ionic bonds are formed when a metal reacts
with a non-metal. When this type of bond is
formed electrons are transferred from the
metal atoms to the non-metal atoms during the
chemical reaction. The atoms that lose electrons become positive ions, called cations,
while the atoms that gain electrons become
negative ions, called anions. Ionic bonds result
from the attraction between these oppositely
charged ions. It should be remembered that
simple ions are charged atoms and that complex ions are groups of charged atoms. Both
types of ion, like atoms, are extremely small.

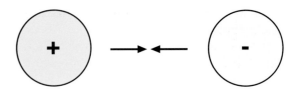

An **ionic bond** is the attractive force
between oppositely charged ions.

Sodium Chloride

Consider the reaction of a sodium atom with a
chlorine atom.

Sodium is in Group I of the periodic table: it
has one valence electron (electrons in the outer
shell are called valence electrons). It needs to
lose this electron to have a full outer shell.

Chlorine is in Group VII of the periodic
table: it has seven valence electrons. It needs
to gain one electron to have a full outer shell.

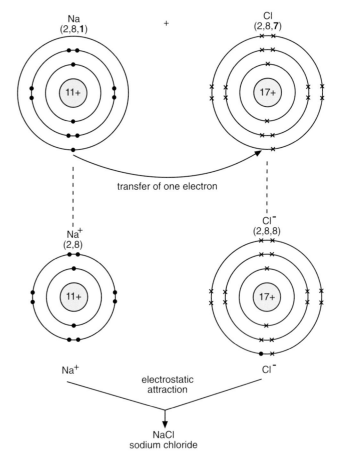

When the two atoms collide, all the electrons
come under the influence of both nuclei.
However, the chlorine atom has the greater
pull on the electrons. This enables the chlorine
atom to pull away the one outer electron from
the sodium atom. This leaves the sodium atom

'one electron short', giving it an overall charge of +1, turning it into a positive sodium ion. The chlorine atom, which gained the electron, has an overall charge of –1, turning it into a negative chloride ion. Because the two ions are oppositely charged they attract each other strongly and form an ionic compound, sodium chloride, NaCl.

Magnesium sulphate – an ionic crystal.

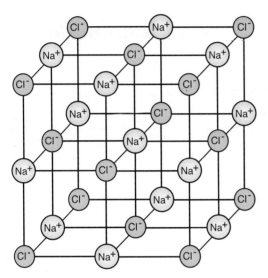
Structure of sodium chloride

Calcium Chloride

When chlorine gas is passed over heated calcium metal, a white substance, calcium chloride, is produced. Calcium is in Group II of the periodic table; thus it has two valence electrons. It needs to lose two electrons to have a full outer shell. The calcium atom transfers these two electrons to two chlorine atoms. This results in the formation of one positive calcium ion, Ca^{2+}, and two negative chloride ions, Cl^-. The calcium and chloride ions combine with each other and form an ionic compound, calcium chloride, $CaCl_2$.

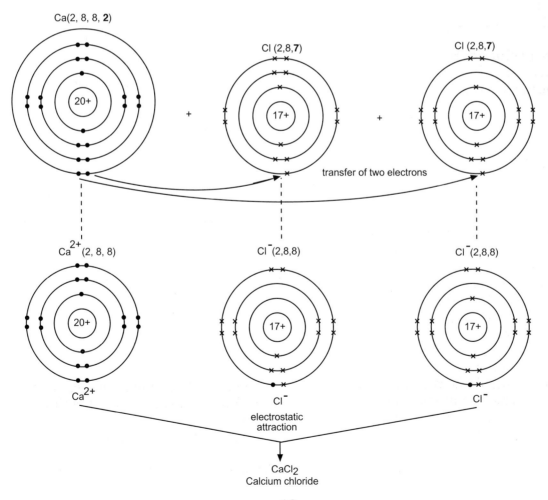

43

Characteristics of Ionic Compounds

(1) Ionic compounds consist of an orderly arrangement of oppositely charged ions which are held together by strong electrostatic forces in a crystalline structure called a lattice.

(2) When an ionic solid is heated, the ions within the crystalline lattice start to vibrate much more vigorously. Eventually as the energy absorbed increases the ions move apart from each other. Consequently, ionic compounds have high melting points and high boiling points. For example, sodium chloride, potassium chloride and calcium chloride all melt at temperatures in excess of 750 °C.

(3) Ionic compounds are usually soluble in water. The attraction between the water and the ions is sufficiently great to pull the crystalline lattice apart.

(4) In the solid state, ionic compounds do not conduct electricity, because the ions are in fixed positions. However, in the molten state or in solution the ions are free to move; consequently an electric current can flow through the molten ions or through the solution.

Ionic Materials in Everyday Life

• Lime, calcium oxide, is added to soil to lower the acidity of the soil.

• Salt tablets, NaCl, are used to replace lost salt during sweating.

• Aluminium sulphate is used in water treatment plants to help small suspended particles in water to coagulate (stick together) and fall to the bottom of the tank.

• Washing soda, Na_2CO_3, is used to soften hard water. Many commercial bath salts contain some washing soda.

Mandatory experiment 3

To test for anions in aqueous solutions

Introduction

Many different negative ions (anions) are found in aqueous solutions. These include chlorides, Cl^-, carbonates, CO_3^{2-}, hydrogencarbonates, HCO_3^-, nitrates, NO_3^-, sulphates, SO_4^{2-}, sulphites, SO_3^{2-}, and phosphates, PO_4^{3-}.

Many of these anions are found in normal concentrations in water and are there to balance the positive ions in the water. However, sometimes these anions can be present in large concentrations and may have a harmful effect. For example, high concentrations of nitrates and phosphates can cause the growth of dense algal blooms in water.

The presence of anions in water needs to tested on a regular basis. Each anion is tested using a particular technique and identified. A sample of each anion is tested by dissolving a salt in aqueous solution and carrying out a series of tests.

When the concentration of a particular ion is very low, it is usually determined using a Colorimeter or an optical comparator (see page 282).

Requirements

Safety glasses, salts of the anions: chloride, carbonate, hydrogencarbonate, nitrate, sulphate and phosphate, deionised water, test-tubes and test-tube rack, dilute sodium hydroxide, dilute hydrochloric acid, dilute nitric acid, concentrated sulphuric acid, silver nitrate solution, barium nitrate or barium chloride solution, ammonia solution and magnesium sulphate solution.

Procedure

(1) The procedure for each of the anions is given in the table. For all the salts, except for the carbonates and hydrogencarbonates, it involves adding the required solutions to an aqueous solution of the particular salt.

(2) Dilute HCl is added to solid samples of the carbonate and hydrogencarbonate salts. The procedure in the table on the next page is then followed.

Anion	Test	Observation	Confirmatory test
Carbonate, CO_3^{2-}	Add dilute HCl	CO_2 evolved turns limewater milky	Add $MgSO_4 \rightarrow$ white precipitate
Hydrogencarbonate, HCO_3^-	Add dilute HCl	CO_2 evolved turns limewater milky	Add $MgSO_4 \rightarrow$ no precipitate
Nitrate, NO_3^-	Make a solution of the salt. Add an equal volume of $FeSO_4(aq)$ and then carefully pour concentrated H_2SO_4 down the side of the test-tube	Brown ring forms	
Chloride, Cl^-	Make a solution of the salt. Add dilute HNO_3 and then silver nitrate solution	White precipitate of AgCl forms	Soluble in NH_3 solution
Sulphate, SO_4^{2-}	Make a solution of the salt. Add dilute HNO_3 and then barium nitrate solution	White precipitate of barium sulphate forms	Precipitate is insoluble in HCl or HNO_3
Sulphite, SO_3^{2-}	Make a solution of the salt. Add dilute HNO_3 and then barium nitrate solution	White precipitate of barium sulphate forms	Precipitate is soluble in HCl or HNO_3
Phosphate, PO_4^{3-}	Make an alkaline solution of the salt. Add barium nitrate solution	White precipitate of barium phosphate forms	

Exercise 2.2
•

(1) When a metal reacts with a non-metal to form an ionic compound, how do the elements attain an octet of electrons?

(2) Potassium fluoride, KF, is an ionic compound. Describe, using diagrams, how potassium combines with fluorine.

(3) (a) Explain how metals normally combine with non-metals.
(b) Describe the formation of the lithium chloride crystal from its constituent atoms.
(c) Why does lithium form an ionic bond with chlorine?

(4) Draw diagrams to illustrate the bonding in each of the following ionic compounds:
(a) $MgCl_2$ (b) CaO and (c) Li_2O.

(5) List some characteristics of ionic compounds.

(6) What particular property (or properties) of ionic compounds enables them to be used in the following everyday applications:
(a) salt tablets, NaCl, to replace salt lost by perspiring
(b) lime, CaO, used to neutralise soil
(c) aluminium sulphate, $Al_2(SO_4)_3$, used in water treatment
(d) sodium carbonate, Na_2CO_3, used in bath salts.

(7) A sample of sea water was tested for the presence of chloride ions, Cl⁻.
(a) What reagents were used to test the salt water?
(b) What was the colour of the precipitate?
(c) What was the name of the precipitate?

(8) An excessive growth of green material was noticed growing in a stream.
(a) What is the possible cause of the excessive growth?
(b) How would you test the water for the cause of the excessive growth?

(9) Many fertilisers contain nitrates and sulphates. Describe how you would test a fertiliser for the presence of these ions.

(10) A paper mill is suspected of dumping sulphite ions, SO_3^{2-}, into a river. Describe how you would test for the presence of sulphites in the water.

(11) A student was given a water sample to test for hardness. The sample was thought to contain ions responsible for temporary hardness. Describe an anion test which would determine if the ions causing temporary hardness were present.

(12) Limestone contains calcium carbonate, $CaCO_3$. How would you test the limestone for the presence of the carbonate ion?

(13) Three test-tubes labelled A, B and C contain sodium chloride, sodium nitrate and sodium sulphate, respectively. How would you discover what is in each test-tube?

(14) Two bags of nitrogen-containing fertilisers have been incorrectly labelled. If they contain ammonium phosphate and ammonium sulphate, describe tests you would use to distinguish between them.

(15) Describe how you would distinguish between carbonate and hydrogencarbonate ions in aqueous solution.

(16) A school ordered five compounds, all of which were soluble in water. When the order was delivered the labels were missing. The delivery contained a nitrate, a sulphate, a carbonate, a phosphate and a chloride. Describe tests you would use to identify the compounds.

2.3 COVALENT BONDING

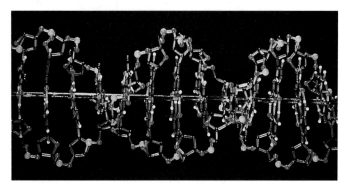

Covalent bonds in the DNA double helix.

Covalent bonds are usually found in non-metals and in the compounds they form with each other. A covalent bond is formed when a pair of electrons is shared between two atoms. By sharing electrons each atom usually obtains a complete outer shell (a noble gas structure).

The Hydrogen Molecule

The hydrogen atom has one electron and one proton. To obtain a full outer shell and have a noble gas structure, it must have two electrons. When one hydrogen atom approaches another, its electron is attracted to the nucleus of the other atom. At the same time, its own nucleus attracts the electron of the other atom. As the atoms draw closer to each other it becomes impossible to distinguish which electron belongs to which atom. The electrons overlap with each other and form an electron pair. This shared pair of electrons is known as a covalent bond.

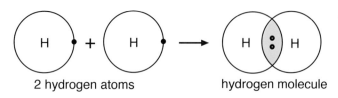

2 hydrogen atoms   hydrogen molecule

Electron-dot structures of this type are often called Lewis structures, named after Gilbert N. Lewis who proposed this theory of covalent bonding in 1916.

The hydrogen molecule can be represented by H : H or by H—H.

The strength of the covalent bond is due to the attraction between the positively charged nucleus and the negative electron cloud surrounding both nuclei.

46

The Hydrogen Chloride Molecule

When a hydrogen atom approaches a chlorine atom, covalent bonding can again take place. The chlorine atom, like the hydrogen atom, requires one electron in order to have a full outer shell. The chlorine atom and the hydrogen atom share a pair of electrons between them.

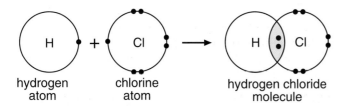

hydrogen atom   chlorine atom   hydrogen chloride molecule

The hydrogen chloride molecule is usually represented by H : Cl or by H—Cl.

Other Covalent Molecules

Some molecules can form more than one covalent bond between two atoms. For example, a double covalent bond is formed when two oxygen atoms combine with each other. The noble gas structure is attained when one oxygen atom shares two electron pairs with the other oxygen atom.

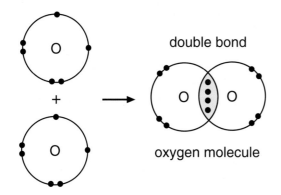

double bond

oxygen molecule

2 oxygen atoms

This double bond is usually represented as O :: O or as O=O.

A triple covalent bond is formed when two nitrogen atoms combine with each other. The inert gas structure is attained when a nitrogen atom shares three electron pairs with the other nitrogen atom.

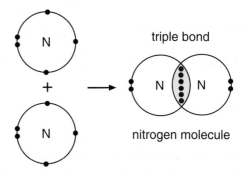

triple bond

nitrogen molecule

2 nitrogen atoms

This triple bond is usually represented as N :: N or as N≡N.

The nitrogen atom in the ammonia molecule, NH_3, does not use all its valence electrons when it bonds to hydrogen atoms. Three of the valence electrons on the nitrogen form bond pairs with the valence electron on each of the three hydrogen atoms. The remaining two valence electrons on the nitrogen atom are paired up with each other to form a non-bonding lone pair.

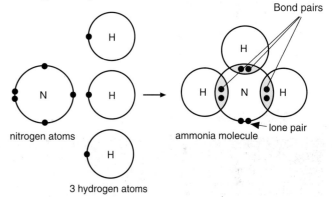

Bond pairs

nitrogen atoms   ammonia molecule

lone pair

3 hydrogen atoms

Covalent Bonding: Sigma and Pi Bonds

Covalent bonds can also be described in terms of the overlap of atomic orbitals. When atomic orbitals overlap they form a molecular orbital. The covalent bond results from the overlap of the unfilled atomic orbitals to form a fully occupied molecular orbital.

The hydrogen molecule, H_2, is formed by the overlap of the 1s electrons of the two hydrogen atoms.

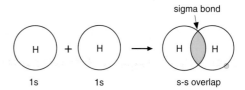

sigma bond

1s   1s   s-s overlap

This 1s—1s overlap results in the formation of a strong covalent bond called a sigma (σ) bond.

47

The hydrogen chloride molecule, HCl, is formed by the overlap of the 1s orbital in hydrogen with the $3p_z$ orbital in chlorine. This 'head-on overlap' of the outer valence electrons, also results in the formation of a strong sigma bond.

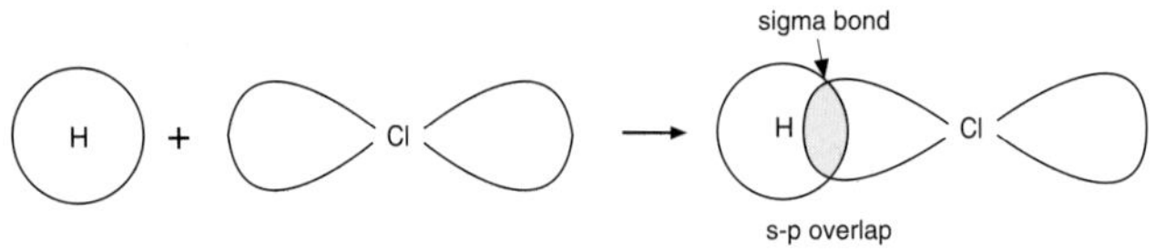

Sigma bonds are formed by the 'head-on overlap' of atomic orbitals.

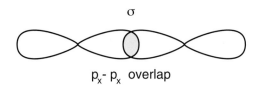

σ

p_x - p_x overlap

sigma bond

s-p overlap

Electron-dot structures and models of some simple molecular substances

Hydrogen H_2	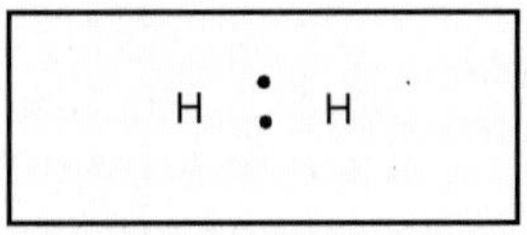 H : H	H - H	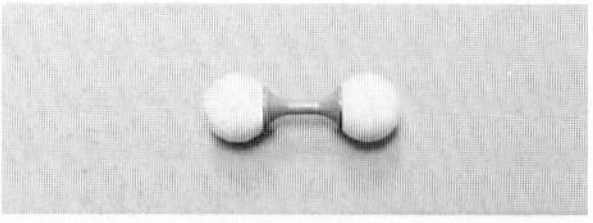
Oxygen O_2	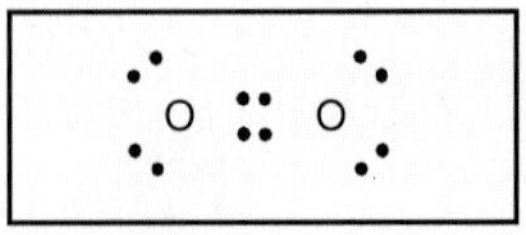 O :: O	O = O	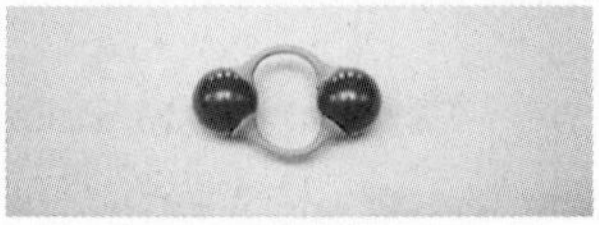
Methane CH_4	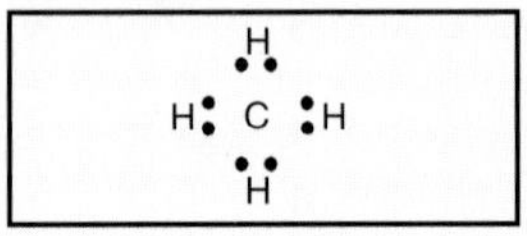 H C H (with H top and bottom)	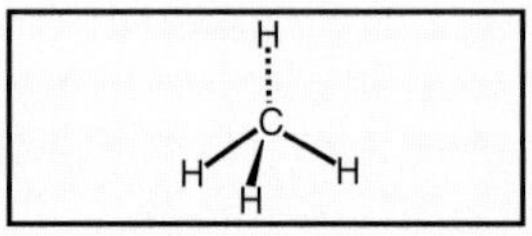 H C H H H	
Carbon dioxide CO_2	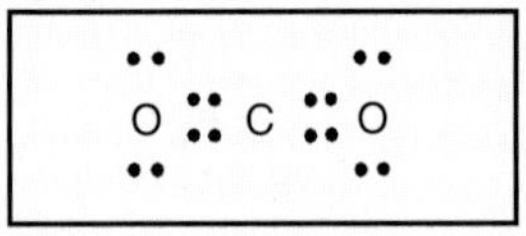 O :: C :: O	O = C = O	
Carbon monoxide CO	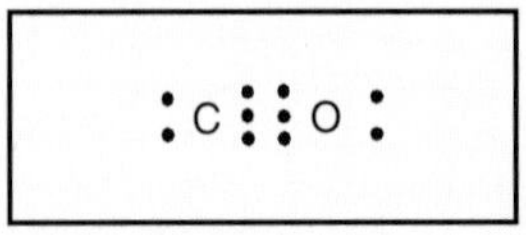 :C ⦂ O:	C ≡ O	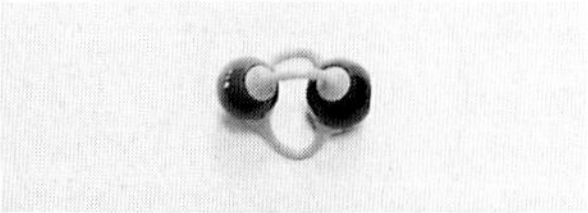
Water H_2O	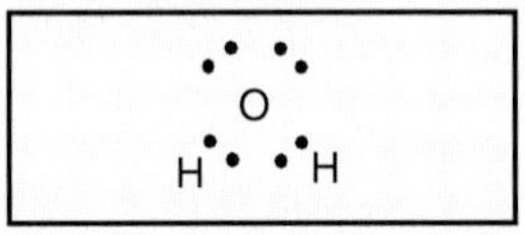 O H H	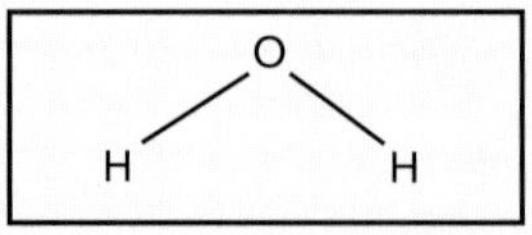 O H H	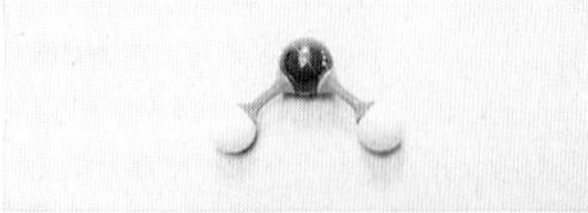
Ammonia NH_3	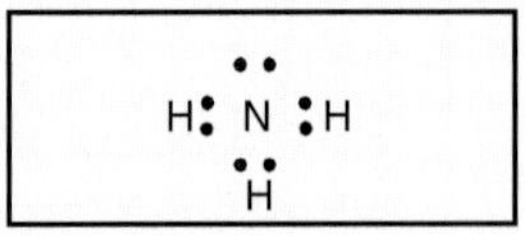 H N H H	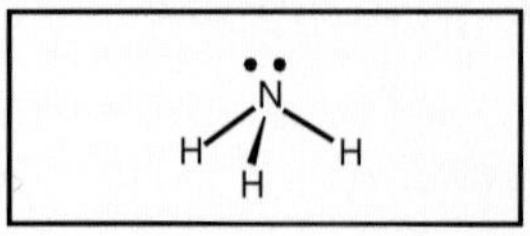 N H H H	
Nitrogen N_2	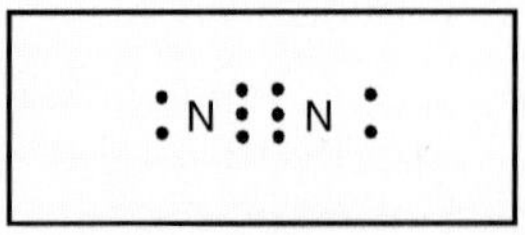 :N ⦂ N:	N ≡ N	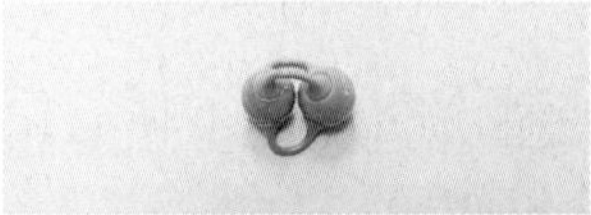
Ethene C_2H_4	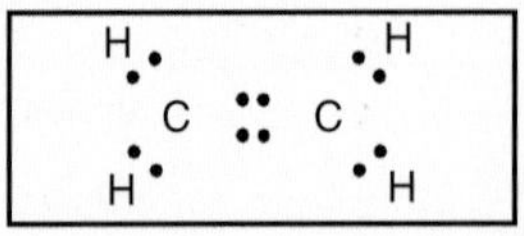 H H C :: C H H	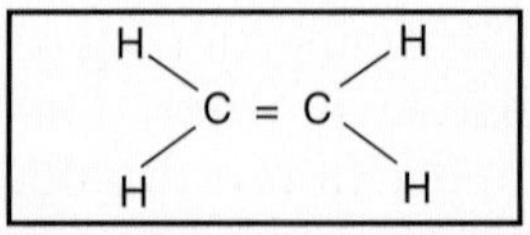 H H C = C H H	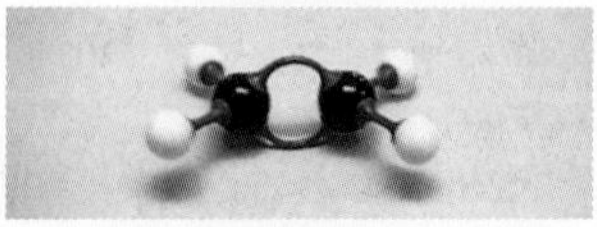
Ethyne C_2H_2	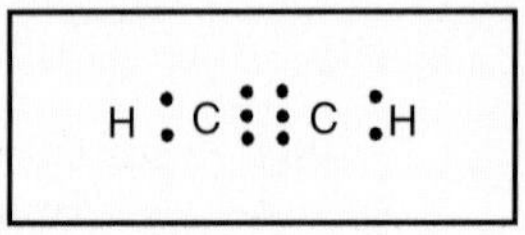 H :C ⦂ C: H	H - C ≡ C - H	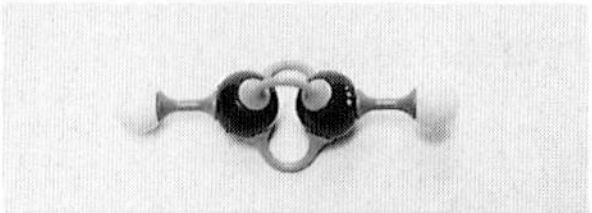

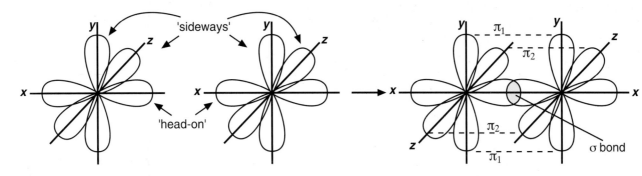

The nitrogen molecule, N_2, is formed by the overlap of the three unfilled atomic orbitals in each nitrogen atom, the $2p_x$ orbital, the $2p_y$ orbital and the $2p_z$ orbital. The orbitals must overlap in different ways if the two atoms are to join together. If the two $2p_x$ orbitals overlap in a 'head-on' manner, the remaining $2p_y$ and $2p_z$ orbitals must then overlap in a 'sideways' manner. This 'sideways overlap' results in the formation of two pi (π) bonds, one bond above and below the plane, the other in front of and behind the plane. These pi bonds are not as strong as sigma bonds.

Pi bonds are formed by the 'sideways overlap' of atomic orbitals.

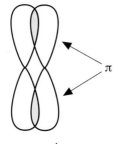

p$_y$- p$_y$ overlap

The bonding in the ethene molecule, C_2H_4, involves the formation of a double bond between the two carbon atoms. This carbon–carbon bond consists of one sigma bond and one pi bond.

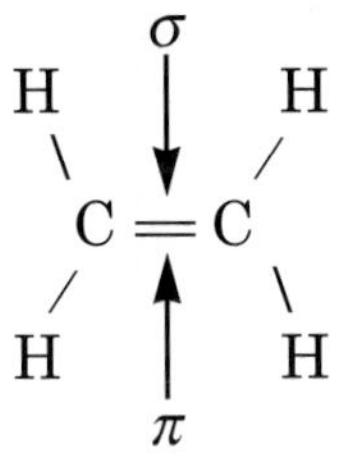

$$
\begin{array}{ccc}
H & \sigma & H \\
\diagdown & \downarrow & \diagup \\
& C = C & \\
\diagup & \uparrow & \diagdown \\
H & \pi & H
\end{array}
$$

The bonding in the ethyne molecule, C_2H_2, involves the formation of a triple bond between the two carbon atoms. This carbon–carbon bond consists of one sigma bond and two pi bonds.

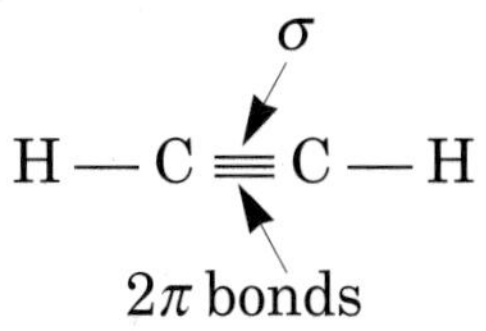

$$
H - C \overset{\sigma}{\equiv} C - H
$$

2π bonds

Polar and Non-polar Covalent Bonding

A covalent bond involves the sharing of electron pairs between two atoms.

When the two atoms are alike, as in the case of the hydrogen molecule, H_2, the shared pair of electrons is equally distributed between the two hydrogen atoms. The shared pair of electrons is under the influence of two similar nuclei. Covalent bonds of this type are termed non-polar covalent bonds.

Electrons occupy equal space around both atoms

Non-polar covalent bonds are bonds where the electrons are shared equally between the atoms.

When the two atoms are not alike, as in the case of the hydrogen chloride molecule, HCl, the shared pair of electrons spend more time nearer the chlorine atom than the hydrogen atom. Covalent bonds of this type are termed polar covalent bonds.

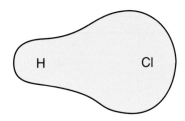

A **polar covalent bond** is a bond where the bonding electrons are shared unequally between the atoms.

A polar covalent bond can be regarded as being intermediate between a non-polar covalent bond, such as in H_2, and an ionic bond, such as in NaCl. A polar covalent bond possesses some percentage of ionic character.

This percentage of ionic character is usually represented by placing a small negative charge, indicated by δ^-, on the atom which has the greater pull on the electron pair and by placing a small positive charge, indicated by δ^+, on the atom which has less pull on the shared pair.

This separation of charge is denoted in HCl as $H^{\delta+}$—$Cl^{\delta-}$.

Polar molecules, such as hydrogen chloride, are often called dipoles.

Some Polar and Non-polar Covalent Compounds

Polar		*Non-polar*	
Hydrogen chloride	HCl	Hydrogen	H_2
Hydrogen iodide	HI	Chlorine	Cl_2
Water	H_2O	Oxygen	O_2

some polar and non-polar molecules

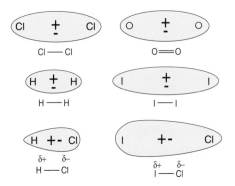

Activity 2.1

Demonstration of the Polarity of Some Liquids

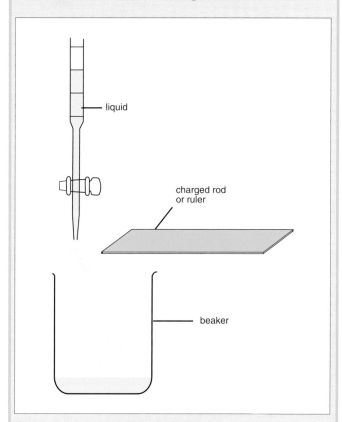

Introduction

A chemical bond is polar when the bonding electrons are unequally distributed between the two atoms which are joined together. Hydrogen chloride, HCl, is an example of a molecule with a polar bond. Molecules which are polar are often called dipoles.

Requirements

Safety glasses, burettes, 250 cm^3 beakers, water, ethanol, cyclohexane, propanone, methylbenzene, 1,1,1-trichloroethane, plastic ruler, woollen cloth.

Procedure

(1) Half fill a clean burette with water. Place a clean beaker under the burette.
(2) Charge the plastic ruler by rubbing it with a dry woollen cloth.
(3) Turn on the burette tap to allow a jet of water to flow.
(4) Bring the charged plastic ruler close to the jet of water but not touching it.

(5) Record the deflection (if any) of the water.
(6) Repeat the procedure for each of the other liquids.

(7) Copy the table below and record your observations in it.
(8) Collect and recycle the solvents.

Observations and Conclusions

Liquid	Formula	Deflection	Polarity
Water			
Ethanol			
Cyclohexane			
Propanone			
Methylbenzene			
1,1,1-trichloroethane			

Activity 2.2

Demonstration of the Solubility of Some Ionic and Covalent Substances

Introduction

Ionic and polar covalent substances are soluble in polar solvents, while non-polar covalent substances are soluble in non-polar solvents. The solubility of a typical ionic compound, sodium chloride, is tested in a range of different solvents, while the solubility of iodine, a covalent solid, is tested in the same solvents.

Requirements

Safety glasses, two test-tube racks, twelve test-tubes, six solvents (water, ethanol, cyclohexane, propanone, methylbenzene, 1,1,1-trichloroethane), two solutes (sodium chloride and iodine), parafilm.

Procedure

(1) Place six test-tubes in a test-tube rack.
(2) Place a very small quantity of sodium chloride in each test-tube.

(3) Add about 3 cm³ of water to the first test-tube. Place some parafilm over the mouth of the test-tube and shake for a few minutes.
(4) Copy the table below and record your observations on it.
(5) Repeat the procedure for the remaining five solvents.
(6) Repeat the procedure for each solvent, but this time placing iodine in each test-tube instead of sodium chloride.
(7) Record your observations in your table.

Observations

Solvent	Solute	Observation
Water		
Ethanol		
Cyclohexane		
Propanone		
Methylbenzene		
1,1,1-trichloroethane		

Conclusion

In general, like dissolves like, that is, ionic compounds dissolve in polar solvents, while covalent compounds dissolve in non-polar solvents.

Polar and Non-Polar Materials in Everyday Life

• Water is perhaps the most important polar chemical compound of all. Practically everything on this earth dissolves to some extent in water.

• Sodium chloride is perhaps the best known ionic compound. It is used by people every day to enhance the taste of food.
• The alcohols, ethanol and propanol, both of which are slightly polar, are very common solvents. Propanol is used in antiseptic wipes and in audio and video cleaning tapes.

51

• 1,1,1-trichloroethane is used to thin Tipp-Ex®. It is a dual solvent. It can dissolve polar and non-polar solutes.
• Chlorofluorocarbons and their ozone-friendly substitutes are non-polar solvents which are used in aerosol sprays and refrigeration units.
• Propanone (acetone), a non-polar solvent, is the main constituent of nail varnish remover.
• Methylbenzene (toluene) is a non-polar solvent and is used to dissolve many compounds.

Characteristics of Covalent Compounds

(1) Covalent compounds usually consist of separate molecules which have very little attractive force between them. Consequently, at room temperature, covalent substances are usually gases or volatile (easily evaporated) liquids.
(2) Because the binding forces between the molecules are extremely weak, covalent compounds have low melting points and boiling points.
(3) Pure covalent compounds do not conduct electricity, which indicates that they do not contain ions. Polar covalent compounds, such as hydrogen chloride and ammonia, do conduct electricity when dissolved in water.
(4) Many covalent compounds do not dissolve in water. Most are soluble in covalent compounds such as trichloroethane and benzene.

Exercise 2.3

•

(1) (a) How small are molecules?
(b) What is a covalent bond?
(c) Describe, using dot and cross diagrams, how the following molecules are covalently bound together:
(i) hydrogen, H_2 (ii) hydrogen chloride, HCl
(iii) oxygen, O_2 (iv) chlorine, Cl_2

(2) Draw each of the following molecules using Lewis (dot and cross) structures:
(a) carbon dioxide, CO_2 (b) methane, CH_4
(c) ethene, C_2H_4.

(3) (a) Explain and illustrate (i) sigma bonds and (ii) pi bonds.
(b) Describe the bonding in the following molecules in terms of the overlap of atomic orbitals. In each case, indicate whether the bonds formed are sigma bonds or pi bonds.
(i) H_2 (ii) HCl (iii) Br_2 (iv) N_2 (v) C_2H_4

(4) (a) Explain the difference between the covalent bond in H_2 and the covalent bond in HCl.
(b) What is meant by a polar covalent bond?

(5) Which of the following molecules would you expect to be polar? In each case explain your reason.
(a) H_2 (b) NO (c) Cl_2 (d) O_2 (e) HCHO
(f) C_2H_4

(6) Using the periodic table, explain why the bond between H and F is more polar than that between H and N.

(7) Describe some of the characteristics of covalent compounds. Explain why they have those particular properties.

(8) Describe, briefly, how you would show polarity in liquids.

(9) Polar and non-polar materials have widespread applications in everday life. For each type of material, give two examples where they are used as solvents.

(10) List all the non-metallic atoms which occur as diatomic molecules.

(11) Lithium aluminium hydride, $LiAlH_4$, is an important reducing agent. Write the Lewis electron-dot formula for the AlH_4^- ion.

(12) Draw Lewis electron-dot structures for each of the following molecules:
(a) H_2S (b) PH_3 (c) SiH_4 (d) HCN (e) ICl
(f) CH_2H_2 (g) H_2O_2 (h) BrCl

(13) Explain, using diagrams, why ethanol is soluble in water.

2.4 ELECTRONEGATIVITY

Electronegativity is a measure of the relative ability of an atom in a molecule to attract electrons to itself.

In 1932, **Linus Pauling** proposed an electronegativity scale based on bond energies. The Pauling scale ranges from 0 to 4.0. Fluorine is the most electronegative atom and is given the value 4.0. Other atoms are given values relative to fluorine depending on their ability to attract electrons.

Electronegativities for the main-group elements

H 2.1							He —
Li 1.0	Be 1.5	B 2.0	C 2.5	N 3.0	O 3.5	F 4.0	N —
Na 0.9	Mg 1.2	Al 1.5	Si 1.8	P 2.1	S 2.5	Cl 3.0	Ar
K 0.8	Ca 1.0	Ga 1.6	Ge 1.8	As 2.0	Se 2.4	Br 2.8	Kr —
Rb 0.8	Sr 1.0	In 1.7	Sn 1.8	Sb 1.9	Te 2.1	I 2.5	Xe —
Cs 0.7	Ba 0.9	Tl 1.8	Pb 1.8	Bi 1.9	Po 2.0	At 2.2	Rn —
Fr 0.7	Ra 0.9						

Polarity and Electronegativity – Prediction of Bond Type

Electronegativity values can be used to predict bond type.
A difference in electronegativity between two atoms of **> 1.7** indicates that the bond is **ionic.**
A difference in electronegativity between two atoms of **< 1.7** indicates that the bond is **covalent**.

The polarity of the HCl bond is due to the difference between the electronegativity of the Cl atom and the electronegativity of the H atom.

The electronegativity of chlorine is 3.0, while the electronegativity of hydrogen is 2.1. The absolute value of the electronegativity difference between the two bonded atoms is $3.0 - 2.1 = 0.9$.

This value gives a rough measure of the polarity of the bond. In this case, the value is relatively large and tells us that the H—Cl bond is polar covalent. This polar covalent bond is said to have some percentage of ionic character.

The electronegativity difference for the H—H bond is $2.1 - 2.1 = 0$; this means that the bond is 100% non-polar covalent.

The electronegativity difference for the Na—Cl bond is $3.0 - 0.9 = 2.1$. This very large value tells us that the bond is ionic (if the difference is greater than 1.7 it is assumed to be an ionic bond).

Bond types in other substances can be predicted from electronegativity differences, which tell us whether the bond type is non-polar covalent, polar covalent or ionic.

General Trends in Electronegativity Values

(1) Electronegativity increases from left to right across any period as the number of valence electrons increases.
(2) Electronegativity values decrease from top to bottom in any group as the atomic size increases.
(3) Metals are the least electronegative elements (they are electropositive), while non-metals are the most electronegative.

Difference in electronegativity	Bond type	% Ionic character	Examples
0	Non-polar covalent	0	H—H O=O Cl—Cl
0.9	Polar covalent	17	H—Cl
1.0	Polar covalent	22	C—O
1.8	Ionic	55	Li—Br
2.1	Ionic	66	Na—Cl
2.4	Ionic	76	Mg—O
3.2	Ionic	92	K—F

Polar Bonds: Hydrogen Bonding

Hydrogen bonding is the attractive force which exists between a hydrogen atom covalently bonded to a very electronegative atom, X, and another small electronegative atom, Y.

Polar molecules containing a hydrogen atom attached to an electronegative element, such as fluorine, oxygen or nitrogen, have a tendency to unite together in a special way by means of hydrogen bonds. The hydrogen atom acts as a bridge between the two strongly electronegative atoms.

The attraction of the highly electronegative element exerts such a strong pull on the bonding electrons in such molecules that the hydrogen atom is left with an appreciable δ^+ charge. In fact, the hydrogen atom acts almost like an exposed proton in such situations.

Some hydrogen-bonded molecules are illustrated below, where the hydrogen bonds are indicated by -----.

$$-----H^{\delta+}\!\!-\!F^{\delta-}\!-----H^{\delta+}\!\!-\!F^{\delta-}\!-----H^{\delta+}\!\!-\!F^{\delta-}\!-----$$

$$-----H^{\delta+}\!\!-\!N^{\delta-}\!-----H^{\delta+}\!\!-\!N^{\delta-}\!-----H^{\delta+}\!\!-\!N^{\delta-}\!-----$$

(each N additionally bonded to two H atoms)

The most familiar hydrogen-bonded structure is water. The individual water molecules are held together by a vast three-dimensional network of hydrogen bonds.

These weak hydrogen bonds (10–40 kJ mol^{-1}) exert a large influence on the physical and chemical properties of the molecules which form them.

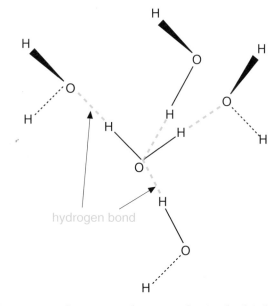

hydrogen bond

For example, water has a relatively high melting point and boiling point for such as a small molecule. The water molecules, because they are joined together by hydrogen bonds, seem to behave as one large molecule.

The boiling points of some hydrogen bonded compounds containing the elements of groups IV, V, VI and VII are illustrated below.

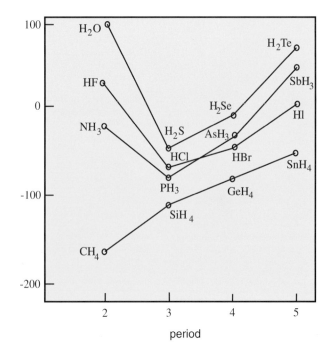

The main points to notice are:

(1) The boiling point of the first member of each series where hydrogen bonding occurs (H_2O, HF and NH_3) is extremely high in comparison to the other members of the group.

(2) There is no hydrogen bonding in the group IV A series.

(3) Really effective hydrogen bonds are only formed by fluorine, oxygen and nitrogen compounds. For instance, although chlorine has the same electronegativity as nitrogen (3.0), H—Cl forms weaker hydrogen bonds than NH_3. The chlorine atom is larger than nitrogen and accordingly has a more diffuse electron cloud and cannot form strong hydrogen bonds.

(4) The boiling point of H_2O is higher than that of HF, although the hydrogen bonding in HF is stronger. There are, however, twice as many hydrogen bonds per molecule in water as there are in hydrogen fluoride.

Hydrogen bonding also accounts for the high solubilities of some compounds containing oxygen, nitrogen and fluorine in hydrogen-bonded solvents like water. Thus, methanol and many other molecules dissolve readily in water due to the formation of hydrogen bonds.

Methanol   Water

Exercise 2.4

•

(1) (a) What is meant by the electronegativity of an element?

(b) Using an electronegativity table, calculate the electronegativity differences between the following pairs of atoms:
(i) H and F (ii) H and S (iii) H and N
(iv) C and O (v) H and H (vi) Cl and Cl
(vii) Na and Br (viii) Mg and O
(ix) K and F.

(c) In each example in (b) above, indicate the bond type you expect to form between each pair of atoms.

(d) If polar covalent bonds are formed between the atoms listed in part (b), assign partial charges to the atoms, using the symbols δ^+ and δ^-.

(2) Discuss the general trends in electronegativity values.

(3) For each of the following pairs of elements, state whether the compound formed is likely to be ionic, non-polar covalent or polar covalent:
(a) H and F (b) Li and Cl (c) Br and Br
(d) Sr and O

(4) (a) What is a hydrogen bond?
(b) Illustrate hydrogen bonding in each of the following compounds:
(i) water, H_2O (ii) ammonia, NH_3

(5) (a) Hydrogen bonds influence the chemical and physical properties of many molecules. Explain why.
(b) The boiling point of water is extremely high for such a small molecule. Explain why this is so.
(c) Ammonia, NH_3, has stronger hydrogen bonds than hydrogen chloride. Explain the reason for this.
(d) Explain why the boiling point of water is higher than that of hydrogen fluoride.

(6) What type of bond exists:
(a) between the atoms in a molecule of hydrogen chloride
(b) between molecules of hydrogen chloride?

55

2.5 SHAPES OF MOLECULES AND INTERMOLECULAR FORCES

Shapes of Simple Molecules

The electron-dot formulae to show how molecules are combined does not give us any information about the shape of a molecule.

Some shapes are easy to predict. All molecules containing only two atoms are linear in shape.

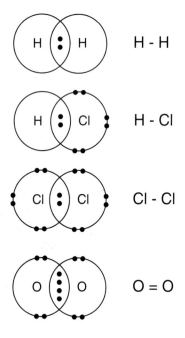

H - H

H - Cl

Cl - Cl

O = O

The shapes of most other molecules cannot be predicted using electron-dot structures.

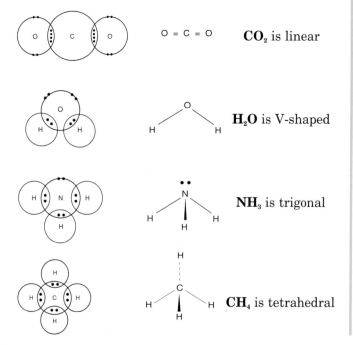

O = C = O   **CO$_2$** is linear

H$_2$O is V-shaped

NH$_3$ is trigonal

CH$_4$ is tetrahedral

Shapes of Molecules – Electron Pair Repulsions

The geometric arrangement of atoms in molecules may be predicted by considering the repulsions between the pairs of electrons about an atom.

Each electron pair is equally important and the repulsions between these pairs determine the shape of the molecule.

One theory which is used to predict molecular shape is called 'the valence shell electron pair repulsion theory' and is written in shorter form as **VSEPR**.

The VSEPR model is based on the fact that electron pairs repel one another and take up positions as far apart from each other as possible.

The valence shell electron pairs around a particular central atom may be either pairs which form covalent bonds (called bond pairs) or they may be unshared pairs (called lone pairs).

The main difference between a lone pair and a bond pair is the amount of space they take up. The space taken up by a lone pair is greater than that taken up by a bond pair. A lone pair is more diffuse (spreads out more) than a bond pair because it is under the influence of the nuclear charge of one atom, whereas a bond pair is under the influence of the nuclear charges of two atoms.

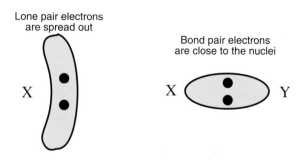

Lone pair electrons are spread out

Bond pair electrons are close to the nuclei

Lone pairs are localised near the central atom and are in a better position to push away other lone pairs. Bond pairs are closer to the nucleus and cannot exert the same influence.

Consequently the expected order of repulsion between electron pairs decreases as shown:

**Lone pair : Lone pair
> Lone pair : Bond pair
> Bond pair : Bond pair**

The shape of a molecule is determined by:
(1) The number of pairs of electrons around the central atom.
(2) The type of electron pairs around the central atom (i.e. whether they are bond pairs or lone pairs).

Two Pairs of Electrons
A molecule in which the central atom has two bonding pairs in its valence shell is always **linear**.

Consider the beryllium hydride molecule, BeH_2.

Beryllium has the electronic configuration $1s^2$, $2s^2$; thus it has two valence electrons.

Hydrogen has the electronic configuration $1s^1$, which means it has one valence electron.

The two valence electrons in beryllium are used to form two covalent bonds with two electrons from two hydrogen atoms.

The BeH_2 molecule formed is linear in shape, because the two bond pairs repel each other as far as possible.

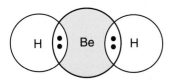 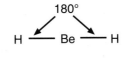

Three Pairs of Electrons
A molecule in which the central atom has three bonding pairs in its valence shell is always triangular **planar**.

Consider the boron fluoride molecule, BF_3.

Boron has the electronic configuration $1s^2$, $2s^2$, $2p_x^1$; thus it has three valence electrons.

Fluorine has the electronic configuration $1s^2$, $2s^2$, $2p_x^2$, $2p_y^2$, $2p_z^1$, which means it has one

electron available for bonding.

In BF_3, boron uses each of its three valence electrons to form a bond pair with an electron from each fluorine atom.

The BF_3 molecule formed is triangular and planar, because this arrangement provides the best separation between the bond pairs.

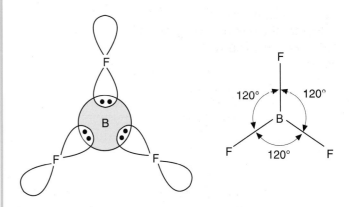

Four Pairs of Electrons – No Lone Pairs
A molecule in which the central atom has four bonding pairs and no lone pairs in its valence shell is always **tetrahedral**.

Consider the methane molecule, CH_4.

Carbon has the electronic configuration $1s^2$, $2s^2$, $2p_x^1$, $2p_y^1$; thus it has four valence electrons.

These valence electrons can bond with each valence electron from four hydrogen atoms and form four bond pairs.

The methane molecule is tetrahedral in shape because this arrangement enables the four bond pairs to repel each other as far apart as possible.

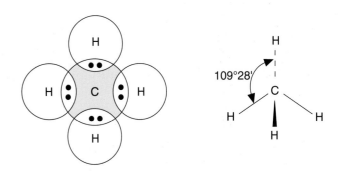

Four Pairs of Electrons – Some Lone Pairs

For a molecule in which the central atom has four bonding pairs in its valence shell:
• if the four pairs consist of one lone pair and three bond pairs, the molecule will be **trigonal pyramidal** in shape
• if the four pairs consist of two lone pairs and two bond pairs, the molecule will be **planar V-shaped**.

Consider the ammonia molecule, NH_3.

Nitrogen has the electronic configuration $1s^2$, $2s^2$, $2p_x^1$, $2p_y^1$, $2p_z^1$; thus it has five valence electrons, three of which can form bond pairs with each valence electron from the three hydrogen atoms, while the other two electrons on the central nitrogen atom remain as a non-bonding lone pair.

The non-bonding lone pair of electrons are more diffuse than the three bond pairs and repel the bond pairs and squeeze them closer together. As a result the H—N—H bond angle is 107° and not the perfect tetrahedral angle of 109° 28′. Because the lone pair is close to the central atom, the molecular shape is a trigonal pyramid.

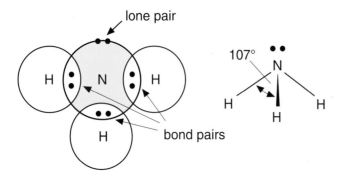

Consider the water molecule, H_2O.

Oxygen has the electronic configuration $1s^2$, $2s^2$, $2p_x^2$, $2p_y^1$, $2p_z^1$; thus it has six valence electrons, two of which can form bond pairs with the valence electrons from the two hydrogen atoms, while the other four electrons on the central oxygen atom remain as two non-bonding lone pairs.

The non-bonding lone pairs of electrons repel the bond pairs and squeeze them even closer together than in ammonia. As a result the H—O—H bond angle is 105° and because of the electron pair repulsions the molecule adopts a planar V-shaped arrangement.

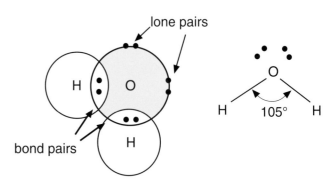

Relationship Between Symmetry and Bond Polarity

Methane, CH_4, is a totally symmetrical molecule: each polar C—H bond is the same as the others. Methane is a non-polar molecule where each corner of the tetrahedron is occupied by a hydrogen atom.

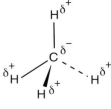

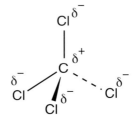

Tetrachloromethane, CCl_4, is also a totally symmetrical molecule. Each corner of the tetrahedron is occupied by a similar atom.

In both cases the centre of positive and negative charges coincide: there is no permanent dipole.

Number of electron pairs in the valence shell of the central atom and the molecular shape

Number of electron pairs			Shape of molecule	Examples
Bond pairs	Lone pairs	Total		
2	0	2	Linear	BeH_2, BeF_2, $BeCl_2$
3	0	3	Planar	BH_3, BF_3, AlH_3
4	0	4	Tetrahedral	CH_4, CCl_4, SiH_4
3	1	4	Trigonal pyramid	NH_3, PH_3
2	2	4	Planar V-shaped	H_2O, H_2S

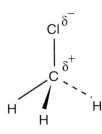

Monochloromethane, CH_3Cl, is not a symmetrical molecule. One corner of the tetrahedron is occupied by a chlorine atom, while the other three corners are occupied by hydrogen atoms. This disturbs the perfect symmetry of the molecule: the centres of positive and negative charges do not coincide and this gives the molecule a permanent dipole. CH_3Cl is a polar molecule.

Bonding Within and Between Molecules: Intramolecular Bonding and Intermolecular Forces

The individual bonds between the central atom and the surrounding atoms were examined previously. The individual bond which exists between the atoms within a molecule is called an **intramolecular bond**.

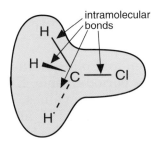

The attractive (or repulsive) forces which exist between two or more molecules are called **intermolecular forces**.

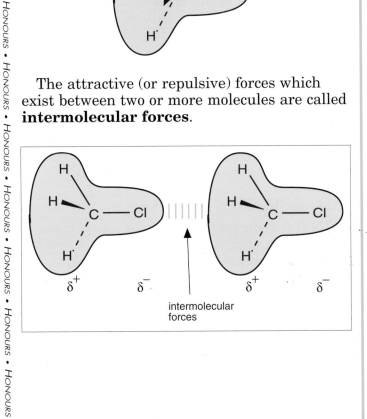

The forces which exist between collections of molecules are:
• van der Waal's forces: these are the weak forces which exist between all forms of matter
• dipole–dipole interactions: these are fairly weak forces which attract polar molecules to each other

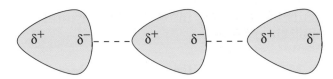

dipole-dipole interactions

• hydrogen bonding: hydrogen bonds (see page 54) are a special type of dipole–dipole interaction.

Boiling Points and Covalent and Intermolecular Forces

The stronger the intermolecular force within a substance, the higher is that substance's boiling point.

The attractive force which exists between hydrogen molecules, H_2, is the weak van der Waal's force. Consequently, hydrogen molecules tend to stay apart from each other; as a result the boiling point of hydrogen is quite low (–252 °C).

The force of attraction between nitric oxide, NO, molecules is a dipole–dipole interaction which is a relatively strong intermolecular force. As a result nitric oxide has a higher boiling point than hydrogen (–152 °C).

The force which holds water molecules, H_2O, together is the hydrogen bond which is a stronger bond than the dipole–dipole interaction. As a result water has a higher boiling point (100 °C) than either hydrogen or nitric oxide.

Intermolecular force	Boiling point
Van der Waal's	Very low
Dipole–dipole	Low
Hydrogen bond	High

Activity 2.4

Comparison of Boiling Points

Find out the boiling points of the pairs of molecules in the table below and record them in a copy of the table. Compare the pairs of molecules in terms of their inter-molecular forces.

Molecule	Boiling point	Intermolecular force
O_2, NO		
C_2H_4, HCHO		
H_2O, H_2S		

• Nitrogen(II) oxide, NO, has a higher boiling point than oxygen, O_2, because NO is a polar molecule, while O_2 is non-polar.
• Chlorine, Cl_2, has a higher boiling point than fluorine, F_2, because Cl_2 is larger than F_2 and consequently the van der Waal's forces are greater.
• Water, H_2O, has a higher boiling point than hydrogen sulphide, H_2S, because water can form strong intermolecular hydrogen bonds, while hydrogen sulphide cannot form hydrogen bonds.

Exercise 2.5

(1) Explain what you understand by the valence shell electron pair repulsion (VSEPR) theory.

(2) Use the VSEPR theory to predict the shape of each of the following molecules:
(a) BeF_2 (b) BH_3 (c) CCl_4

(3) (a) Explain why lone pairs repel each other more than bond pairs.
(b) List the order of repulsion of electron pairs according to the VSEPR theory.
(c) Explain why BH_3 is triangular and planar in shape, while NH_3 is trigonal pyramidal in shape.

(4) Give examples of molecules (two in each case) which have the following shapes:
(a) triangular and planar (b) linear
(c) tetrahedral (d) V-shaped and planar.

(5) (a) Which of the following molecules is linear in shape: CH_4, H_2S, N_2, O_2, HCl, NH_3?
(b) Why do you think they are linear ?

(6) Some molecular shapes are given below

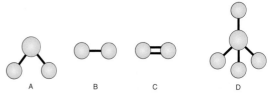

Identify the shape which could represent
(a) oxygen (b) chlorine (c) water
(d) methane

(7) The shape of hydrogen chloride can be represented as H–Cl. Use similar diagrams to represent the shapes of the following molecules
(a) hydrogen, H_2 (b) methane, CH_4
(c) ammonia, NH_3, (d) carbon dioxide, CO_2
(e) oxygen, O_2

(8) Discuss the bonding and the shape of the molecule in each of the following:
(a) H_2 (b) H_2O (c) HCl (d) PH_3 (e) CH_4 (f) BF_3
(g) $SiCl_4$ (h) O_2 (i) $BeCl_2$ (j) N_2 (k) SF_4.

(9) (a) Nitrogen and phosphorus are both in group V of the periodic table. Explain, using electronegativity values, why the bond angles in NH_3 and PH_3 are different.
(b) Explain why the H—O—H bond angle in water is approximately 104°, while the H—S—H bond angle in hydrogen sulphide, H_2S, is approximately 92°.

(10) If A represents a central atom, while B represents an atom bonded by an electron pair to A, determine the shape of each of the following molecules using the VSEPR theory:
(a) AB_3 (b) AB_4 (c) AB (d) AB_2.

(11) If A represents a central atom, B represents an atom bonded by an electron pair to A and E represents a lone pair on A, determine the shape of each of the following molecules using the VSEPR theory:
(a) AB_3E (b) AB_2E_2.

(12) Describe how you would predict the shape of (a) ethene and (b) ethyne, using the valence shell electron pair repulsion theory.

(13) The bond angles in methane, ammonia and water are 109° 28′, 107° 30′ and

104° 30′, respectively. Explain the reason for this trend.

(14) Water, H_2O, and beryllium hydride, BeH_2, may both be represented by the formula AB_2. The B—A—B bond angle in water is 104° 30′, while the B—A—B angle in beryllium hydride is 180°. Explain clearly, using diagrams, why the bond angles differ.

(15) The bond distance in nitrogen, N_2, is 109 pm, while the bond distance in oxygen, O_2, is 112 pm. Suggest a reason for the difference.

(16) Explain the meaning of the following terms:
(a) intramolecular bonding
(b) intermolecular forces (c) van der Waal's forces (d) dipole–dipole interactions
(e) hydrogen bonding.

(17) The intermolecular forces which exist between molecules can affect the boiling point of a substance. Use examples to explain why molecules which can form hydrogen bonds between them have higher boiling points than molecules which are held together by van der Waal's forces.

(18) Explain each of the following in terms of the intermolecular forces involved.
(a) Water has a higher boiling point than hydrogen sulphide.
(b) Oxygen gas has a higher boiling point than hydrogen gas.
(c) Methanal has a higher boiling point than ethene.

(19) For each of the following pairs, which substance is more soluble in water:
(a) CH_3OH or CH_3CH_3 (b) CCl_4 or NaCl. Explain why in each case.

(20) The physical and chemical properties of a compound are effected by hydrogen bonding. Which of the following compounds are capable of hydrogen bonding? Explain why.
(a) CH_4 (b) HCl (c) CO_2 (d) CF_4 (e) CH_3OH
(f) CH_3CHO

(21) Pure ethanoic acid, CH_3COOH, exists as a dimer (double molecule). Explain why.

(22) Sodium chloride, NaCl, is soluble in water; ethanol, C_2H_5OH, is also soluble in water, while iodine, I_2, is practically insoluble in water. Discuss the reasons for this trend.

(23) Iodine is soluble in tetrachloromethane and practically insoluble in water. Give reasons for this.

(24) Explain why the C—Cl bond is a polar bond, while the CCl_4 molecule is a non-polar molecule.

2.6 OXIDATION NUMBERS

Many chemical reactions do not involve ionic species, where the gain or loss of electrons can be easily seen. Many oxidation reduction reactions involve covalent compounds; this makes it difficult to explain redox reactions. The use of oxidation numbers overcomes that difficulty.

The **oxidation number** (or state) of each single atom or radical in an ionic compound is the same as its electrical charge, while in a covalent compound the oxidation number is a hypothetical (imaginary) number invented which assumes that the covalent compound is in fact an ionic compound.

For example, in ionic compounds the oxidation number is the ionic charge. Sodium chloride is Na^+Cl^-, so sodium has an oxidation number of +1 and chlorine has an oxidation number of –1. Calcium bromide is $CaBr_2$, so calcium has an oxidation number of +2 and each bromine has an oxidation number of –1.

For covalent compounds a set of hypothetical (imaginary) numbers is used, which assumes that all compounds are ionic.

For example, when hydrogen forms an ion it is usually the H^+ ion, so in the covalent compound, HCl, the oxidation number of hydrogen is +1 and the oxidation number of chlorine is –1.

In sulphuric acid, H_2SO_4, if we assume that hydrogen is +1 and that oxygen is –2 we find that sulphur is +6, as the sum of the oxidation numbers in the molecule must equal 0.

The rules for assigning oxidation numbers are given below.
(1) The oxidation number of an uncombined atom or an atom in a molecule of an element is 0.

For example, the oxidation numbers of Na, Fe, Ag are all 0 and the oxidation numbers of Cl in Cl_2, O in O_2 and P in P_4 are all 0.
(2) The sum of the oxidation numbers in a compound is 0, since all compounds are electrically neutral.

For example, the oxidation numbers in HNO_3 add up to 0. As H has an oxidation number of +1 and O is –2, this means that N has an oxidation number of +5.

(3) For simple ions the oxidation number is the same as the charge on the ion.

For example, the oxidation number of Br in Br^- is –1, that of K in K^+ is +1, while that of O in O^{2-} is –2.

(4) The oxidation number of group I elements is always +1, and the oxidation number of group II elements is always +2.

The oxidation number of fluorine is always –1, while the oxidation number of the other group VII elements is –1 in all binary compounds where the halogen is combined with a less electronegative element.

In a compound such as BrCl, the oxidation number of the bromine is +1 while the oxidation number of the more electronegative chlorine is –1, i.e. the more electronegative halogen is assigned a minus number.

(5) The oxidation number of oxygen in a compound is –2 except for peroxides, such as H_2O_2, where it is –1, and superoxides, such as OF_2, where it is +2.

(6) The oxidation number of hydrogen is +1, except in the metallic hydrides, such as LiH and KH, where it is –1.

• • • • • **Example 2.1** • • • • •

Find the oxidation numbers of each of the following:
(a) chromium in potassium chromate(VI), $K_2Cr_2O_7$
(b) sulphur in the sulphate ion, SO_4^{2-}
(c) manganese in manganese(IV) oxide.

(a) The oxidation number for K is +1 and for O is –2.
The overall oxidation number of $K_2Cr_2O_7 = 0$.
Then, inserting oxidation numbers into the formula (letting x = oxidation number of chromium)

$2(+1) + 2(x) + 7(-2) = 0$
$2 + 2x - 14 = 0$
$2x = +12$
$x = +6$, i.e. the oxidation number of chromium in $K_2Cr_2O_7$ is +6.

(b) The oxidation number for O is –2.
The overall oxidation number of $SO_4^{2-} = -2$.
Then, inserting oxidation numbers into the formula (letting x = oxidation number of sulphur)

$x + 4(-2) = -2$
$x - 8 = -2$
$x = +6$, i.e. the oxidation number of sulphur in SO_4^{2-} is +6.

(c) The oxidation number for O is –2.
The overall oxidation number for $MnO_2 = 0$.
Then, inserting the oxidation numbers into the formula (letting x = oxidation number of manganese)

$x + 2(-2) = 0$
$x - 4 = 0$
$x = +4$, i.e. the oxidation number of manganese in MnO_2 is +4.

• •

Redox Reactions

Oxidation and reduction reactions can now be defined more clearly using oxidation numbers.

> Oxidation is an increase in oxidation number.
> Reduction is a decrease in oxidation number.

Consider the important industrial reaction where iron oxide, Fe_2O_3, is reduced by carbon monoxide to metallic iron.

Inserting the oxidation numbers for each atom into the stoichiometric equation,

$$Fe_2O_3 \;+\; 3CO \longrightarrow 2Fe \;+\; 3CO_2$$

2(+3), 3(-2)   +2, -2   0   +4, 2(-2)

└──────Decrease in O.N. = Reduction──────┘
└──────Increase in O.N. = Oxidation──────┘

Here, the iron(III) is reduced to iron(0), while the carbon(II) is oxidised to carbon(IV).

In the reaction we say the reducing agent, CO, reduces the iron(III) oxide to metallic iron(0), while at the same time the oxidising agent, Fe_2O_3, oxidises the carbon(II) oxide to carbon(IV) oxide.

Today, the term oxidant is sometimes used instead of oxidising agent. The word anti-oxidant is often seen on food labels where a chemical has been added to prevent oxidation taking place.

• • • • • **Example 2.2** • • • • •

Determine which substance is oxidised, which is reduced and which are the oxidising and the reducing agents in the following equation.

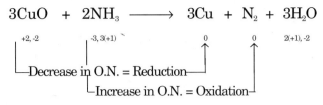

$$3CuO + 2NH_3 \longrightarrow 3Cu + N_2 + 3H_2O$$

Decrease in oxidation number of 2 = reduction

Increase in oxidation number of 3 = oxidation

The copper(II) oxide is reduced to metallic copper(0), while simultaneously the ammonia, NH_3, is oxidised to nitrogen. The copper(II) oxide is the oxidising agent (oxidant) and the ammonia is the reducing agent.

● ● ● ● ● ● ● ● ● ● ● ● ● ● ● ● ● ● ●

Social and Applied Aspects of Oxidising and Reducing Agents

• Sodium chlorate, NaOCl, is used extensively as a domestic bleach. It oxidises biological and other substances making materials whiter and killing bacteria. NaOCl can also be used as a disinfectant in swimming pools where it kills the bacteria present in the water.

• Sulphur dioxide, SO_2, is widely used as a preservative in many foods and drinks. It works by killing bacteria and by preventing oxidation. In foodstuffs, such as flour, sulphur dioxide improves the appearance by bleaching the flour. It is also used in the paper industry to bleach paper. Sulphur dioxide gas and its solution in water act as reducing agents.

Exercise 2.6
•

(1) (a) What are the oxidation numbers of the following elements:
(i) Na in Na (ii) Cl in Cl^- (iii) Na in Na^+
(iv) Mg in Mg^{2+} (v) O in O^{2-} (vi) H in H^+
(b) Assign oxidation numbers to each of the underlined atoms:
(a) $\underline{Mn}_2O_7$ (b) $K\underline{Mn}O_4$ (c) $Na_2\underline{Cr}_2O_7$
(d) $Na_2\underline{S}_2O_3$ (e) $Na_2\underline{S}_4O_6$ (f) $\underline{Xe}F_2$ (g) $H\underline{N}O_3$
(h) $K\underline{Cl}O_3$ (i) $\underline{Cr}Cl_3$ (j) $\underline{N}H_4NO_3$

(2) (a) What is an oxidation number?
(b) If a compound contains two elements, which element would be expected to have a positive oxidation number?
(c) What oxidation number would you expect an ion to have?

(d) What are the expected oxidation numbers of hydrogen and oxygen in their compounds?

(3) What is the oxidation number of P in:
(a) PCl_3 (b) $H_2PO_4^-$?

(4) Use oxidation numbers to determine whether or not each of the following reactions is a redox reaction and, if so, state the species reduced.

(a) $S_2O_3^{2-} + 2HCl \rightarrow H_2O + SO_2 + S + 2Cl^-$

(b) $2O_2^{2-} + 2H_2O \rightarrow 4OH^- + O_2$

(c) $BiCl_4^- + H_2O \rightarrow BiOCl + 2HCl + Cl^-$

(5) (a) What is the oxidation number of chromium in the following:
(i) $HCrO_4^-$ (ii) CrO_3 (iii) $Cr_2O_7^{2-}$?
(b) What is the oxidation number of oxygen in OF_2? Explain why it is given this oxidation number.

(6) Find the oxidation number of phosphorus in each of the following compounds:
(a) H_3PO_4 (b) PCl_3 (c) $Ca(H_2PO_4)_2$

(7) (a) Define oxidation, reduction, oxidising agent and reducing agent in terms of oxidation numbers.
(b) For each of the following equations show which substances are oxidised and which substances are reduced. In each case name the oxidising agent and the reducing agent.

(i) $Mg + CuCl_2 \rightarrow MgCl_2 + Cu$

(ii) $Fe_2O_3 + 2Al \rightarrow Al_2O_3 + 2Fe$

(iii) $OF_2 + H_2O \rightarrow O_2 + 2HF$

(iv) $2NO + O_2 \rightarrow 2NO_2$

(v) $2SO_2 + O_2 \rightarrow 2SO_3$

(8) By assigning oxidation numbers, decide whether or not each of the following is a redox reaction.

(a) $2KBr + Cl_2 \rightarrow 2KCl + Br_2$

(b) $Fe + CuSO_4 \rightarrow FeSO_4 + Cu$

(c) $C_2H_4 + H_2 \rightarrow C_2H_6$

(d) $CH_4 + 2O_2 \rightarrow 2H_2O + CO_2$

(e) $SO_2 + NO_2 \rightarrow SO_3 + NO$

(f) $Ca + 2H_2O \rightarrow Ca(OH)_2 + H_2$

(g) $Al^{3+} + 3OH^- \rightarrow Al(OH)_3$

(h) $Cl_2 + 2I^- \rightarrow I_2 + 2Cl^-$

63

(9) In each of the following industrial redox reactions, state which substance is oxidised and which substance is reduced:

(a) $N_2 + 3H_2 \rightarrow 2NH_3$

(b) $Fe_2O_3 + 3CO \rightarrow 2Fe + 3CO_2$

(c) $CH_4 + 2H_2O \rightarrow CO_2 + 4H_2$

(d) $2H_2 + CO \rightarrow CH_3OH$

(10) Bleaches are oxidising agents. Write a brief note on some common everyday oxidising agents and reducing agents.

▼▼▼ Key Terms ▼▼▼

Some of the more important terms are listed below. Other terms not listed may be located by means of the index.

1. Compounds: A compound is a substance made by combining two or more elements.

2. Chemical bond: A chemical bond is a strong attractive force that exists between certain atoms in a substance.
(a) Ionic bonding occurs when chemical bonds are held together by oppositely charged ions, such as in sodium chloride.
(b) Covalent bonding occurs when chemical bonds are held together by sharing of electron pairs, such as in methane.
(c) Metallic bonding occurs when chemical bonds are held together by a sea of mobile electrons surrounding an array of positive metal ions, such as in sodium metal.

3. Valency: The valency of an atom tells us the number of chemical bonds that an atom can form – i.e. it is the combining power of the atom.

4. Polar and non-polar covalent bonds: A non-polar covalent bond is a bond where the electrons are shared equally between two atoms. A polar covalent bond is a bond where the bonding electrons are unequally distributed throughout the molecule. A polar covalent bond possesses some percentage of ionic character. Polar molecules are often called dipoles.

5. Electronegativity: Electronegativity is a measure of the relative ability of an atom in a molecule to attract electrons to itself.

6. Hydrogen bonding: Hydrogen bonding is the attractive force which exists between a hydrogen atom covalently bonded to a very electronegative atom, X, and another small electronegative atom, Y. Although hydrogen bonds are much weaker (10–40 kJ mol^{-1}) than ordinary covalent bonds (100–1000 kJ mol^{-1}), they exert a large influence on the physical and chemical properties of the molecules which form them.

7. The forces which exist between collections of molecules are:
• van der Waal's forces: these are the weak forces which exist between all forms of matter
• dipole–dipole interactions: these are fairly weak forces which attract non-polar molecules to each other.
• hydrogen bonding: hydrogen bonds are a special type of dipole–dipole interaction.

8. Intramolecular bonds: The individual bonds which exist between the atoms within a molecule are called intramolecular bonds.

9. Sigma and pi bonds: Sigma bonds are are formed by the 'head-on overlap' of atomic orbitals. Pi bonds are formed by the 'sideways overlap' of atomic orbitals.

10. VSEPR: VSEPR is short for valence shell electron pair repulsion theory. The VSEPR model is based on the fact that electron pairs repel one another and consequently try to take positions as far apart from each other as possible.

11. Shapes of covalent molecules: The shape of a molecule is determined by:
(a) the number of pairs of electrons around the central atom
(b) the type of electron pairs around the central atom (i.e. whether they are bond pairs or lone pairs).

12. Oxidation numbers: The oxidation number (or state) of each single atom or radical in an ionic compound is the same as its electrical charge, while in a covalent compound the oxidation number is a hypothetical (imaginary) number invented assuming that the covalent compound is in fact an ionic compound.

13. Oxidation and reduction:
Oxidation is increase in oxidation number.
Reduction is decrease in oxidation number.

The glass contains 1 mole of red wine.

3.1 STATES OF MATTER

When you started to study science you found out that 'Matter occupies space'. The space is occupied by solids, liquids or gases, and these are called the three states of matter.

The three states of matter: solid, liquid and gas.

Change of State

Solids, liquids and gases can undergo a change of state when the temperature is changed. Pure water is the only substance that normally exists on earth in all three states.

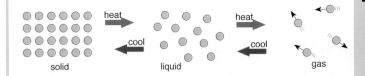

States of Matter and the Particle Theory

Solids

Solids may exist as crystals or as glasses.

Crystals are well-ordered arrangements of fixed particles. The size and shape (volume) of a solid are affected only slightly by a change in temperature (liquid crystals are an exception) and are relatively unaffected by a change in pressure.

The particles in a solid are very close to each other, due to the large attractive forces between them. Solids have low kinetic energy and move by vibrating from side to side. The individual particles which make up a solid can also move if they are dissolved in a liquid or if they are converted into a gas.

When a small crystal such as potassium manganate(VII) is allowed to dissolve in a large volume of water, the purple colour of the potassium manganate(VII) can still be seen. This indicates that solids are made up of particles which can spread throughout the water.

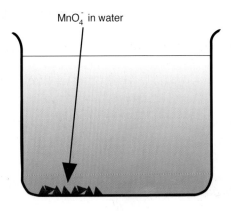

MnO_4^- in water

Crococite crystals, Tasmania.

Liquids

Liquids are mobile and almost struc-tureless fluids.

The particles in a liquid are much further apart than in a solid and are constantly moving about with high kinetic energy. The high kinetic energy allows the particles in a liquid not only to vibrate and rotate, but to move from place to place.

Liquids have some degree of order due to their attractive forces, and this enables them to have a definite volume. However, since the particles are free to move, a liquid does not have any characteristic shape.

The main characteristic of liquids is their ability to flow. Very viscous liquids such as syrups and molten polymers flow very slowly because their large molecules become entangled with each other. Mobile liquids such as tetrachloromethane have low viscosities because of the weak attractive forces between the molecules, whereas liquids such as water are less mobile because the molecules are more tightly bound together.

Ink moving through water.

Gases

Gases are like a swarm of high energy molecules in constant chaotic motion.

The word 'gas' is derived from the word 'chaos'. The particles travel in straight lines at high speeds until they bump into each other, and are deflected all over the place.

The speed at which gases move is greatly affected by temperature. A large decrease in temperature can slow the gas down so much that the particles can come so close together that they may form a liquid.

The space between the gas particles is very large relative to their size; this implies that the attractive forces are very small between the particles. Gases are easily compressed and when the pressure is high enough the gas may be turned into a liquid.

Diffusion

Diffusion is the process by which a substance spreads out throughout another substance in a uniform manner.

Gases diffuse quite easily and quickly. If a substance with a strong odour is released at one end of a room, minutes or even seconds later it will be detected at the other end of the room. The smell of an apple tart from an oven is a familiar example. The gas particles move at speeds of over a thousand miles an hour, constantly colliding with one another along the way and zig-zagging off in different directions.

The process where a gas flows through a small hole in a container is called **effusion**.

Effusion of gas particles through a small hole

Liquids that are miscible with one another can diffuse through each other. For example, Coca Cola and other soft drinks which are used as mixers in alcoholic drinks move through the alcohol by diffusion.

Solids can diffuse through liquids if the forces holding the solid together are weaker than the forces holding the liquid together. Sodium chloride diffuses very quickly through water in body fluids.

Blugas: butane gas used for heating and cooking.

Activity 3.1

To demonstrate that matter consists of moving particles

Introduction
Solid particles, liquid particles and gaseous particles can all move by diffusing through one another.

The following can be examined: ammonia and hydrogen chloride, ink and water, and smoke and air.

Requirements
Safety glasses, dilute hydrochloric acid, dilute ammonia, long glass tube, cotton wool, ink, water, piece of rope.

Procedure
(a) Ammonia and hydrogen chloride (This must be done in a fume cupboard)

(1) Using tweezers, carefully dip a ball of cotton wool into some concentrated hydrochloric acid. Allow the excess liquid to drip off. Insert the cotton wool at one end of the long glass tube. Stopper the tube.

white fumes

cotton wool soaked in ammonia solution

cotton wool soaked in concentrated hydrochloric acid solution

(2) As quickly as possible repeat the procedure, only this time dip another ball of cotton wool into the concentrated ammonia solution. Insert the cotton wool at the other end of the glass tube. Stopper the tube.
(3) Observe what happens when the two gases meet. Note approximately where they meet.
(4) Write an equation for the reaction and explain why the reaction occurs nearer to one end of the tube than the other end.
(b) Ink and water
(1) Using a teat pipette place a drop of ink in the bottom of a beaker of water.
(2) Observe what happens.
(c) Smoke and air
Light a small piece of rope and observe what happens to the smoke as it moves through the air.

Exercise 3.1
•

(1) Matter exists in three states, solids, liquids and gases. Explain, in terms of the particles of which they are made up of, how they differ from one another. What evidence is there to show that matter is made up of particles?

(2) Describe how gases, liquids and solids differ in terms of the motion of the particles which make them up and also in terms of the arrangement of the particles in space.

(3) Explain, briefly, why water can exist in all three states of matter under normal conditions.

(4) (a) What is a vapour?
(b) What is a volatile liquid?
(c) List five chemicals that are gases at normal conditions.
(d) Explain diffusion in molecular terms.
(e) What causes atmospheric pressure?
(f) How is the density of a gas affected by a change in pressure?

(5) Explain the meaning of the terms diffusion and effusion. Describe some simple experiments to illustrate the following:
(a) diffusion in gases
(b) diffusion in liquids.

3.2 THE MOLE – THE AMOUNT OF SUBSTANCE

Everyday items are counted using definite units, such as a pair of gloves, a dozen eggs, a litre of oil, a ream of paper and many other items. In each case we buy a certain amount of the goods which we require. Very small items, such as atoms, molecules and ions also need to be counted.

A mole of ethanol.

A mole of sodium chloride.

The unit used to count the amount of atoms, molecules or ions is the mole (abbreviation: mol).

A **mole** (**mol**) is defined as the amount of substance which contains as many elementary particles as there are carbon atoms in 12 g of carbon-12.

One mole of water contains the same number of water molecules as there are carbon atoms in 12 g of carbon-12.

The mass of an atom of carbon-12 is 1.99252×10^{-23} g.

Therefore, the number of atoms in 12 g of carbon-12 is

$$\frac{12 \text{ g}}{1.99252 \times 10^{-23} \text{ g}} = 6.022 \times 10^{23} \text{ atoms}$$

The number of molecules in one mole of water is also 6.022×10^{23}.

This number, 6.022×10^{23}, called the Avogadro number, N_A, tells us the number of particles (ions, atoms or molecules) in one mole of the substance.

$$N_A = 6.022 \times 10^{23} \text{ particles mol}^{-1}$$

Relative Atomic Mass, Relative Molecular Mass and Molar Mass

Relative Atomic Mass

The relative atomic mass, A_r, is defined as the ratio of the average mass of that atom compared to $\frac{1}{12}$ of the mass of an atom of carbon-12.

The relative atomic mass of fluorine, $A_r(F)$, is 19. This is also called the atomic weight and is found under the symbol of each element in the periodic table.

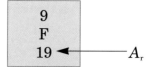

Relative Molecular Mass

The relative molecular mass, M_r, is defined as the average mass of the molecule compared to $\frac{1}{12}$ of the mass of an atom of carbon-12.

The relative molecular mass of methane, CH_4, is calculated by adding up the atomic masses of all of the atoms in CH_4.

$$M_r(CH_4) = 12 + (1 \times 4) = 16$$

• • • • • • **Example 3.1** • • • • • •

Calculation of relative molecular mass

Find the relative molecular mass of each of the following:
(a) oxygen, O_2 (b) carbon dioxide, CO_2
(c) ammonia, NH_3 (d) sulphuric acid, H_2SO_4,
(e) ethanol, C_2H_5OH.

(a) $M_r(O_2) = 16 \times 2 = 32$

(b) $M_r(CO_2) = 12 + (16 \times 2) = 44$

(c) $M_r(NH_3) = 14 + (1 \times 3) = 17$

(d) $M_r(H_2SO_4) = (1 \times 2) + 32 + (16 \times 4) = 98$

(e) $M_r(C_2H_5OH) = (12 \times 2) + (1 \times 5) + 16 + 1$
$= 46$

Always be careful when using the term 'oxygen'.
This is not precise enough. You should always state whether you mean an oxygen molecule, O_2, or an oxygen atom, O.

It should be noted that M_r can be used for entities that are not strictly molecular. Ionic compounds such as KCl are not molecular, but have a simple empirical formula.

For example, the relative molecular mass of potassium chloride, $M_r(KCl) = 39 + 35.5 = 74.5$.

∙∙∙∙∙∙∙∙∙∙∙∙∙∙∙∙∙∙∙∙∙∙∙∙∙∙∙∙∙∙∙∙

Molar Mass

Molar mass, M, is an extremely useful term because it can be applied to any chemical entity. It is the mass of one mole of the particular entity. It is expressed in g mol^{-1} (grams per mole).

The molar mass is calculated by adding the atomic masses of all the atoms in the particular entity together.

For example, the molar mass of fluorine, F, is written as $M(F) = 19$ g mol^{-1}. The molar mass of methane is written as $M(CH_4) = 16$ g mol^{-1}.

∙∙∙∙∙∙ **Example 3.2** ∙∙∙∙∙∙

Calculation of molar mass

Find the molar mass of each of the following:
(a) sodium atom, Na (b) hydrogen, H_2 (c) carbon monoxide, CO (d) sodium chloride, NaCl (e) magnesium ion, Mg^{2+} (f) nitrate ion, NO_3^-

(a) $M(Na) = 23$ g mol^{-1}

(b) $M(H_2) = (1 \times 2)$ g mol^{-1} = 2 g mol^{-1}

(c) $M(CO) = (12 + 16)$ g mol^{-1} = 28 g mol^{-1}

(d) $M(NaCl) = (23 + 35.5)$ g mol^{-1}
$\qquad = 58.5$ g mol^{-1}

(e) $M(Mg^{2+}) = 24$ g mol^{-1}

(f) $M(NO_3^-) = (14 + (16 \times 3))$ g mol^{-1}
$\qquad = 62$ g mol^{-1}

∙∙∙∙∙∙∙∙∙∙∙∙∙∙∙∙∙∙∙∙∙∙∙∙∙∙∙∙∙∙∙∙

Counting in Moles

It is far easier to buy 1000 reams of paper, rather than 500 000 sheets (1 ream = 500 sheets). Chemists find it convenient to use moles when they are counting chemicals.

The number of moles (amount) of a substance is found by dividing the mass of the substance by the molar mass of the substance.

Number of moles (amount) of a substance	$= \dfrac{\text{mass}}{\text{molar mass}}$	$n = \dfrac{m}{M}$

The amount (n) is measured in moles (mol), the mass (m) in grams (g) and the molar mass (M) in grams per mole (g mol^{-1}).

∙∙∙∙∙∙ **Example 3.3** ∙∙∙∙∙∙

Converting grams to moles

Sulphuric acid sales are used as a market guide for the chemical industry because vast quantities of it are produced annually by the large multinational chemical companies. 2.45 g of pure H_2SO_4 was weighed in a beaker. How many moles of sulphuric acid is this?

The molar mass of sulphuric acid is found using a table of relative atomic masses,

$$M(H_2SO_4) = ((1 \times 2) + 32 + (16 \times 4)) \text{ g mol}^{-1}$$
$$= 98 \text{ g mol}^{-1}$$

Number of moles (amount) of sulphuric acid $= \dfrac{\text{mass}}{\text{molar mass}}$

$$n = \frac{m}{M} = \frac{2.45 \text{ g}}{98 \text{ g mol}^{-1}} = 0.025 \text{ mol}$$

∙∙∙∙∙∙∙∙∙∙∙∙∙∙∙∙∙∙∙∙∙∙∙∙∙∙∙∙∙∙∙∙

∙∙∙∙∙∙ **Example 3.4** ∙∙∙∙∙∙

Converting moles to masses

Ammonium nitrate, NH_4NO_3 is a white crystalline solid used as a fertiliser. Calculate the mass of NH_4NO_3 in 1.5 mol of ammonium nitrate.

$$M(NH_4NO_3) = (14 + 1 \times 4 + 14 + 16 \times 3) \text{ g mol}^{-1}$$
$$= 80 \text{ g mol}^{-1}$$

$n = \dfrac{m}{M}$ is rearranged to $m = nM$

$$m = 1.5 \text{ mol} \times 80 \text{ g mol}^{-1}$$
$$= 120 \text{ g } NH_4NO_3$$

∙∙∙∙∙∙∙∙∙∙∙∙∙∙∙∙∙∙∙∙∙∙∙∙∙∙∙∙∙∙∙∙

Avogadro number, $N_A = 6.022 \times 10^{23}$ = 1 mol of particles

•••••• **Example 3.5** ••••••

Converting masses to number of particles

Calculate:
(a) the number of HCl molecules in 3.65 g HCl
(b) the number of atoms in 18 g of carbon, C.

$$(a)\ n\ (\text{HCl}) = \frac{m}{M} = \frac{3.65\ \text{g}}{36.5\ \text{g mol}^{-1}} = 0.1\ \text{mol}$$

Using Avogadro's number:

1.0 mol HCl = 6.022×10^{23} molecules of HCl

0.1 mol HCl = $0.1 \times 6.022 \times 10^{23}$ molecules of HCl

$$= 6.022 \times 10^{22}\ \text{molecules of HCl.}$$

Alternatively this calculation can be done by simply using the formula

$$N = nN_A$$

where N = number of particles
n = amount in moles
N_A = Avogadro number

$$(b)\ n(\text{C}) = \frac{m}{M} = \frac{18\ \text{g}}{12\ \text{g mol}^{-1}} = 1.5\ \text{mol}$$

The number of particles,

$$N = nN_A = 1.5\ \text{mol} \times 6.022 \times 10^{23}\ \text{mol}^{-1}\ \text{C atoms}$$

$$= 9.033 \times 10^{23}\ \text{atoms of carbon.}$$

•••••• **Example 3.6** ••••••

Converting number of particles to moles

A silicon chip in a computer circuit contains 1.22×10^{20} silicon atoms. What is the mass of the silicon in the computer chip?

The number of moles of silicon atoms

$$n = \frac{N}{N_A} = \frac{1.22 \times 10^{20}\ \text{atoms}}{6.022 \times 10^{23}\ \text{atoms mol}^{-1}}$$

$$= 2.026 \times 10^{-4}\ \text{mol}$$

Mass of silicon

$$m = nM = 2.026 \times 10^{-4}\ \text{mol} \times 28\ \text{g mol}^{-1}$$

$$= 5.67 \times 10^{-3}\ \text{g.}$$

•••••• **Example 3.7** ••••••

Converting moles to litres (gases only)

Calculate the number of litres of the gas hydrogen sulphide, H_2S, in 1.25 moles of the gas at STP.

At STP, the molar volume, V_m = 22.4 L (see page 75)

$$V = nV_m = 1.25\ \text{mol} \times 22.4\ \text{L mol}^{-1} = 28\ \text{L}$$

Chemistry in Action: Mass Spectrometry

A quick and accurate method of determining the relative molecular mass of a compound is mass spectrometry.

When a compound is passed into a mass spectrometer it is bombarded with high energy electrons. The compound is broken up into fragments,

$$M \rightarrow M^+$$

For example, a molecule of ethanol, C_2H_5OH, can break up into several fragments such as $C_2H_5O^+$, CH_3^+ and so on. The mass spectrum of the compound is detected and analysed by computer. Many peaks are noticed in the mass spectrum, but the one with the heaviest mass corresponds to the ethanol; in this case M_r = 58.

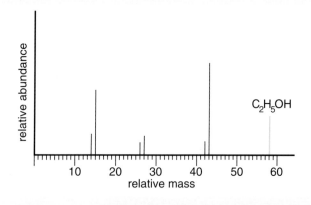

The fragments produced by a mass spectrum often give useful information about the structure of a compound. However, the main method of determining the molecular structure is spectroscopy.

Exercise 3.2

•

(1) (a) Define each of the following terms:
(i) Avogadro's number (see p.74) (ii) relative atomic mass (iii) relative molecular mass (iv) molar mass (v) the amount of a substance.
(b) State clearly the unit (if any) in which each of the terms in part (a) is measured. In each case use a suitable example to show that you fully understand the particular definition.

(2) Calculate the relative molecular mass of each of the following:
(i) nitrogen, N_2 (ii) hydrogen sulphide, H_2S (iii) carbon monoxide, CO (iv) carbon dioxide, CO_2 (v) nitric acid, HNO_3 (vi) potassium manganate(VII), $KMnO_4$ (vii) sodium chlorite, $NaOCl$ (viii) calcium hydrogencarbonate, $Ca(HCO_3)_2$.

(3) How can a mass spectrometer be used to determine the relative molecular mass of a molecular species?

(4) Calculate the molar mass of each of the following entities:
(a) $CHCl_3$ (b) KCl (c) Na^+ (d) HCO_3^-
(e) $Ca(OH)_2$ (f) H (g) H_2O_2.

(5) Calculate the molar mass of each of the following compounds:
(a) glucose, $C_6H_{12}O_6$ (b) urea, NH_2CONH_2
(c) Epsom salts, $MgSO_4.7H_2O$ (d) ethanoic acid, CH_3COOH (e) sodium thiosulphate, $Na_2S_2O_3$ (f) magnesium nitrate, $Mg(NO_3)_2$ (g) iron (III) oxide, Fe_2O_3.

(6) Calculate the number of moles (amount) in each of the following:
(a) 2 g NaOH (b) 1.1 g CO_2 (c) 25 g $CaCO_3$
(d) 14 g N (e) 1 g H_2 (e) 16 g CH_3OH
(f) 0.8 g SO_2.

(7) Calculate the amount of substance in each of the following:
(a) 3 g C (b) 76 g C_4H_{10} (c) 4.8 g N_2H_4
(d) 58.8 g $K_2Cr_2O_7$.

(8) (a) Which has more mass, 0.5 mol H_2O or 2 mol He?
(b) Which has more mass, 2 mol C_2H_5OH or 5 mol H_2O?
(c) Which has more mass, 2.5 mol NaCl or 1.25 mol $CaCO_3$?

(9) Calculate the mass of each of the following:

(a) 0.1 mol Cu, 5 mol Na, 4.5 mol C, 6 mol S
(b) 2 mol Na^+, 5 mol O^{2-}, 0.25 mol SO_4^{2-}, 2.5 mol HCO_3^-
(c) 2.2 mol CO, 0.5 mol NH_3, 16 mol SO_2, 1.5 mol C_2H_5OH
(d) 2.5 mol $Fe_2(SO_4)_3$, 4.0 mol Al_2O_3, 1.5 mol $K_2Cr_2O_7$.

(10) Write out the number of moles of each element in 1 mole of the following substances:
(a) NH_3 (b) NH_2CONH_2 (c) $NaNO_3$ (d) N_2, (e) N_2H_4.

(11) Write out the number of moles of each element in one mole of the following substances:
(a) $CoCl_2.6H_2O$ (b) $CuSO_4.5H_2O$
(c) $MgSO_4.7H_2O$.

(12) (a) Calculate the molar mass of aspirin, $C_9H_8O_4$.
(b) A typical aspirin tablet contains 500 mg of $C_9H_8O_4$. How many moles of $C_9H_8O_4$ are there in a 500 mg tablet?

(13) Calculate the number of moles present in a 10 g sample of each of the following:
(a) helium, He (b) ethanol, C_2H_5OH (c) calcium carbonate, $CaCO_3$ (d) carbon, C.

(14) (a) What is Avogadro's number?
(b) Calculate the number of particles in each of the following:
(i) 2 mol HCl (ii) 0.1 mol NH_3 (iii) 1.25 mol C
(iv) 2 g H_2 (v) 7 g N_2 (vi) 1.6 g SO_2.
(c) Calculate the number of ions in:
(i) 3.5 mol H_2O (ii) 2 mol NaCl (iii) 4.5 mol $Mg(OH)_2$ (iv) 2 mol $Ca_3(PO_4)_2$ (v) 5.85 g NaCl
(vi) 37 g $CaCl_2$.

(15) How many carbon atoms are present in 2.0 g of each of the following:
(a) CO_2 (b) C_2H_4 (c) C_2H_5OH (d) $C_6H_{12}O_6$
(e) CH_3Cl (f) CH_4?

(16) How many atoms of hydrogen and oxygen are needed to form 3.6 g of water, H_2O?

(17) Vitamin C, or ascorbic acid, has the chemical formula $C_6H_8O_6$. The recommended daily intake for an adult is 90 mg. How many molecules of ascorbic acid are there in the recommended intake?

(18) How many molecules are present in:
(a) 22 g carbon dioxide, CO_2 (b) 20 g of glucose, $C_6H_{12}O_6$ (c) 17.1 g of cane sugar, $C_{12}H_{22}O_{11}$?

71

(19) Convert each of the following numbers of particles to moles:
(a) 2.0×10^{22} molecules of O_2 (b) 1.0×10^{27} molecules of N_2 (c) 1.5×10^{12} molecules of CO_2 (d) 1.75×10^{15} atoms of O (e) 3.0×10^{18} atoms of He (f) 4.0×10^{22} atoms of S
(g) 1.75×10^{15} of Na^+ ions (h) 2.0×10^{20} Mg^{2+} ions.

(20) (a) What is STP? (see page 73)
(b) What is the molar volume of a gas at STP?
(c) How many moles of methane, CH_4, are there in 3.36 L of methane at STP?
(d) How many molecules of methane, CH_4, are there in 3.36 L of methane at STP?
(e) How many atoms are there in 3.36 L of methane at STP?

(21) (a) If it takes 2.5×10^{22} atoms of neon to completely fill a flask at STP, what is the capacity of the flask?
(b) The total number of molecules in a certain volume of oxygen at STP is 1.5×10^{21}. Calculate the volume.
(c) What is the volume occupied by 4.4 g of CO_2 gas at STP?
(d) How many atoms are there in 1.4 g of CO gas at STP?
(e) The mass of 2.8 L of a gaseous hydrocarbon at STP is 3.75 g. What is the molar mass of the hydrocarbon?

3.3 THE GAS LAWS

Cushioning the impact: an airbag in a car can save a life.

Most substances which are made up of small molecules are gases under normal conditions of temperature and pressure. Liquids, like ethanol, which are made from small molecules are easily vaporised (volatile). The physical appearance of gases and volatile liquids are easily changed when the temperature and pressure is changed. For example, gases can be compressed into a smaller volume when pressure is applied. Changes such as this can be interpreted by using the gas laws.

Robert Boyle 1627–1691
Robert Boyle was born in Lismore, Ireland. He was regarded as one of the foremost experimental scientists of his time. It is thought that he was the first to collect gases by displacing water in an inverted flask. He discovered the relationship between the pressure and the volume of a gas in 1660. The relationship, $P \times V$ = constant, is known as Boyle's law and was one of the first attempts to express a scientific principle in a mathematical form. Boyle separated chemistry from the realm of alchemy and established it as a science. On the basis of experiment he defined an element as something that cannot be broken up into smaller substances. Robert Boyle devoted his life to experimental science, taking careful notes of each experiment, enabling other scientists to learn from his work. He is regarded as the father of experimental science.

Boyle's Law

One of the main properties of a gas is its ability to be compressed and squeezed into a smaller volume. This is done by applying pressure. **Robert Boyle** noticed that when he poured mercury into the open end of a J-tube containing air, the volume of the air decreased as more mercury was added.

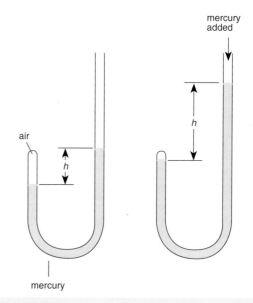

Boyle's law: At constant temperature, the volume of a fixed mass of gas is inversely proportional to the pressure.

Boyle's law is usually written as an equation:

$$PV = k \text{ (constant)} \quad \text{or as } P_1V_1 = P_2V_2$$

Boyle's law can be explained in terms of the kinetic theory of gases as follows. If the volume of a gas is reduced at constant temperature, the gas molecules moving at a constant average speed collide more frequently with each other and with the walls of the container. The increase in frequency of the collisions causes an increase in pressure.

Charles' Law

Jacques Charles was a French scientist who did many experiments on hot air and hydrogen-filled air balloons. He noticed that temperature affected the volume of a gas: a gas contracts when cooled and expands when heated.

> **Charles' law:** At constant pressure, the volume of a fixed mass of gas is directly proportional to the absolute temperature (K).

Charles' law is usually written as an equation:

$$V / T = k \text{ (constant)}$$

Charles' law can be explained in terms of the kinetic theory as follows. As the temperature of the gas is raised, the average speed of the gas increases, causing more and more collisions. If the pressure is constant, the volume must increase so that the number of particles per unit volume decreases and consequently fewer collisions occur.

The Combined Gas Law

Chemical reactions occur under various conditions of temperature, pressure and volume. Boyle's law and Charles's law can be combined to give the combined gas law as follows:

$$\frac{P_1 V_1}{T_1} = \frac{P_2 V_2}{T_2}$$

This can be used to calculate changes in volume, temperature or pressure as conditions change in a chemical reaction.

> The temperature of a gas must be converted to the absolute scale (K) in all calculations involving the combined gas law.
> **0 °C = 273 K**

Ballooning in the Rocky mountains.

Standard Temperature and Pressure (STP)

The volume of a gas changes with changes in temperature and pressure. It is important to measure gas volumes using the same standard conditions.

> **Standard temperature = 273 K;**
> **Standard pressure = 1.01×10^5 Pa.**

• • • • • • **Example 3.8** • • • • • •

A balloon is filled with helium gas, He, which occupies a volume of 460 cm^3 at a temperature of 35 °C and a pressure of 1.2×10^5 Pa. Calculate the volume of the helium in the balloon at a pressure of 1.0×10^5 Pa and a temperature of 25 °C.

Initial conditions:

$V_1 = 460 \text{ cm}^3 \quad T_1 = (273 + 35) \text{ K} = 308 \text{ K}$

$P_1 = 1.2 \times 10^5 \text{ Pa}$

Final conditions:

$V_2 = ? \qquad T_2 = (273 + 25) \text{ K} = 298 \text{ K}$

$P_2 = 1.0 \times 10^5 \text{ Pa}$

$$\frac{P_1 V_1}{T_1} = \frac{P_2 V_2}{T_2}$$

$$\frac{1.2 \times 10^5 \text{ Pa} \times 460 \text{ cm}^3}{308 \text{ K}} = \frac{1.0 \times 10^5 \text{ Pa} \times V_2}{298 \text{ K}}$$

$$V_2 = \frac{1.2 \times 10^5 \text{ Pa} \times 460 \text{ cm}^3 \times 298 \text{ K}}{1.0 \times 10^5 \text{ Pa} \times 308 \text{ K}}$$

$$= 534 \text{ cm}^3$$

• • • • • **Example 3.9** • • • • •

The volume of a sample of gas is 600 cm³ at 15 °C and 1.5×10^5 Pa. Calculate the volume of the gas at STP.

Initial conditions:

$V_1 = 600$ cm³ : $T_1 = (273 + 15)$ K = 288 K

$P_1 = 1.5 \times 10^5$ Pa

Final conditions:

$V_2 = ?$: $T_2 = 273$ K : $P_2 = 1.01 \times 10^5$ Pa

$$\frac{P_1 V_1}{T_1} = \frac{P_2 V_2}{T_2}$$

$$\frac{1.5 \times 10^5 \,\text{Pa} \times 600 \,\text{cm}^3}{288 \,\text{K}} = \frac{1.01 \times 10^5 \,\text{Pa} \times V_2}{273 \,\text{K}}$$

$$V_2 = \frac{1.5 \times 10^5 \,\text{Pa} \times 600 \,\text{cm}^3 \times 273 \,\text{K}}{1.01 \times 10^5 \,\text{Pa} \times 288 \,\text{K}}$$

$$= 844.7 \,\text{cm}^3$$

• •

Gay-Lussac's Law of Combining Volumes

In 1808 **Gay-Lussac** published the results of some experiments on gases. He concluded that the volumes of reactant and product gases at a given temperature and pressure are in the ratios of small whole numbers.

For example, he observed the following

1. Two volumes of ammonia gas break down to give one volume of nitrogen gas and three volumes of hydrogen gas.

2 volumes ammonia → 1 volume nitrogen + 3 volumes hydrogen

We can write this as the equation

$$2NH_3(g) \rightarrow N_2(g) + 3H_2(g)$$

2. Two volumes of hydrogen gas and one volume of oxygen gas combine to form one volume of steam.

2 volumes hydrogen + 1 volume oxygen → 2 volumes steam

We can write this as the equation

$$2H_2(g) + O_2(g) \rightarrow 2H_2O(g)$$

Gay-Lussac's law: Gases always react with each other in simple whole-number relationships.

• •

Avogadro's Law

Three years after Gay-Lussac published his findings, **Amedeo Avogadro** interpreted Gay-Lussac's law as follows:

Equal volumes of gases, under the same conditions of temperature and pressure, contain equal numbers of molecules.

This law enables us to change over directly from volume of gas to molecules of gas and vice versa. This rule applies to gases only.

When this rule is applied to the decomposition of ammonia gas:

According to Gay-Lussac
Ammonia → Hydrogen + Nitrogen
2 volumes → 3 volumes + 1 volume

According to Avogadro
Ammonia → Hydrogen + Nitrogen
2 molecules → 3 molecules + 1 molecule
2 mol → 3 mol + 1 mol (Avogadro's number)

We can now substitute, almost at will, the terms volume, molecule and mole in all calculations involving gases.

Amedeo Avogadro
1776–1856
Amedeo Avogadro was a professor of physics in the University of Turin, but is best known for his contributions to chemistry. He followed the work of Gay–Lussac closely and realised early on the difference between atoms and molecules. Avogadro suggested that equal volumes of gases under the same conditions of temperature and pressure contained equal numbers of particles. The number of particles in a mole, 6.022×10^{23}, is called Avogadro's number in his honour.

••••• **Example 3.10** •••••

Calculate the volume of gases produced when 50 cm³ N_2O decomposes to produce N_2 and O_2 in accordance with the equation given below.

$2N_2O \rightarrow 2N_2 + O_2$
2 mol → 2 mol + 1 mol (Avogadro)
2 volumes → 2 volumes + 1 volume (Gay-Lussac)

then,
50 cm³ → 50 cm³ + 25 cm³

The volume of gases produced is 75 cm³.

•••••••••••••••••••••••••••••••

Avogadro's Number and Molar Volume

The volume of a gas is easier to measure than its mass. Avogadro's law tells us that equal volumes of any two gases at the same temperature and pressure contain the same number of particles. This means that one mole of any gas contains the same number of particles as one mole of any other gas. The number of particles in one mole of gas is called Avogadro's number and is equal to 6.022×10^{23}.

Avogadro's number,

$N_A = 6.022 \times 10^{23}$ **particles mol⁻¹.**

Molar Volume

One mole of any gas contains the same number of molecules (6.022×10^{23}) and by Avogadro's law must occupy the same volume at a given temperature and pressure. The volume of one mole of gas is called the molar gas volume.

The molar volume, V_m, varies with temperature and pressure. Its value at standard temperature and pressure (STP) is the reference standard for gases.

At STP, the molar volume, $V_m = 22.4$ L.

Standard temperature = 273 K;
Standard pressure = 1.01×10^5 Pa.

The number of moles (amount) of gas can be calculated from its volume using

$$\text{Number of moles} = \frac{\text{volume of gas}}{\text{molar volume of gas}}$$

$$n = \frac{V}{V_m}$$

••••• **Example 3.11** •••••

Calculate the number of moles (amount) of carbon dioxide in 250 cm³ of the gas at STP.
Note: 250 cm³ = 0.25 L

$$n(CO_2) = \frac{\text{volume of } CO_2}{\text{molar volume}}$$

$$n = \frac{V}{V_m} = \frac{0.25 \text{ L}}{22.4 \text{ L mol}^{-1}} = 0.011 \text{ mol } CO_2$$

•••••••••••••••••••••••••••••••

The Kinetic Theory of Gases

The kinetic theory of gases is based on the idea that a gas consists of particles in constant random motion. These particles may be molecules such as oxygen, O_2, or atoms such as helium, He. The theory provides a model which explains the regular behaviour observed in all gases.

Many major scientists of the eighteenth and nineteenth centuries, such as Boyle, Bernoulli, Clausius, Maxwell and Boltzman, developed the present theory.

The kinetic theory of gases is based on the following assumptions:

1. Gases consist of particles widely separated in space.

Most of the volume occupied by the gas is empty space, which means that the volume occupied by the gas particles can be ignored. Gases can then be regarded as points in space.

2. The gas particles move randomly in straight lines in all directions.

Gas particles can collide with each other and the walls of the container which hold them with no overall loss of energy. This is shown by Brownian motion.

3. The forces of attraction (or repulsion) between the particles is weak except when they collide.

Evidence for this is shown by the fact that a gas can fill any container that it is put into.

4. The average kinetic energy of a molecule is proportional to the absolute temperature.

This shows the relationship between molecular motion and temperature; the higher the temperature, the greater the kinetic energy.

An **ideal gas** is one which obeys the gas laws at all temperatures and pressures.

The Ideal Gas Law – The Equation of State

The gas laws can be combined into one equation, called the **equation of state** for an ideal gas. The equation $PV = nRT$, which combines all the gas laws, is called the ideal gas law. The equation includes all the information contained in Boyle's law, Charles's law and Avogadro's law.

Ideal gas law $PV = nRT$

P is the pressure of the gas measured in Pascals, Pa
V is the volume of the gas measured in m^3
n is the number of moles of the gas
R is a constant, called the gas constant,
$R = 8.314$ J K^{-1} mol^{-1}
T is the temperature of the gas measured in degrees Kelvin, K

• • • • • Example 3.12 • • • • •

0.5 moles of hydrogen gas has a volume of 8.4 L at 0 °C. Calculate the pressure exerted by the gas, assuming ideal behaviour.

Pressure of gas: $P = ?$

Volume of gas: $V = 8.4$ L $= 8.4 \times 10^{-3}$ m^3

Amount of gas in moles: $n(H_2) = 0.5$ mol

Gas constant: $R = 8.314$ J K^{-1} mol^{-1}

Temperature of gas: $T = 0$ °C $= 273$ K

The equation of state, $PV = nRT$ is rearranged to

$$P = \frac{nRT}{V}$$

Therefore,

$$P = \frac{0.5 \text{ mol} \times 8.314 \text{ J K}^{-1} \text{ mol}^{-1} \times 273 \text{ K}}{8.4 \times 10^{-3} \text{ m}^3}$$

$= 135102.5$ J m^{-3}

$= 135102.5$ N m m^{-3} ← 1 J = 1 N m

$= 135102.5$ N m^{-2}

$= 135102.5$ Pa ← 1 Pa = 1 N m^{-2}

$= 1.35 \times 10^5$ Pa

• • • • • Example 3.13 • • • • •

Calculate the volume occupied by 4 g of carbon dioxide at 20 °C and 50 kPa pressure, assuming ideal behaviour.

Amount of gas in moles:

$$n(CO_2) = \frac{m}{M} = \frac{4 \text{ g}}{44 \text{ g mol}^{-1}} = 0.091 \text{ mol}$$

Pressure of gas:
$P = 50$ kPa $= 50\,000$ N m^{-2} ← 1 Pa = 1 N m^{-2}

Temperature of gas:
$T = 20$ °C $= (273 + 20)$ K $= 293$ K

Gas constant:
$R = 8.314$ J K^{-1} mol^{-1}

Volume of gas:
$V = ?$

The equation of state,
$PV = nRT$ is rearranged to

$$V = \frac{nRT}{P}$$

Therefore,

$$V = \frac{0.091 \text{ mol} \times 8.314 \text{ J K}^{-1} \text{ mol}^{-1} \times 293 \text{ K}}{50\,000 \text{ N m}^{-2}}$$

$= 4.4335 \times 10^{-3}$ J N^{-1} m^2

$= 4.4335 \times 10^{-3}$ N m N^{-1} m^2 ← 1 J = 1 N m

$= 4.4335 \times 10^{-3}$ m^3

HONOURS • Honours • Honours • Honours • Honours • Honours • Honours • Honours • Honours • Honours • Honours • Honours • Honours • Honours • Honours • Honours • Honours • Honours • Honours • Hono

Honours • Honours • Honours • Honours • Honours • Honours • Honours • Honou

• • • • • Example 3.14 • • • • •

Calculate the mass of oxygen in a 50 L tank at 29 °C and a pressure of 2×10^5 Pa.

Pressure:
$P = 2 \times 10^5 \, \text{Pa} = 2 \times 10^5 \, \text{N m}^{-2}$

Volume:
$V = 50 \, \text{L} = 50 \times 10^{-3} \, \text{m}^3$

Number of moles of oxygen:
$n = ?$

Gas constant:
$R = 8.314 \, \text{J K}^{-1} \, \text{mol}^{-1}$

Temperature:
$T = 29 \, °\text{C} = (273 + 29) \, \text{K} = 302 \, \text{K}$

The equation of state
$PV = nRT$ is rearranged to

$$n = \frac{PV}{RT}$$

$$n(O_2) = \frac{2 \times 10^5 \, \text{N m}^{-2} \times 50 \times 10^{-3} \, \text{m}^3}{8.314 \, \text{J K}^{-1} \, \text{mol}^{-1} \times 302 \, \text{K}}$$

$$= \frac{2 \times 10^5 \times 50 \times 10^{-3} \, \text{N m}}{8.314 \times 302 \, \text{J mol}^{-1}}$$

$$= \frac{2 \times 10^5 \times 50 \times 10^{-3} \, \text{N m}}{8.314 \times 302 \, \text{N m mol}^{-1}} \quad \leftarrow 1 \, \text{J} = 1 \, \text{N m}$$

$$= 3.983 \, \text{mol}$$

Mass of oxygen,
$m = nM = 3.983 \, \text{mol} \times 32 \, \text{g mol}^{-1}$

$$= 127.46 \, \text{g} \, O_2$$

• •

Real Gases Deviate from Ideal Behaviour

The ideal gas law describes the behaviour of real gases quite well at high temperatures and low pressures.

Real gases do not obey the equation $PV = nRT$ at low temperatures and high pressures.

Two key assumptions are made by the kinetic theory model of gases:
• the volume of space occupied by the molecules is small in comparison to the total volume
• the molecules are so far apart that intermolecular forces can be disregarded.

However, as the molecules become more densely packed at low temperatures and high pressures, there are attractive forces (polar and van der Waal's) between the molecules as they are closer to each other.

Mandatory experiment 4

To determine the relative molecular mass, M_r, of a volatile liquid such as ethanol.

Introduction

This method may be used to determine the relative molecular mass of many volatile liquids. The method involves heating a liquid to a temperature high enough to convert it into a gas. The liquid is vaporised by using either a steam jacket or a small furnace.

If a steam jacket is used to vaporise the liquid sample, the boiling temperature of the volatile liquid must be low (<80 °C).

If a furnace is used, the relative molecular masses of liquids with boiling temperatures of less than 300 °C can be determined.

Some ethanol is drawn into a hypodermic syringe and weighed. Some of the ethanol is then injected into a glass gas syringe. The mass of ethanol injected is found by reweighing the hypodermic syringe.

The ethanol injected into the glass gas syringe is heated in a furnace or a steam jacket. The liquid vaporises into a gas, the volume of which can be measured in the glass gas syringe.

The ideal gas law is used to determine the relative molecular mass.

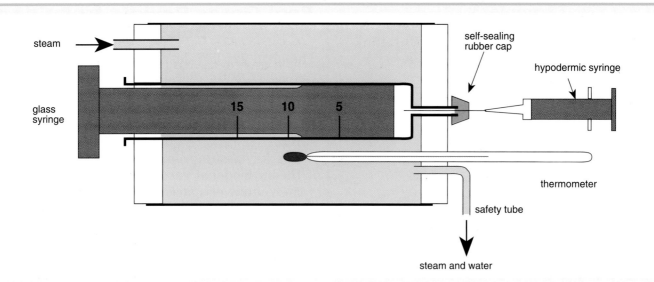

steam

glass syringe

15 10 5

self-sealing rubber cap

hypodermic syringe

thermometer

safety tube

steam and water

Requirements

Safety glasses, glass gas syringe, self-sealing rubber cap, hypodermic syringe and needle, steam jacket and steam generator (or furnace), filter paper, barometer, thermometer.

Procedure

(1) Set up a steam generator, as shown in the diagram, making sure that it has a safety tube.

(2) Place the glass gas syringe into the steam jacket. Draw 5 cm^3 of air into the syringe and fit the self-sealing rubber cap.

(3) Pass steam through the steam jacket until the temperature reading and the volume of air in the syringe are at equilibrium (i.e. steady values).

(4) Draw about 1 cm^3 of ethanol into the hypodermic syringe and rinse it out by flushing the barrel up and down. When the syringe has been rinsed thoroughly, expel the ethanol and draw in another 1 cm^3 of ethanol. Make sure that there are no air bubbles by expelling some ethanol very slowly from the syringe.

(5) Dry the outside of the hypodermic syringe and then weigh it. Record the mass in your copy of the table.

(6) Carefully inject about 0.2 cm^3 of the ethanol through the self-sealing rubber cap in the glass gas syringe in the steam jacket.

(7) Remove the hypodermic syringe carefully and reweigh it. Record the mass.

(8) Make sure the pressure inside the glass syringe is at atmospheric pressure by gently rotating the plunger; it is possible to 'feel' that the pressure is at atmospheric pressure.

Record the atmospheric pressure.

(9) Note when the temperature and the volume of air and ethanol are at steady values. Record these values.

Results

Copy and complete the table. Use your results to determine the relative modecular mass of ethanol (see the specimen calculation).

Mass of hypodermic syringe and ethanol before injection

Mass of hypodermic syringe and ethanol after injection

Steady volume of air in glass gas syringe

Steady volume of air and ethanol vapour in glass gas syringe

Atmospheric pressure

Temperature of ethanol vapour

Specimen Calculation

Mass of ethanol = 0.15 g

Volume of ethanol vapour
= 112 cm^3 = 112 × 10^{-6} m^3

Atmospheric pressure
= 9.6 × 10^{-4} Pa = 9.6 × 10^4 N m^{-2}

Temperature of ethanol vapour
= 100 °C = 373 K

Gas constant, R = 8.314 J K^{-1} mol^{-1}

We can use the ideal gas equation directly to determine the molar mass, M, of the ethanol by rearranging it into the form below.

As $n = \dfrac{m}{M}$ then $PV = nRT = \dfrac{m}{M}RT$

Therefore, $M = \dfrac{mRT}{PV}$

$$= \frac{0.15\ \text{g} \times 8.314\ \text{J K}^{-1}\ \text{mol}^{-1} \times 373\ \text{K}}{9.6 \times 10^{4}\ \text{N m}^{-2} \times 112 \times 10^{-6}\ \text{m}^{3}}$$

$$= \frac{0.15\ \text{g} \times 8.314\ \text{N m K}^{-1}\ \text{mol}^{-1} \times 373\ \text{K}}{9.6 \times 10^{4}\ \text{N m}^{-2} \times 112 \times 10^{-6}\ \text{m}^{3}} (1\,\text{J} = 1\,\text{N m})$$

$= 43.26\ \text{g mol}^{-1}$

The relative molecular mass, M_r, is then 43.26 (no units)

Correct answer = 46

Exercise 3.3
•

(1) Explain the following everyday occurrences in terms of the kinetic theory of gases.
(a) The air in a bicycle tyre leaks out slowly through a tiny hole.
(b) The hole is repaired in the shade. When the tyre is moved out into the sun, the tyre expands.
(c) The volume of air in the other tyre increases when you pump it up.
(d) When you sit on the bicycle, the volume of the tyre decreases.

(2) (a) State: (i) Boyle's law (ii) Charles's law (iii) Avogadro's law (iv) Gay-Lussac's law.
(b) A gas has a volume of 200 cm^3 at a pressure of 97 kPa. Calculate its volume at 104 kPa at the same temperature.
(c) A gas occupies 450 cm^3 at a temperature of 27 °C. Calculate its volume at 37 °C at the same pressure.

(3) (a) List the main points of the kinetic theory. Give some evidence for each point you have listed.

(b) How does kinetic theory explain:
(i) Boyle's law (ii) Charles's law?

(4) A gas contained in a flexible container in an aeroplane exerts a pressure of 10^5 Pa and has a volume of 2.5 L. Suppose that due to a sudden change in pressure the gas expands to 10.5 L. What is the pressure in the aeroplane, assuming that the temperature has remained constant?

(5) The air in the cylinder of a diesel engine occupies 1.0 L at 25 °C at a pressure of 10^5 Pa.
What is the pressure in the cylinder when air is compressed to a compression ratio of 14 : 1 (0.0714 L) and the engine has heated to 500 °C?

(6) A rigid gas cylinder contains 3.5 L of gas at 15 °C and 10^5 Pa.
Determine the pressure when:
(a) the temperature is doubled
(b) the temperature is halved.

(7) Use the combined gas law to calculate the following:
(a) the volume occupied by a gas at a temperature of 20 °C and a pressure of 99 kPa, if it occupies a volume of 40 cm^3 at a temperature of 30 °C and a pressure of 1.03×10^5 Pa.
(b) the pressure exerted by a gas if its volume is 250 cm^3 at a temperature of 21 °C, if it occupies a volume of 280 cm^3 at a pressure of 1.02×10^5 Pa and a temperature of 27 °C.

(8) A sample of biogas, consisting of methane, was produced from sewage. If 40 L of methane was produced at a temperature of 21 °C and a pressure of 9.5×10^4 Pa, calculate the volume of methane which would be produced at STP.

(9) A bottled gas cylinder contains 4.5 L of gas at a temperature of 20 °C and a pressure of 4.0×10^5 Pa. Calculate the volume of gas it would release at STP.

(10) A balloon containing 6.0 L of gas at 15 °C and 100 kPa rises to a height of 2 km, where the temperature is 20 °C and the pressure of gas in the balloon is 80 kPa. What is the volume of gas in the balloon at this altitude?

(11) A volume of air is taken from the earth's surface, at 14 °C and 98 kPa and

sent up into the stratosphere where the temperature is –21 °C and the pressure is 1 kPa. Calculate the percentage increase in the volume of the gas at those conditions.

(12) (a) What is meant by: (i) the molar volume of a gas (ii) the ideal gas equation?
(b) Calculate the amount of oxygen, O_2, in 200 cm³ of the gas at STP.
(c) Calculate the number of moles of nitrogen in 11.2 L at STP.

(13) (a) What is an ideal gas?
(b) Give some reasons why gases deviate from ideal behaviour.
(c) What is the value of R in the ideal gas equation?
(d) Calculate the volume occupied by 2 g of nitrogen, N_2, at 10 °C and 70 kPa, assuming ideal behaviour.
(e) Calculate the mass of hydrogen, H_2, contained in a 10 L vessel at 27 °C and 1.5×10^5 Pa.

(14) A flask containing liquid marked 'alcohol' is analysed to determine whether it is ethanol, an alcoholic drink, or methanol, a poison. When a 0.63 g sample of the alcohol was taken and vaporised at 100 °C and at 10^5 Pa the volume of the alcohol vapour was 415 cm³. What was the molar mass of the alcohol? Was the alcohol ethanol or methanol?

(15) (a) Describe, using diagrams of the apparatus used, how you would determine the relative molecular mass of a volatile liquid.
(b) In an experiment to determine the relative molecular mass of propanone, the following results were tabulated.
Mass of propanone injected = 0.30 g
Temperature of propanone vapour = 100 °C
Atmospheric pressure = 101 kPa
Volume of propanone vapour = 61.5 cm³
Determine the relative molecular mass of propanone from these results.
What is the correct value for the relative molecular mass?
Suggest possible reasons for the difference in this value from the experimental value.

(16) In an experiment, a student collected 190 cm³ of clean dry oxygen, O_2, at a pressure of 101.7 kPa and 12 °C. If the gas weighed 0.27 g, determine the value of R, the gas constant, from the data given.

(17) (a) What is an ideal gas?
(b) Give two reasons why gases deviate from ideal behaviour. Under what conditions of temperature and pressure do real gases come closest to ideal behaviour?
(c) In an experiment to measure the relative molecular mass of a volatile liquid a mass of 0.18 g of the liquid was vaporised by heating it to 100 °C in suitable apparatus. If the volume of the vapour was 94 cm³ and the pressure was 1.02×10^5 Pa, use the ideal gas law to find the relative molecular mass of the liquid.

3.4 CHEMICAL FORMULAE

Polarised light micrograph of vitamin B₁₂. Vitamin B₁₂ deficiency affects almost all body tissues. The most serious effects are pernicious anaemia and a degeneration of the nervous system. Vitamin B₁₂ is contained only in foods of animal origin – liver, fish and eggs.

When a chemist discovers a new compound, the first question he asks is 'what is the formula of this compound?' If the formula of a compound is not known it is not possible to write a chemical equation for a reaction.

Empirical Formulae

The **empirical formula** tells us the simplest whole number ratio of the various atoms in a compound.

The empirical formula is the simplest formula of a compound. It tells us the simplest whole number ratio of the atoms in a compound. Sometimes the empirical formula is the same as the molecular formula, sometimes it is not.

Some examples

- The molecular formula of hydrogen peroxide is H_2O_2, whereas its empirical formula is HO.
- The molecular formula of phosphorus(V) oxide is P_4O_{10}, whereas its empirical formula is P_2O_5.
- The molecular formula of ethene is C_2H_4, whereas its empirical formula is CH_2.
- The molecular formula of carbon dioxide is CO_2, and its empirical formula is CO_2.
- In the case of ionic compounds, the empirical formula is also the formula of the compound. For example, the empirical formula of sodium chloride is NaCl.

Determination of Empirical Formulae

Empirical formulae are determined by experiment by calculating the amounts of the elements present from either
(a) the masses of the constituents, or
(b) the percentage composition by mass.

· · · · · · Example 3.15 · · · · · ·

Mass of Constituents

Analysis of the common oxidising agent, sodium chromate(VII), showed that it contained 1.75 g Na, 3.97 g Cr and 4.28 g O. What is the empirical formula of this compound?

It is suggested that the results are set out in tabular form and that the steps below are followed.

Steps

(a) List the mass of each constituent and then list the molar mass of each constituent.
(b) Calculate the amount (number of moles) of each constituent by dividing the mass by the molar mass.

	Na	Cr	O
Mass/g	1.75	3.97	4.28
Molar mass /g mol^{-1}	23.0	52.0	16.0
Amount /mol	0.0761	0.0763	0.268
Amount / Smallest amount	$\dfrac{0.0761}{0.0761} = 1$	$\dfrac{0.0761}{0.0763} = 1$	$\dfrac{0.268}{0.0761} = 3.5$
Simplest ratio	2	2	7

(c) Divide each amount by the smallest amount.
(d) Express this ratio as a simple whole number ratio. This is the empirical formula of the compound.
Therefore, the empirical formula is $Na_2Cr_2O_7$.

· ·

· · · · · Example 3.16 · · · · ·

Percentage Composition by Mass

A compound contains 40% of carbon, 6.6% hydrogen and 53.4% oxygen. Calculate its empirical formula.
In this case we assume that a 100 g sample is being determined. Thus, the masses of the constituents are 40 g C, 6.6 g H and 53.4 g O, respectively.

A table is then constructed as before.

	C	H	O
Mass/g	40.0	6.6	53.4
Molar mass /g mol^{-1}	12.0	1.0	16.0
Amount/mol	3.33	6.6	3.34
Amount / Smallest amount	$\dfrac{3.33}{3.33} = 1$	$\dfrac{6.6}{3.3} = 2$	$\dfrac{3.34}{3.33} = 1$
Simplest ratio	1	2	1

Therefore, the empirical formula is CH_2O.

· ·

Molecular Formulae from Empirical Formulae

The molecular formula is a multiple of the empirical formula. For example, the empirical formula of glucose is CH_2O but its molecular formula is $C_6H_{12}O_6$,
i.e. the molecular formula is six times the empirical formula

$$C_6H_{12}O_6 = 6(CH_2O)$$

Molecular formula = x(empirical formula) where x is a whole number.

81

· · · · · · **Example 3.17** · · · · · ·

Molecular Formula from Empirical Formula

The ester, ethyl butyrate has an odour of pineapples and is used as an artificial flavour. It has a relative molecular mass of 116 and is composed of 62.0% C, 10.4% H and 27.6% O. Determine its molecular formula.

In this case we assume that a 100 g sample is used in the analysis.

Therefore, the masses of the constituents are 62.0 g C, 10.4 g H and 27.6 g O, respectively.

A table is then constructed as before.

	C	H	O
Mass/g	62.0	10.4	27.6
Molar mass /g mol^{-1}	12.0	1.0	16.0
Amount/mol	5.17	10.4	1.73
Amount / Smallest amount	$\frac{5.17}{1.73} = 3$	$\frac{10.4}{1.73} = 6$	$\frac{1.73}{1.73} = 1$
Simplest ratio	3	6	1

Therefore, the empirical formula is C_3H_6O.

Molecular formula = x(empirical formula)

$$= x(C_3H_6O)$$

Inserting the relative molecular mass of each term gives the following equation:

$116 = x(12 \times 3 + 1 \times 6 + 16)$
$116 = 58x$

Therefore $x = 2$
and the molecular formula = $2(C_3H_6O)$

$$= C_6H_{12}O_2.$$

· ·

· · · · · · **Example 3.18** · · · · · ·

Molecular Formula from Empirical Formula

Urea is used as a fertiliser and as an animal feed. It has a relative molecular mass of 60 and is composed of 46.66% N, 26.66% O, 20% C and 6.66 % H. Determine its molecular formula.

In this case we assume that a 100 g sample is used in the analysis.

Therefore, the masses of the constituents are 46.66 g N, 26.66% O, 20.0 g C, and 6.66 g H, respectively.

A table is then constructed as before.

Constituent	N	O	C	H
Mass/g	46.66	26.66	20.0	6.66
Molar mass /g mol^{-1}	14.0	16.0	12.0	1.0
Amount/mol	3.33	1.67	1.25	6.66
Amount / Smallest amount	$\frac{3.33}{1.25}$	$\frac{1.67}{1.25} = 2$	$\frac{1.25}{1.25} = 1$	$\frac{6.66}{1.25} = 1$ ... $= 4$
Simplest ratio	2	1	1	4

Therefore, the empirical formula is N_2OCH_4.

Molecular formula = x(empirical formula)

$$= x(N_2OCH_4)$$

Inserting the relative molecular mass of each term gives the following equation:

$60 = x(14 \times 2 + 16 + 12 + 1 \times 4)$
$60 = 60x$

Therefore $x = 1$
and the molecular formula = N_2OCH_4.

· ·

• • • • • Example 3.19 • • • • •

Empirical Formula from Masses of Reactants and Products

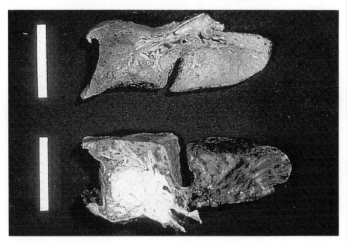

Normal lung and bronchial carcinoma

Nicotine is a compound containing carbon, hydrogen and nitrogen only. When a 5.00 g sample of nicotine is burned in oxygen, 13.56 g of CO_2, 3.88 g of H_2O and 0.42 g of N_2 are the resulting combustion products.

Calculate the empirical formula of nicotine.

5.00 g nicotine + oxygen →
13.56 g CO_2 + 3.88 g H_2O + 0.42 g N_2

Number of moles of CO_2 formed

$$n(CO_2) = \frac{m}{M} = \frac{13.56 \text{ g}}{44.0 \text{ g mol}^{-1}} = 0.308 \text{ mol } CO_2$$

One mole of CO_2 contains one mole of C; this means that the amount of C in the nicotine sample is the same as the amount of CO_2, i.e it is also 0.308 mol.

Mass of C,
$m = nM = 0.308 \text{ mol} \times 12 \text{ g mol}^{-1} = 3.696 \text{ g}$

Number of moles of H_2O formed

$$n(H_2O) = \frac{m}{M} = \frac{3.88 \text{ g}}{18.0 \text{ g mol}^{-1}} = 0.216 \text{ mol } H_2O$$

One mole of H_2O contains two moles of H, which means that the amount of H in the nicotine sample is 2×0.216 mol or 0.432 mol.

Mass of H,
$m = nM = 0.432 \text{ mol} \times 1 \text{ g mol}^{-1} = 0.432 \text{ g}$

Number of moles of N formed

As the nitrogen did not combine with the oxygen in this combustion process and evolves as N_2, then,

Mass of N in sample =
5.00 g – (3.696 + 0.432) g = 0.872 g N

$$n(N) = \frac{m}{M} = \frac{0.0872 \text{ g}}{14.0 \text{ g mol}^{-1}} = 0.062 \text{ mol}$$

Ratio of moles

The nicotine contains the following amounts:
0.308 mol C, 0.432 mol H and 0.062 mol N.

Dividing each by the smallest amount, we get the simplest ratio

$$\frac{0.308}{0.062} = 5 \text{ moles C}$$

$$\frac{0.432}{0.062} = 7 \text{ moles H}$$

$$\frac{0.062}{0.062} = 1 \text{ mole N}$$

Therefore, the empirical formula of nicotine is C_5H_7N.

• •

Percentage Composition of Compounds

The percentage composition of a compound can be readily calculated from the formula of the compound.

The subscripts in the formula of a compound tell us the number of moles of each element in one mole of the compound.

For example, the formula of ethanol is C_2H_5OH and this tells us that in one mole of ethanol there are two moles of C, six moles of H and one mole of O.

From this information and from the relative atomic masses of the elements the percentage composition can be calculated.

$$\text{Mass } \% \text{ A} = \frac{\text{Mass of A in compound}}{\text{Molar mass of compound}} \times 100\%$$

• • • • • • **Example 3.20** • • • • • • •

Percentage Composition by Mass

Methanal, HCHO, is used in the manufacture of plastics and as a preservative for biological specimens. Calculate the percentage mass of each element in methanal.

1 mol of HCHO contains
2 mol H + 1 mol C + 1 mol O

$$M(\text{HCHO}) = (1 + 12 + 1 + 16) \text{ g mol}^{-1}$$
$$= 30 \text{ g mol}^{-1}$$

$$\text{Mass \% H} = \frac{2 \times M(\text{H})}{M(\text{HCHO})} \times 100\%$$

$$= \frac{2 \times 1 \text{ g mol}^{-1}}{30 \text{ g mol}^{-1}} \times 100\% = 6.67\%$$

$$\text{Mass \% C} = \frac{M(\text{C})}{M(\text{HCHO})} \times 100\%$$

$$= \frac{12 \text{ g mol}^{-1}}{30 \text{ g mol}^{-1}} \times 100\% = 40.0\%$$

$$\text{Mass \% O} = \frac{M(\text{O})}{M(\text{HCHO})} \times 100\%$$

$$= \frac{16 \text{ g mol}^{-1}}{30 \text{ g mol}^{-1}} \times 100\% = 53.33\%$$

• •

• • • • • • **Example 3.21** • • • • •

Percentage Composition by Mass

Spearmint oil gets its characteristic smell from carvone, $C_{10}H_{14}O$. Calculate the percentage mass of each element in carvone.

1 mol of $C_{10}H_{14}O$ contains
10 mol C + 14 mol H + 1 mol O

$$M(C_{10}H_{14}O) = (12 \times 10 + 1 \times 14 + 16) \text{ g mol}^{-1}$$
$$= 150 \text{ g mol}^{-1}$$

$$\text{Mass \% C} = \frac{10 \times M(\text{C})}{M(C_{10}H_{14}O)} \times 100\%$$

$$= \frac{10 \times 12 \text{ g mol}^{-1}}{150 \text{ g mol}^{-1}} \times 100\% = 80.0\%$$

$$\text{Mass \% H} = \frac{14 \times M(\text{H})}{M(C_{10}H_{14}O)} \times 100\%$$

$$= \frac{14 \times 1 \text{ g mol}^{-1}}{150 \text{ g mol}^{-1}} \times 100\% = 9.33\%$$

$$\text{Mass \% O} = \frac{M(\text{O})}{M(C_{10}H_{14}O)} \times 100\%$$

$$= \frac{16 \text{ g mol}^{-1}}{150 \text{ g mol}^{-1}} \times 100\% = 10.66\%$$

• •

Structural Formula

The structural formula gives us more information about molecules. It tells us how the atoms are arranged in a molecule.

Sometimes, the structural formula only gives the order of atoms in a molecule.
For example, the structural formula of water, H_2O is written as H—O—H and tells us that two hydrogen atoms are joined to one oxygen atom.

Sometimes the structural formula is written so that we can tell the shape of the molecule.

$$\overset{\displaystyle O}{\underset{\text{H} \quad 105° \quad \text{H}}{\diagup \diagdown}}$$ tells us that the water molecule is V-shaped.

(Shapes of molecules are studied in detail in Unit 2.5.)

Some molecules have the same molecular formula, but different structural formulas. For example, ethanol and methoxy methane have the same molecular formula, C_2H_6O, but different structural formulae.

```
      H  H
      |  |
  H—C—C—H
      |  |
      H  OH
    Ethanol
```

```
      H      H
      |      |
  H—C—O—C—H
      |      |
      H      H
   Methoxy methane
```

84

Exercise 3.4

(1) (a) A substance has the molecular formula $C_6H_{12}O_2$. What is its empirical formula?
(b) Explain the difference between the empirical formula and the molecular formula of a compound.

(2) What is the empirical formula for each of the following compounds:
(a) glucose, $C_6H_{12}O_6$ (b) methane, CH_4 (c) vitamin C, $C_6H_8O_6$ (d) ethanoic acid, CH_3COOH
(e) benzene, C_6H_6 (f) ethyne, C_2H_2?

(3) On analysis, a compound was found to contain 4.6 g of sodium, Na, and 1.6 g of oxygen, O. Calculate the empirical formula of the compound.

(4) A compound was analysed and was found to contain 48.8% carbon, 13.5% hydrogen and 37.7% nitrogen by mass. Determine the empirical formula of the compound.

(5) A hydrocarbon of relative molecular mass 80 was analysed and found to contain 90% C and 10% H by mass. Determine the molecular formula of the hydrocarbon.

(6) A sample of urea contains 2.24 g of N, 0.32 g of H, 0.96 g of C and 1.28 g of O. What is the empirical formula of urea?

(7) Calculate the percentage composition of nitrogen, N, in each of the following:
(a) NH_3 (b) HNO_3 (c) NH_4NO_3 (d) NO_2
(e) N_2O_4 (f) $(NH_4)_2SO_4$ (g) NH_2CONH_2

(8) Calculate the percentage composition of the elements in each of the following compounds:
(a) NaCl (b) KOH (c) H_2SO_4 (d) $Ca(OH)_2$
(e) CH_3COOH (f) $CaCO_3$ (g) $Na_2Cr_2O_7$.

(9) Calculate the percentage by weight of water in each of the following:
(a) washing soda, $Na_2CO_3.10H_2O$
(b) bluestone, $CuSO_4.5H_2O$
(c) Epsom salts, $MgSO_4.7H_2O$
(d) alum, $K_2SO_4.Al_2(SO_4)_3.24H_2O$.

(10) The empirical formula of 'angel dust' is $C_{17}H_{25}N$. A forensic scientist tested a suspected sample of this substance and found that it contained 83.7% C, 10.4% H and 5.6% N. Do the scientist's data match the theoretical data for 'angel dust'?

(11) A forensic scientist analysed a sample of plastic found at the scene of a crime. He found that it contained 86% C and 14% H. What is the empirical formula of the plastic?

(12) Calculate the empirical formula of each of the following from their percentage composition:
(a) benzoic acid: 69% C, 5% H and 26% O
(b) aspirin: 60% C, 4.5% H and 35.5% O
(c) Teflon: 24% C and 76% F.

(13) Calculate the molecular formula of each of the following, given the empirical formula and the relative molecular mass:
(a) benzene, CH, 78 (b) glucose, CH_2O, 180
(c) ethanoic acid, CH_2O, 60 (d) nicotine, C_5H_7N, 162.

(14) (a) The female sex hormone, oestrogen, has a relative molecular mass of 270 and the empirical formula $C_9H_{11}O$. What is its molecular formula?
(b) The hallucinogenic drug LSD contains 74.3% C, 7.74% H, 4.95% O and 13% N. What is its empirical formula?
(c) Oxalic acid occurs in rhubarb and contains 27% C, 2.2% H and 71% O. What is its empirical formula?

(15) (a) What is the difference between a chemical formula and a structural formula?
(b) What information does a structural formula tell us?
(c) Draw the structural formulae of O_2, CO_2, CH_4 and C_2H_5OH.

3.5 CHEMICAL EQUATIONS

Offshore oil rig burning off excess natural gas.

A chemical equation is a short way of presenting a lot of information about a chemical reaction.

When natural gas (methane) burns in oxygen it forms carbon dioxide and water. This information is written in a chemical equation as:

$$CH_4(g) + 2O_2(g) \rightarrow CO_2(g) + 2H_2O(g)$$

The equation tells us:

• The chemical formulae of reactants and products: CH_4 is the chemical formula for methane.

• The names of reactants and products: CH_4 is chemical shorthand for methane.

• The physical states of reactants and products: $CH_4(g)$ means methane is in gaseous form.

• The number of moles of reactants and products: 1 mole of methane reacts with 2 moles of oxygen to form 1 mole of carbon dioxide and 2 moles of water.

In the Haber process, nitrogen gas combines with hydrogen gas to form ammonia gas. The equation is written as

$$N_2(g) + 3H_2(g) \rightarrow 2NH_3(g)$$

This tells us that 1 mole of nitrogen gas combines with 3 moles of hydrogen gas to form 2 moles of ammonia gas.

Balancing Chemical Equations

Chemical equations can be balanced if the reactants and products are known. The rules for balancing equations are based on the law of conservation of mass: atoms are conserved during a chemical reaction.

· · · · · **Example 3.22** · · · · ·

Balancing chemical equations

Write a balanced equation for the reaction of magnesium with oxygen to form magnesium oxide.

Before you start, make sure that you know what the reactants and products are.

Step 1: write a word equation for the reaction.

Magnesium + oxygen $\rightarrow$ magnesium oxide

Step 2: write in the correct formula for each element or compound.

$$Mg + O_2 \rightarrow MgO$$

Step 3: balance the number of each type of atoms on each side.

There are 2 oxygen atoms on the left but only 1 oxygen atom on the right. To balance the number of oxygen atoms we need to double the number of oxygen atoms on the right. This is done by doubling the amount of magnesium oxide.

$$Mg + O_2 \rightarrow 2MgO$$

Now there are 2 magnesium atoms on the right but only 1 magnesium atom on the left. So we need to double the amount of magnesium on the left.

$$2Mg + O_2 = 2\ MgO$$

The equation is now balanced, with equal numbers of each type of atom on the right and on the left.

· ·

Remember: equations cannot be balanced by changing the formulae of the compounds.

Balancing chemical equations

Write a balanced equation for the combustion of ethanol in oxygen to form carbon dioxide and water.

Before you start, make sure that you know what the reactants and products are.

Step 1: write a word equation for the reaction.

Ethanol + oxygen → carbon dioxide + water

Step 2: write in the correct formula for each element or compound.

$$C_2H_5OH + O_2 \rightarrow CO_2 + H_2O$$

Step 3: balance the number of each type of atoms on each side.

It is always best to start with the most complicated molecule (the one containing the greatest number of atoms). In this case, C_2H_5OH is the most complicated molecule.

There are 2 carbon atoms on the left but only 1 carbon atom on the right. To balance the number of carbon atoms we need to double the number of carbon atoms on the right. This is done by doubling the amount of carbon dioxide.

$$C_2H_5OH + O_2 \rightarrow 2CO_2 + H_2O$$

Now there are 6 hydrogen atoms on the left but only 2 hydrogen atoms on the right. So we need to multiply by three the amount of water on the right.

$$C_2H_5OH + O_2 \rightarrow 2CO_2 + 3H_2O$$

We balance the oxygen atoms last. The left-hand side has 3 oxygen atoms (1 + 2), but the right has 7 oxygen atoms ($2 \times 2 + 3 \times 1$). We balance this by multiplying by three the amount of oxygen, O_2, on the left.

$$C_2H_5OH + 3O_2 = 2CO_2 + 3H_2O$$

The equation is now balanced, with equal numbers of each type of atom on the right and on the left.

● ●

Stoichiometry (pronounced stoy-key-om-i-tree), derived from the Greek, stoichion, meaning 'element' and metron, meaning 'to measure', is the branch of chemistry which deals with all the quantitative relationships in chemical reactions.

The balanced chemical equation is called the stoichiometric equation.

Masses of Reactants and Products

We have seen from the previous examples that the balanced chemical equation tells us the number of moles of reactants and products in a chemical reaction. However, in a laboratory or in a chemical plant the amount of reactants used is determined by weighing. In this section, we will see how chemical equations are used to deal with masses of reactants and products.

Calculations involving masses are carried out using the following steps:
1. Write a balanced equation for the reaction.
2. Change the known masses of reactants and products into moles (amounts).
3. Use the information given in the equation.
4. Change from moles back into masses.

Remember: $n = \dfrac{m}{M}$

The iron ore, haematite, Fe_2O_3, is reduced to iron by reacting it with carbon monoxide in a blast furnace. How much iron can theoretically be produced from 1.0 kg of Fe_2O_3?

Step 1 $Fe_2O_3 + 3CO = 2Fe + 3CO_2$
$\qquad$ 1 mol $\quad$ 3 mol $\quad$ 2 mol $\quad$ 3 mol

The balanced chemical equation tells us that 1 mol of Fe_2O_3 produces 2 mol of Fe.

Step 2 Number of moles (amount) of Fe_2O_3 used

$$n(Fe_2O_3) = \frac{m}{M} = \frac{1000 \text{ g}}{160 \text{ g mol}^{-1}}$$

$$= 6.25 \text{ mol } Fe_2O_3$$

Loading iron ore.

Step 3 The number of moles (amount) of Fe formed is twice the number of moles (amount) of Fe_2O_3,

i.e. 2×6.25 mol Fe = 12.50 mol of Fe is produced

Step 4 Change this amount (number of moles) to mass using the equation

$m = nM$
$= 12.50$ mol $\times 56$ g mol^{-1} Fe = 700 g Fe

● ● ● ● ● ● ● ● ● ● ● ● ● ● ● ● ● ● ● ●

● ● ● ● ● ● **Example 3.25** ● ● ● ● ● ●

Aluminium is produced by electrolysis of pure aluminium oxide. What mass of aluminium can theoretically be produced from 2.04 kg of Al_2O_3?

Step 1 $\quad 2Al_2O_3 = 4Al + 3O_2$
$\qquad\qquad$ 2 mol $\quad$ 4 mol $\quad$ 3 mol

The balanced chemical equation tells us that 2 mol of Al_2O_3 produces 4 mol of Al.

Step 2 Number of moles (amount) of Al_2O_3 used

$$n(Al_2O_3) = \frac{m}{M} = \frac{2040 \text{ g}}{102 \text{ g mol}^{-1}}$$

$$= 20.0 \text{ mol } Al_2O_3$$

Step 3 The number of moles (amount) of Al formed is twice the number of moles (amount) of Al_2O_3,

i.e. 2×20.0 mol Al = 40.0 mol of Al is produced

Step 4 Change this amount (number of moles) to mass using the equation

$m = nM$
$= 40.0$ mol $\times 27$ g mol^{-1} Al = 1080 g Al

● ●

● ● ● ● ● ● **Example 3.26** ● ● ● ● ● ●

'Epsom salts' have the formula $MgSO_4.xH_2O$, where x represents the number of water molecules in the crystal. 24.6 g of the crystalline 'Epsom salts' were heated to remove all the water of crystallisation. If 12.0 g of anhydrous $MgSO_4$ was formed after heating the crystals until there was no change in mass, calculate the value of x.

● $MgSO_4.xH_2O \rightarrow MgSO_4 + xH_2O$
$\quad$ 24.6 g $\qquad\qquad$ 12.0 g $\quad$ (24.6 – 12.0) g

● Number of moles (amounts)

Number of moles of $MgSO_4$

$$n(MgSO_4) = \frac{m}{M} = \frac{12.0 \text{ g}}{120 \text{ g mol}^{-1}} = 0.1 \text{ mol}$$

Number of moles of H_2O

$$n(H_2O) = \frac{m}{M} = \frac{12.6 \text{ g}}{18 \text{ g mol}^{-1}} = 0.7 \text{ mol}$$

● This tells us that for every 0.1 mol of anhydrous $MgSO_4$ formed, 0.7 mol of water is given off. This means that for every 1 mol of $MgSO_4$ formed, 7 mol of water is given off, i.e. $x = 7$. The formula of 'Epsom salts' is $MgSO_4.7H_2O$.

● ●

Calculations Involving Volumes of Gases

Calculations involving gases are carried out using the following steps:
● Write a balanced equation for the reaction.
● Change the known volumes and/or masses of reactants and products into moles (amounts).
● Use the information given in the equation.
● Change from moles back into volumes.

Remember: $n = \dfrac{V}{V_m}$

$V_m = 22.4$ L mol^{-1} at STP

Example 3.27

Sulphur dioxide emissions from waste gases are reduced using calcium carbonate (limestone). What volume of sulphur dioxide (measured at STP) could be removed from the waste gases using 1.5 kg of calcium carbonate?

Desulphurisation plant at Bayer, Uerdingen, Germany.

• $2CaCO_3 + 2SO_2 + O_2 = 2CaSO_4 + 2CO_2$

   2 mol   2 mol   1 mol   2 mol   2 mol

The balanced chemical equation tells us that 2 mol of $CaCO_3$ absorbs 2 mol of SO_2.

• Number of moles (amount) of $CaCO_3$ used

$$n(CaCO_3) = \frac{m}{M} = \frac{1500 \text{ g}}{100 \text{ g mol}^{-1}}$$

$$= 15 \text{ mol } CaCO_3$$

• The number of moles (amount) of SO_2 absorbed is equal to the number of moles (amount) of $CaCO_3$,

i.e. 15 mol of SO_2 is absorbed

• Change this amount (number of moles) to volume using the equation

$V = nV_m$
   $= 15 \text{ mol} \times 22.4 \text{ L mol}^{-1} SO_2 = 336 \text{ L } SO_2$

Example 3.28

Information Given by a Balanced Equation

The Haber process is a very important reaction in making fertilisers. In the reaction, nitrogen reacts with hydrogen. Write a balanced chemical equation for the reaction. What information does the balanced chemical reaction tell you?

$$N_2(g) + 3H_2(g) = 2NH_3(g)$$

This balanced chemical equation tells us the following information.

Reactants	$\rightarrow$	Products
$N_2(g) + 3H_2(g)$	$\rightarrow$	$2NH_3(g)$
1 molecule N_2 + 3 molecules H_2	$\rightarrow$	2 molecules NH_3
1 volume N_2 + 3 volumes H_2	$\rightarrow$	2 volumes NH_3
1 mol N_2 + 3 mol H_2	$\rightarrow$	2 mol NH_3
28 g N_2 + 3(2 g) H_2	$\rightarrow$	2(17 g) NH_3
34 g reactants	$\rightarrow$	34 g products
6.022×10^{23} molecules N_2 + 3(6.022×10^{23}) H_2 molecules	$\rightarrow$	2(6.022×10^{23}) molecules NH_3

Limiting Reactant – Theoretical and Percentage Yields

The **limiting reactant** is the reactant that is fully consumed when a reaction goes to completion.

Suppose you are a shoe manufacturer and you have a supply of 300 right shoes and 200 left shoes. How many pairs of shoes can you produce? As you have fewer left shoes than right shoes, this **limits** the number of pairs of shoes you can produce. Left shoes are your **limiting component**; they determine the number of pairs of shoes you can produce. You can only produce 200 pairs of shoes, while at the same time you have an excess of right shoes.

300 right shoes	+	200 left shoes	=	200 pairs of shoes
excess component		**limiting** component		

Usually when chemicals are reacted with each other in a reaction vessel they are added in amounts which are different to the molar ratios given in the balanced chemical equation. This means that only one reactant is completely consumed during the reaction; this reactant is called the **limiting reactant**. A reactant that is not fully consumed is called the **excess reactant**.

The limiting reactant is the reactant that is fully consumed during a chemical reaction. It is this reactant which determines the amount of product formed during the reaction. If the amount of the limiting reactant is known, then the amount of product formed is also known.

Consider the balanced equation for the burning of hydrogen in oxygen to form water.

$$2H_2(g) + O_2(g) = 2H_2O(g)$$

This equation tells us that if 2 mol of H_2 is burned with 1 mol of O_2, then 2 mol of H_2O can be produced.

Suppose you mix together 1 mol of H_2 and 1 mol of O_2. How many moles of H_2O will be produced? The answer is 1 mol of H_2O, because hydrogen, H_2, is the limiting reactant. When all of the hydrogen has been used up the reaction stops. Oxygen, O_2, is the excess reactant.

• • • • • **Example 3.29** • • • • •

Limiting reactant

Methanol, CH_3OH, is used as a fuel in racing cars. Methanol is produced by combining carbon monoxide, CO, and hydrogen. Calculate the theoretical amount of methanol formed when 420 g of carbon monoxide, CO, is reacted with 34 g of hydrogen, H_2.

- **Balanced equation**

$$2H_2(g) + CO(g) = CH_3OH(l)$$
$$\text{2 mol} \qquad \text{1 mol} \qquad \text{1 mol}$$

- **Convert masses to moles**

  Number of moles of H_2

$$n(H_2) = \frac{m}{M} = \frac{34\text{ g}}{2\text{ g mol}^{-1}} = 17\text{ mol}$$

  Number of moles of CO

$$n(CO) = \frac{m}{M} = \frac{420\text{ g}}{28\text{ g mol}^{-1}} = 15\text{ mol}$$

- **Limiting reactant**

According to the chemical equation, 2 moles of H_2 react exactly with 1 mole of CO. This means that 17 moles of H_2 react exactly with 8.5 moles of CO. But 15 moles of CO was used for the reaction, which means that CO is the excess reactant and 17 moles of H_2 is the limiting reactant.

- **Theoretical amount of CH_3OH formed**

17 moles of H_2 is the limiting reactant
2 moles $H_2 \rightarrow$ 1 mole CH_3OH
17 moles $H_2 \rightarrow$ 8.5 moles CH_3OH

- **Convert moles to mass**

Mass of CH_3OH,
$m = nM = 8.5\text{ mol} \times 32\text{ g mol}^{-1} = 272\text{ g}$

• •

The amount of product formed depends on the limiting reactant.

Theoretical Yield and Percentage Yield

> The **theoretical yield** is the amount of a product formed when the limiting reactant is completely used up.

The actual yield obtained from a chemical reaction is rarely the same as the theoretical yield. For example, some of the product may be lost during the separation and purification procedures. It is important to know the actual yield in a reaction, mainly in order to know if the process is economically viable. The actual yield is often given as a percentage of the theoretical yield.

$$\% \text{ yield} = \frac{\text{Actual experimental yield}}{\text{Theoretical yield}} \times 100\%$$

$\bullet \bullet \bullet \bullet \bullet$ **Example 3.30** $\bullet \bullet \bullet \bullet \bullet$

Percentage yield

In an experiment 12 g of ethanoic acid, CH_3COOH, reacts with 4.6 g of ethanol, C_2H_5OH, to produce 6.0 g of ethyl ethanoate, $CH_3COOC_2H_5$, and water, H_2O. Calculate the percentage yield of the ester, ethyl ethanoate, formed.

- **Balanced equation**

$$CH_3COOH + C_2H_5OH = CH_3COOC_2H_5 + H_2O$$
$$\;\;\; 1 \text{ mol} \qquad\quad 1 \text{ mol} \qquad\quad 1 \text{ mol} \qquad\qquad 1 \text{ mol}$$

- **Convert masses to moles**

Number of moles of CH_3COOH used

$$n(CH_3COOH) = \frac{m}{M} = \frac{12 \text{ g}}{60 \text{ g mol}^{-1}} = 0.2 \text{ mol}$$

Number of moles of C_2H_5OH used

$$n(C_2H_5OH) = \frac{m}{M} = \frac{4.6 \text{ g}}{46 \text{ g mol}^{-1}} = 0.1 \text{ mol}$$

- **Limiting reactant**

The balanced equation tells us that 1 mole of CH_3COOH reacts exactly with 1 mole of C_2H_5OH. This means that 0.1 mole of CH_3COOH reacts with exactly 0.1 mole of C_2H_5OH.

As 0.2 mol (the amount used) > 0.1 mol (the amount needed) there is an excess of ethanoic acid being used and therefore, **0.1 mol of ethanol is the limiting reactant.**

- **Theoretical amount of $CH_3COOC_2H_5$ formed**

0.1 mol of ethanol is the limiting reactant. This determines the amount of product formed.

$$0.1 \text{ mol } C_2H_5OH \rightarrow 0.1 \text{ mol } CH_3COOC_2H_5$$

- **Convert moles to mass**

The maximum mass of $CH_3COOC_2H_5$ formed

$$m = nM = 0.1 \text{ mol} \times 88 \text{ g mol}^{-1} = 8.8 \text{ g}$$

- **Percentage yield**

As the actual mass of ethyl ethanoate formed was 6.0 g, then

$$\% \text{ yield} = \frac{\text{Actual experimental yield}}{\text{Theoretical yield}} \times 100\%$$

$$= \frac{6.0 \text{ g}}{8.8 \text{ g}} \times 100\% = 68.2\%$$

Balancing Equations Using Oxidation Numbers

Ionic equations are used when full equations are not necessary. Ionic equations ignore the presence of spectator ions.

When silver nitrate reacts with sodium chloride to form a precipitate of silver chloride, the full equation is:

$$AgNO_3(aq) + NaCl(aq) = AgCl(s) + NaNO_3(aq)$$

When ionic substances like $AgNO_3$ and $NaCl$ dissolve in water they break up into hydrated ions. The full equation can be represented as:

$$Ag^+(aq) + NO_3^-(aq) + Na^+(aq) + Cl^-(aq) = AgCl(s) + Na^+(aq) + NO_3^-(aq)$$

The nitrate ion, NO_3^-, and the sodium ion, Na^+, do not take part in the reaction: they are **spectator ions**.

The full equation can now be simplified and written as an ionic equation:

$$Ag^+(aq) + Cl^-(aq) = AgCl(s)$$

Ionic equations are used mainly for oxidation and reduction reactions.

･ ･ ･ ･ ･ ･ Example 3.31 ･ ･ ･ ･ ･

Potassium manganate(VII) solutions can be used to determine the concentration of Fe^{2+} ions in iron tablets. Balance the redox equation given below using oxidation numbers.

$$MnO_4^-(aq) + Fe^{2+}(aq) + H^+(aq) \rightarrow Mn^{2+}(aq) + Fe^{3+}(aq) + H_2O(l)$$

- **Write in the oxidation numbers for each atom in the equation**

$$MnO_4^- \ + \ Fe^{2+} + \ H^+ \Rightarrow Mn^{2+} + Fe^{3+} + H_2O$$
$$+7, 4(-2) \quad +2 \quad\quad +1 \quad\quad +2 \quad\quad +3 \quad\quad 2(1), -2$$

- **Find out which atoms change oxidation number**

Manganese changes from +7 to +2: a decrease of 5
Iron changes from +2 to +3: an increase of 1

- **Balance each of the atoms which change oxidation number**

The oxidation number for each manganese atom decreases by 5 and for each iron atom increases by 1. The decrease in oxidation number of the manganese must be equal to the increase in oxidation number of the iron. This is achieved by multiplying the iron by 5.

Manganese: decrease of 5×1 = decrease of 5
Iron: increase of 1×5 = increase of 5

- **Insert the changes into the equation**

$$MnO_4^- + Fe^{2+} + H^+ \rightarrow Mn^{2+} + 5Fe^{3+} + H_2O$$

- **Balance the manganese and iron on the left-hand side of the equation**

$$MnO_4^- + 5Fe^{2+} + H^+ \rightarrow Mn^{2+} + 5Fe^{3+} + H_2O$$

- **Balance the overall electrical charge by multiplying the H^+ ion by the appropriate number**

$$MnO_4^- + 5Fe^{2+} + 8H^+ \Rightarrow Mn^{2+} + 5Fe^{3+} + H_2O$$
$$-1 \quad\quad 5(+2) \quad 8(+1) \quad +2 \quad\quad 5(+3) \quad\quad 0$$

- **Balance the remaining atoms by multiplying by the appropriate numbers**

$$MnO_4^- + 5Fe^{2+} + 8H^+ \rightarrow Mn^{2+} + 5Fe^{3+} + 4H_2O$$

･ ･

· · · · · **Example 3.32** · · · · ·

Chlorine, Cl_2, can be prepared by oxidation with potassium manganate(VII) solution. Balance the redox equation given below using oxidation numbers.

$$MnO_4^-(aq) + Cl^-(aq) + H^+(aq) \rightarrow Mn^{2+}(aq) + Cl_2(aq) + H_2O(l)$$

• **Write in the oxidation numbers for each atom in the equation**

$$MnO_4^- \quad + \quad Cl^- + H^+ \rightarrow Mn^{2+} + Cl_2 + H_2O$$
$$\text{+7, 4(−2)} \quad \text{−1} \quad \text{+1} \quad \text{+2} \quad 0 \quad \text{2(1), −2}$$

• **Find out which atoms change oxidation number**

Mn changes from +7 to +2: a decrease of 5. Cl changes from −1 to 0: an increase of 1

• **Balance the chlorines on the right and left**

$$MnO_4^- + 2Cl^- + H^+ \rightarrow Mn^{2+} + Cl_2 + H_2O$$
Mn changes from +7 to +2: a decrease of 5. Cl changes from 2(−1) to 0: an increase of 2

• **Balance each of the atoms which change oxidation number**

The decrease in oxidation number of the manganese must be equal to the increase in oxidation number of the chlorine. This is achieved by multiplying the manganese by 2 and the chloride ion, Cl^-, by 5.

Mn: decrease of 5×2 = decrease of 10
Cl: increase of 2×5 = increase of 10

• **Insert the changes into the equation**

$$2MnO_4^- + 10Cl^- + H^+ \rightarrow Mn^{2+} + Cl_2 + H_2O$$

• **Balance the manganese and chlorine on the right-hand side of the equation.**

$$2MnO_4^- + 10Cl^- + H^+ \rightarrow 2Mn^{2+} + 5Cl_2 + H_2O$$

• **Balance the overall electrical charge by multiplying the H⁺ ion by the appropriate number.**

$$2MnO_4^- + 10Cl^- + 16H^+ \rightarrow 2Mn^{2+} + 5Cl_2 + H_2O$$
$$\text{2(−1)} \quad \text{10(−1)} \quad \text{16(+1)} \quad \text{2(+2)} \quad \text{5(0)} \quad 0$$

• **Balance the remaining atoms by multiplying by the appropriate numbers**

$$2MnO_4^- + 10Cl^- + 16H^+ \rightarrow 2Mn^{2+} + 5Cl_2 + 8H_2O$$

· ·

Exercise 3.5

•

(1) What information does the chemical equation below tell us?

$$2H_2(g) + O_2(g) = 2H_2O(g)$$

(2) What information does the chemical equation below tell us?

$$2Mg(s) + O_2(g) = 2MgO(s)$$

(3) Balance each of the following reactions, which occur in the blast furnace during the production of iron:

(a) $Fe_2O_3 + CO \rightarrow Fe + CO_2$

(b) $Fe_2O_3 + C \rightarrow Fe + CO_2$

(4) Balance the following combustion reactions:

(a) $CH_4 + O_2 \rightarrow CO_2 + H_2O$

(b) $C_2H_4 + O_2 \rightarrow CO_2 + H_2O$

(c) $C_3H_6 + O_2 \rightarrow CO_2 + H_2O$

(d) $C_4H_8 + O_2 \rightarrow CO_2 + H_2O$

(5) Balance the following equations:

(a) $SO_2 + O_2 \rightarrow SO_3$

(b) $Ca + H_2O \rightarrow Ca(OH)_2 + H_2$

(c) $KClO_3 \rightarrow KCl + O_2$

(d) $Mg + HCl \rightarrow MgCl_2 + H_2$

(6) Describe, in words, the meaning of the equation:
$$C_2H_4 + O_2 = CO_2 + H_2O$$
in terms of (a) molecules, (b) moles and (c) masses.

(7) What mass of iron(III) oxide, Fe_2O_3, will be produced when 50 g of iron is oxidised to iron(III) oxide?
$$4Fe + 3O_2 = 2Fe_2O_3$$

(8) When cane sugar, $C_{12}H_{22}O_{11}$, is burned in air it produces carbon dioxide and water. What are the masses of water and carbon dioxide formed when 17.1 g of cane sugar is burned?
$$C_{12}H_{22}O_{11} + 12O_2 = 12CO_2 + 11H_2O$$

(9) Nitric acid is manufactured using the Ostwald process, where nitrogen dioxide, NO_2, reacts with water and oxygen.
$$4NO_2 + 2H_2O + O_2 = 4HNO_3$$
How many grams of nitrogen dioxide are required to produce 12.6 kg of HNO_3?

(10) (a) Calculate the mass of carbon dioxide formed when 1.25 kg of calcium carbonate are heated.
$$CaCO_3 = CaO + CO_2$$
(b) Calculate the mass of aluminium produced when 20.4 kg of alumina, Al_2O_3, is reduced to aluminium by electrolysis.
$$2Al_2O_3 = 4Al + 3O_2$$
(c) What mass of carbon is needed to produce 1.4 kg of iron from iron(III) oxide?
$$2Fe_2O_3 + 3C = 4Fe + 3CO_2$$

(11) 21g of iron, Fe, reacts with 8 g of oxygen gas, O_2, to produce an oxide of iron, Fe_3O_4.
(a) Write a balanced equation for the reaction.
(b) How many moles of the oxide are formed?

(12) The main reactions involved in the production of sulphuric acid, H_2SO_4, by the contact process are:
$$S(s) + O_2(g) = SO_2(g)$$
$$2SO_2(g) + O_2(g) = 2SO_3(g)$$
$$SO_3(g) + H_2O(l) = H_2SO_4(aq)$$
If all the sulphur were successfully converted into sulphuric acid, (a) what amount and (b) what mass of sulphur would be needed to produce 2.0 kg of sulphuric acid?

(13) (a) Aluminium reacts with iodine, I_2, as follows:
$$2Al + 3I_2 = 2AlI_3$$
What mass of iodine, I_2, will react completely with 10 g of aluminium?
(b) Iron and sulphur react forming iron sulphide, FeS. Calculate the mass of sulphur which reacts with 14.0 g of iron.

(14) Silica, SiO_2, is removed from iron by reacting it with quicklime, CaO, to produce a molten slag $CaSiO_3$. What mass of slag is produced from 1 kg of silica?
$$SiO_2 + CaO = CaSiO_3$$

(15) (a) Calculate the volume of carbon dioxide formed at STP when 250 g of calcium carbonate is heated.
$$CaCO_3 = CaO + CO_2$$
(b) Calculate the volume of oxygen produced when 102 g of alumina, Al_2O_3, is reduced to aluminium by electrolysis at STP.
$$2Al_2O_3 = 4Al + 3O_2$$
(c) What volume of carbon dioxide is produced at STP when 640 g of iron(III) oxide is reduced to iron?
$$2Fe_2O_3 + 3C = 4Fe + 3CO_2$$

(16) Ethyne gas is produced by the reaction of calcium dicarbide and water as follows:
$$CaC_2 + 2H_2O = C_2H_2 + Ca(OH)_2$$
If 12.8 g of CaC_2 were used in the reaction, calculate:

(a) the mass of C_2H_2 produced, and (b) the volume of C_2H_2 produced at STP.

(17) Carbon dioxide is produced in the laboratory by reacting dilute hydrochloric acid with calcium carbonate.

$$CaCO_3(s) + 2HCl(aq)$$
$$= CaCl_2(aq) + CO_2(g) + H_2O(l)$$

Determine the volume of CO_2 produced at STP from 150 g of $CaCO_3$.

(18) Coal contains many sulphur impurities, such as iron pyrites, FeS_2, which releases undesirable sulphur dioxide into the atmosphere when it is burned:

$$4FeS_2(s) + 11O_2(g) = 2Fe_2O_3(s) + 8SO_2(g)$$

Calculate the volume of SO_2 produced at STP when 616 g of FeS_2 is burned.

(19) 0.8 g of calcium reacted with 0.32 g of oxygen, O_2, to form a compound X, which has a relative molar mass of 56.
(a) How many moles of calcium reacted?
(b) How many moles of oxygen, O_2, reacted?
(c) What is the empirical formula of X?
(d) What is the molecular formula of X?

(20) Hydrogen sulphide, H_2S, is released during the refining of crude oil. It can be treated by combining it with oxygen:

$$2H_2S(g) + O_2(g) = 2S(s) + 2H_2O(l)$$

(a) How many kilograms of oxygen are needed to react with 1700 g of H_2S?
(b) How many grams of sulphur are formed from 153 kg of H_2S?

(21) For the reaction given, identify the limiting reactant in each of the following cases:

$$2H_2 + O_2 = 2H_2O$$

(a) 1 mol H_2 and 1 mol O_2
(b) 0.5 mol H_2 and 2 mol O_2
(c) 1g H_2 and 16 g O_2
(d) 10 g H_2 and 160 g O_2
(e) 1.5 kg H_2 and 4.0 kg O_2

(22) Ammonia is burned in the presence of a catalyst to form nitric oxide, NO, in the first stage in the production of nitric acid.

$$4NH_3 + 5O_2 = 4NO + 6H_2O$$

Suppose the reaction vessel contains 240 moles of ammonia and 280 moles of oxygen. Which is the limiting reactant?

(23) Ammonia is produced by the Haber process as shown:

$$N_2 + 3H_2 = 2NH_3$$

(a) Which is the limiting reactant when 2 kg of nitrogen, N_2, reacts with 1 kg of hydrogen, H_2?
(b) What is the mass of the unreacted starting material?

(24) Zinc reacts with sulphur to form zinc sulphide:

$$Zn(s) + S(s) = ZnS(s)$$

(a) Determine which is the limiting reactant when 130 g of zinc and 72 g of sulphur are reacted together to form zinc sulphide.
(b) How many grams of the excess reactant will be left over?

(25) Ethyne gas, C_2H_2, and oxygen, O_2, react together, giving off intense heat. This heat is used in 'oxyacetylene' burners to weld and cut metals. The products formed are carbon dioxide and water.

$$C_2H_2 + \tfrac{5}{2}O_2 = 2CO_2 + H_2O$$

(a) How many moles of oxygen are needed to react with 52 g of C_2H_2?
(b) Determine which is the limiting reactant when 52 g of C_2H_2 and 240 g of O_2 react together to form carbon dioxide and water.

(26) (a) What is meant by (i) the actual yield of a reaction, (ii) the theoretical yield and (iii) the percentage yield?
(b) Why does the actual yield differ from the theoretical yield?

(27) The theoretical yield in the production of ammonia in an industrial process was 55 kg, but only 42 kg was obtained. What is the percentage yield of ammonia?

(28) (a) What is a limiting reactant? Is there a limiting reactant in all chemical reactions?
(b) What is the percentage yield? Why is it useful?
(c) In a particular chemical reaction the theoretical yield of product was calculated as 34.54 g. If the actual yield obtained was 24.1 g, determine the percentage yield.

(29) In a laboratory experiment the following reaction was used to prepare hydrogen gas, H_2.

$$Zn(s) + 2HCl(aq) = ZnCl_2(aq) + H_2(g)$$

If 0.15 mol of Zn were added to 1.04 mol HCl, determine which of the two reactants was the limiting reactant and then determine how many moles of H_2 were produced in the reaction.

(30) Aluminium chloride, $AlCl_3$, is an important industrial catalyst. It can be prepared by the following reaction:

$$2Al(s) + 6HCl(g) = 2AlCl_3(s) + H_2(g)$$

Calculate the amount of $AlCl_3$ formed when 0.30 mol Al and 0.70 mol HCl react in a reaction vessel.

(31) Aluminium is made from bauxite ore, which is an impure oxide of aluminium. The aluminium(III) oxide, Al_2O_3, is reduced by carbon to form aluminium and carbon dioxide.

$$2Al_2O_3 + 3C = 4Al + 3CO_2$$

(a) How many grams of aluminium oxide are needed to produce 540 g of aluminium metal?
(b) If it is assumed that bauxite is 50% Al_2O_3, how many grams of aluminium can be produced from 1 kg of ore?
(c) Determine how many kilograms of aluminium are formed when 816 kg of aluminium(III) oxide reacts with 252 kg of carbon.

(32) Hydrogen chloride can be prepared from natural gas by burning it in a mixture of chlorine and air by the following reaction:

$$2CH_4(g) + O_2(g) + 4Cl_2 = 8HCl(g) + 2CO(g)$$

(a) If the reaction vessel contains 26.0 g CH_4, 142 g Cl_2 and excess O_2, calculate (i) the limiting reactant (ii) the theoretical yield of HCl.
(b) If the actual yield of HCl obtained was 116 g, then determine the percentage yield of HCl.

(33) Explain why ionic equations are sometimes used in preference to full equations.

(34) Balance the following equations using oxidation numbers:

(a) $Fe^{2+} + H^+ + ClO_3^- \rightarrow Fe^{3+} + Cl^- + H_2O$

(b) $MnO_4^- + I^- \rightarrow MnO_2 + IO_3^-$

(c) $IO_3^- + H_2O + SO_2 \rightarrow I_2 + SO_4^{2-} + H^+$

(d) $MnO_4^- + H^+ + C_2O_4^{2-} \rightarrow Mn^{2+} + H_2O + CO_2$

(e) $Cr_2O_7^{2-} + H^+ + Fe^{2+} \rightarrow Cr^{3+} + Fe^{3+} + H_2O$

(f) $S_2O_3^{2-} + I_2 \rightarrow S_4O_6^{2-} + I^-$

Some of the more important terms are listed below. Other terms not listed may be found by means of the index.

1. States of matter:
Solids may exist as crystals or as glasses. Solids have low kinetic energy and move by vibrating from side to side.
Liquids are mobile and almost structureless fluids. The particles are constantly moving about with high kinetic energy. The high kinetic energy allows the particles in a liquid to vibrate, rotate and translate.
Gases can be pictured as a swarm of molecules of extremely high kinetic energy in constant chaotic motion.

2. Diffusion: Diffusion is the process by which a substance spreads out throughout another substance in a uniform manner. The particles move from a region of high concentration to one of low concentration.

3. The mole: The unit used to count atoms, ions or molecules is called the mole. One mole (mol) is defined as the amount of the substance which contains as many elementary particles as there are carbon atoms in 0.012 kg (12 g) of carbon-12.

4. Avogadro's number: Avogadro's number, N_A, tells us the number of particles (ions, atoms or molecules) in one mole of the substance.

$$N_A = 6.022 \times 10^{23} \text{ particles mol}^{-1}$$

5. Relative atomic mass: The relative atomic mass, A_r, is defined as the ratio of the average mass of that atom compared to $\frac{1}{12}$ of the mass of an atom of carbon-12. It is a pure number, and is also called the atomic weight.

6. Relative molecular mass: The relative molecular mass, M_r, is defined as the average mass of the molecular entity compared to $\frac{1}{12}$ of the mass of an atom of carbon-12.

7. Molar mass: The molar mass, M, is the mass of one mole of the particular entity. It is expressed in grams per mole (g mol^{-1}).

8. Amount (number of moles):

$$\text{Amount of substance} = \frac{\text{Mass}}{\text{Molar mass}}$$

$$n = \frac{m}{M}$$

9. Mass:

$$\text{Mass, } m = nM$$

10. Number of particles in one mole:

$N = n\, N_A$ where N = number of particles
n = amount in moles
N_A = Avogadro's number

11. The kinetic theory of gases:

• Gases consist of particles widely separated in space.
• The gas particles move randomly in straight lines in all directions.
• The forces of attraction (or repulsion) between the particles is weak except when they collide.
• The average kinetic energy of a molecule is proportional to the absolute temperature.

12. The gas laws:

Ideal gas

An ideal gas is one which obeys the gas laws at all temperatures and pressures.

Boyle's law

At constant temperature, the volume of a fixed mass of gas varies inversely with pressure.

$$PV = k \text{ (constant)}$$

Charles' law

At constant pressure, the volume of a fixed mass of gas is directly proportional to the absolute temperature.

$$\frac{V}{T} = k \text{ (constant)}$$

The combined gas law

Chemical reactions occur under various conditions of temperature, pressure and volume. Boyle's law and Charles' law can be combined to give the combined gas law as follows:

$$\frac{P_1 V_1}{T_1} = \frac{P_2 V_2}{T_2}$$

Gay-Lussac's law of combining volumes

The volumes of reactant and product gases at a given temperature and pressure are in the ratios of small whole numbers.

Avogadro's law

Equal volumes of gases, under the same conditions of temperature and pressure, contain equal numbers of molecules.

The ideal gas law: the equation of state

$$PV = nRT$$

13. Standard temperature and pressure (STP):

Standard temperature = 273 K;
standard pressure = 1.01×10^5 Pa.

14. Molar volume: The volume of one mole of gas is called is called the molar gas volume. At STP, the molar volume, $V_m = 22.4$ L.

15. Converting moles to litres (gases only):

$$V = nV_m$$

16. Empirical formula: The empirical formula is the simplest whole number ratio of the various atoms in a compound.

17. Molecular formula from empirical formula:

Molecular formula = x(empirical formula) where x is a whole number.

18. Percentage composition of compounds:

$$\text{Mass \% A} = \frac{\text{Mass of A in compound}}{\text{Molar mass of compound}} \times 100\%$$

19. Structural formula: The structural formula tells us how the atoms are arranged in a compound.

20. Limiting reactant: The limiting reactant is the reactant that is fully consumed when a reaction goes to completion.

21. Theoretical yield: The theoretical yield is the amount of a product formed when the limiting reactant is completely used up.

22. Percentage yield:

$$\text{\% yield} = \frac{\text{Actual experimental yield}}{\text{Theoretical yield}} \times 100\%$$

VOLUMETRIC ANALYSIS

Water drops on surface.

$$\text{Molar concentration} = \frac{\text{Amount of solute in moles}}{\text{Volume of solution}}$$

The unit of concentration is moles per litre (mol L^{-1}).

Moles per litre is written as mol L^{-1}.

It is often useful to use the simple formula

$$c = \frac{n}{V}$$

where c is the molar concentration, n is the number of moles of solute and V is the volume of the solution, in order to calculate moles, concentrations, or volumes in calculations involving solutions.

The substance which is dissolved is called the solute.
The liquid used to dissolve this substance is the solvent.
The solute and solvent mix together to form a solution.

4.1 CONCENTRATIONS OF SOLUTIONS

Reactions in Solutions

Reactions between solids proceed extremely slowly or sometimes not at all. Reactions in solution are much quicker. For this reason most reactions are carried out in solution. It is, therefore, important to know the amount of reactants in the solution.

We must know the concentration of the solution and the volume of the solution in order to determine the amount (number of moles) of a reactant in solution.

There are many ways of expressing the concentration of a solution. The most convenient unit of concentration is **molar concentration** or **molarity (M)**.

The molar concentration of a solution is the amount (number of moles) of solute dissolved in a stated volume of solution.

For instance, a 1.0 mol L^{-1} solution would contain one mole of solute dissolved in one litre of solution.

• • • • • **Example 4.1** • • • • •

To calculate concentration from amount and volume

Calculate the concentration of a solution which is made up by dissolving 0.4 mol of sodium carbonate in 200 cm³ of solution.

Amount of solute in moles,
$$n(\text{Na}_2\text{CO}_3) = 0.4 \text{ mol}$$

Volume of solution in litres,
$$V = 200 \text{ cm}^3 = 0.2 \text{ L}$$
then:

$$c = \frac{n}{V} = \frac{0.4 \text{ mol}}{0.2 \text{ L}} = 2.0 \text{ mol L}^{-1} \text{ Na}_2\text{CO}_3$$

$2.0 \text{ mol L}^{-1} \text{Na}_2\text{CO}_3$ is sometimes written as $2.0 \text{ M Na}_2\text{CO}_3$.

• •

•••••• **Example 4.2** ••••••

To calculate concentration in grams per litre (g L⁻¹)

Calculate the concentration of sodium hydroxide solution made by dissolving 20.0 g of pure sodium hydroxide, NaOH, in 500 cm³ of solution. Give the concentration in grams per litre (g L⁻¹).

$M(NaOH) = (23 + 16 + 1)$ g mol⁻¹ $= 40$ g mol⁻¹

Number of moles of NaOH:

$$n(NaOH) = \frac{m}{M} = \frac{20.0 \text{ g}}{40.0 \text{ g mol}^{-1}} = 0.5 \text{ mol}$$

Molar concentration,

$$c = \frac{n}{V} = \frac{0.5 \text{ mol}}{0.5 \text{ L}} = 1.0 \text{ mol L}^{-1} \text{ NaOH}$$

Concentration in g L⁻¹,

$$c = 1.0 \text{ mol L}^{-1} \times 40 \text{ g mol}^{-1} \text{ NaOH}$$
$$= 40.0 \text{ g L}^{-1} \text{ NaOH}$$

• • • • • • • • • • • • • • • • • • •

• • • • • **Example 4.3** • • • • •

To calculate amount (number of moles) from concentration and volume

Calculate the amount of sodium hydroxide in 250 cm³ of 2.0 mol L⁻¹ NaOH.
Rearranging the formula

$$c = \frac{n}{V} \text{ to } n = cV$$

$n(NaOH) = cV = 2.0 \text{ mol L}^{-1} \times 0.250 \text{ L}$
$$= 0.50 \text{ mol}$$

Note: it is convenient to write the number of moles of any substance such as NaOH as $n(NaOH)$.

• • • • • • • • • • • • • • • • • • •

•••••• **Example 4.4** ••••••

To calculate concentration from mass of solute and volume

Calculate the concentration of sodium nitrate solution made by dissolving 4.25 g NaNO₃ in 250 cm³ of solution.

$M(NaNO_3)$
$= (23 + 16 + 16 \times 3)$ g mol⁻¹ $= 85$ g mol⁻¹

$$n(NaNO_3) = \frac{m}{M} = \frac{4.25 \text{ g}}{85.0 \text{ g mol}^{-1}}$$

$= 0.05 \text{ mol NaNO}_3$

Now,

$$c = \frac{n}{V} = \frac{0.05 \text{ mol}}{0.250 \text{ L}} = 0.2 \text{ mol L}^{-1}$$

• • • • • • • • • • • • • • • • • • • •

Concentrations Expressed in Different Ways

Precise concentrations are usually expressed in moles per litre (mol L⁻¹) of solution. Sometimes, other ways of expressing concentration are used, such as the following.

Weight per volume (w/v): this tells us the concentration expressed as the weight of solute in a specified volume of the solution. Hydrogen peroxide solution is usually made up by weight dissolved in a volume of solution, e.g. 30% w/v H₂O₂ = 30 g H₂O₂ in 100 cm³ of solution.

Volume per volume (v/v): this tells us the concentration expressed as the volume of solute in a specified volume of solution, e.g. 5% v/v ethanol in water = 5 cm³ C₂H₅OH in 100 cm³ of solution.

Weight per weight (w/w): this tells us the concentration expressed as the weight of solute in a specified weight of solution, e.g. 5% w/w starch solution = 5 g starch in 95 g water

Parts per million (ppm): this is used to describe small concentrations which are often expressed as milligrams per litre (mg L⁻¹), that is 1/1000 g solute in 1 L of solution. Parts per million are usually used to describe small concentrations of impurities in water, e.g. 30 ppm of $Ca(OH)_2$ = 30 mg $Ca(OH)_2$ in 1 L of solution.

$$ppm = \frac{mass\ of\ solute\ /\ g}{mass\ of\ solution\ /\ g} \times 10^6$$

• • • • • • **Example 4.5** • • • • •

Describe how you would make up 300 cm^3 of a 30% (v/v) solution of ethanol.
30% w/v ≡ 30 cm^3 in 100 cm^3 of solution ≡ 90 cm^3 in 300 cm^3 of solution
(1) Measure out 90 cm^3 of ethanol in a 500 cm^3 graduated cylinder.
(2) Add water until the volume of liquid in the graduated cylinder is 300 cm^3.

• • • • • • • • • • • • • • • • • • • •

• • • • • • **Example 4.6** • • • • •

The molar concentration of oxygen dissolved in clean water is approx. 0.0003125 mol L^{-1}. Express this in parts per million, ppm.

Concentration of oxygen,
c = 0.0003125 mol L^{-1}
 = 0.0003125 mol L^{-1} × 32 g mol^{-1}
 = 0.01 g L^{-1} (1 litre of water = 1000 g)
 = 10 ppm

• • • • • • • • • • • • • • • • • • • •

Colour Intensity

Many substances which react in solution are highly coloured. Potassium manganate(VII), $KMnO_4$, forms an intense purple solution when it is dissolved in water. As more and more water is added to the solution the colour of the solution becomes lighter and lighter. The intensity of the colour depends on the concentration of the potassium manganate(VII). The colour intensity can be measured using an instrument called a colorimeter.

Dilution

Most solutions are made in the school laboratory either by dissolving solids in water or by adding water to a concentrated solution.

These solutions are stored in the laboratory as concentrated solutions in order to save time and space: these solutions are called stock solutions. When a particular solution is required it is prepared by diluting the concentrated stock solution.

The common acids are purchased as concentrated solutions. For example, concentrated sulphuric acid is approximately 16 mol L^{-1}.

As only water is added during a dilution, the amount of solute in the final dilute solution must equal the amount of solute in the concentrated stock solution.
Number of moles of solute after dilution = number of moles of solute before dilution

• • • • • • **Example 4.7** • • • • •

Effect of dilution on concentration

What volume of 16.0 mol L^{-1} nitric acid is required to prepare 2.5 L of a 0.1 mol L^{-1} nitric acid solution?
Number of moles before dilution = number of moles after dilution

$n = cV$

c_1V_1 (before) = c_2V_2 (after)

16 mol L^{-1} × V_1 = 0.1 mol L^{-1} × 2.5 L

$$V_1 = \frac{0.1\ mol\ L^{-1} \times 2.5\ L}{16.0\ mol\ L^{-1}} = 0.015625\ L$$

= 15.625 cm^3 = 15.6 cm^3

• • • • • • • • • • • • • • • • • • • •

Standard Solutions

A standard solution is a solution of known concentration.

Some standard solutions can be prepared accurately by dissolving a known amount of solute in a solution, if the solute is in a pure state. Anhydrous sodium carbonate is one such example, which can be used to make a standard solution directly. Standard solutions such as these are often called primary standards.

Most standard solutions are made up by titrating a known standard solution with an unknown solution. For instance, a standard solution of sodium hydroxide cannot be made up directly because sodium hydroxide is hygroscopic (absorbs water) and also absorbs carbon dioxide from the atmosphere.

Requirements for a substance used as a primary standard:
(1) It must be available in a pure state.
(2) It must be stable in air.
(3) It must dissolve easily in water.
(4) Ideally, it should have a high molar mass which ensures greater accuracy when weighing.
(5) When used in titration it should react quickly and completely.

Making up a solution.

Mandatory experiment 5
To prepare a standard solution of sodium carbonate.

Introduction
Anhydrous sodium carbonate can be bought in an extremely pure state. However, it absorbs water from the atmosphere very quickly. It must be heated in a crucible to 250 °C to remove any water present and then cooled in a dessicator.

In this experiment 250 cm^3 of an exact 0.1 mol L^{-1} Na$_2$CO$_3$ solution is prepared in a volumetric flask.

Amount of Na$_2$CO$_3$

in 0.25 L of 0.1 mol L^{-1} Na$_2$CO$_3$

$n(\text{Na}_2\text{CO}_3) = cV$

$= 0.1 \text{ mol L}^{-1} \times 0.25 \text{ L} = 0.025 \text{ mol}$

The molar mass of Na$_2$CO$_3$ is 106 g mol^{-1} then,

mass of Na$_2$CO$_3$,

$m = nM = 0.025 \text{ mol} \times 106 \text{ g mol}^{-1} = 2.65 \text{ g}$

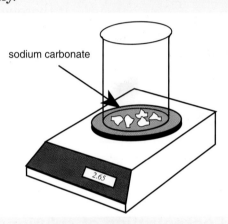

Required
Safety glasses, 250 cm^3 volumetric flask and label, 100 cm^3 beaker, filter funnel, stirrer, wash bottle, deionised water, pipette, balance and anhydrous sodium carbonate.

Procedure
(1) Rinse the 250 cm^3 volumetric flask, filter funnel, 100 cm^3 beaker and glass stirrer thoroughly with deionised water.
(2) Weigh out accurately 2.65 g of the pure dry anhydrous sodium carbonate in the 100 cm^3 beaker. Add about 50 cm^3 of deionised water to the beaker. Stir constantly.

sodium carbonate

2.65

(3) When the sodium carbonate has fully dissolved transfer the solution to the volumetric flask. Make sure that all the sodium carbonate solution is transferred to the volumetric flask by constantly rinsing the beaker with deionised water and adding the rinsings to the volumetric flask.

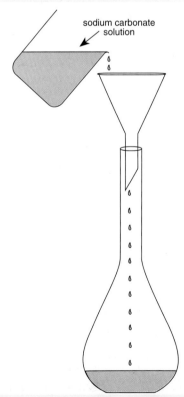

sodium carbonate solution

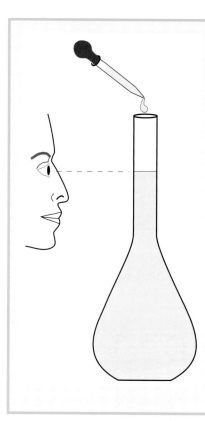

(4) Make the solution in the volumetric flask almost up to the engraved mark with deionised water and then add more deionised water drop by drop from a pipette until the meniscus sits on the engraved mark. (5) Stopper the flask and shake it thoroughly 9–10 times to ensure correct mixing. The solution is now an exact

0.1 mol L⁻¹ sodium carbonate solution.
(6) Label the flask with the exact concentration, your name and the date on which the solution was made up. Place aside if it is required for another experiment.

Results and Calculations

Mass of Na_2CO_3,
$m = 2.65$ g

Molar mass of Na_2CO_3,
$M = 106$ g mol⁻¹

Amount of Na_2CO_3,

$$n = \frac{m}{M} = \frac{2.65 \text{ g}}{106 \text{ g mol}^{-1}} = 0.025 \text{ mol}$$

Concentration of Na_2CO_3,

$$c = \frac{n}{V} = \frac{0.025 \text{ mol}}{0.25 \text{ L}} = 0.1 \text{ mol L}^{-1}$$

Exercise 4.1

•

Concentration of solutions

(1) Define each of the following terms using a suitable example:
(a) molar concentration
(b) standard solution and primary standard.

(2) Calculate the concentrations of each of the following solutions:
(a) 1 mol NaOH in 2 L of solution
(b) 0.1 mol Na_2CO_3 in 250 cm³ of solution
(c) 0.1 mol HCl in 500 cm³ of solution
(d) 0.25 mol H_2SO_4 in 5 L of solution
(e) 0.2 mol $NaHCO_3$ in 400 cm³ of solution.

(3) Calculate the concentrations of each of the following solutions:
(a) 20 g NaOH in 250 cm³ of solution
(b) 13.1 g $Na_2Cr_2O_7$ in 500 cm³ of solution
(c) 3.65 g HCl in 1 L of solution
(d) 4.9 g H_2SO_4 in 500 cm³ of solution
(e) 4.2 g $NaHCO_3$ in 200 cm³ of solution.

(4) Describe how you would prepare 1 L of each of the following solutions:
(a) 0.2 mol L⁻¹ Na_2CO_3 solution from pure anydrous Na_2CO_3

(b) 0.1 mol L⁻¹ Na_2CO_3 solution from 2.5 mol L⁻¹ Na_2CO_3 solution
(c) 0.3 mol L⁻¹ $NaHCO_3$ solution from 1.5 mol L⁻¹ $NaHCO_3$
(d) 0.2 mol L⁻¹ HCl solution from 0.5 mol L⁻¹ HCl
(e) 0.1 mol L⁻¹ H_2SO_4 solution from 1.0 mol L⁻¹ H_2SO_4.

(5) (a) Describe, briefly, how you would prepare 1.0 L of a 0.1 mol L⁻¹ solution of each of the following:
(i) HCl from concentrated HCl (12.0 mol L⁻¹)
(ii) H_2SO_4 from concentrated H_2SO_4 (16.0 mol L⁻¹)
(iii) HNO_3 from concentrated HNO_3 (18.0 mol L⁻¹).
(b) Describe, briefly, how you would prepare 2.0 L of a 0.2 mol L⁻¹ solution of each of the following:
(i) HCl from concentrated HCl (12.0 mol L⁻¹)
(ii) H_2SO_4 from concentrated H_2SO_4 (18.0 mol L⁻¹)
(iii) HNO_3 from concentrated HNO_3 (16.0 mol L⁻¹).

(6) (a) What volume of solution is required to obtain 2 mol NaCl from a 0.1 mol L⁻¹ NaCl solution?
(b) What volume of solution is required to

obtain 0.25 mol NaOH from a 1.5 mol L^{-1} NaOH solution?

(c) What volume of 0.5 mol L^{-1} H_2SO_4 solution contains 4.9 g of H_2SO_4?

(7) You are asked to make up a 0.8 mol L^{-1} HNO_3 solution from a 16.0 mol L^{-1} stock solution of HNO_3. What volume of the stock solution should you use to make up 1.0 L of 0.8 mol L^{-1} HNO_3?

(8) A bottle of stock HCl solution is 12.5 mol L^{-1}. How would you make up 2.0 L of 0.5 mol L^{-1} HCl?

(9) Calculate the amount (number of moles) of solute in each of the following:
(a) 2.0 L of 2.5 mol L^{-1} NaOH solution
(b) 300 cm^3 of 0.5 mol L^{-1} HCl solution
(c) 50 cm^3 of 2.0 mol L^{-1} H_2SO_4 solution.

(10) How would you make up each of the following solutions:
(a) 250 cm^3 of approximately 0.25 mol L^{-1} NaOH solution using sodium hydroxide pellets
(b) 100 cm^3 of approximately 0.5 mol L^{-1} HCl solution using 12 mol L^{-1} HCl solution
(c) 200 cm^3 of exactly 0.1 mol L^{-1} Na_2CO_3 solution using pure anhydrous sodium carbonate?

(11) Describe, briefly, how you would make up 200 cm^3 of a 25% (v/v) solution of ethanol in water.

(12) Describe, briefly, how you would make up 200 g of a 25% (w/w) solution of sodium chloride in water.

(13) How would you make up each of the following solutions:
(a) 20% w/v of H_2O_2 solution from 100% w/v of H_2O_2 solution
(b) 10% w/v of sucrose solution
(c) 25% v/v of methanol in ethanol
(d) 25 ppm of $CaCO_3$ solution?

(14) Sea water contains 1.29 g of magnesium ions, Mg^{2+}, on average, in every kilogram of water. Determine the concentration of magnesium ions expressed in ppm in sea water.

(15) The public health standard for some heavy metals such as cadmium is 0.01 ppm of dissolved cadmium ions, Cd^{2+}. If a 5.0 L water sample contains 0.021 mg of Cd^{2+} ions, calculate whether the concentration of cadmium in the water contravenes the health regulations.

(16) Fish die if the concentration of carbon dioxide in the water exceeds 200 ppm. When a 2.5 L sample of river water was taken and analysed it was found to contain 0.40 g of carbon dioxide. Is this level high enough to kill the fish?

(17) (a) An ethanol and water solution contains 20% w/w of ethanol. Express this concentration in mol L^{-1} of ethanol.
(b) An ethanol and water solution contains 40% w/v of ethanol. Express this concentration in mol L^{-1} of ethanol.

(18) Does the amount of solute in a solution change when the solution is diluted? Explain your answer.

(19) Calculate the molar concentration of a 20% (v/v) ethanol in water solution, given that the density of ethanol is 0.8 g cm^{-3}.

(20) Sodium carbonate, Na_2CO_3, can be used as a primary standard. Give four reasons why it is suitable for use as a primary standard.

(21) Sodium hydroxide, NaOH, cannot be used as a primary standard. Give reasons why it is unsuitable for use as a primary standard.

(22) What is a standard solution? Describe, in detail, how you would make up a standard solution of sodium carbonate.

4.2 ACIDS AND BASES

Acids and bases can be recognised by simple properties such as taste, how they react with indicators and how they react with each other. It would be foolish to try to identify an acid by its taste as most acids used in the laboratory are extremely corrosive.

Bee stinging.

Properties of acids

1. Acids have a sour taste: vinegar and citrus fruits are acids which have a stinging sour taste.
2. Acids are corrosive: acids can eat away fabric, metals and buildings.
3. Acids turn blue litmus paper and solution red.
4. Acids have a pH less than 7: strong acids like hydrochloric acid have a pH close to 0 while weak acids like sour milk have values close to 7.
5. Most acids react with most metals to form salts and release hydrogen gas:

$$metal + acid \rightarrow salt + hydrogen$$

For example:
(a) Magnesium reacts with dilute sulphuric acid forming magnesium sulphate and releasing hydrogen gas.

$$Mg + H_2SO_4 = MgSO_4 + H_2$$

(b) Zinc reacts with dilute hydrochloric acid to form zinc chloride and releases hydrogen gas.

$$Zn + 2HCl = ZnCl_2 + H_2$$

6. Acids react with carbonates: they form a salt, release carbon dioxide gas and form water.

$$carbonate + acid \rightarrow salt + carbon\ dioxide + water$$

For example:
(a) Calcium carbonate reacts with dilute hydrochloric acid as follows:

$$CaCO_3 + 2HCl = CaCl_2 + CO_2 + H_2O$$

7. Acids neutralise bases: when an acid is added to a base they cancel each other out (neutralise each other) by forming a salt and water.

$$acid + base \rightarrow salt + water$$

For example:
(a) Hydrochloric acid reacts with sodium hydroxide as follows:

$$HCl + NaOH = NaCl + H_2O$$

(b) Sulphuric acid reacts with sodium hydroxide as follows:

$$H_2SO_4 + 2NaOH = Na_2SO_4 + 2H_2O$$

Formulae of Some Common Acids and Bases

Acid	Formula	Base	Formula
hydrochloric acid	HCl	sodium hydroxide	$NaOH$
nitric acid	HNO_3	potassium hydroxide	KOH
sulphuric acid	H_2SO_4	sodium hydrogen carbonate	$NaHCO_3$
carbonic acid	H_2CO_3	magnesium oxide	MgO
ethanoic acid	CH_3COOH	calcium oxide	CaO

Bases

Bases are the opposites of acids. Some bases are soluble in water: these bases are called alkalis. Alkalis are soapy to touch: this is because they react with the natural oils on the skin to make soap.

Neutralisation

When an acid is mixed with a base they react with each other and cancel each other out. This is called neutralisation because the substances you end up with are neither acidic or basic: they are neutral.

The following are examples of everyday neutralisation reactions:
· When a bee stings you it injects an acid into you. Bee stings can be treated with a base, such as baking soda or calamine lotion, which neutralises the sting.
· Wasp stings and nettle stings contain the base histamine. They are neutralised by anti-histamines, which are acids. Vinegar, another acid, is often used to treat wasp stings.
· Medicines, such as milk of magnesia, which relieve pain caused by excess stomach acid are called antacids (anti-acids) because they neutralise the excess acid.
· We use toothpaste to neutralise the acids produced in our mouths when food is eaten because excess acid causes tooth decay.

Household Acids and Bases

Acid	Where found
Ethanoic acid	Vinegar
Citric acid	Citrus fruits such as oranges and lemons
Carbonic acid	Aerated drinks such as Coca Cola
Oxalic acid	Rhubarb
Lactic acid	Sour milk and yoghurt

Base	Where found
Ammonia solution	Household cleaners
Sodium hydroxide (caustic soda)	Drain cleaner
Lime	Building materials
Magnesium hydroxide	Milk of magnesia – an antacid
Zinc oxide	Skin creams

Arrhenius Concept of Acids and Bases

Svante August Arrhenius 1859–1927
Svante Arrhenius was born in Sweden. He learned to read at the age of three and became interested in mathematics and physics at an early age. He proposed in his doctoral thesis that electrolytes split into ions in water. For his efforts he was awarded the barest of passes. Fortunately, William Ostwald and Jacobus van't Hoff promoted his work on electrolytic theory. He was awarded the 1903 Nobel Prize for Chemistry for roughly the same thesis that had been nearly rejected nineteen years previously. He had universal interests in science and proposed the greenhouse effect.

In 1884, Arrhenius suggested that acids were substances which dissociated in water to produce hydrogen ions, H^+.

$$HCl \rightarrow H^+ + Cl^-$$

It was thought that the hydrogen ion, H^+, was unlikely to exist independently as a bare proton in aqueous solution, but existed as a proton chemically bound to a polar water molecule – that is, $H_3O^+(aq)$. This is called the oxonium ion (formerly called the hydronium ion).

The dissociation of acids in aqueous solution can be regarded as

$$HCl(aq) + H_2O(l) \rightarrow H_3O^+(aq) + Cl^-(aq)$$

The role of water molecules in the dissociation of acids explains why substances such as hydrogen chloride, HCl, and ethanoic acid, CH_3COOH, are regarded as acidic in water. Neither of these substances dissociates to form H^+ ions or H_3O^+ ions in organic solvents such as 1,1,1-trichloroethane. When they are not dissolved in water, they do not react with dry litmus paper or release carbon dioxide from carbonates.

Arrhenius regarded bases as substances which reacted with H^+ ions and formed water.

$$CuO + 2H^+ \rightarrow Cu^{2+} + H_2O$$

Soluble bases, called alkalis, were substances which dissociated in aqueous solution to produce OH^- ions.

$$NaOH(s) + H_2O(l) \rightarrow Na^+(aq) + OH^-(aq)$$

Evidence for the Arrhenius theory comes from the molar heat of neutralisation of any strong acid by any strong base. The neutralisation process is essentially the combination of $H_3O^+(aq)$ and $OH^-(aq)$ ions to form water. However, despite its success, the Arrhenius theory is unsatisfactory because it considers reactions only in aqueous solutions and restricts the definition of a base to those substances which react with H^+ ions to form water.

Bronsted–Lowry Theory of Acids and Bases

Bronsted and Lowry independently proposed a theory of acids and bases.

According to Bronsted–Lowry theory,

An acid is any substance (atom, ion or molecule) that can donate a proton. A base is any substance, (atom, ion, or molecule) that can accept a proton.

In the reaction

$$CH_3COOH(l) + H_2O(l) \rightleftharpoons H_3O^+(aq) + CH_3COO^-(aq)$$

| B/L acid | B/L base | Conjugate acid | Conjugate base |

the ethanoic acid, CH_3COOH, is acting as a Bronsted–Lowry acid and donates a proton to the water molecule, which is acting as a Bronsted–Lowry base by accepting the proton.

The reaction is reversible ($\rightleftharpoons$) and exists at equilibrium. In the reverse reaction, the H_3O^+

ion donates a proton to the ethanoate ion, CH_3COO^-, which accepts the proton.

The H_3O^+ ion is called the **conjugate acid** of the **base**, H_2O, because the water molecule changed into that acid **having accepted a proton.**

The CH_3COO^- ion is called the **conjugate base** of the **acid** CH_3COOH, because the acid changed into that base **having donated a proton.**

H_3O^+ is the conjugate acid of the base H_2O. CH_3COO^- is the conjugate base of the acid CH_3COOH.

Hence, H_3O^+ and H_2O and CH_3COO^- and CH_3COOH are called **conjugate pairs**.

There are many molecules and ions that can function in one reaction as acids and in other reactions as bases; such substances are termed **amphoteric** or **amphiprotic**.

In the previous example, water is acting as a base, but in the following reaction with ammonia, water is acting as an acid.

$$H_2O(l) + NH_3(aq) \rightleftharpoons NH_4^+(aq) + OH^-(aq)$$

| B/L acid | B/L base | Conjugate acid | Conjugate base |

As a result of this dual behaviour, water is classified as amphoteric.

Neutralisation can be interpreted in Bronsted–Lowry terms as the combination of the conjugate pairs H_3O^+ and OH^- to form water.

$$H_3O^+(aq) + OH^-(aq) \rightleftharpoons H_2O(l) + H_2O(l)$$

Exercise 4.2

•

(1) Describe some simple properties of acids. In your answer refer to some everyday examples of acids.

(2) Write the names and formulae of three common strong acids and of three common strong bases. In the case of one strong acid and of one strong base, mention an everyday use of each.

(3) Write out four properties of a base. In your answer refer to sodium hydroxide.

(4) Describe, in words and using equations, each of the following:
(a) the reaction of magnesium with sulphuric acid
(b) the reaction of calcium carbonate with dilute hydrochloric acid
(c) the reaction of sulphuric acid with potassium hydroxide.

(5) Write balanced equations for each of the following reactions:
(a) hydrochloric acid and sodium hydroxide solution
(b) hydrochloric acid and zinc metal
(c) hydrochloric acid and magnesium hydroxide solid
(d) sulphuric acid and aqueous ammonia
(e) nitric acid and magnesium oxide solid
(f) sulphuric acid and aluminium hydroxide solid
(e) ethanoic acid and sodium metal

(f) sulphuric acid and calcium carbonate solid.

(6) Complete and balance each of the following reactions:
(a) $H_2SO_4(aq) + Al(OH)_3 \rightarrow$
(b) $CH_3COOH(aq) + NaOH(aq) \rightarrow$
(c) $HCl(aq) + Na_2CO_3(aq) \rightarrow$

(7) Magnesium oxide, MgO, magnesium hydroxide, $Mg(OH)_2$, aluminium hydroxide, $Al(OH)_3$, and sodium hydrogen carbonate, $NaHCO_3$, are often used as antacids to relieve stomach acidity. Write balanced equations showing how each of these compounds neutralises stomach acid.

(8) (a) What are the Arrhenius definitions of an acid and a base?
(b) Why is HCl regarded as acidic in water?
(c) What evidence is there for Arrhenius's theory?
(d) Why is the theory unsatisfactory?

(9) Define each of the following terms in relation to Bronsted–Lowry theory:
(a) acid (b) base (c) conjugate acid (d) conjugate base (e) conjugate acid–base pairs.

(10) Identify all the Bronsted–Lowry acids and bases in the following equations:
(a) $NH_3 + H_2O \rightarrow NH_4^+ + OH^-$
(b) $HNO_3 + H_2O \rightarrow H_3O^+ + NO_3^-$
(c) $NH_3 + NH_3 \rightarrow NH_2^- + NH_4^+$
(d) $O^{2-} + H_2O \rightarrow OH^- + OH^-$
(e) $H_2SO_4 + HCN \rightarrow HSO_4^- + H_2CN^+$
(f) $CH_3COOH + H_2O \rightarrow CH_3COO^- + H_3O^+$

(11) (a) What are the conjugate bases of each of the following: (i) $H_2PO_4^-$ (ii) HNO_2 (iii) HS^- (iv) NH_3 (v) H_3PO_4?

(b) What are the conjugate acids of each of the following: (i) PH_3 (ii) S_2^- (iii) PO_4^{3-} (iv) NH_2^- (v) H_2O (vi) HF?

(12) Explain clearly, using at least two relevant examples, the meaning of the term 'amphoteric (amphiprotic)'.

(13) Identify the conjugate acid–base pairs in each of the following reactions:
(a) $HSO_4^- + SO_3^{2-} \rightarrow HSO_3^- + SO_4^{2-}$
(b) $CH_3COOH + H_2O \rightarrow CH_3COO^- + H_3O^+$
(c) $CN^- + H_3O^+ \rightarrow HCN + H_2O$
(d) $HOCl + H_2O \rightarrow H_3O^+ + OCl^-$

4.3(a)
VOLUMETRIC ANALYSIS: ACIDS AND BASES

Laboratory research. Analysts working on crop protection, BASF.

Chemists who identify and classify materials are called analysts. Analysts work in medical laboratories, industrial laboratories, forensic laboratories and many others. They use a variety of techniques when they are trying to analyse something. One of the techniques which is often used is volumetric analysis.

Volumetric analysis is a technique in which one solution is used to analyse another. Acid–base titrations and redox titrations are common forms of volumetric analysis.

A **titration** is the process by which a known volume of one solution is added to a measured volume of another solution until the reaction is complete.

The solution used to carry out the analysis is usually delivered from an instrument called a burette, while the solution to be analysed is measured into a conical flask using a pipette.

The titration is complete at a point called the end-point or the equivalence point. The end-point is usually determined by means of a colour change of an indicator.

Guidelines for Correct Titration Procedure

Correct titration procedure involves very careful technique which requires proper use of the main glassware involved.

1. Pipette: a specially calibrated piece of glassware which delivers an exact volume of liquid. The most common pipette used delivers exactly 25 cm³ of solution.

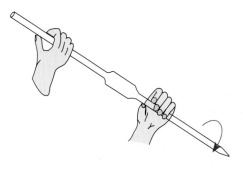

• Wash with deionised water and then with some of the solution to be used in it. It should be constantly rotated when being rinsed. Constant rotation washes the walls of the pipette thoroughly with deionised water.

• Use a pipette filler or pipette pump at all times to avoid sucking up dangerous material.

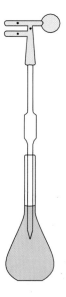

• Never blow the solution down the pipette because the pipette is specially calibrated for a set rate of flow and for some retention of solution at the tip.

• Blowing can also cause contamination of the solution by addition of carbon dioxide or water vapour.

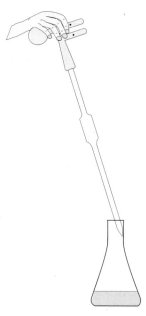

• The tip of the pipette should be held just touching the side of the conical flask for 10–15 seconds to allow the solution to drain out.

2. Burette: a specially graduated piece of glass tubing, marked in increments, usually of 0.1 cm³, which delivers a measured volume of solution.

• Wash well with deionised water and then with some of the solution to be used in it. Rotate the burette constantly when it is being washed. Make sure that some solution flows through the jet below the stop cock.

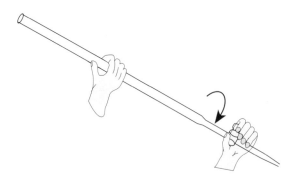

• Fill the burette using a funnel which has been washed with deionised water. Remember to remove the funnel before the start of the titration.

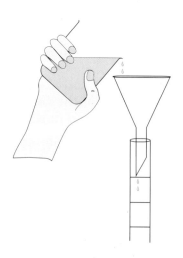

• Remove any air bubbles by allowing some solution to run through the tap and then recheck your initial reading. This also fills the burette tip with solution.

• Rinse the tip of the burette with deionised water into the conical flask near the end-point.

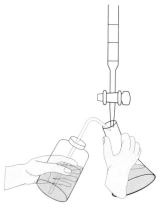

• Remember that you can read a burette accurately to 0.05 cm³ and that three consecutive titres with an error of less than 0.1 cm³ are required.

3. Conical flask: a piece of glassware, shaped like a cone, which allows the reaction mixture to be shaken constantly without spilling.

• Rinse well several times with deionised water only.

• Place a white tile underneath the conical flask in order to see clearly when the indicator changes colour.

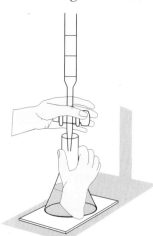

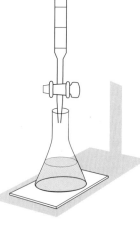

• During the titration swirl the conical flask constantly and also wash the walls with deionised water from time to time in case any of the solution from the burette has stuck to the neck of the conical flask.

Mandatory experiment 5A
Ordinary level only

To prepare a salt by neutralising sodium hydroxide with dilute hydrochloric acid.

$$HCl + NaOH = NaCl + H_2O$$

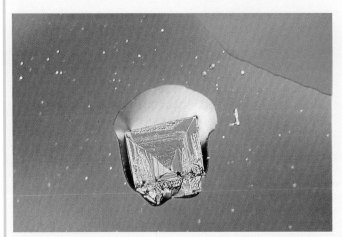

Polarised light micrograph of a crystal of salt forming in a drop of sea water.

Requirements

Safety glasses, retort stand, burette, pipette, pipette filler, conical flasks, filter funnel, wash bottle with deionised water, methyl orange indicator, white tile, 0.1 mol L^{-1} hydrochloric acid solution, 0.1 mol L^{-1} sodium hydroxide solution.

Procedure

Students should practise using pipettes, pipette fillers and burettes before carrying out this experiment.

(1) Rinse the pipette, the burette, the funnel and the conical flask thoroughly with deionised water.

(2) Using the funnel, rinse the burette at least three times with some of the 0.1 mol L^{-1} HCl. Make sure that you fill and rinse the tip of the burette.

(3) Fill the burette with the 0.1 mol L^{-1} HCl making sure that there are no air bubbles. Record your initial reading. Place a white tile on the base plate of the stand underneath the burette. This allows you to see clearly any colour changes during the reaction.

(4) Rinse out the pipette at least three times with the 0.1 mol L^{-1} NaOH solution.

(5) Fill the 25 cm^3 pipette with 0.1 mol L^{-1} NaOH solution so that the bottom of the meniscus sits on the engraved mark.

(6) Transfer the measured 25 cm^3 of 0.1 mol L^{-1} NaOH solution to a clean conical flask.

(7) Add 2–3 drops of methyl orange indicator.

(8) Run the sodium hydroxide solution from the burette into the conical flask, constantly swirling the flask, until the solution just turns pink. Record your final reading and use this titre as a guide for your next titres.

(9) Refill your burette (if necessary) with the hydrochloric acid solution and record your initial reading.

(10) Refill your pipette with the sodium hydroxide solution and transfer it to a clean conical flask. Add 2–3 drops of indicator.

(11) Add the hydrochloric acid solution very carefully from the burette. **As the acid is added the orange colour of the methyl orange turns red-pink in the middle of the flask but initially disappears quickly. When the pink colour does not disappear quickly, add the acid dropwise until the solution remains permanently pink.** Record your final reading to the nearest **0.1 cm^3**.

(12) Repeat the procedure until at least three consecutive titres agree to within 0.1 cm^3 of each other.

(13) When you have an average titration figure, repeat the procedure without using an indicator, that is, add the average volume of hydrochloric acid used to exactly 25 cm^3 of the sodium hydroxide solution.

(14) Place a sample of this neutral solution on an evaporating dish and evaporate it to dryness. Sodium chloride crystals will be seen on the evaporating dish.

Results

Complete a table similar to that below.

Pipette solution:
25 cm^3 of 0.1 mol L^{-1} NaOH

Burette solution: 0.1 mol L^{-1} HCl

Indicator: Methyl orange

Burette readings / cm^3

	Titre 1	Titre 2	Titre 3	Titre 4
Final	25.7	25.1	25.2	25.3
Initial	0	0	0	0
Volume used	25.7	25.1	25.2	25.3

• • • • • • **Example 4.8** • • • • • •

Acid–base neutralisation

During a titration to determine the exact concentration of hydrochloric acid, a student used 25.0 cm^3 of exactly 0.2 mol L^{-1} NaOH solution to neutralise 20.0 cm^3 of the hydrochloric acid solution. Calculate the concentration of the hydrochloric acid solution (a) in mol L^{-1} and (b) in g L^{-1}.
The balanced chemical equation is

$$HCl + NaOH \rightarrow NaCl + H_2O$$

$$\frac{c_A V_A}{a} = \frac{c_B V_B}{b}$$

c_A and c_B are the concentrations of HCl and NaOH, respectively.
V_A and V_B are the volumes of the HCl and NaOH, respectively.
a and b are the number of moles of reactants in the balanced chemical equation.
(a) Now

$c_A = ?$	$c_B = 0.2$ mol L^{-1}
$V_A = 20.0$ cm^3	$V_B = 25.0$ cm^3
$a = 1$	$b = 1$

$$\frac{c_A V_A}{a} = \frac{c_B V_B}{b}$$

$$\frac{c_A \times 20.0 \text{ cm}^3}{1} = \frac{0.2 \text{ mol L}^{-1} \times 25.0 \text{ cm}^3}{1}$$

$$c_A = \frac{0.2 \text{ mol L}^{-1} \times 25.0 \text{ cm}^3}{24.0 \text{ cm}^3} = 0.25 \text{ mol L}^{-1}$$

(b) The concentration of the hydrochloric acid expressed in g L^{-1} is obtained by multiplying the molar concentration by the molar mass of HCl (36.5 g mol^{-1}).

$$c = 0.25 \text{ mol L}^{-1} \times 36.5 \text{ g mol}^{-1}$$

$$= 9.125 \text{ g mol}^{-1} \text{ HCl solution}$$

• •

Mandatory experiment 6

To standardise an approximately 0.2 mol L^{-1} HCl solution using a standard solution of sodium carbonate.

Introduction

The purpose of this experiment is to standardise the HCl solution using an exact 0.1 mol L^{-1} Na$_2$CO$_3$ solution.
 Sodium carbonate solution reacts with hydrochloric solution as follows:

$$Na_2CO_3 + 2HCl \rightarrow 2NaCl + CO_2 + H_2O$$

 This means that 1 mol Na$_2$CO$_3$ reacts with 2 mol HCl.
 A suitable indicator such as methyl orange is used to determine the end-point. The addition of one drop of HCl changes the solution from orange to pink and shows that the reaction is just complete. This is called the end-point.
 Concentrated hydrochloric acid is approximately 10–12 mol L^{-1} HCl. If 20 cm^3 of concentrated acid are diluted in 1 L of solution, the concentration of the resulting solution is approximately 0.2 mol L^{-1} HCl.
 For safety reasons, make sure when you are diluting the acid that the acid is added to the water, rather than the other way round.

⌐∪⌐ Requirements

Safety glasses, filter funnel, 2 beakers, 4 conical flasks, pipette, burette and stand, methyl orange indicator, 0.1 mol L^{-1} sodium carbonate solution, 250 cm^3 volumetric flask, concentrated hydrochloric acid solution, white tile and wash-bottle of deionised water.

Procedure

The procedure is described in detail for this particular experiment and should be used as a guide for all further titrations.

(1) Rinse the pipette, the burette, the funnel and the conical flask thoroughly with deionised water.

(2) Using a funnel rinse the burette at least three times with some of the approximately 0.2 mol L^{-1} HCl (made as described above). Make sure that you fill and rinse the tip of the burette.

(3) Fill the burette with the 0.2 mol L^{-1} HCl (approx.) making sure that there are no air bubbles. Record your initial reading. Place a white tile on the base plate of the stand underneath the burette. This allows you to see clearly any colour changes during the reaction.

(4) Rinse out the pipette at least three times with the 0.1 mol L^{-1} Na$_2$CO$_3$ solution.

(5) Fill the 25 cm^3 pipette with 0.1 mol L^{-1} Na$_2$CO$_3$ solution so that the bottom of the meniscus sits on the engraved mark.

(6) Transfer the 25 cm^3 of 0.1 mol L^{-1} Na$_2$CO$_3$ solution to a clean conical flask.

(7) Add 2–3 drops of methyl orange indicator.

(8) Run the sodium carbonate solution from the burette into the conical flask, constantly swirling the flask, until the solution just turns pink. Record your final reading and use this titre as a guide for your next titres.

(9) Refill your burette (if necessary) with the hydrochloric acid solution and record your initial reading.

(10) Refill your pipette with the sodium carbonate solution and transfer it to a clean conical flask. Add 2–3 drops of indicator.

(11) Add the hydrochloric acid solution very carefully from the burette, adding it drop by drop near the end-point. Record your final reading to the nearest 0.1 cm^3.

(12) Repeat the titre until at least three consecutive titres agree to within 0.1 cm^3 of each other.

(13) Rinse your burette and pipette well with water.

Results

Complete a table similar to that below.

Pipette solution:
25 cm^3 of 0.1 mol L^{-1} Na$_2$CO$_3$

Burette solution: Approx. 0.2 mol L^{-1} HCl

Indicator: Methyl orange

Burette readings / cm^3

	Trial	Titre 1	Titre 2	Titre 3
Final	23.60	46.80	23.40	46.60
Initial	0	23.50	0	23.40
Volume used	23.60	23.30	23.40	23.20

Specimen calculations

$$\text{Mean titre /cm}^3 = \frac{23.30 + 23.40 + 23.20}{3}$$

$$= 23.30 \text{ cm}^3 \text{ HCl}$$

The balanced chemical equation is

$$2HCl + Na_2CO_3 = 2NaCl + H_2O + CO_2$$

Now

$c_A = ?$ $c_B = 0.1$ mol L^{-1}
$V_A = 23.30$ cm^3 $V_B = 25.00$ cm^3
$a = 2$ $b = 1$

$$\frac{c_A V_A}{a} = \frac{c_B V_B}{b}$$

c_A and c_B are the concentrations of HCl and Na$_2$CO$_3$, respectively.
V_A and V_B are the volumes of HCl and Na$_2$CO$_3$, respectively.
a and b are the number of moles of reactants in the balanced chemical equation.

$$\frac{c_A \times 23.30 \text{ cm}^3}{2} = \frac{0.1 \text{ mol } L^{-1} \times 25.00 \text{ cm}^3}{1}$$

$$c_A = \frac{0.1 \text{ mol } L^{-1} \times 25.00 \text{ cm}^3 \times 2}{23.30 \text{ cm}^3 \times 1}$$

Therefore, c_A, the concentration of HCl, is exactly 0.21 mol L^{-1}.

Mandatory experiment 7

To determine the concentration of ethanoic acid in vinegar.

Introduction

Vinegar is a weak solution of ethanoic acid. It is often made by fermentation of malt, which gives the vinegar formed a more subtle flavour than vinegar made from industrial ethanoic acid.

Ethanoic acid and phenolphthalein solution changes from colourless to pink as sodium hydroxide is added.

The amount of ethanoic acid in the vinegar can be determined by titration with sodium hydroxide.

$$CH_3COOH + NaOH = CH_3COONa + H_2O$$

Phenolphthalein is used as an indicator to determine the end-point because the reaction is between a weak acid and a strong base. This indicator is a pink colour in alkaline solution and colourless in acidic solution. In this experiment the base is added using the burette because it is preferable to add the indicator to the acidic vinegar solution and not to the sodium hydroxide solution. In this way we can observe a change in colour from colourless to pink and not from pink to colourless. This colour change is easier to see.

Requirements

Safety glasses, filter funnel, beakers, conical flasks, 25 cm³ volumetric flask, pipette, burette and stand, phenolphthalein, 0.1 mol L⁻¹ sodium hydroxide solution, vinegar, white tile and wash bottle of deionised water.

Procedure

(1) Rinse the pipette, the burette, the funnel, the conical flask and the volumetric flask thoroughly with deionised water.

(2) Commercial vinegar is too concentrated and should be diluted. Transfer exactly 25 cm³ of the vinegar into the volumetric flask. Make the solution up to the 250 cm³ mark by following the correct procedure as given in experiment 5.
(3) Wash out the burette and the pipette with the solutions which you are going to use in each of them.
(4) Fill the burette up to the mark in the usual manner with the 1.0 mol L⁻¹ sodium hydroxide solution.
(5) Pipette 25 cm³ of the diluted vinegar into a conical flask.
(6) Add 1–2 drops of phenolphthalein indicator. The vinegar solution is colourless.
(7) Titrate the sodium hydroxide solution against the dilute vinegar solution until you have three consecutive readings within the accepted experimental error.
(8) Record your results as shown before.
(9) Wash all glassware thoroughly and in particular the burette. If the burette is not washed out with water the sodium hydroxide may fix the stopcock if it is made of glass.

Specimen calculations

The balanced chemical equation is

$$CH_3COOH + NaOH = CH_3COONa + H_2O$$

Now

$c_A = ?$	$c_B = 0.1$ mol L⁻¹
$V_A = 27.2$ cm³	$V_B = 25.00$ cm³
$a = 1$	$b = 1$

$$\frac{c_A V_A}{a} = \frac{c_B V_B}{b}$$

c_A and c_B are the concentrations of CH_3COOH and $NaOH$, respectively.
V_A and V_B are the volumes of CH_3COOH and $NaOH$, respectively.
a and b are the number of moles of reactants in the chemical equation.

$$\frac{c_A \times 27.2 \text{ cm}^3}{1} = \frac{0.1 \text{ mol L}^{-1} \times 25.00 \text{ cm}^3}{1}$$

Therefore, c_A, the concentration of diluted CH_3COOH is 0.092 mol L⁻¹.

As the original vinegar sample was diluted by a factor of ten (25 cm³ in 250 cm³) then the concentration of the commercial vinegar was 0.92 mol L⁻¹.

Mandatory experiment 8

To determine the amount of water of crystallisation in a compound.

Introduction

Many common compounds contain water of crystallisation. Bluestone, $CuSO_4.5H_2O$, Epsom salts, $MgSO_4.7H_2O$, oxalic acid, $C_2H_2O_4.2H_2O$ and washing soda, $Na_2CO_3.10H_2O$, are familiar examples.

Washing soda is crystalline sodium carbonate, $Na_2CO_3.10H_2O$. It can be analysed quite easily to determine the amount of water of crystallisation by titration with hydrochloric acid solution.

$$Na_2CO_3 + 2HCl = 2NaCl + CO_2(g) + H_2O$$

It should be remembered that $10H_2O$ in the formula $Na_2CO_3.10H_2O$ means that there are 10 moles of water bound to 1 mole of anhydrous sodium carbonate.

This can be written simply as

$$10 = \frac{n(H_2O)}{n(Na_2CO_3)}$$

Requirements

Safety glasses, laboratory balance, filter funnel, 250 cm³ beaker, conical flasks, pipette, burette and stand, methyl orange indicator, sodium carbonate crystals, 250 cm³ volumetric flask, 0.2 mol L⁻¹ hydrochloric acid solution, white tile and wash-bottle of deionised water.

Procedure

(1) Rinse the pipette, the burette, the funnel, the conical flask and the volumetric flask thoroughly with deionised water.

(2) Dry the 250 cm³ beaker and place it on the balance. Weigh approximately 4.3 g of crystalline sodium carbonate in the beaker. Note the exact mass.

(3) Add about 100 cm³ of deionised water to the crystalline sodium carbonate in the beaker. After a while the Na_2CO_3 crystals will dissolve.

(4) Transfer the solution and washings to the 250 cm³ volumetric flask. Make the solution up to 250 cm³ by following the usual procedure.

(5) Titrate 25 cm³ samples of this solution against the 0.2 mol L⁻¹ hydrochloric acid solution.

Results

Complete a table similar to that below.

Mass of crystalline Na_2CO_3 = 4.2 g

Pipette solution: 25 cm³ of Na_2CO_3 solution

Burette solution: 0.2 mol L⁻¹ HCl

Burette readings / cm³

	Trial	Titre 1	Titre 2	Titre 3
Final				
Initial				
Volume used				

Specimen calculations

Determination of (a) % water of crystallisation and (b) the number of moles of water of crystallisation.

Mean titre = 15.0 cm³ HCl

The balanced chemical equation is

$$2HCl + Na_2CO_3 = 2NaCl + H_2O + CO_2$$

113

Now

$$c_A = 0.2 \text{ mol L}^{-1} \qquad c_B = ?$$
$$V_A = 15.0 \text{ cm}^3 \qquad V_B = 25.0 \text{ cm}^3$$
$$a = 2 \qquad b = 1$$

$$\frac{c_A V_A}{a} = \frac{c_B V_B}{b}$$

c_A and c_B are the concentrations of HCl and Na_2CO_3, respectively.
V_A and V_B are the volumes of HCl and Na_2CO_3, respectively.
a and b are the number of moles of reactants in the balanced chemical equation.

$$\frac{0.2 \text{ mol L}^{-1} \times 15.0 \text{ cm}^3}{2} = \frac{c_B \times 25.00 \text{ cm}^3}{1}$$

$$c_B = \frac{0.2 \text{ mol L}^{-1} \times 15.0 \text{ cm}^3 \times 1}{2 \times 25.0 \text{ cm}^3}$$

$c_B = 0.06 \text{ mol L}^{-1}$ anhydrous Na_2CO_3

Amount of anhydrous Na_2CO_3, $n = cV$
$= 0.06 \text{ mol L}^{-1} \times 0.25 \text{ L} = 0.015 \text{ mol}$

Mass of anhydrous Na_2CO_3, $m = nM$
$= 0.015 \text{ mol} \times 106 \text{ g mol}^{-1} = 1.62 \text{ g}$

Mass of water of crystallisation = mass of crystalline Na_2CO_3 – anhydrous Na_2CO_3
$= 4.2 \text{ g} - 1.62 \text{ g} = 2.58 \text{ g}$

(a) % Water of crystallisation

$$= \frac{\text{Mass of water}}{\text{Mass of crystalline } Na_2CO_3} \times 100\%$$

$$= \frac{2.58 \text{ g}}{4.2 \text{ g}} \times 100\% = 61.4\%$$

(b) Amount of water

$$n(H_2O) = \frac{m}{M} = \frac{2.78 \text{ g}}{18.0 \text{ g mol}^{-1}} = 0.1433 \text{ mol}$$

[handwritten: 2.58]

Number of moles of water of crystallisation

$$x = \frac{n(H_2O)}{n(Na_2CO_3)} = \frac{0.1433 \text{ mol}}{0.015 \text{ mol}} = 9.6 \text{ mol}$$

The correct answer is 10 mol. This indicates that the washing soda must have lost some water of crystallisation by drying out during storage.

• • • • • • **Example 4.9** • • • • • •

Calculation of the relative molecular mass of a compound from titration data

1.92 g of a group I metal carbonate was dissolved in 250 cm³ of water and analysed by titrating the solution with 0.2 mol L⁻¹ hydrochloric acid solution. If 15.0 cm³ of the hydrochloric acid solution was required to neutralise 25 cm³ of the carbonate solution, determine the relative molecular mass of the metal carbonate.

The balanced chemical equation is given as

$$2HCl + M_2CO_3 = 2MCl + H_2O + CO_2$$

where M is the group I metal

Now

$$c_A = 0.2 \text{ mol L}^{-1} \qquad c_B = ?$$
$$V_A = 15.0 \text{ cm}^3 \qquad V_B = 25.00 \text{ cm}^3$$
$$a = 2 \qquad b = 1$$

$$\frac{c_A V_A}{a} = \frac{c_B V_B}{b}$$

c_A and c_B are the concentrations of the HCl and M_2CO_3, respectively.
V_A and V_B are the volumes of HCl and M_2CO_3, respectively.
a and b are the number of moles of reactants in the balanced chemical equation.

$$\frac{0.2 \text{ mol L}^{-1} \times 15.0 \text{ cm}^3}{2} = \frac{c_B \times 25.00 \text{ cm}^3}{1}$$

$$\frac{0.2 \text{ mol L}^{-1} \times 15.0 \text{ cm}^3 \times 1}{2 \times 25.0 \text{ cm}^3}$$

$c_B = 0.06 \text{ mol L}^{-1} M_2CO_3$

Amount of M_2CO_3, $n = cV$

$= 0.06 \text{ mol L}^{-1} \times 0.25 \text{ L} = 0.015 \text{ mol}$

Molar mass of M_2CO_3

$$M = \frac{m}{n} = \frac{1.92 \text{ g}}{0.015 \text{ mol}} = 128 \text{ g mol}^{-1}$$

Relative molecular mass, $M_r = 128$

• •

Exercise 4.3(a)

(1) Explain each of the following terms:
(a) volumetric analysis (b) titration
(c) neutralisation (d) end-point.

(2) (a) Describe each of the following:
(i) pipette (ii) burette (iii) conical flask
(iv) volumetric flask.
(b) What is the correct procedure for using:
(i) a pipette, (ii) a burette and (iii) a conical flask, in a titration?

(3) Burettes are designed to measure precise volumes of chemicals. Explain why they are designed with each of the following features:
(a) a volume at the top which is not graduated
(b) a volume at the bottom which is not graduated
(c) a stopcock and tip.

(4) Explain how you would determine the concentration of an acid solution using a standard solution of a base.

(5) Answer each of the following questions in relation to a titration where 0.1 mol L^{-1} HCl solution neutralises 0.05 mol L^{-1} NaOH solution.
(a) How would you accurately measure out 25 cm^3 of 0.05 mol L^{-1} NaOH solution?
(b) Describe how you would fill the burette with the 0.1 mol L^{-1} HCl solution.
(c) What is the purpose of using a white tile in the titration?
(d) Name an indicator which you could use for this titration.
(e) Describe what happens when the acid is run from the burette into the conical flask.
(f) When do you stop adding the acid to the conical flask?
(g) How would you isolate a pure sample of the salt formed?
(h) Write a balanced equation for the reaction.
(i) What volume of 0.1 mol L^{-1} HCl solution exactly neutralises 25 cm^3 of 0.05 mol L^{-1} NaOH solution?

(6) In an experiment 25 cm^3 of sodium hydroxide solution was neutralised by 21.50 cm^3 of 0.1 mol L^{-1} hydrochloric acid. Calculate the concentration of the sodium hydroxide solution in:
(a) mol L^{-1} (b) g L^{-1} .

(7) During a titration to determine the exact concentration of hydrochloric acid, a student used 25 cm^3 of exactly 0.23 mol L^{-1} NaOH solution for each titration in order to neutralise the acid. If his titre values of hydrochloric acid were as follows:
22.8 cm^3, 22.4 cm^3, 22.5 cm^3 and 22.6 cm^3, determine the exact concentration of the hydrochloric acid.

(8) In an experiment NaHCO$_3$ solution reacts with hydrochloric acid as follows:
$$NaHCO_3(aq) + HCl(aq)$$
$$= NaCl(aq) + CO_2(g) + H_2O(l)$$
If 25 cm^3 of a solution containing 3.60 g L^{-1} NaHCO$_3$ reacted with 22.50 cm^3 of hydrochloric acid, calculate the concentration of the hydrochloric acid solution in:
(a) mol L^{-1} (b) g L^{-1}.

(9) The gastric juice in the stomach is dilute hydrochloric acid. A 10.0 cm^3 sample of gastric juice was neutralised by 22.0 cm^3 of 0.01 mol L^{-1} NaOH solution.
(a) Determine the concentration of HCl in the sample.
(b) If the density of the gastric juice is assumed to be 1.0 g cm^{-3}, calculate the % (w/w) of HCl in the gastric juice.

(10) Indigestion caused by excess hydrochloric acid in the stomach is often relieved by taking some bread soda, NaHCO$_3$.
$$HCl + NaHCO_3 = NaCl + H_2O + CO_2$$
What mass of bread soda must be taken to neutralise 150 cm^3 of stomach acid if it contains 0.05 mol L^{-1} HCl?

(11) Calculate the volume of 0.2 mol L^{-1} NaOH solution which would neutralise each of the following solutions:
(a) 100 cm^3 of 0.1 mol L^{-1} HCl
(b) 300 cm^3 of 0.2 mol L^{-1} H$_2$SO$_4$
(c) 1 L of 0.4 mol L^{-1} HNO$_3$
(d) 25 cm^3 of 2.5 mol L^{-1} HNO$_3$
(e) 50 cm^3 of 1.0 mol L^{-1} CH$_3$COOH
(f) 2.5 L of 0.01 mol L^{-1} HCl.

(12) In a titration, 20.0 cm^3 of 0.1 mol L^{-1} Na$_2$CO$_3$ solution was needed to react completely with 22.3 cm^3 of HCl solution.

$$2HCl + Na_2CO_3 = 2NaCl + H_2O + CO_2$$

Calculate the molar concentration of the hydrochloric acid solution.

(13) A 25 cm^3 sample of vinegar, consisting of ethanoic acid, CH_3COOH, was titrated against 0.5 mol L^{-1} NaOH solution.

$$CH_3COOH(aq) + NaOH(aq)$$
$$= CH_3COONa(aq) + H_2O(l)$$

If the mean titre was 15.2 cm^3 of the sodium hydroxide solution, determine the molar concentration of the ethanoic acid in the vinegar.

(14) 16.5 cm^3 of 0.5 mol L^{-1} NaOH solution is required to exactly neutralise 20 cm^3 of vinegar, consisting of an aqueous solution of ethanoic acid, $CH_3COOH(aq)$.

$$CH_3COOH(aq) + NaOH(aq)$$
$$= CH_3COONa(aq) + H_2O(l)$$

(a) Name the indicator used for this titration.
(b) Why was the sodium hydroxide solution placed in the burette?
(c) Calculate the molar concentration of the ethanoic acid in the vinegar.
(d) If the density of the vinegar is assumed to be 1.0 g cm^{-3}, calculate the % w/w of ethanoic acid in the vinegar.

(15) A solution of washing soda was made up by dissolving 8.6 g of crystalline sodium carbonate in 500 cm^3 of solution. If 25 cm^3 of this solution required 15 cm^3 of 0.2 mol L^{-1} HCl solution for neutralisation, calculate:
(a) the % H$_2$O in the crystals
(b) the value of x in the formula $Na_2CO_3.xH_2O$.

4.3(b)
VOLUMETRIC ANALYSIS: OXIDATION AND REDUCTION

Redox Titrations

A redox titration is similar to an acid–base titration. The main types of redox titrations are based on the use of the following oxidising agents: (a) potassium manganate(VII) and (b) sodium thiosulphate.

Redox titrations are not as straightforward as acid–base titrations. Some essential background theory and some experimental guidelines are necessary in order to perform a redox titration.

Potassium Manganate(VII): Theory

Potassium manganate(VII), $KMnO_4$, is a relatively powerful oxidising agent. Like some other oxidising agents it can be used in volumetric analysis to determine the end-point in a titration. The manganate(VII) ion is an intense purple colour, which can be reduced by a suitable reducing agent to the colourless Mn(II) ion. Because of the dramatic colour change from purple to colourless, an indicator is not necessary in a titration involving potassium manganate(VII).

Acidic potassium manganate(VII) is used chiefly in redox titrations with iron(II) salts, Fe^{2+}, ethanedioate salts and ethanedoic acid, $(COOH)_2$, and hydrogen peroxide solution.

Potassium manganate(VII) sets up the following equilibrium in acidic conditions:

$$MnO_4^- + 8H^+ + 5e^- \rightleftharpoons Mn^{2+} + 4H_2O$$
$$\underset{purple}{+7} \qquad\qquad \underset{colourless}{+2}$$

Here, the purple manganate(VII) ion is reduced to the colourless manganese(II) ion. In the reaction each manganate(VII) ion accepts five electrons.

It is difficult to keep potassium manganate(VII) in a pure state, because it decomposes in the presence of sunlight and being a strong oxidising agent it is easily reduced. Therefore, a standard solution cannot be prepared directly; it must be standardised by titrating it against a primary standard, such as a solution of an iron(II) salt.

Ammonium iron(II) sulphate, $FeSO_4(NH_4)_2SO_4.6H_2O$, is used as a source of Fe^{2+} ions in preference to iron(II) sulphate, $FeSO_4$, because it is more stable. Iron(II) sulphate can be impure due to oxidation and efflorescence. The high relative molecular mass of ammonium iron(II) sulphate ($M_r = 392$) leads to greater accuracy when it is weighed. The ammonium sulphate part of the molecule does not become involved in the transfer of electrons.

The ionic half equation for the oxidation of iron(II) ions is as follows:

$$Fe^{2+} \rightarrow Fe^{3+} + e^-$$
$$\quad +2 \qquad\quad +3$$

The overall ionic equation is given as:

$$MnO_4^- + 8H^+ + 5Fe^{2+} \rightarrow Mn^{2+} + 4H_2O + 5Fe^{3+}$$

and tells us that one mole of MnO_4^- ions oxidises five moles of Fe^{2+} ions in acidic conditions.

It should be noted that sulphuric acid is used as an acidifying agent because it has no oxidising or reducing powers in dilute solution. Hydrochloric acid cannot be used as it is itself oxidised by the manganate(VII) ion, while nitric acid cannot be used as it is a strong oxidising agent. The appearance of a brown colour during the titration is due to the formation of insoluble manganese (III) compounds which indicates that insufficient sulphuric acid has been added.

Sulphuric acid also inhibits the oxidation of Fe^{2+} ions to Fe^{3+} ions by atmospheric oxygen which is dissolved in the water.

Because potassium manganate(VII) is not only a powerful oxidising agent but is also not very soluble in water, a dilute solution (0.02 mol L^{-1}) is used. Owing to the intense purple colour of the manganate(VII) ion it is difficult to read the bottom of the meniscus of a solution of potassium manganate(VII); it is usual to read the top of the meniscus.

- - - - - - **Example 4.10** - - - - - -

A solution of ammonium iron(II) sulphate was made up by dissolving 9.80 g of $FeSO_4(NH_4)_2SO_4.6H_2O$ in 250 cm^3 of acidified solution, in accordance with the correct procedure.
25.0 cm^3 of the solution completely reacted with 24.65 cm^3 of a potassium manganate(VII) solution. Calculate the concentration of the potassium manganate(VII) solution.

It is not necessary to know the complete balanced chemical equation for the reaction. The ionic equation given below gives us all the information we need to know.

$$MnO_4^- + 8H^+ + 5Fe^{2+} \rightarrow Mn^{2+} + 5Fe^{3+} + 4H_2O$$

This tells us that one mole of MnO_4^- ions reacts with five moles of Fe^{2+} ions.

Amount of Fe^{2+} ions,

$$n = \frac{m}{M} = \frac{9.80 \text{ g}}{392 \text{ g mol}^{-1}} = 0.025 \text{ mol}$$

Concentration of Fe^{2+} ions,

$$c = \frac{n}{V} = \frac{0.025 \text{ mol}}{0.25 \text{ L}} = 0.1 \text{ mol L}^{-1}$$

Now using

$$\frac{c_A V_A}{a} = \frac{c_B V_B}{b} \text{ where } a = 1 \text{ and } b = 5 \text{ (from the ionic equation above)}$$

$$\frac{c_A \times 24.6 \text{ cm}^3}{1} = \frac{0.1 \text{ mol L}^{-1} \times 25.0 \text{ cm}^3}{5}$$

$$c_A = \frac{0.01 \text{ mol L}^{-1} \times 25.0 \text{ cm}^3}{24.65 \text{ cm}^3 \times 5}$$

Therefore,
c_A, the concentration of $KMnO_4$
$= 0.0203$ mol L^{-1}

- -

Mandatory experiment 9

To standardise a potassium manganate(VII) solution using ammonium iron(II) sulphate solution.

Introduction

Potassium manganate(VII) is unavailable in a very pure state. Therefore, the concentration of a manganate(VII) solution must be determined by titration with a known standard solution. Ammonium iron(II) sulphate is a stable substance with a high relative molecular mass and is a suitable source of Fe^{2+} ions which can be used as a primary standard in this experiment.

The titration is carried out under acidic conditions in accordance with the ionic reaction:

$$MnO_4^- + 8H^+ + 5Fe^{2+} \rightarrow Mn^{2+} + 5Fe^{3+} + 4H_2O$$

Requirements

Safety glasses, balance, clock glass, beaker, stirrer, filter funnel, pipette, burette, volumetric flask, white tile, wash bottle and deionised water, stand and clamp, ammonium iron(II) sulphate, potassium manganate(VII) and dilute sulphuric acid.

Procedure

(1) Wash all the glassware thoroughly in deionised water.

(2) Make up an exact 0.1 mol L^{-1} ammonium iron(II) sulphate solution by dissolving exactly 9.8 g of ammonium ion(II) sulphate crystals in some deionised water in a beaker to which 20–30 cm^3 of dilute sulphuric acid has been added. Transfer the contents and rinsings into a 250 cm^3 volumetric flask. Make the solution up to the mark by adding more deionised water. Stopper the flask. Invert it and shake the solution thoroughly to ensure complete mixing.

(3) Make up the potassium manganate(VII) solution to be standardised by dissolving approximately 0.8 g KMnO$_4$ crystals in some deionised water in a beaker. Transfer the contents and rinsings into a 250 cm^3 volumetric flask. Make the solution up to the mark with deionised water. Stopper the flask and shake thoroughly to ensure complete mixing.

(4) Pour some potassium manganate(VII) solution through a filter funnel into the burette. Rinse the sides of the walls with the KMnO$_4$ solution making sure to run some out through the stop-cock. Refill the burette, remove the filter funnel and bring the solution to the zero mark by reading the top of the meniscus.

(5) Rinse the pipette out with some ammonium iron(II) sulphate solution. Carefully pipette 25 cm^3 of the ammonium iron(II) solution into a conical flask. Add approximately 20 cm^3 of dilute sulphuric acid and then titrate against the potassium manganate(VII) solution until the pink colour of the manganate(VII) just remains in the conical flask. If a brown precipitate should appear during the titration insufficient sulphuric acid has been added. If this happens, more sulphuric acid should be added.

(6) Repeat the titration until three consecutive readings are within the accepted experimental error of 0.1 cm^3.

(7) Record your results in the usual way and determine the exact concentration of the unknown potassium manganate(VII) solution using:

$$\frac{c_A V_A}{a} = \frac{c_B V_B}{b}$$

where c_A = concentration of KMnO$_4$ and c_B = concentration of Fe^{2+} ions and V_A and V_B are the respective volumes used, noting that $a = 1$ and $b = 5$.

Mandatory experiment 10

To calculate the percentage of iron in iron tablets.

Introduction

Iron plays a vital role in the body. It is found in haemoglobin in blood, in myoglobin in muscles and is stored in organs such as the liver.

Many people, especially young teenagers, suffer from a lack of iron. Iron deficiency (or anaemia) can be remedied by correct diet or by taking iron tablets. Iron tablets are made from iron(II) sulphate, FeSO$_4$.7H$_2$O, and are sold in pharmacies as ferrous sulphate tablets. The amount of iron in an iron tablet can be determined by a titration with potassium manganate (VII) solution.

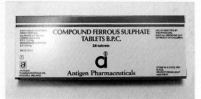

Iron tablets.

The titration is carried out under acidic conditions in accordance with the ionic reaction:

$$MnO_4^{2-} + 8H^+ + 5Fe^{2+} \rightarrow Mn^{2+} + 5Fe^{3+} + 4H_2O$$

Requirements

Safety glasses, balance, clock glass, beaker, stirrer, filter funnel, pipette, burette, 250 cm^3 volumetric flask, white tile, wash bottle and deionised water, stand and clamp, iron tablets, pestle and mortar, 0.02 mol L^{-1} potassium manganate(VII) solution and dilute sulphuric acid.

Procedure

(1) Wash all the glassware thoroughly in deionised water.

(2) Weigh an iron tablet accurately. Grind the tablet up in the mortar with some dilute sulphuric acid.

(3) Using a filter funnel, transfer the paste consisting of the tablet and the acid carefully into the 250 cm^3 volumetric flask. Make sure that you add all the rinsings from the pestle and mortar to the volumetric flask.

(4) Make the solution up to 250 cm^3 in the volumetric flask.

(5) Carefully pipette 25 cm³ of the iron(II) solution into a conical flask. Add approximately 20 cm³ of dilute sulphuric acid and then titrate against the potassium manganate(VII) solution until the pink colour of the manganate(VII) just remains in the conical flask. If a brown precipitate should appear during the titration insufficient sulphuric acid has been added. If this happens more sulphuric acid should be added.

(6) Repeat the titration until three consecutive readings are within the accepted experimental error of 0.1 cm³.

(7) Record your results in the usual way and determine the exact concentration of the iron(II) sulphate solution.

Results

Complete a table similar to that below.

Mass of iron tablet = 1.5 g

Pipette solution: 25 cm³ of $FeSO_4$ solution

Burette solution: 0.02 mol L⁻¹ $KMnO_4$ solution

Burette readings / cm³

	Trial	Titre 1	Titre 2	Titre 3
Final				
Initial				
Volume used				

Specimen calculation

Mean titre = 9.4 cm³ $KMnO_4$

Determine the concentration of the $FeSO_4$ solution using

$$\frac{c_A V_A}{a} = \frac{c_B V_B}{b}$$

where c_A = concentration of $KMnO_4$, c_B = concentration of $FeSO_4$ and V_A and V_B are the respective volumes used, noting that $a = 1$ and $b = 5$.

$$\frac{0.02 \text{ mol L}^{-1} \times 9.4 \text{ cm}^3}{1} = \frac{c_B \times 25.0 \text{ cm}^3}{5}$$

$$c_B = \frac{0.02 \text{ mol L}^{-1} \times 9.4 \text{ cm}^3 \times 5}{25.0 \text{ cm}^3}$$

Therefore, c_A, the concentration of $FeSO_4$ = 0.0376 mol L⁻¹.

Amount of iron(II) sulphate in 250 cm³,

$$n = cV = 0.0376 \text{ mol L}^{-1} \times 0.25 \text{ L}$$
$$= 0.0094 \text{ mol}$$

Since 1 mol of $FeSO_4$ contains 1 mol of Fe, then,

mass of iron, $m = nM$

$$= 0.0094 \text{ mol} \times 56.0 \text{ g mol}^{-1}$$

$$= 0.5264 \text{ g}$$

Percentage of iron in tablet

$$= \frac{\text{mass of iron}}{\text{mass of tablet}} \times 100\%$$

$$= \frac{0.5264 \text{ g}}{1.5 \text{ g}} \times 100\% = 35\%$$

Remember: amount of solute after dilution = amount of solute before dilution

Sodium Thiosulphate: Theory

Sodium thiosulphate, $Na_2S_2O_3$, reacts as a reducing agent towards iodine, producing sodium iodide and sodium tetrathionate.

$$2Na_2S_2O_3 + I_2 \rightarrow 2NaI + Na_2S_4O_6$$

Here, **2 mol $S_2O_3^{2-}$ ions reduces 1 mol of I_2.**

The colour change from the brown-red colour of iodine, I_2, through a straw-yellow colour to the colourless iodide ion, I^-, is slow and difficult to monitor. Therefore, an indicator, starch, is used to detect the end-point. Starch forms an intense blue complex with iodine which rapidly turns colourless when the iodine, I_2, is converted to iodide, I^-. The starch indicator is added towards the end of the titration (when the solution is a straw-yellow colour) because it changes colour so abruptly.

Sodium thiosulphate cannot be obtained in a sufficiently pure state to make a standard solution. It must be standardised with potassium manganate(VII). When potassium manganate(VII) reacts with potassium iodide, the potassium iodide is oxidised to free iodine, I_2. The free iodine can then be titrated against the thiosulphate solution using starch as an indicator.

The two reactions are as follows:

$$2KMnO_4 + 10KI + 8H_2O =$$
$$6K_2SO_4 + 2MnSO_4 + 8H_2O + 5I_2$$

i.e. 2 mol $KMnO_4 \rightarrow$ 5 mol I_2

$$2Na_2S_2O_3 + I_2 = 2NaI + Na_2S_4O_6$$

i.e. 2 mol $Na_2S_2O_3 \rightarrow$ 1 mol of I_2

Overall, since

2 mol $KMnO_4 \rightarrow$ 5 mol I_2

and

2 mol $Na_2S_2O_3 \rightarrow$ 1 mol of I_2

then

2 mol $KMnO_4 \equiv$ 5 mol $I_2 \equiv$ 10 mol $Na_2S_2O_3$

It should also be remembered that iodine, I_2, cannot be used as a primary standard because it does not dissolve readily in water.

Sodium thiosulphate is standardised by reacting a standard solution of potassium manganate(VII) with an excess of acidified potassium iodide solution. The liberated iodine, I_2, is then titrated against the unknown thiosulphate solution. It is not necessary to know the amount of iodine produced; the iodine merely acts as a link between the manganate(VII) and the thiosulphate solutions.

Mandatory experiment 11

To standardise a sodium thiosulphate solution.

Introduction

Sodium thiosulphate is unsuitable for use as a primary standard. It is standardised using potassium manganate(VII) solution, which first reacts with acidified potassium iodide solution to produce free iodine. The free iodine is then reduced by the thiosulphate solution to form iodide again. In the reactions the iodine merely acts as a bridge between the manganate(VII) and thiosulphate solutions.

Requirements

Safety glasses, filter funnel, pipette, burette, volumetric flask, white tile, wash bottle and deionised water, stand and clamp, standardised potassium manganate(VII) solution (0.02 mol L^{-1}) sodium thiosulphate crystals, 10% potassium iodide solution, dilute sulphuric acid and starch indicator.

Procedure

(1) Wash all glassware to be used thoroughly in deionised water. Remember to rinse out the pipette, the burette and the filter funnel with the solutions which you are going to use in each of them.
(2) Pipette exactly 25 cm^3 of the 0.02 mol L^{-1} $KMnO_4$ solution into a conical flask and then add approximately 10 cm^3 of potassium iodide solution and about 20 cm^3 of dilute sulphuric acid into the conical flask.
(3) Titrate the above solution against the sodium thiosulphate solution to be standardised using starch as an indicator. Remember to add the starch as the end-point (straw-yellow colour) approaches.
(4) Repeat the titration until you have three consecutive readings within the accepted experimental error.
(5) Tabulate your results and determine the concentration of the thiosulphate solution using

$$\frac{c_A V_A}{a} = \frac{c_B V_B}{b}$$

where c_A = concentration of $KMnO_4$

c_B = concentration of $Na_2S_2O_3$

V_A and V_B are the respective volumes used, noting that $a = 2$ and $b = 10$.

⊙ ⊙ ⊙ ⊙ ⊙ **Example 4.11** ⊙ ⊙ ⊙ ⊙ ⊙

Sodium thiosulphate was standardised as follows. 25 cm^3 of 0.02 mol L^{-1} potassium manganate(VII) solution was added to excess acidified potassium iodide solution. The iodine liberated as a result of the reaction required 22.40 cm^3 of sodium thiosulphate to reduce it. Determine the concentration of the thiosulphate solution.

$$2KMnO_4 + 10KI + 8H_2SO_4 =$$
$$6K_2SO_4 + 2MnSO_4 + 8H_2O + 5I_2$$

$$2Na_2S_2O_3 + I_2 = 2NaI + Na_2S_4O_6$$

Overall,
2 mol $KMnO_4 \equiv$ 5 mol $I_2 \equiv$ 10 mol $Na_2S_2O_3$

$$\frac{c_A V_A}{a} = \frac{c_B V_B}{b}$$

where c_A = concentration of $KMnO_4$,
c_B = concentration of $Na_2S_2O_3$ and

V_A and V_B are the respective volumes used, noting that $a = 2$ and $b = 10$.

$$\frac{0.02 \text{ mol L}^{-1} \times 25.00 \text{ cm}^3}{2} = \frac{c_B \times 22.40 \text{ cm}^3}{10}$$

$$c_B = \frac{0.02 \text{ mol L}^{-1} \times 25.00 \text{ cm}^3 \times 10}{22.40 \text{ cm}^3 \times 2}$$

Therefore, c_B, the concentration of thiosulphate = 0.112 mol L^{-1}.

• •

Mandatory experiment 12

To determine the % (w/v) of hypochlorite in bleach by means of an iodine/sodium thiosulphate titration.

Introduction

Most commercial and domestic bleaches are aqueous solutions of sodium hypochlorite, NaOCl(aq). The concentration of bleach can be indirectly determined by an iodine–sodium thiosulphate titration.

Sodium hypochlorite reacts with iodide ions, I$^-$, in dilute acidic solution to produce iodine, I$_2$.

$$NaOCl + 2I^- + 2H^+ = I_2 + NaCl + H_2O$$

Here, the hypochlorite (chlorate(I)) ions oxidise the iodide ions, I$^-$, to iodine, I$_2$.

The concentration of the sodium hypochlorite can be determined by titrating the liberated iodine, I$_2$, with a standard solution of sodium thiosulphate.

$$2Na_2S_2O_3 + I_2 = 2NaI + Na_2S_4O_6$$

Overall, 1 mol NaOCl ≡ 2 mol Na$_2$S$_2$O$_3$

Sodium thiosulphate is unsuitable for use as a primary standard. It should first be standardised using potassium manganate(VII) solution.

Requirements

Safety glasses, filter funnel, pipette, burette, volumetric flask, white tile, wash bottle and deionised water, stand and clamp, domestic bleach, standardised (0.1 mol L^{-1}) sodium thiosulphate solution, 10% potassium iodide solution, dilute sulphuric acid and starch indicator.

Procedure

(1) Wash all the glassware thoroughly in deionised water. Remember to rinse out the pipette, the burette and the filter funnel with the solutions which you are going to use in each of them.

(2) Pipette exactly 25 cm^3 of the domestic bleach solution into a 250 cm^3 volumetric flask. Add deionised water and make the solution up to the 250 cm^3 mark following the usual procedure.

(3) Using a clean pipette, transfer exactly 25 cm^3 of the diluted bleach to a conical flask. Acidify this solution by adding about 20 cm^3 of dilute sulphuric acid. Add about 10 cm^3 of potassium iodide solution.

(4) Fill the burette with the 0.1 mol L^{-1} sodium thiosulphate solution.

(5) Titrate the above solution against the 0.1 mol L^{-1} sodium thiosulphate solution using starch as an indicator. Remember to add the starch when a straw-yellow colour is presented.

(6) Repeat the titration until you have three consecutive readings within the accepted experimental error.

(7) Tabulate your results as shown and determine the concentration of the hypochlorite.

Results

Complete a table similar to that below.

Pipette solution: 25 cm^3 of NaOCl solution

Burette solution: 0.1 mol L^{-1} Na$_2$S$_2$O$_3$ solution

Burette readings / cm^3

	Trial	Titre 1	Titre 2	Titre 3
Final				
Initial				
Volume used				

Domestic bleach.

Specimen Calculation

Mean titre = 24.5 cm^3 Na$_2$S$_2$O$_3$

Determine the concentration of the diluted sodium hypochlorite solution using

$$\frac{c_A V_A}{a} = \frac{c_B V_B}{b}$$

where c_A = concentration of NaOCl,
c_B = concentration of Na$_2$S$_2$O$_3$ and
V_A and V_B are the respective volumes used,
noting that $a = 1$ and $b = 2$.

$$\frac{c_A \times 25.00 \text{ cm}^3}{1} = \frac{0.1 \text{ mol L}^{-1} \times 24.5 \text{ cm}^3}{2}$$

$$c_A = \frac{0.1 \text{ mol L}^{-1} \times 24.5 \text{ cm}^3}{25.0 \text{ cm}^3 \times 2}$$

Therefore, c_A, the concentration of sodium hypochlorite = 0.049 mol L^{-1}.

Amount of sodium hypochlorite in 250 cm^3,
$n = cV = 0.049$ mol L$^{-1} \times 0.25$ L
= 0.01225 mol

Mass of sodium hypochlorite, $m = nM$
= 0.01225 mol $\times$ 74.5 g mol^{-1}
= 0.912625 g
Percentage (w/v) of sodium hypochlorite in bleach

$$= \frac{\text{mass of sodium hypochlorite}}{\text{original volume of bleach}} \times 100\%$$

$$= \frac{0.912625 \text{ g}}{25 \text{ cm}^3} \times 100\% = 3.65 \text{ w/v\%}$$

Remember: amount of solute after dilution = amount of solute before dilution
The amount of sodium hypochlorite in 250 cm^3 is the same amount as that in the bottle of domestic bleach.

Exercise 4.3(b)

(1) Potassium manganate(VII) is a relatively powerful oxidising agent.
(a) Indicate the changes in oxidation number which occur when it reacts with an iron(II) salt.
(b) Describe the experimental conditions under which the titration is carried out. In your answer indicate all possible colour changes and what causes these colour changes.

(2) Potassium manganate(VII) reacts in acidic solution with ammonium iron(II) sulphate solution as follows:

$$MnO_4^- + 8H^+ + 5Fe^{2+} \rightarrow Mn^{2+} + 5Fe^{3+} + 4H_2O$$

Calculate the concentration of a potassium manganate(VII) solution, if 25 cm^3 of the solution was standardised against 24.5 cm^3 of a 0.1 mol L^{-1} ammonium iron(II) sulphate solution.

(3) Potassium manganate(VII) solution was standardised against an exactly 0.1 mol L^{-1} ammonium iron(II) sulphate solution.
25 cm^3 of the iron(II) solution was oxidised by 24.8 cm^3 of the manganate(VII) solution.
(a) Describe briefly how you would make up an exact 0.1 mol L^{-1} ammonium iron(II) solution.
(b) Explain why ammonium iron(II) sulphate solution is used as a source of Fe^{2+} ions in preference to iron(II) sulphate.
(c) Explain why an acidifying agent is used.
(d) During the titration a brown colour is sometimes noticed. Explain how this is caused and how you would remedy the situation.
(e) Determine the concentration of the manganate solution.

(4) 5.2 g of iron(II) sulphate crystals was dissolved in 250 cm^3 of aqueous solution in the usual manner. 25 cm^3 samples of this solution were completely oxidised by 0.02 mol L^{-1} potassium manganate(VII) solution. If the mean titre obtained was 19.8 cm^3, determine the amount of water of crystallisation in the iron(II) sulphate crystals.

(5) A 2.90 g sample of iron wire was dissolved in excess sulphuric acid and the solution was made up to 500 cm^3, following the usual procedure. The iron reacted as follows:

$$Fe(s) + H_2SO_4(aq) = FeSO_4(aq) + H_2(g)$$

25 cm^3 of this iron(II) solution required 24.8 cm^3 of 0.02 mol L^{-1} KMnO$_4$ for oxidation.

$$MnO_4^- + 8H^+ + 5Fe^{2+} \rightarrow Mn^{2+} + 5Fe^{3+} + 4H_2O$$

(a) Calculate the molar concentration of the Fe^{2+} ions.
(b) Determine the percentage of iron in the iron wire sample.

(6) The iron content of an iron ore can be determined by titration with potassium manganate (VII) solution. The iron ore is first reduced to Fe^{2+} ions by reaction with hydrochloric acid. If 25 cm^3 of the Fe^{2+} solution required 24.6 cm^3 of 0.02 mol L^{-1} KMnO$_4$ for oxidation, determine the molar concentration of the Fe^{2+} solution.

(7) 25 cm^3 of 0.02 mol L^{-1} potassium manganate(VII) solution was added to excess potassium iodide, the solution was acidified and then titrated against sodium thiosulphate solution. It was found that 19.5 cm^3 of the sodium thiosulphate was required to react with 25 cm^3 of the mixture.
(a) Why was the potassium manganate(VII) first reacted with excess potassium iodide?
(b) Explain how you would determine the end-point in the titration.
(c) Explain why sulphuric acid is used as an acidifying agent.
(d) Determine the concentration of the sodium thiosulphate solution.

(8) 25 cm^3 of a solution of iodine required 22.2 cm^3 of 0.12 mol L^{-1} sodium thiosulphate solution to react completely with it. Determine the concentration of the sodium thiosulphate solution expressed in
(a) mol L^{-1} (b) g L^{-1}

(9) The formula of crystalline iron(II) sulphate may be written as FeSO$_4$.xH$_2$O where x is the number of moles of water of crystallisation. In an experiment to determine x a student dissolved 6.25 g of crystalline iron(II) sulphate in deionised water and made the solution up to 250 cm^3 following the usual procedure. 20 cm^3 of this solution required on average 18.1 cm^3 of potassium manganate(VII) solution for oxidation.

(a) Write a balanced equation for the reaction.
(b) What is the usual procedure for making up an iron(II) sulphate solution for oxidation with potassium manganate(VII)?
(c) Describe briefly how the average titre of 18.1 cm^3 was determined.
(d) Calculate the molarity of the iron(II) solution.
(e) Calculate the percentage water of crystallisation.
(f) Determine the value of x.

(10) 20 cm^3 of 0.02 mol L^{-1} potassium manganate(VII) solution was oxidised in acidic solution by 25 cm^3 of iron(II) sulphate solution. If the iron(II) sulphate solution was made up by dissolving 5.5 g of iron(II) sulphate crystals in 250 cm^3 of solution, calculate the percentage water of crystallisation in the crystals.

(11) Potassium manganate(VII) is used in acidic solution as an oxidising agent as shown:

$$MnO_4^- + 8H^+ + 5Fe^{2+} \rightarrow Mn^{2+} + 5Fe^{3+} + 4H_2O$$

(a) Explain what is meant by (i) oxidising agent and (ii) reducing agent.
(b) Why is it not necessary to use an indicator when titrating Fe^{2+} ions against MnO$_4^-$ ions?
(c) Why is sulphuric acid and not hydrochloric acid used as an acidifying agent?
(d) What would you notice if insufficient concentrated sulphuric acid were added as an acidifying agent?

(12) Sodium thiosulphate solution is often used in volumetric analysis to estimate oxidising agents. Household bleaches are aqueous solutions of sodium hypochlorite, NaOCl(aq). 25 cm^3 of domestic bleach was diluted with deionised water in a 250 cm^3 volumetric flask following the usual procedure. Exactly 25 cm^3 of the diluted bleach was transferred to a conical flask and about 20 cm^3 of dilute sulphuric acid and 10 cm^3 of potassium iodide solution were added. The mixture in the conical flask was titrated with 0.1 mol L^{-1} sodium thiosulphate solution. The procedure was repeated several times to give a mean titre of 14.7 cm^3.

$$NaOCl + 2I^- + 2H^+ \rightarrow I_2 + NaCl + H_2O$$

$$2Na_2S_2O_3 + I_2 = 2NaI + Na_2S_4O_6$$

(a) Why is sodium thiosulphate not suitable for use as a primary standard?

(b) How are sodium thiosulphate solutions usually standardised?
(c) Why was the household bleach diluted?
(d) Describe, briefly, how the diluted bleach was transferred to the conical flask.
(e) Name the indicator used and its colour at the end-point.
(f) At what stage in the titration is the indicator added? Why should it not be added at the beginning?
(g) Calculate the concentration of the domestic bleach in (i) mol L^{-1} and
(ii) g L^{-1} NaOCl.

(13) 25 cm^3 of commercial bleach was accurately measured into a 250 cm^3 volumetric flask and diluted to the mark with deionised water. 25 cm^3 of this solution was placed in a conical flask, was acidified and excess potassium iodide was added. The iodine liberated was titrated against 0.1 mol L^{-1} sodium thiosulphate solution. The titres obtained were 24.8 cm^3, 24.5 cm^3, 24.4 cm^3 and 24.6 cm^3.

$$NaOCl + 2I^- + 2H^+ \rightarrow I_2 + NaCl + H_2O$$

$$2Na_2S_2O_3 + I_2 = 2NaI + Na_2S_4O_6$$

(a) Describe how the 25 cm^3 of bleach was measured accurately.
(b) How was the solution diluted to 250 cm^3 in the volumetric flask?
(c) Why was an excess of potassium iodide added?
(d) What was observed when the potassium iodide was added?
(e) Describe the colour changes which take place in the conical flask during the titration.
(f) What is the mean titre?
(g) Calculate:
(i) the concentration of NaOCl in the diluted bleach
(ii) the concentration of NaOCl in the commercial bleach
(iii) the % (w/v) of NaOCl in the commercial bleach.

(14) 2.5 g of bleaching powder was made into a paste with deionised water and made up to 250 cm^3 of solution in a volumetric flask. 25 cm^3 samples were acidified and excess potassium iodide was added. The 25 cm^3 samples required 24.1 cm^3 of 0.01 mol L^{-1} sodium thiosulphate solution to completely oxidise the liberated iodine.

Calculate the percentage of chlorine liberated by the action of the dilute acid on the bleaching powder.

$$Cl_2 + 2I^- \rightarrow 2Cl^- + I_2$$

$$2Na_2S_2O_3 + I_2 = 2NaI + Na_2S_4O_6$$

▼▼▼▼ Key Terms ▼▼▼▼

Some of the more important key terms are listed below. Other terms not listed may be located by means of the index.

1. Molar concentration: The molar concentration of a solution is the amount (number of moles) of solute dissolved in a stated volume of solution.

$$\text{Molar concentration} = \frac{\text{Amount of solute in moles}}{\text{Volume of solution}}$$

$$c = \frac{n}{V}$$

The unit of concentration is moles per litre (mol L^{-1}).

2. Number of moles from concentration and volume:

Amount in moles, $n = cV$

3. Concentration in grams per litre:

Concentration in grams per litre
= molar concentration × molar mass

$$\text{g L}^{-1} = cM$$

4. Concentrations expressed in different ways:

Weight per volume (w/v): This tells us the concentration expressed as the weight of solute in a specified volume of solution.

Volume per volume (v/v): This tells us the concentration expressed as the volume of solute in a specified volume of solution.

Weight per weight (w/w): This tells us the concentration expressed as the weight of solute in a specified weight of solution.

Parts per million (ppm): This refers to small concentrations which are often expressed as mg L^{-1}, i.e. 1/1000 g solute in 1 L of solution.

5. Dilution:

Number of moles of solute after dilution = number of moles of solute before dilution.

6. Standard solution: A standard solution is a solution of known concentration.

7. Acids:

Acids have a sour taste.
Acids are corrosive.
Acids turn blue litmus paper and solution red.
Acids have a pH less than 7.
Most acids react with most metals to form salts and release hydrogen gas.
Acids react with carbonates.
Acids neutralise bases.

8. Bases and alkalis: Bases are the opposite of acids. Some bases are soluble in water: these bases are called alkalis. Alkalis are soapy to touch; this is because they react with the natural oils on the skin to make soap.

9. Neutralisation: When an acid is mixed with a base they react with each other and cancel each other out. This is called neutralisation because the substances formed are neither acidic or basic: they are neutral.

10. Arrhenius concept of acids and bases: Arrhenius suggested that acids were substances which dissociated in water to produce hydrogen ions, H^+. Arrhenius regarded bases as substances which reacted with H^+ ions and formed water.

11. Bronsted–Lowry acids and bases: An acid is any substance (atom, ion or molecule) that can donate a proton.
A base is any substance, (atom, ion, or molecule) that can accept a proton.

12. Titration: A titration is a process by which a known volume of one solution is added to a measured volume of another solution until the reaction is complete. The titration is complete at a point called the end-point or the equivalence point.

13. Determination of concentration or volume via titration:

$$\frac{c_A V_A}{a} = \frac{c_B V_B}{b}$$

• Unit 5 •

FUELS AND HEATS OF REACTION

5.1 Sources of Hydrocarbons

5.2 Structure of Aliphatic Hydrocarbons

5.3 Aromatic Hydrocarbons

5.4 Exothermic and Endothermic Reactions

5.5 Oil Refining and its Products

5.6 Other Chemical Fuels

U n i t 5 •

Fossil – Cretaceous ammonite.

5.1 SOURCES OF HYDROCARBONS

Decomposition of Plant and Animal Waste

Oil rig at sea.

The petroleum and gas products which we use today were formed millions of years ago. Much of the Earth's surface was covered by sea, in which an abundance of marine plants and animals lived. As these organisms died, normal decay was prevented by rapid burial. The dead material was converted by heat, pressure and possibly by bacteria into the many kinds of hydrocarbon molecules which make up coal, crude oil and natural gas. It is important to realise that oil and gas do not accumulate in giant underground lakes but are contained in the pores of rocks such as sandstone, like water in a sponge.

Hydrocarbons are compounds formed by the combination of carbon and hydrogen.

Petroleum (Crude Oil)

Crude oil contains many types of hydrocarbon molecules. Some are aliphatic hydrocarbons: these form straight chain and branched chain molecules. Others are aromatic hydrocarbons (given that name because of their characteristic smell): these contain a ring structure.

Natural Gas

Natural gas rig, offshore, near Kinsale.

Natural gas is also a mixture of hydrocarbons. Natural gas is found in Ireland under the sea near Kinsale. It is composed of approximately 95% methane, CH_4. Methane is produced when dead or decaying material is converted into hydrocarbon molecules. It is found in slurry pits, coal mines and refuse dumps. When methane is produced in such areas it can be hazardous, and has caused many explosions.

Coal

Power station.

Coal is a hard black solid, composed mainly of carbon. The higher the carbon content the more efficient the coal is as a fuel. Coal is used mainly to generate electricity. Burning of coal can cause air pollution (see p. 311).

Hydrocarbons as Fuels

Today, we are familiar with many fuels, such as coal, natural gas, petrol and liquid petroleum gas (LPG). All are hydrocarbons which release energy when they are burnt in oxygen.

$$\text{Fuel} + O_2 \rightarrow \text{energy}$$

A **fuel** is a substance which can be used as an energy source.

Hydrocarbons from coal, natural gas and petroleum are one of the main sources of chemicals. Many complicated chemical compounds are synthesised from simple hydrocarbon molecules such as methane, ethene, propene and others. Without the natural resources of hydrocarbon molecules many of the compounds used today would not be available.

Some Applications of Crude Oil

Chemical feedstocks	Drugs, detergents, dyes, plastics
Solvents	Dry-cleaning liquids, paint removers
Lubricating oils	Medicinal, machinery, cosmetics
LPG	Propane, butane
Petrol	Leaded, unleaded
Fuels	Jet fuel, diesel oil, heating oil

Exercise 5.1

•

(1) How do hydrocarbon molecules form?

(2) What types of products are made from hydrocarbons?

(3) (a) What is a fuel?
(b) Name four fuels. Have they anything in common with each other?
(c) Write down the chemical formulae of any two fuels.
(d) Write a balanced equation for the combustion of any fuel.

(4) (a) What is a hazardous material?
(b) When methane is produced in the ground it can be hazardous. Why?
(c) How is methane produced in the ground?
(d) Name three places where methane is produced.

(5) (a) List some chemicals which are made from crude oil.
(b) Write a brief note on the importance of crude oil.

5.2 STRUCTURE OF ALIPHATIC HYDROCARBONS

Organic Chemistry

Organic chemistry is the study of carbon compounds, with the exception of the oxides of carbon and the metallic carbonates and related compounds. The scale of the modern organic chemical industry is so vast that it affects almost every aspect of our lives. The list of organic chemicals is endless: it includes drugs and medicines, clothing, farm products, domestic products, sports equipment and many more chemical products.

A data storage disk made of Macrolon.

Automotive underseal provides protection against stone chipping.

Hairdrier components made from polycarbonate which is heat resistant to 205 °C.

Pharmaceuticals to improve health care.

The bonds which carbon forms are covalent; that is, each bond is formed by the sharing of an electron pair. Sometimes covalent bonds are formed by sharing a pair of electrons with another carbon atom, while at other times covalent bonds are formed by sharing a pair of electrons with a different atom such as hydrogen, oxygen or nitrogen.

$$\begin{array}{ccc} & H & \\ & | & \\ H - & C & - H \\ & | & \\ & H & \end{array} \qquad \begin{array}{cccc} H & & H & \\ | & & | & \\ H - C & - & C - H \\ | & & | & \\ H & & H & \end{array}$$

Covalent Carbon

Carbon has four electrons available for sharing; this means that it can form four bonds. It is usual to represent these bonds by —; for example the bond between a carbon atom and a hydrogen atom is shown as C—H.

It should be remembered that in most organic compounds carbon always has four bonds around itself, whether they are single, double or treble bonds.

$$\begin{array}{ccc} & H & \\ & | & \\ H - & C & - H \\ & | & \\ & H & \end{array}$$

Four single

$$\begin{array}{cc} H \quad\quad H \\ \backslash \quad\quad / \\ C = C \\ / \quad\quad \backslash \\ H \quad\quad H \end{array}$$

Two single and one double per carbon

$$H - C \equiv C - H$$

One single and a triple per carbon

The ability of carbon atoms to form strong bonds to other atoms, including other carbon atoms, results in a vast number of carbon compounds.

Use of Models

The compounds studied in organic chemistry, in common with all compounds, vary in size and in shape. For simplicity, we often draw compounds in two dimensions only. Methane is usually represented as

$$\begin{array}{ccc} & H & \\ & | & \\ H - & C & - H \\ & | & \\ & H & \end{array}$$

This two-dimensional illustration conceals the three-dimensional tetrahedral shape of the methane molecule.

Methane is usually represented in three dimensions as

$$\begin{array}{c} H \\ \vdots \\ C \\ / \; \blacktriangle \; \backslash \\ H \quad H \quad H \end{array}$$

where — indicates bonds on the plane,
--- indicates bonds going behind the plane
and ◄ indicates bonds coming out from the plane.

Methane and the many other organic compounds you will come across should be studied using molecular models in order to get a better picture of the structure or the shape of the molecule. The simplest models include the ball and spring models and space-filling models illustrated below.

Methane – ball and spring model.

Methane – space-filling model.

Three-dimensional representation of molecules using computer modelling.

Alkanes, Alkenes and Alkynes

The alkanes, the alkenes and the alkynes are called **aliphatic hydrocarbons**, in which the carbon atoms are joined together in chains, whereas the arenes are called **aromatic hydrocarbons** and they contain a special ring structure, called a benzene ring.

• **Alkanes:** compounds with single bonds
• **Alkenes:** compounds with double bonds
• **Alkynes:** compounds with triple bonds

The simplest hydrocarbons are the alkanes. Alkanes are termed **saturated hydrocarbons** because all the chemical bond are single bonds. Compounds with double and triple bonds are said to be unsaturated.

The Alkanes

Alkanes are saturated hydrocarbons because each molecule contains carbon atoms bonded to four other atoms by single covalent bonds. The general formula of the alkanes is C_nH_{2n+2} where n = 1, 2, 3 … and the structural formula is

$$
\begin{array}{c}
H \\
| \\
R - C - H \\
| \\
H
\end{array}
$$
where R = H or an alkyl group (CH_3, C_2H_5 etc.).

Homologous Series

The molecular formula of each member of the alkanes follows the pattern or general formula given earlier. The alkanes are known as a homologous series, that is, a series or family of compounds in which the molecular formula of each member differs from that of the previous member by a definite number of carbon and hydrogen atoms. In this case, each alkane differs from the next one by **$-CH_2-$**.

The **first eight members** of the alkane family are given in the table on page 130.

Systematic Naming of Organic Compounds

Organic compounds are often known by their common names. However, today, most compounds are named in accordance with the IUPAC (International Union of Pure and Applied Chemistry) system.

The rules for naming the alkanes are as follows:

(1) The simplest member of the alkanes is called methane, CH_4. The next three members are called by their original names, ethane, propane and butane. The other straight chain alkanes are named by combining a prefix from the Greek for the number of carbon atoms in the chain with the suffix **-ane**.

(2) **The part with the longest chain is the parent alkane.**

(3) Substituted groups are named and their positions are indicated by the lowest possible number of the carbon atom to which they are attached.

(4) If there is more than one substituted alkyl group (an alkane with a H atom missing), they are named in order of increasing size, i.e. methyl before ethyl, ethyl before propyl etc., or else alphabetically.

The first eight members of the alkane family

Alkane	Formula	Structural formula		Physical state at room temperature
Methane	CH_4	$H-\underset{\underset{H}{\vert}}{\overset{\overset{H}{\vert}}{C}}-H$		Gas
Ethane	C_2H_6	$H-\underset{\underset{H}{\vert}}{\overset{\overset{H}{\vert}}{C}}-\underset{\underset{H}{\vert}}{\overset{\overset{H}{\vert}}{C}}-H$		Gas
Propane	C_3H_8	$H-\underset{\underset{H}{\vert}}{\overset{\overset{H}{\vert}}{C}}-\underset{\underset{H}{\vert}}{\overset{\overset{H}{\vert}}{C}}-\underset{\underset{H}{\vert}}{\overset{\overset{H}{\vert}}{C}}-H$		Gas
Butane	C_4H_{10}	$H-\underset{\underset{H}{\vert}}{\overset{\overset{H}{\vert}}{C}}-\underset{\underset{H}{\vert}}{\overset{\overset{H}{\vert}}{C}}-\underset{\underset{H}{\vert}}{\overset{\overset{H}{\vert}}{C}}-\underset{\underset{H}{\vert}}{\overset{\overset{H}{\vert}}{C}}-H$		Gas
Pentane	C_5H_{12}	$H-\underset{\underset{H}{\vert}}{\overset{\overset{H}{\vert}}{C}}-\underset{\underset{H}{\vert}}{\overset{\overset{H}{\vert}}{C}}-\underset{\underset{H}{\vert}}{\overset{\overset{H}{\vert}}{C}}-\underset{\underset{H}{\vert}}{\overset{\overset{H}{\vert}}{C}}-\underset{\underset{H}{\vert}}{\overset{\overset{H}{\vert}}{C}}-H$		Liquid
Hexane	C_6H_{14}	$H-\underset{\underset{H}{\vert}}{\overset{\overset{H}{\vert}}{C}}-\underset{\underset{H}{\vert}}{\overset{\overset{H}{\vert}}{C}}-\underset{\underset{H}{\vert}}{\overset{\overset{H}{\vert}}{C}}-\underset{\underset{H}{\vert}}{\overset{\overset{H}{\vert}}{C}}-\underset{\underset{H}{\vert}}{\overset{\overset{H}{\vert}}{C}}-\underset{\underset{H}{\vert}}{\overset{\overset{H}{\vert}}{C}}-H$		Liquid
Heptane	C_7H_{16}	$H-\underset{\underset{H}{\vert}}{\overset{\overset{H}{\vert}}{C}}-\underset{\underset{H}{\vert}}{\overset{\overset{H}{\vert}}{C}}-\underset{\underset{H}{\vert}}{\overset{\overset{H}{\vert}}{C}}-\underset{\underset{H}{\vert}}{\overset{\overset{H}{\vert}}{C}}-\underset{\underset{H}{\vert}}{\overset{\overset{H}{\vert}}{C}}-\underset{\underset{H}{\vert}}{\overset{\overset{H}{\vert}}{C}}-\underset{\underset{H}{\vert}}{\overset{\overset{H}{\vert}}{C}}-H$		Liquid
Octane	C_8H_{18}	$H-\underset{\underset{H}{\vert}}{\overset{\overset{H}{\vert}}{C}}-\underset{\underset{H}{\vert}}{\overset{\overset{H}{\vert}}{C}}-\underset{\underset{H}{\vert}}{\overset{\overset{H}{\vert}}{C}}-\underset{\underset{H}{\vert}}{\overset{\overset{H}{\vert}}{C}}-\underset{\underset{H}{\vert}}{\overset{\overset{H}{\vert}}{C}}-\underset{\underset{H}{\vert}}{\overset{\overset{H}{\vert}}{C}}-\underset{\underset{H}{\vert}}{\overset{\overset{H}{\vert}}{C}}-\underset{\underset{H}{\vert}}{\overset{\overset{H}{\vert}}{C}}-H$		Liquid

Alkanes and Alkyl Groups

Alkane	Formula	Alkyl group	Formula
Methane	CH_4	Methyl	CH_3
Ethane	C_2H_6	Ethyl	C_2H_5
Propane	C_3H_8	Propyl	C_3H_7
Butane	C_4H_{10}	Butyl	C_4H_9
Pentane	C_5H_{12}	Pentyl	C_5H_{11}
Hexane	C_6H_{14}	Hexyl	C_6H_{13}
Heptane	C_7H_{16}	Heptyl	C_7H_{15}
Octane	C_8H_{18}	Octyl	C_8H_{17}

It should be noted carefully at this point that the IUPAC system for naming all organic compounds cannot be used if the rules for naming the alkanes are not understood. The following example will help you to understand these rules.

• • • • • Example 5.1 • • • • •

Use the IUPAC system to name the following alkane.

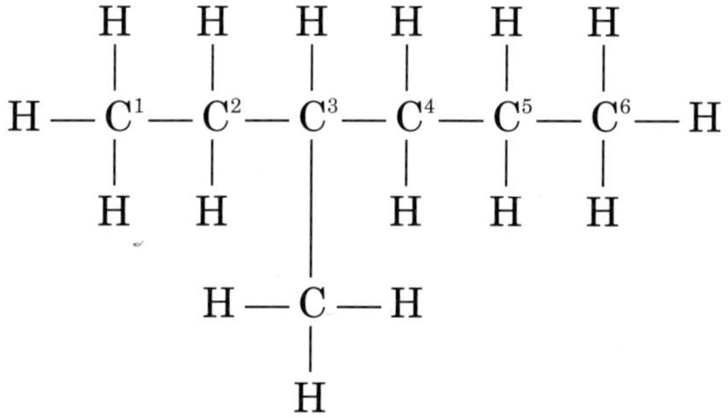

(1) Parent is hexane: longest chain has six carbons.
(2) Substituted group is methyl, CH_3.
(3) Methyl is at carbon-3.
(4) The name of the compound is, therefore, 3-methylhexane.

Isomerism

The fourth member of the alkane series is butane, C_4H_{10}. There are two different ways of arranging the atoms in C_4H_{10}.
The structure can be given in either of the following ways:

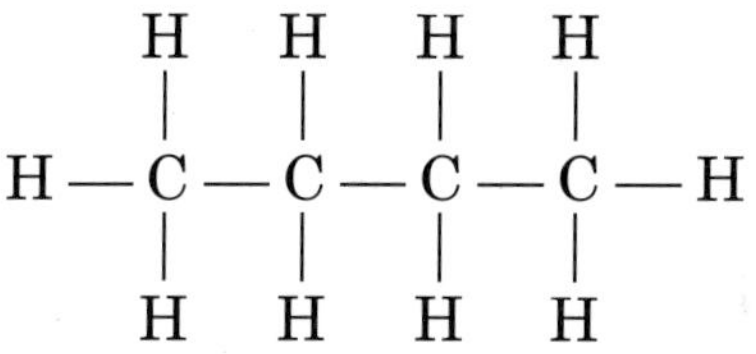

Butane

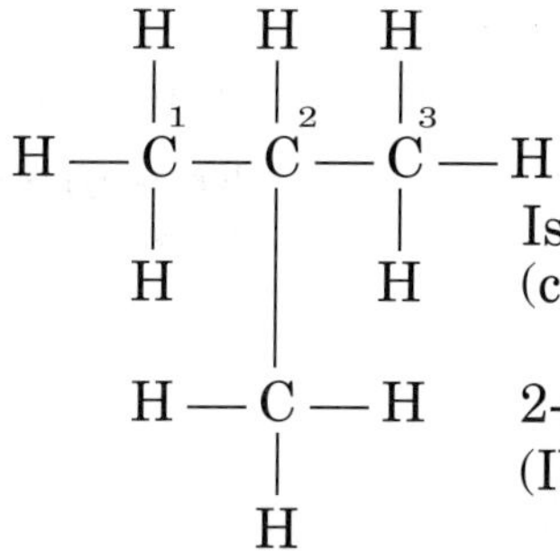

Isobutane
(common)

2-methylpropane
(IUPAC)

The fifth member of the alkane series is pentane, C_5H_{12}. There are three different ways of arranging the atoms in C_5H_{12}.
The structure can be given in any of the following ways:

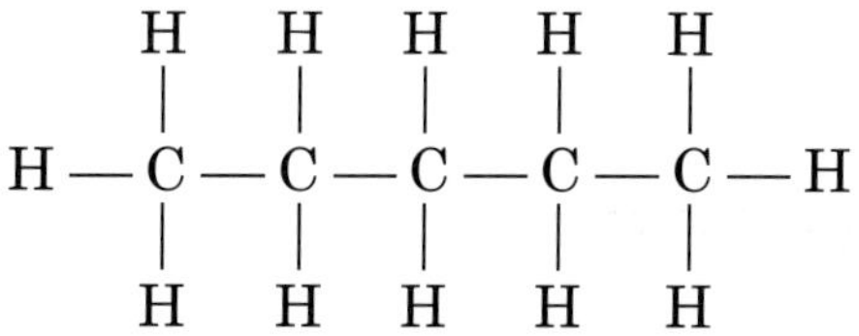

Pentane

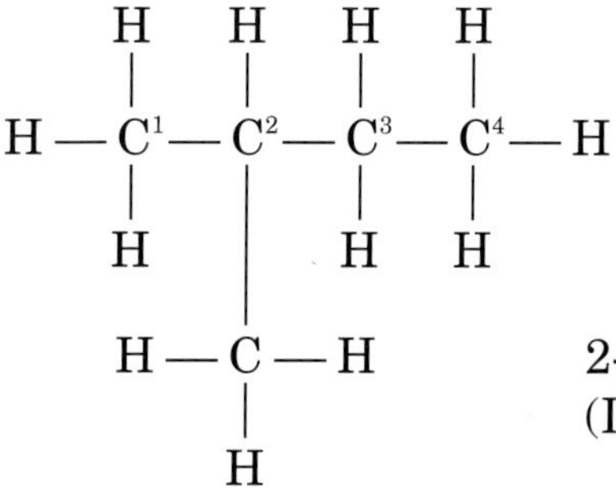

Isopentane
(common)

2-methylbutane
(IUPAC)

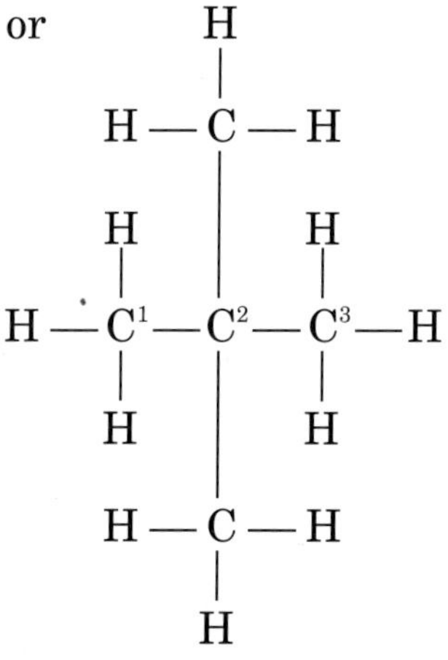

2,2-dimethylpropane
(IUPAC)

Compounds like those above are called **isomers**.

Isomers are substances with the same molecular formula but different structural formulae.

Octane, C_8H_{18}, has some isomers. The most familiar isomer is 2,2,4-trimethylpentane (isooctane).

$$
\begin{array}{ccc}
& H & H \\
& | & | \\
& H-C-H & H-C-H \\
& | & | \\
H & H & H \\
| & | & | \\
H-C^5-C^4-C^3-C^2-C^1-H \\
| & | & | & | \\
H & H & H & H \\
& & H-C-H \\
& & | \\
& & H
\end{array}
$$

Cyclic Structures

Alkanes with three or more carbon atoms can be arranged in cyclic structures. Cyclohexane, C_6H_{12}, is a typical example.

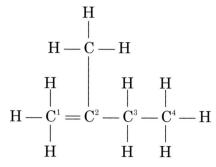

cyclohexane

Physical Properties of Alkanes

Methane and the other alkanes are symmetrical covalent molecules. Because they are symmetrical molecules they are non-polar. The attraction between non-polar molecules is the weak van der Waal's attraction. Consequently, the forces between the alkane molecules are tiny compared with, say, ionic forces. These very small forces can be easily overcome and so it is not surprising that ethane is a gas at room temperature. Ethane, propane and butane are also gases at ordinary temperatures. The remaining alkanes are liquids or solids.

Because of the weak non-polar forces between them, alkanes not only have very low boiling points but are also practically insoluble in water and other polar solvents. They are, however, soluble in non-polar solvents such as 1,1,1-trichloroethane (see p. 184).

Alkenes

The alkenes are another hydrocarbon family. They are characterised by the presence of a carbon–carbon double bond in the molecule. Because they have a double bond they are said to be unsaturated compounds.

The general formula of the alkenes is C_nH_{2n}, where n = 2, 3, 4 ...
and the structural formula is

$$
\begin{array}{ccc}
R & & H \\
\backslash & & / \\
& C=C & \\
/ & & \backslash \\
H & & H
\end{array}
$$
where R = H or any alkyl group.

The **first three members** of the family are given on page 133.

Systematic Naming of Alkenes

(1) The parent alkene is named by counting the maximum number of carbons in the longest chain containing the double bond. The prefixes used are the same as those for the alkanes, but this time they are combined with the ending **-ene**.
(2) The position of the double bond is indicated by the lowest possible number.
(3) Other substituted groups are then indicated in the same way as in alkanes.

· · · · · · **Example 5.2** · · · · · ·

Name the following compound using the IUPAC system.

$$
\begin{array}{c}
H \\
| \\
H-C-H \\
| \\
\begin{array}{cccc}
H & H & H \\
| & | & | \\
H-C^1=C^2-C^3-C^4-H \\
| & | & | \\
H & H & H
\end{array}
\end{array}
$$

(1) Longest straight chain has four carbons: parent is butene.
(2) Double bond is at carbon-2: parent is but-2-ene.
(3) Methyl is at carbon-2.
(4) The name of the compound is, therefore, 2-methylbut-2-ene.

· · · · · · · · · · · · · · · · · ·

132

The first three members of the alkene family

Alkene	Formula	Structural formula		Physical state at room temperature
Ethene	C_2H_4	$\overset{\displaystyle H}{\underset{\displaystyle H}{{\Large \diagdown}\diagup}}C = C\overset{\displaystyle H}{\underset{\displaystyle H}{\diagup {\Large \diagdown}}}$		Gas
Propene	C_3H_6	$H - \overset{\displaystyle H}{\underset{\displaystyle H}{\overset{\displaystyle |}{\underset{\displaystyle |}{C}}}} - \overset{\displaystyle H}{\overset{\displaystyle |}{C}} = \overset{\displaystyle H}{\overset{\displaystyle |}{C}} - H$		Gas
Butene	C_4H_8	$H - \overset{\displaystyle H}{\underset{\displaystyle H}{\overset{\displaystyle |}{\underset{\displaystyle |}{C}}}} - \overset{\displaystyle H}{\underset{\displaystyle H}{\overset{\displaystyle |}{\underset{\displaystyle |}{C}}}} - \overset{\displaystyle H}{\overset{\displaystyle |}{C}} = \overset{\displaystyle H}{\overset{\displaystyle |}{C}} - H$		Gas

· · · · · · **Example 5.3** · · · · · ·

Name the following compound using the IUPAC system.

$$H - \overset{\displaystyle H}{\underset{\displaystyle H}{\overset{\displaystyle |}{\underset{\displaystyle |}{C^1}}}} - \overset{\displaystyle H}{\underset{\displaystyle H}{\overset{\displaystyle |}{\underset{\displaystyle |}{C^2}}}} = \overset{\overset{\displaystyle H}{\overset{\displaystyle |}{\overset{\displaystyle H-C-H}{\overset{\displaystyle |}{\,}}}}}{C^3} - \overset{\displaystyle H}{\underset{\displaystyle H}{\overset{\displaystyle |}{\underset{\displaystyle |}{C^4}}}} - \overset{\displaystyle H}{\underset{\displaystyle H}{\overset{\displaystyle |}{\underset{\displaystyle |}{C^5}}}} - H$$

(1) Longest straight chain has five carbons: parent is pentene.
(2) Double bond is at carbon-2: parent is pent-2-ene.
(3) Methyl is at carbon-3.
(4) The name of the compound is, therefore, 3-methylpent-2-ene.

· ·

Physical Properties of Alkenes

The boiling points of the lower alkenes are similar to those of the alkanes with the same number of carbon atoms. Ethene, propene and butene are gases at room temperature, while the other members of the homologous series are liquids.

Some alkenes, such as propene, are very slightly polar because they are not symmetrical like ethene. The slight polarity of some alkenes does not dramatically alter their physical properties.

Alkenes, like alkanes, are insoluble in water, but soluble in non-polar solvents such as trichloroethane or benzene.

Structural Isomerism

Alkenes with more than three carbon atoms can form structural isomers. For instance, butene can have the double bond in two possible positions, i.e. at carbon-1 or at carbon-2.

$$H - \overset{\displaystyle H}{\overset{\displaystyle |}{C^1}} = \overset{\displaystyle H}{\overset{\displaystyle |}{C^2}} - \overset{\displaystyle H}{\underset{\displaystyle H}{\overset{\displaystyle |}{\underset{\displaystyle |}{C^3}}}} - \overset{\displaystyle H}{\underset{\displaystyle H}{\overset{\displaystyle |}{\underset{\displaystyle |}{C^4}}}} - H$$

But-1-ene

$$H - \overset{\displaystyle H}{\underset{\displaystyle H}{\overset{\displaystyle |}{\underset{\displaystyle |}{C^1}}}} - \overset{\displaystyle H}{\overset{\displaystyle |}{C^2}} = \overset{\displaystyle H}{\overset{\displaystyle |}{C^3}} - \overset{\displaystyle H}{\underset{\displaystyle H}{\overset{\displaystyle |}{\underset{\displaystyle |}{C^4}}}} - H$$

But-2-ene

Alkynes

Alkynes are unsaturated hydrocarbons which are even more unsaturated than alkenes as they contain a carbon–carbon triple bond.

The general formula of the alkynes is C_nH_{2n-2} and the structural formula is R—C≡C—R, where R = H or an alkyl group.

They form a homologous series, the first member of which is ethyne, C_2H_2. The common name of ethyne is acetylene.

Systematic Naming of Alkynes

The compounds are named as for the alkenes but with the suffix **-yne**.
For example: H—C≡C—H is ethyne.

Physical Properties of Alkynes

The boiling points of the alkynes are similar to those of the alkenes with the same number of carbon atoms. The alkynes are also insoluble in water but soluble in non-polar solvents.

Activity 5.1

Demonstration of the Solubility of Methane, Ethene and Ethyne

Requirements

Safety glasses, water, hexane, ethanol, propanone, methane, ethene and ethyne, disposable Pasteur pipettes, test-tubes and rack.
Natural gas can used as a source of methane. Ethene and ethyne can be tested when they are prepared in Experiments 19 and 14.

Procedure

(1) Fill four test-tubes with methane. Stopper each test-tube.
(2) In turn, place about 2 cm³ of water, of hexane, of ethanol and of propanone into the test-tubes.
(3) Shake each test-tube and observe what happens.
(4) Repeat the procedure for ethene and for ethyne.
(5) Copy and complete the table by indicating whether or not the gas is soluble in the solvent. How did you know if the gas was soluble?
(6) Discuss your conclusions.

Results

Solute in solvent	Water	Hexane	Ethanol	Propanone
Methane				
Ethene				
Ethyne				

Exercise 5.2

(1) (a) What is a hydrocarbon?
(b) Explain the meaning of each the following terms:
(i) aliphatic (ii) aromatic (iii) saturated (iv) unsaturated.

(2) (a) What is the general formula of the alkanes?
(b) Write the molecular formula and name of each of the first six members of the alkanes.
(c) Draw the structural formulae of the first four alkanes.

(3) (a) What is the difference between the structural formula of a molecule and the real molecular structure of the molecule?
(b) How can models be used to illustrate molecular structures?

(4) (a) What is a homologous series?
(b) Describe how each successive member of the alkane family differs from the next one.
(c) What is an alkyl group?

(5) Name each of the following alkanes using the IUPAC system.

(a)

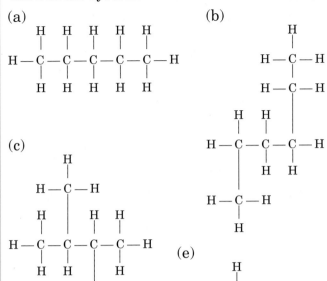

(b)

(c)

(d)

(e)

(f)

(g)

(6) Draw the structural formula of each of the following alkanes:
(a) pentane (b) hexane (c) heptane (d) octane
(e) 2-methylpropane (f) butane
(g) cyclohexane (h) 2,2,4-trimethylpentane.

(7) (a) What is an isomer?
(b) Draw the structures of the isomers of formula C_4H_8.

(8) (a) Why would you expect the boiling points of the alkanes to increase in the order methane, ethane, propane and so on?
(b) Methane and the other members of the homologous series are practically insoluble in water, yet soluble in non-polar solvents. Explain why this is so.

(9) (a) What is the general formula of the alkenes?
(b) Give the names and the molecular formulae of the first four alkenes.
(c) Draw the structural formulae of the first four alkenes.

(10) Name the following alkenes using the IUPAC system.

(a)

(b)

(c)

(11) Draw the structure of each of the following alkenes:
(a) ethene (b) propene (c) but-1-ene
(d) but-2-ene.

(12) The alkenes have similar physical properties to the alkanes. Explain why this is so.

(13) (a) What are the alkynes?
(b) Ethyne, the first member of the alkynes, is a gas which is insoluble in water and soluble in non-polar solvents. Explain why this is so.

(14) (a) What is acetylene?
(b) Draw the structure of ethyne.

(15) Describe an experiment to demonstrate the solubility of methane, of ethene and of ethyne in different solvents.

(16) Write structural formulae for the simplest possible:
(a) alkane (b) alkene (c) alkyne.

5.3 AROMATIC HYDROCARBONS

Aromatic hydrocarbons are **cyclic unsaturated hydrocarbons**: the most common one found in crude oil is benzene.

The term aromatic was first used to describe a group of compounds which had a pleasant aroma or smell. This group of compounds included benzene and its many derivatives. Today, the term aromatic is used to describe a range of compounds which have a similar structure to benzene. Aromatic compounds have many similar chemical properties to benzene, but have many completely different chemical properties to the aliphatic compounds.

Benzene was discovered in 1825 by **Michael Faraday**. Analysis showed that it had a relative molecular mass of 78 and that its empirical formula was CH. The molecular formula was found to be C_6H_6.

Experiments have shown that benzene is a cyclic compound, with a six-membered ring of carbon atoms and with one hydrogen atom attached to each carbon atom. As a result, it was thought that benzene had a structure consisting of alternate single and double carbon–carbon bonds. The following structure was proposed by **Friedrich Kekulé** in 1865.

The benzene structure is now represented as shown, where the hexagonal corners represent H—C bonds and the inner circle represents the six equivalent carbon–carbon bonds.

The complete benzene structure is indicated as shown.

Methylbenzene and ethylbenzene have similar structures.

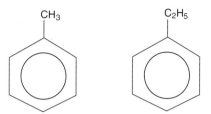

Physical Properties

The physical properties of the aromatic hydrocarbons are similar to those of the aliphatic hydrocarbons. They are non-polar compounds which are insoluble in water but soluble in other non-polar compounds. Benzene is an excellent solvent, but because it has cancer-causing (carcinogenic) properties, other less carcinogenic solvents such as methylbenzene and ethylbenzene are used in preference to benzene.

Activity 5.2

Demonstration of the Solubility of Methylbenzene

Requirements
Safety glasses, water, hexane, ethanol, propanone, methylbenzene, disposable Pasteur pipettes, test-tubes and rack.

Procedure
(1) Place about 5 cm³ of methylbenzene into four different test-tubes.
(2) In turn, place about 1 cm³ of water, hexane, ethanol and propanone into the test-tubes.
(3) Shake each test-tube and observe what happens.
(4) Copy and complete the table.
(5) Discuss your conclusions.

Results

Solute	Type	Result
Water	Polar	
Hexane	Non-polar	
Ethanol	Polar + Non-polar	
Propanone	Polar + Non-polar	

136

Social and Applied Aspects

• Many aromatic compounds are used as solvents in the paint industry.
• Others form the basis of dyestuffs, detergents, herbicides and many pharmaceutical compounds.
• Many aromatic hydrocarbons, such as benzene, are carcinogenic. Proper care must be taken when using aromatic compounds.

Exercise 5.3

•

(1) (a) Why is benzene termed an unsaturated hydrocarbon?
(b) Draw the structure of benzene.

(2) (a) What is a cyclic compound?
(b) Draw the structures of two named cyclic compounds.

(3) (a) Which famous scientist discovered benzene?
(b) What is the empirical formula of benzene?
(c) What did Kekulé discover about the structure of benzene?

(4) (a) What type of solvent is benzene?
(b) Why is benzene not often used as a solvent?
(c) What aromatic hydrocarbons are used as solvents instead of benzene?

(5) How would you demonstrate that benzene is a non-polar solvent?

(6) The structural formulae of three aromatic hydrocarbons are shown as follows:

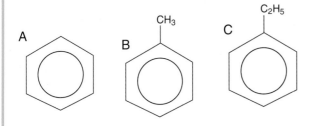

(a) Give the systematic name and the common name for A.
(b) Which is not commonly used as a solvent?
(c) Why is it not used?

5.4 EXOTHERMIC AND ENDOTHERMIC REACTIONS

The majority of the world's sources of energy are chemical in nature. The chemical sources of energy include coal, oil, wood and farm and domestic waste, while the non-chemical sources include wind, water and nuclear energy.

Thermochemistry is the study of the heat changes which occur during a chemical reaction.

Heat is produced in many chemical reactions and this causes the temperature to rise. In other chemical reactions heat is absorbed and this causes the temperature to fall.

The study of the amount of energy taken in or given out during a chemical reaction is important to engineers, chemists, biologists, doctors, geologists, nutritionists and many others.

Energy Changes

Almost every chemical reaction involves an energy change. When a reaction gives out heat it is said to be **exothermic**. A reaction which absorbs heat is said to be **endothermic**.

In an exothermic reaction, heat energy is released to the surroundings (the atmosphere). The heat change causes a rise in the temperature of the reaction system.

In an endothermic reaction, heat energy is taken in from the surroundings. The heat change causes a fall in temperature.

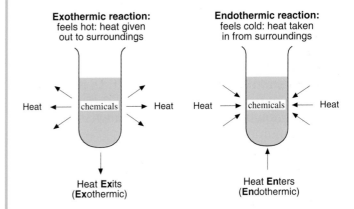

The chemical energy in glucose is converted into heat, mechanical and electrical energy.

Changes of State

Heat changes occur when solids are converted into liquids. For example, when ice melts, it absorbs heat from its surroundings (mainly the atmosphere): this is an endothermic process. When steam condenses into water, it releases energy to its surroundings (mainly the atmosphere): this is an exothermic process.

When sulphuric acid, H_2SO_4, dissolves in water, the temperature of the water rises. The process

$$H_2SO_4(l) + H_2O(l) \rightarrow H_2SO_4(aq)$$

is exothermic.

When ammonium nitrate, NH_4NO_3, dissolves in water, the temperature of the water falls. The process

$$NH_4NO_3(s) + H_2O(l) \rightarrow NH_4NO_3(aq)$$

is endothermic.

Activity 5.3

Demonstration of exothermic and endothermic reactions

Exothermic reaction

Add a small piece of magnesium to about 10 cm³ of dilute hydrochloric acid in a test-tube. Measure the rise in temperature with a thermometer.

Endothermic reaction

(1) Place a drop of water on a wooden block.
(2) Put about 10 g of ammonium thiocyanate and 20 g of barium hydroxide into a glass beaker. Stir the mixture with a glass rod.

---

(3) Place the beaker on the drop of water on the wooden block. Continue stirring the mixture.

Barium hydroxide and ammonium thiocyanate are mixed together in a glass beaker. The reaction is endothermic. When the beaker is set in a puddle of water on a board, the water freezes and attaches the beaker solidly to the board.

(4) After some time lift up the beaker. What do you observe?

Heat of Reaction

The amount of energy given out or taken in can be measured when a temperature change takes place. The energy change is normally measured in kilojoules, kJ, when one mole of a particular chemical reacts.

This change is written as ΔH, called 'delta H', meaning 'change in' energy.

ΔH is the change in energy of the reactants and products and is written as

$$\Delta H = H(\text{products}) - H(\text{reactants})$$

In exothermic reactions, heat is evolved and the product formed has less energy than the reactants. The heat change is written, by convention, as $-\Delta H$ (heat lost from system).

In endothermic reactions, heat is absorbed and the product formed has more energy than the reactants. Here, the heat change is written, by convention, as $+\Delta H$ (heat gained by system).

The **standard conditions** for measuring heat changes are a temperature of 298 K and a pressure of 10^5 Pa.

Examples of heat changes:
(1) When one mole (16 g) of methane combines with two moles (64 g) of oxygen to produce one mole (44 g) of carbon dioxide and two moles (36 g) of water, 890 kJ of heat is evolved under standard conditions. For this reaction, $\Delta H = -890$ kJ mol⁻¹.

$$CH_4(g) + 2O_2(g) = CO_2(g) + 2H_2O(l);$$
$$\Delta H = -890 \text{ kJ mol}^{-1}$$

This is shown on an energy level diagram as

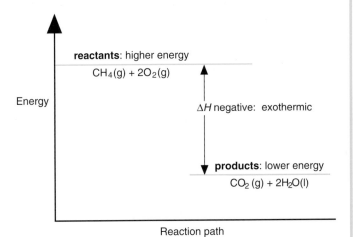

(2) When one mole (100 g) of calcium carbonate, $CaCO_3$, is decomposed by heating, 178 kJ of energy must be supplied to convert it into one mole (56 g) of calcium oxide, CaO, and one mole (44 g) of carbon dioxide, CO_2. For this reaction, $\Delta H = +178$ kJ mol^{-1}.

$$CaCO_3(s) = CaO(s) + CO_2(g);$$
$$\Delta H = +178 \text{ kJ mol}^{-1}$$

This is shown on an energy level diagram as

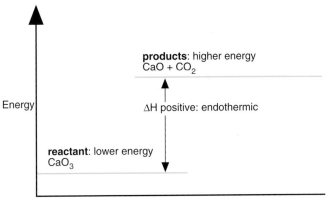

Different chemical reactions are compared with each other when all the reactants and products in the chemical reactions are in their standard states and at standard conditions.

Standard state refers to the normal state the reactants and products are in at the stated standard conditions. In other words, it must be clearly stated whether the reactants and products are solids (s), liquids (l), gases (g), allotropes, such as graphite or diamond, or in aqueous solution (aq) at the **standard conditions of temperature and pressure**.

Combustion of Alkanes and Other Fuels

Most fuels are 'fossil fuels', such as coal, oil and natural gas. Fossil fuels are organic compounds, containing carbon and hydrogen atoms bonded together by strong covalent bonds. When these bonds are broken energy is released. Bonds are broken when fuels are burnt in air or oxygen (undergo combustion). Combustion reactions are exothermic reactions.

The complete combustion of a fuel can be represented in the following ways:

Coal + air/oxygen → carbon dioxide + water + heat

Oil + air/oxygen → carbon dioxide + water + heat

Alcohol + air/oxygen → carbon dioxide + water + heat

Most of the common fuels in use today are alkanes: methane in natural gas, propane and butane in LPG, octane in petrol, and many, many more.

When alkanes burn in oxygen they form carbon dioxide and water and release heat energy.

$$CH_4(g) + 2O_2(g) = CO_2(g) + 2H_2O(l); \Delta H = -890 \text{ kJ mol}^{-1}$$

$$C_2H_6(g) + \tfrac{7}{2}O_2(g) = 2CO_2(g) + 3H_2O(l); \Delta H = -1560 \text{ kJ mol}^{-1}$$

$$C_3H_8(g) + 5O_2(g) = 3CO_2(g) + 4H_2O(l); \Delta H = -2220 \text{ kJ mol}^{-1}$$

$$C_4H_{10}(g) + \tfrac{13}{2}O_2(g) = 4CO_2(g) + 5H_2O(l); \Delta H = -2877 \text{ kJ mol}^{-1}$$

Standard Heat of Combustion of a Substance

This is the amount of heat in joules liberated when one mole of a substance is burned in excess oxygen, under standard conditions, e.g.

$$C_2H_6(g) + \tfrac{7}{2}O_2(g) = 2CO_2(g) + 3H_2O(l); \Delta H = -1554 \text{ kJ mol}^{-1}$$

Bond Energies and Heats of Combustion

During a chemical reaction many bonds are broken and new bonds are made. The energy change which takes place is caused by the making and breaking of chemical bonds.

When methane, CH_4, is burnt in oxygen, O_2, to form carbon dioxide, CO_2, and water, H_2O, some chemical bonds are broken and new ones are made.

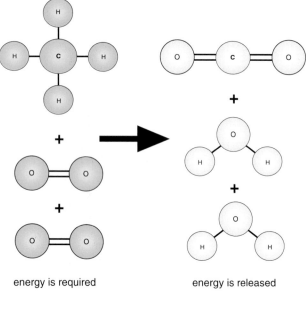

energy is required   energy is released

Overall energy change   = energy required - energy released

= (2376 - 3266) kJ mol^{-1}

= - 890 kJ mol^{-1}

$$H—\overset{\displaystyle H}{\underset{\displaystyle H}{C}}—H \quad + 2(O{=}O) \rightarrow O{=}C{=}O + 2(H—O—H)$$

$$\Delta H = -890 \text{ kJ mol}^{-1}$$

In the reaction, the bonds in methane and in oxygen are broken: this requires energy. Carbon dioxide and water are formed by making new bonds: this releases energy. When the energy released is greater than the energy taken in, the reaction is exothermic.

When a bigger hydrocarbon molecule, like ethane, is burnt in oxygen, the reaction is more exothermic (ΔH is negative): as the hydrocarbon molecule gets bigger more covalent bonds are broken and more heat energy is released.

$$H—\overset{\displaystyle H\ \ H}{\underset{\displaystyle H\ \ H}{C—C}}—H \ + \tfrac{7}{2}(O{=}O) \rightarrow 2(O{=}C{=}O) + 3(H—O—H)$$

Ethane $\Delta H = -1560 \text{ kJ mol}^{-1}$

Other compounds also release energy when burnt in oxygen. In each case, energy is released when bonds are made with oxygen.

$$H—H + \tfrac{1}{2}(O{=}O) \Rightarrow H—O—H; \Delta H = -286 \text{ kJ mol}^{-1}$$

$$H—\overset{\displaystyle H}{\underset{\displaystyle H}{C}}—O—H \ + \tfrac{3}{2}(O{=}O) \rightarrow O{=}C{=}O + 2(H—O—H)$$

Methanol $\Delta H = -726 \text{ kJ mol}^{-1}$

Methanol, CH_3OH, does not produce as much energy as methane, CH_4, when it burns, because methanol already has a C—O bond.

Average Bond Energies

Bond energy is the heat change required to convert one mole of gaseous molecules into its constituent gaseous atoms.

When a covalent bond is broken, energy is required to break the bond into atoms.

$$A{\overset{.}{=}}B(g) \rightarrow A(g) + B(g)$$

For simple diatomic molecules like H_2, Cl_2 and others, the bond dissociation energy is the energy required to convert one mole of the gaseous molecule into gaseous atoms.

$$H{\overset{.}{=}}H \ (g) \rightarrow H(g) + H(g); \Delta H = +436 \text{ kJ mol}^{-1}$$

$$Cl{\overset{.}{=}}Cl \ (g) \rightarrow Cl(g) + Cl(g); \Delta H = +242 \text{ kJ mol}^{-1}$$

For polyatomic molecules, like methane, we have to take an average bond energy.

$$H—\overset{\displaystyle H}{\underset{\displaystyle H}{C}}—H \ (g) \rightarrow C(g) + 4H(g); \Delta H= +1648 \text{ kJ mol}^{-1}$$

140

The bond energy of each C—H bond is taken as $\frac{1}{4}$ of 1648 kJ, because four C—H bonds were broken. The bond energy for the following reactions is as follows.

$$CH_4 \rightarrow CH_3 + H; \; \Delta H = +461 \text{ kJ mol}^{-1}$$

$$CH_3 \rightarrow CH_2 + H; \; \Delta H = +403 \text{ kJ mol}^{-1}$$

$$CH_2 \rightarrow CH + H; \; \Delta H = +424 \text{ kJ mol}^{-1}$$

$$CH \rightarrow C + H; \; \Delta H = +374 \text{ kJ mol}^{-1}$$

When values from other similar molecules are taken into account, the average bond energy of a C—H bond is given as $E(\text{C—H}) = +412 \text{ kJ mol}^{-1}$.

Kilogram Calorific Values

The heat of combustion of a substance, that is the energy released when one mole of a fuel is burnt in excess oxygen, does not always give a true picture of the value of a fuel. A better picture is obtained from the energy released when a kilogram of the fuel is burnt. Remember, fuels have to be transported so the energy per unit of weight is important. The kilogram calorific values of fuels are calculated from their molar masses.

> The energy produced by the burning of a quantity of fuel is described as its **calorific value**, even though chemists no longer use the calorie as a unit of energy. The most common unit is the kilojoule.

Kilogram calorific values of some common fuels

Fuel	Kilogram calorific value / kJ kg^{-1}
Anthracite	31 400
Bituminous coal	28 000
Petrol	35 000
Ethanol	29 800
Natural gas	55 000
LPG	49 500

Social and Applied Aspects

INGREDIENTS
MAIZE, SUGAR, MALT FLAVOURING, SALT, NIACIN, IRON, VITAMIN B6, RIBOFLAVIN (B2), THIAMIN (B1), FOLIC ACID, VITAMIN D, VITAMIN B12.

NUTRITION INFORMATION

		Typical value per 100g	Per 30g Serving with 125ml of Semi-Skimmed Milk
ENERGY	kJ	1550	700 *
	kcal	370	170
PROTEIN	g	7	6
CARBOHYDRATE	g	83	31
(of which sugars)	g	(8)	(9)
(starch)	g	(75)	(22)
FAT	g	0.7	2.5 *
(of which saturates)	g	(0.2)	(1.5)
FIBRE	g	1.0	0.3
SODIUM	g	1.1	0.4
VITAMINS:		(%RDA)	(%RDA)
VITAMIN D	µg	4.2 (85)	1.3 (25)
THIAMIN (B$_1$)	mg	1.2 (85)	0.4 (30)
RIBOFLAVIN (B$_2$)	mg	1.3 (85)	0.6 (40)
NIACIN	mg	15 (85)	4.6 (25)
VITAMIN B$_6$	mg	1.7 (85)	0.6 (30)
FOLIC ACID	µg	333 (165)	110 (55)
VITAMIN B$_{12}$	µg	0.85 (85)	0.75 (75)
IRON	mg	7.9 (55)	2.4 (17)

* For whole milk increase energy by 100kJ (25kcal) and fat by 3g.
* For skimmed milk reduce energy by 70kJ (20kcal) and fat by 2g.
Contribution provided by 125ml of semi-skimmed milk:-
250kJ (60kcal) of energy, 4g of protein, 6g of carbohydrates (sugars), 2g of fat.

The calorific values of foods is determined using a bomb calorimeter (see page 144).

Kilogram calorific values and uses of some other fuels

Fuel	Molar heat of combustion / kJ mol^{-1}	Molar mass /g mol^{-1}	Kilogram calorific value / kJ kg^{-1}	Use
Hydrogen	−286	2	−143 000	Rocket fuel
Methane	−890	16	−55 600	Natural gas heaters
Butane	−2877	58	−49 603	LPG
Methanol	−726	32	−22 688	Petrol substitute

Measurement of Heat of Reaction

Heat released or absorbed during a chemical reaction is measured using a technique called calorimetry. The apparatus used is called a calorimeter.

> A **calorimeter** is an instrument which is used to measure the heat changes which accompany chemical reactions.

A simple calorimeter usually consists of a vessel of negligible heat capacity which acts as an insulator and prevents heat exchange between the system and the surroundings.

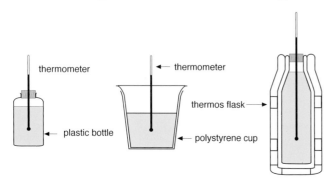

The calorimeters illustrated here are suitable for measuring heat changes in the laboratory because the heat exchange between the system and the surroundings is small enough to be ignored.

> **The unit of energy is the joule (J).** It is a rather small amount of energy. For this reason, most chemical reactions are measured in kilojoules (kJ).

Mandatory experiment 13

To determine the heat of reaction of hydrochloric acid solution with sodium hydroxide solution.

Introduction

Hydrochloric acid and sodium hydroxide solution are mixed together using a calorimeter. Temperature readings are taken before, during and after the reaction, at regular time intervals. The initial temperature and the final maximum temperature of the reaction are determined from a graph of the temperature changes plotted against time.

Safety glasses, 2 burettes, 2 beakers, polystyrene cup and lid, 1.0 mol L^{-1} hydrochloric acid solution, 1.0 mol L^{-1} sodium hydroxide solution, thermometer, stop-watch and balance.

Procedure

(1) Weigh the empty calorimeter (polystyrene cup).
(2) Using a burette, transfer 50 cm^3 of the 1.0 mol L^{-1} hydrochloric acid solution into a clean dry beaker.
(3) Using a second burette, transfer 50 cm^3 of the 1.0 mol L^{-1} sodium hydroxide solution into the calorimeter.
(4) Record the temperature of both solutions. If the temperature of the two solutions is not the same, the average temperature should be taken as the initial temperature.
(5) Add the 50 cm^3 of dilute hydrochloric acid into the 50 cm^3 of sodium hydroxide solution in the calorimeter. Insert the thermometer into the calorimeter and put the cover on the calorimeter.
(6) Stir constantly and record the temperature, to the nearest 0.1 °C, every 30 seconds for about 5–10 minutes.
(7) Copy the table below and use it to record your results.

Results

Time / minutes	Temperature / °C
0.5	
1.0	
1.5	
2.0	
2.5	
3.0	
3.5	
4.0	
4.5	
5.0	
5.5	
6.0	
6.5	
7.0	
7.5	
8.0	
8.5	
9.0	
9.5	
10.0	

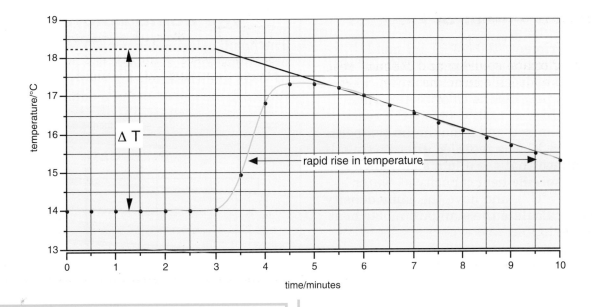

Calculations

(1) Plot a graph of temperature against time. Make a cooling correction (as shown) in order to determine the maximum rise in temperature.

(2) Reweigh the calorimeter and contents in order to determine the exact mass of reactants used. Alternatively you may assume that if exactly 50 cm³ of each reactant were used, the mass of each reactant is 50 g; the density of the dilute solutions is assumed to be equal to the density of water.

(3) The heat change for the reaction = heat gained by the solution

$$\Delta H = -mc\Delta T$$

where m = mass of reactants in kilograms
c = specific heat capacity of solutions
= heat capacity of water, because the solutions are dilute
ΔT = change in temperature (maximum temperature – initial temperature)

(4) When the heat change involved for 50 cm³ of 1.0 mol L⁻¹ acid has been determined, the standard molar heat of neutralisation is then calculated.

Possible Errors

Some assumptions were made during the experiment which are the causes of error.

1. The specific heat capacity of the calorimeter was assumed to be zero.
2. The specific heat capacity of both acid and base was assumed to be equal to the specific heat capacity of water.
3. The densities of both acid and base were assumed to be equal to the density of water.

The **specific heat capacity**, c, of a substance is the amount of energy needed to raise the temperature of 1 kg of a substance by 1 degree Kelvin.

The specific heat capacity of water is 4.2 kJ kg⁻¹ K⁻¹ or, in words, 4.18 kilojoules per kilogram per degree Kelvin.

Standard Heat Change for a Reaction

This is a general definition, particular examples of which are (a) heat of combustion, (b) heat of neutralisation and (c) heat of formation.

The heat change for a reaction refers to the amounts of the substances shown in the particular equation, at a temperature of 298 K and at a pressure of 10⁵ Pa, with the substances in the physical states normal under these conditions.

$$HCl(aq) + NaOH(aq) = NaCl(aq) + H_2O(l);$$
$$\Delta H = -57 \text{ kJ mol}^{-1}$$

······ **Example 5.4** ······

In an experiment, 100 cm³ of 1.0 mol L⁻¹ sodium hydroxide solution and 100 cm³ of 1.0 mol L⁻¹ hydrochloric acid solution were mixed in a polystyrene calorimeter. The temperature was 14.0 °C before the reactants were mixed and a maximum temperature of 20.8 °C was recorded after mixing. If it is assumed that the dilute solutions have the same specific heat capacity as water (4.2 kJ kg⁻¹ °C⁻¹), and have the same density as water, calculate the standard molar heat of reaction.

Heat change

$$= \Delta H = -mc\Delta T$$

$$= -0.2 \text{ kg} \times 4.2 \text{ kJ kg}^{-1} \text{ °C}^{-1} \times 6.8 \text{ °C}$$

$$= -5.712 \text{ kJ}$$

The number of moles used, $n = cV$ (concentration in moles per litre × volume in litres)

$$n(\text{HCl}) = cV = 1.0 \text{ mol L}^{-1} \times 0.1 \text{ L HCl}$$

$$= 0.1 \text{ mol HCl}$$

$$n(\text{NaOH}) = cV = 1.0 \text{ mol L}^{-1} \times 0.1 \text{ L NaOH}$$

$$= 0.1 \text{ mol NaOH}$$

The heat change for 1 mol of reactants,

$$\Delta H_m = \frac{\Delta H(\text{reaction})}{n}$$

$$\Delta H_m = \frac{-5.712 \text{ kJ}}{0.1 \text{ mol}} = -57.12 \text{ kJ mol}^{-1}$$

···

James Dewar
1842–1923

James Dewar was born in the little Scottish town of Kincardine-on-Forth. He was a small man with a quick temper, difficult to work with, but was an immensely gifted experimenter and demonstrator, able to entrance the most critical audience. He invented the vacuum flask to help him in his low temperature studies. The flask is still known as a 'Dewar flask', many forms of which are used today by thermochemists everywhere. As a child he was a keen flautist. Due to a skating accident he was unable to attend school and decided to spend his time making violins. Dewar always believed that his exceptional laboratory skills were based on his early training in a carpenter's shop. He achieved prominence as a scientist in describing possible arrangements of the atoms in benzene. This impressed the German chemist, Kekulé, who invited Dewar to work with him. He became Resident Professor at the Royal Institution and devoted most of his time to endeavouring to liquefy gases. He successfully liquefied hydrogen but was unable to liquefy helium, partly because he could not obtain enough supplies of the gas.

The Bomb Calorimeter

The bomb calorimeter is used to measure heats of combustion accurately. The calorific values of foods and the kilogram calorific values of fuels are measured using a bomb calorimeter.

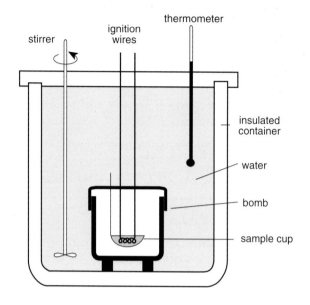

Bomb calorimeter

The calorimeter consists of a steel-walled container or 'bomb' in which a weighed sample of the substance to be tested is burnt in excess oxygen under stated conditions of temperature and pressure.

Oxygen is pumped through a valve. The bomb is placed in an insulated calorimeter which contains water. The water in the calorimeter is stirred constantly with an electrical stirrer. A small electrical heater is used to ignite the sample in the bomb to start the combustion process.

The energy change is determined by measuring the increase in the temperature of the water and the calorimeter parts.

The amount of heat produced by the reaction can be calculated if the following are known:
• amount of substance combusted
• temperature rise of the water and calorimeter
• heat capacity of the water and the calorimeter.

The heat change is given as

$$\Delta H = -mc\Delta T$$

where m is the mass, c is the specific heat capacity and ΔT is the change in temperature.

Calculation of Heat of Reaction Using Heats of Formation

We use the term 'above sea level' without knowing its actual distance from the centre of the Earth. However, we use the term 'above sea level' as a standard for comparison for calculating heights of aeroplanes and so on.

In chemistry, we cannot measure the energy content of a substance, H. However, we can calculate the heat change of a reaction, ΔH, from values given for the heats of formation of substances.

Standard heats of formation, like sea level, are a standard used for comparing heats of reaction.

Standard Heat of Formation of a Substance

This is the amount of heat in joules evolved or absorbed when one mole of a substance is formed in its standard state from its elements in their standard states and under standard conditions.

$$H_2(g) + \tfrac{1}{2}O_2(g) = H_2O(l); \Delta H_f = -286 \text{ kJ mol}^{-1}$$

The heat of formation of an element is zero.

Calculation of heat of reaction using heats of formation

Calculate the standard heat of reaction for the reaction

$$2NO_2(g) = N_2O_4(g) \; \Delta H = ?$$

using the heats of formation given below:

[1]
$$\tfrac{1}{2}N_2(g) + O_2(g) = NO_2(g); \Delta H_f = +33.2 \text{ kJ mol}^{-1}$$

[2]
$$N_2(g) + 2O_2(g) = N_2O_4(g); \Delta H_f = +9.2 \text{ kJ mol}^{-1}$$

Heat of reaction = heat of formation of products – heat of formation of reactants

$$\Delta H = \Delta H_f(\text{products}) - \Delta H_f(\text{reactants})$$

$$= \Delta H_f(N_2O_4) - 2 \times \Delta H_f(NO_2) \longleftarrow \text{remember there are 2 moles of } NO_2$$

$$= +9.2 \text{ kJ mol}^{-1} - 2 \times (+33.2) \text{ kJ mol}^{-1}$$

$$= 9.2 \text{ kJ mol}^{-1} - 66.4 \text{ kJ mol}^{-1}$$

$$= -57.2 \text{ kJ mol}^{-1}$$

$$\Delta H = \Delta H_f(\text{products}) - \Delta H_f(\text{reactants})$$

●●●●●●●●●●●●●●●●●●●●●●●●

Heat of formation / kJ mol^{-1}	
Aluminium(III) oxide, Al_2O_3 (s)	–1676
Ammonia, NH_3 (g)	–46
Butane, C_4H_{10} (g)	–126
Calcium carbonate, $CaCO_3$ (s)	–1207
Carbon dioxide, CO_2 (g)	–394
Carbon monoxide, CO (g)	–110
Carbon, C (s)	0
Chloroethane, C_2H_5Cl (l)	–136
Ethane, C_2H_6 (g)	–85
Ethanol, C_2H_5OH (l)	–278
Hydrogen chloride, HCl (g)	–92
Hydrogen sulphide, H_2S (g)	–21

Heat of formation / kJ mol$^-$	
Hydrogen, H_2 (g)	0
Iron(III) oxide, Fe_2O_3 (s)	–822
Methane, CH_4 (g)	–75
Methanol, CH_3OH (l)	–215
Nitrogen(IV) oxide, NO_2 (g)	33.2
Nitrogen(IV) oxide, N_2O_4 (g)	9.2
Nitrogen, N_2 (g)	0
Oxygen, O_2 (g)	0
Phosphorus(V) chloride, PCl_5 (s)	–444
Phosphoryl chloride, $POCl_3$ (l)	–597
Propane, C_3H_8 (g)	–104
Sulphur dioxide, SO_2 (g)	–297
Water, H_2O (l)	–286

····· **Example 5.6** ·····

Calculation of heat of reaction using heats of formation

Calculate the standard heat of reaction for the reaction

$$CH_3OH(l) + \tfrac{3}{2}O_2(g) = CO_2(g) + 2H_2O(l)$$

given that the standard heats of formation of methanol, carbon dioxide, water and oxygen are -215 kJ mol^{-1}, -393 kJ mol^{-1}, -286 kJ mol^{-1} and 0, respectively.

[1]
$$C(graphite) + 2H_2(g) + \tfrac{1}{2}O_2(g) = CH_3OH(l);$$
$$\Delta H_f = -215 \text{ kJ mol}^{-1}$$

[2]
$$C(graphite) + O_2(g) = CO_2(g); \Delta H_f = -394 \text{ kJ mol}^{-1}$$

[3]
$$H_2(g) + \tfrac{1}{2}O_2(g) = H_2O(l); \Delta H_f = -286 \text{ kJ mol}^{-1}$$

Heat of reaction = heat of formation of products – heat of formation of reactants

$$\Delta H = \Delta H_f(\text{products}) - \Delta H_f(\text{reactants})$$

$$= \Delta H_f(CO_2) + 2 \times \Delta H_f(H_2O) - \Delta H_f(CH_3OH)$$

$$- \tfrac{3}{2}\Delta H_f(O_2) \longleftarrow \text{remember the number of moles}$$

$$= (-394 + 2 \times (-286)) \text{ kJ mol}^{-1}$$
$$- (-215 + 3/2(0)) \text{ kJ mol}^{-1}$$

$$= -751 \text{ kJ mol}^{-1}$$

····· **Example 5.7** ·····

Calculation of heat of reaction using heats of formation

Calculate the standard heat of reaction for the reaction

$$C_2H_5OH(l) + PCl_5(s) = C_2H_5Cl(l) + POCl_3(l) + HCl(g)$$

given that the standard heats of formation are:

$$\Delta H_f(C_2H_5OH) = -278 \text{ kJ mol}^{-1}$$
$$\Delta H_f(PCl_5) = -444 \text{ kJ mol}^{-1}$$
$$\Delta H_f(C_2H_5Cl) = -136 \text{ kJ mol}^{-1}$$
$$\Delta H_f(POCl_3) = -597 \text{ kJ mol}^{-1}$$
$$\Delta H_f(HCl) = -92 \text{ kJ mol}^{-1}$$

Heat of reaction = heat of formation of products – heat of formation of reactants

$$\Delta H = \Delta H_f(\text{products}) - \Delta H_f(\text{reactants})$$

$$= \Delta H_f(C_2H_5Cl) + \Delta H_f(POCl_3) + \Delta H_f(HCl)$$
$$- \Delta H_f(C_2H_5OH) - \Delta H_f(PCl_5)$$

$$= (-136 - 597 - 92) \text{ kJ mol}^{-1}$$
$$- (-278 - 444) \text{ kJ mol}^{-1}$$

$$= -103 \text{ kJ mol}^{-1}$$

Law of conservation of energy: Energy cannot be created or destroyed but can be converted from one form into another.

····· **Example 5.8** ·····

Calculation of heat of formation

Calculate the standard heat of formation of butane, $\Delta H_f(C_4H_{10})$, $4C(s) + 5H_2(g) = C_4H_{10}(g)$
given that the standard heats of formation of carbon dioxide and water are

$$C(s) + O_2(g) = CO_2(g); \Delta H_f = -393 \text{ kJ mol}^{-1}; \quad H_2(g) + \tfrac{1}{2}O_2(g) = H_2O(l); \Delta H_f = -286 \text{ kJ mol}^{-1}$$

and that the heat of combustion of butane is -2878 kJ mol^{-1}.

$$C_4H_{10}(g) + \tfrac{13}{2}O_2(g) = 4CO_2(g) + 5H_2O(l); \Delta H_c = -2878 \text{ kJ mol}^{-1}$$

Using the heat of combustion of butane as the heat of reaction, insert the known heats of formation and the heat of combustion of butane into the equation below.

Heat of combustion = heat of formation of products – heat of formation of reactants

$$\Delta H_c = \Delta H_f(\text{products}) - \Delta H_f(\text{reactants})$$

$$\Delta H_c(C_4H_{10}) = 4 \times \Delta H_f(CO_2) + 5 \times \Delta H_f(H_2O) - \Delta H_f(C_4H_{10}) - \tfrac{13}{2}\Delta H_f(O_2)$$

$$-2878 \text{ kJ mol}^{-1} = (4 \times -393 + 5 \times -286) \text{ kJ mol}^{-1} - [\Delta H_f(C_4H_{10}) - \tfrac{13}{2} \times 0] \text{ kJ mol}^{-1} \longleftarrow \text{remember}$$
$$\text{the heats of formation of elements} = 0$$

$$-2878 \text{ kJ mol}^{-1} = -3002 \text{ kJ mol}^{-1} - \Delta H_f(C_4H_{10})$$

$$\Delta H_f(C_4H_{10}) = -3002 \text{ kJ mol}^{-1} + 2878 \text{ kJ mol}^{-1} = -124 \text{ kJ mol}^{-1}$$

· · · · · **Example 5.9** · · · · ·

Calculation of heat of formation

Calculate the standard heat of formation of iron(III) oxide, $\Delta H_f(Fe_2O_3)$,

$$2Fe(s) + \tfrac{3}{2}O_2(g) = Fe_2O_3(g)$$

given the standard heat of formation of aluminium(III) oxide, Al_2O_3,

$$2Al(s) + \tfrac{3}{2}O_2(g) = Al_2O_3(s); \ \Delta H_f = -1676 \text{ kJ mol}^{-1}$$

and the following heat of reaction:

$$2Al(s) + Fe_2O_3(s) = 2Fe(s) + Al_2O_3(s);$$
$$\Delta H = -854 \text{ kJ mol}^{-1}$$

Using the heat of reaction, insert the known heats of formation and the heat of the reaction into the equation below.

Heat of reaction = heat of formation of products − heat of formation of reactants

$$\Delta H = \Delta H_f(\text{products}) - \Delta H_f(\text{reactants})$$

$$\Delta H = 2 \times \Delta H_f(Fe) + \Delta H_f(Al_2O_3) - 2 \times \Delta H_f(Al) - \Delta H_f(Fe_2O_3)$$

$$-854 \text{ kJ mol}^{-1} = (2 \times 0 - 1676) \text{ kJ mol}^{-1} - [2 \times 0 + \Delta H_f(Fe_2O_3)] \text{ kJ mol}^{-1}$$

$$-854 \text{ kJ mol}^{-1} = -1676 \text{ kJ mol}^{-1} - \Delta H_f(Fe_2O_3)$$

$$\Delta H_f(Fe_2O_3) = -1676 \text{ kJ mol}^{-1} + 854 \text{ kJ mol}^{-1}$$

$$= -822 \text{ kJ mol}^{-1}$$

· ·

Hess's Law of Constant Heat Summation

Many heats of reaction cannot be measured directly.

The reaction between carbon and hydrogen to form the gas methane may seem a simple process. However, during the reaction many other compounds may form as products. This makes it difficult to measure the heat of formation of methane.

Hess's law, another form of the law of conservation of energy, provides a means of measuring many heats of reaction.

Hess's law

The total energy change accompanying a chemical change is independent of the route by which the chemical change takes place, provided that the initial and final conditions are the same.

Suppose that, on the one hand, A reacts with B to form C directly or that, on the other hand, A and B react to form D and E which then form C. In this case, whichever pathway or route is chosen, the energy change, ΔH, remains the same.

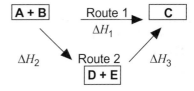

The overall energy for routes 1 and 2 is the same:

$$\Delta H_1 = \Delta H_2 + \Delta H_3$$

· · · · · · **Example 5.10** · · · · ·

Heat of formation using Hess's law

Hess's law provides a method of calculating the heat of formation of methane gas from the following heats of reaction:

[1]
$$CH_4(g) + 2O_2(g) = CO_2(g) + 2H_2O(l);$$
$$\Delta H_c = -890 \text{ kJ mol}^{-1}$$

[2]
$$C(\text{graphite}) + O_2(g) = CO_2(g); \ \Delta H_f = -394 \text{ kJ mol}^{-1}$$

[3]
$$H_2(g) + \tfrac{1}{2}O_2(g) = H_2O(l); \ \Delta H_f = -286 \text{ kJ mol}^{-1}$$

The equation required is:

$$C(\text{graphite}) + 2H_2(g) = CH_4(g); \ \Delta H_f = ?$$

Equations [1], [2] and [3] can be rearranged using the basic arithmetical rules for multiplication, addition and subtraction. It is suggested that the required reactants and products be highlighted as shown. The reactants and products not required can then be easily deleted using the basic arithmetical rules.

[1] − [2]

$$CH_4(g) + 2O_2(g) = CO_2(g) + 2H_2O(l);$$
$$\Delta H = -890 \text{ kJ mol}^{-1}$$

$$C(\text{graphite}) + O_2(g) = CO_2(g); \ \Delta H = -394 \text{ kJ mol}^{-1}$$

───────────────────────

$$CH_4(g) + O_2(g) = C(\text{graphite}) + 2H_2O(l);$$
$$\Delta H = -496 \text{ kJ mol}^{-1} \hspace{2cm} [4]$$

[4] – [3] × 2

$CH_4(g) + O_2(g) = C(graphite) + 2H_2O(l);$
$$\Delta H = -496 \text{ kJ mol}^{-1}$$

$2H_2(g) + O_2(g) = 2H_2O(l); \Delta H = -572 \text{ kJ mol}^{-1}$

$CH_4(g) = C(graphite) + 2H_2(g); \Delta H = +76 \text{ kJ mol}^{-1}$

This equation can be reversed to give the required equation:

$C(graphite) + 2H_2(g) = CH_4(g); \Delta H = -76 \text{ kJ mol}^{-1}$

• •

• • • • • • **Example 5.11** • • • • • •

Heat of formation using Hess's law

Calculate the standard heat of formation of ammonia, NH_3, from the following data.

[1]
$4NH_3(g) + 3O_2(g) = 2N_2(g) + 6H_2O(l);$
$$\Delta H = -1532 \text{ kJ mol}^{-1}$$

[2]
$H_2(g) + \frac{1}{2}O_2(g) = H_2O(l); \Delta H = -286 \text{ kJ mol}^{-1}$

The required equation is:

$\frac{1}{2}N_2(g) + \frac{3}{2}H_2(g) = NH_3(g); \Delta H = ?$

[1] – [2] × 6

$4NH_3(g) + 3O_2(g) = 2N_2(g) + 6H_2O(l);$
$$\Delta H = -1532 \text{ kJ mol}^{-1}$$

$6H_2(g) + 3O_2(g) = 6H_2O(g); \Delta H = -1716 \text{ kJ mol}^{-1}$

$4NH_3(g) = 2N_2(g) + 6 H_2(l); \Delta H = 184 \text{ kJ mol}^{-1}$

The above reaction is reversed and divided by 4 to give the required equation:

$\frac{1}{2}N_2(g) + \frac{3}{2}H_2(g) = NH_3(g); \Delta H = -46 \text{ kJ mol}^{-1}$

• •

Exercise 5.4

•

(1) List some examples of chemical reactions which cause a change in temperature.

(2) (a) Explain the meaning of the following terms:
(i) exothermic (ii) endothermic reactions.
(b) Give an example of (i) an exothermic reaction (ii) an endothermic reaction, where a change of state is involved.

(3) 'During a chemical reaction, heat is transferred from the surroundings to the system.'
(a) What is the meaning of the underlined terms?
(b) Use diagrams to illustrate how heat is transferred from the surroundings to the system.

(4) (a) Give two examples of exothermic reactions and two examples of endothermic reactions.
(b) Explain why all combustion reactions are exothermic.
(c) Write a thermochemical equation showing how methane, CH_4, burns in oxygen.

(5) (a) What is meant by the 'standard heat change' for a reaction?
(b) Give an example of such a heat change.

(6) (a) What is the meaning of ΔH?
(b) Use energy level diagrams to illustrate the difference between an exothermic reaction and an endothermic reaction.

(7) (a) Define the following terms:
(i) bond energy (ii) molar heat of combustion (iii) calorimeter (iv) joule (v) specific heat capacity.
(b) Describe the energy changes involved when chemical bonds are broken and new bonds made during a chemical reaction.
(c) What is meant by 'average bond energy of methane'?

(8) When hydrogen gas is burnt in chlorine gas, hydrogen chloride gas is formed.

$$H_2(g) + Cl_2(g) \rightarrow 2HCl(g)$$

Use structural formulae to show which bonds are broken and which bonds are formed.

(9) Alkanes form a homologous series of general formula C_2H_{2n+2}, where n is the number of carbon atoms present in each respective alkane. The standard molar heats of combustion of the first four members of that series are:
Methane, CH_4, -890 kJ mol^{-1}
Ethane, C_2H_6, $-1560 \text{ kJ mol}^{-1}$
Propane, C_3H_8, $-2220 \text{ kJ mol}^{-1}$
Butane, C_4H_{10}, $-2877 \text{ kJ mol}^{-1}$

(a) Write a balanced thermochemical equation for the combustion of each alkane.

(b) Plot a graph of the heat of combustion of each alkane against the number of carbon atoms in each alkane.

(c) From your graph estimate the molar heat of combustion of pentane, C_5H_{12}.

(10) (a) Describe how a simple calorimeter works.

(b) How is the heat change in a chemical reaction calculated?

(11) (a) What is a bomb calorimeter?

(b) Draw a labelled diagram of a bomb calorimeter.

(c) How is a bomb calorimeter more accurate than a polystyrene cup used as a calorimeter?

(d) Give two examples of how a bomb calorimeter is used to determine kilogram calorific values.

(12) Define each of the following:

(a) law of conservation of energy (b) Hess's law (c) standard molar heat of formation.

(13) Write thermochemical equations that correspond to the standard molar heats of formation of the following:

(a) $\Delta H_f [CH_4(g)] = -75$ kJ mol^{-1}

(b) $\Delta H_f [C_2H_6(g)] = -85$ kJ mol^{-1}

(c) $\Delta H_f [C_2H_5OH(l)] = -278$ kJ mol^{-1}

(d) $\Delta H_f [H_2O(g)] = -241$ kJ mol^{-1}

(e) $\Delta H_f [NH_3(g)] = -46$ kJ mol^{-1}

(f) $\Delta H_f [CaCO_3(s)] = -1207$ kJ mol^{-1}

(14) Write thermochemical equations that correspond to the standard molar heats of formation of the following:

(a) $\Delta H_f(C_2H_5OH) = -278$ kJ mol^{-1}

(b) $\Delta H_f(PCl_5) = -444$ kJ mol^{-1}

(c) $\Delta H_f(C_2H_5Cl) = -136$ kJ mol^{-1}

(d) $\Delta H_f(POCl_3) = -597$ kJ mol^{-1}

(e) $\Delta H_f(HCl) = -92$ kJ mol^{-1}

(15) Calculate the standard molar heat change for the reaction:

$$2H_2S(g) + 3O_2(g) = 2H_2O(l) + 2SO_2(g)$$

given the following standard molar heats of formation:

$H_2(g) + S(s) = H_2S(g); \Delta H_f = -21$ kJ mol^{-1}

$H_2(g) + \frac{1}{2}O_2(g) = H_2O(l); \Delta H_f = -286$ kJ mol^{-1}

$S(s) + O_2(g) = SO_2(g); \Delta H_f = -297$ kJ mol^{-1}

(16) Calculate the standard molar heat change for the reaction:

$$2H_2S(g) + SO_2(g) = 3S(s) + 2H_2O(l)$$

given that the standard molar heats of formation of hydrogen sulphide, $H_2S(g)$, sulphur dioxide, $SO_2(g)$, and water, $H_2O(l)$, are 21, -297 and -286 kJ mol^{-1}, respectively.

(17) Calculate the standard molar heat of formation of carbon disulphide, $CS_2(l)$, given the following standard molar heat changes

$CS_2(s) + 3O_2(g) = CO_2(g) + 2SO_2(g);$
$\Delta H_c = -1075$ kJ mol^{-1}

$C(s) + O_2(g) = CO_2(g); \Delta H_f = -393$ kJ mol^{-1}

$S(s) + O_2(g) = SO_2(g); \Delta H_f = -297$ kJ mol^{-1}

(18) Calculate the standard molar heat of formation of ethanol, $C_2H_5OH(l)$, from the following standard heat changes:

$C(s) + O_2(g) = CO_2(g); \Delta H_f = -394$ kJ mol^{-1}

$H_2(g) + \frac{1}{2}O_2(g) = H_2O(l); \Delta H_f = -286$ kJ mol^{-1}

$C_2H_5OH(l) + \frac{3}{2}O_2(g) = 2CO_2(g) + 3 H_2O(l);$
$\Delta H_c = -1370$ kJ mol^{-1}

(19) Calculate the standard heat changes for the following reactions from the heats of formation given below:

(a) $2CO(g) + O_2(g) = 2CO_2(g)$

(b) $CH_3OH(l) + \frac{3}{2}O_2(g) = CO_2(g) + 2H_2O(l)$

(c) $H_2(g) + \frac{1}{2}O_2(g) = H_2O(l)$

(d) $2SO_2(g) + O_2(g) = 2SO_3(g)$

(e) $N_2(g) + 3H_2(g) = 2NH_3(g)$

(f) $3C(s) + 4H_2(g) = C_3H_8(g)$

$\Delta H_f[CO(g)] = -110$ kJ mol^{-1}
$\Delta H_f[O_2(g)] = 0$
$\Delta H_f[CO_2(g)] = -393$ kJ mol^{-1}
$\Delta H_f[CH_3OH(l)] = -234$ kJ mol^{-1}
$\Delta H_f[H_2(g)] = 0$
$\Delta H_f[SO_2(g)] = -297$ kJ mol^{-1}
$\Delta H_f[SO_3(g)] = -395$ kJ mol^{-1}
$\Delta H_f[N_2(g)] = 0$
$\Delta H_f[H_2O(g)] = -286$ kJ mol^{-1}
$\Delta H_f[NH_3(g)] = -46$ kJ mol^{-1}
$\Delta H_f[C(s)] = 0$
$\Delta H_f[C_3H_8(g)] = -104$ kJ mol^{-1}

(20) In an experiment to determine the molar heat of neutralisation of sodium hydroxide solution by nitric acid solution, 250 cm^3 of 0.2 mol L^{-1} NaOH was mixed with 250 cm^3 of 0.2 mol L^{-1} HNO$_3$. The maximum temperature change observed was 1.35 °C. If the specific heat capacity of water is 4.2 kJ kg^{-1} °C^{-1}

(a) calculate the molar heat change

(b) state clearly any assumptions you have made in your calculations.

5.5 OIL REFINING AND ITS PRODUCTS

Approximately 90% by weight of all organic chemicals come from crude oil (often called petroleum) and natural gas. The chemicals made from these sources are called petrochemicals. Many inorganic chemicals, such as sulphur, helium and ammonia, are derived from crude oil or petroleum. Most petrochemicals are derived from relatively simple molecules, such as methane, ethene, propene, the butenes and butadiene. The main feedstocks for the chemical industry are the naphtha and gas oil fractions from crude oil and natural gas; these consist predominantly of hydrocarbons.

Composition of Crude Oil

Crude oil is a complex mixture of hydrocarbon molecules, whose composition varies depending on where in the world the oil is found. The hydrocarbons in crude oil range from small simple molecules such as methane and ethene, which contain only a few atoms, to large complex molecules containing a large number of atoms. The molecules containing few atoms are volatile and are used as fuels and as chemical feedstocks, while those with many atoms are heavy and viscous and are often used for making road surfaces and other similar uses.

Fractionation: Separation by Fractional Distillation

The first step in the treatment of crude oil at a refinery is to sort out the molecules according to their size by distillation. This process sorts out the mixture into various fractions by making use of the different boiling points of the components.

The crude oil is first heated in a large furnace to about 400 °C. The resulting mixture is fed into large fractionating towers which may be up to 80 m tall.

The heaviest vapours (those with high relative molecular masses) condense near the bottom of the tower, while only the lightest and most volatile components (those with low relative molecular masses) reach the top. The tower is fitted with thirty or forty trays at different heights in the column to collect the various fractions as they condense. The trays are fitted with bubble caps or jet trays which enable the rising vapour to pass up to the next tray and at the same time allow the condensed liquid to fall through onto the lower tray. The products in the required boiling range are then drawn from selected trays.

Major Fractions and their Uses

(1) **Gas** is the uncondensed vapour from the top of the column. Some of this is used as refinery fuel, while the remainder is liquefied and sold as liquid propane and butane (LPG).
(2) **Naphtha** is the fraction which boils between 40 and 180 °C. It is used for blending into petrol (gasoline) and is the main feedstock for the petrochemical industry.
(3) **Kerosene** is used for jet fuels and for domestic heating.
(4) **Gas oil and heavy gas oil** are used as diesel oil or as feedstocks for catalytic cracking.
(5) **Residue** is used as a heavy industrial fuel or is further refined by vacuum distillation for catalytic cracking and for lubricating oils.

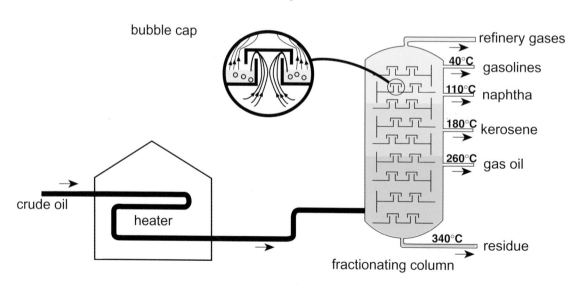

fractionating column

Natural Gas

Natural gas, like crude oil, is a mixture of hydrocarbons. As with crude oil, the composition of natural gas varies depending on where it is found. Its main constituents are methane, ethane, propane and butane. The North Sea and Kinsale natural gas fields contain approximately 95% methane, whereas the natural gas fields in the USA contain large amounts of propane and butane. Some natural gas fields, such as Lacq in France, contain a large amount of sulphur dioxide, which is an important source of sulphur. Certain American fields contain up to 5% helium, and are an important source of this gas.

Uses of Natural Gas Components

• **Methane** is used mainly in the formation of synthesis gas. It is used to make methanol, ethyne and hydrogen cyanide. It is also used as a domestic and industrial fuel. Natural gas is fairly odourless. For safety reasons, before it is distributed to homes and factories, compounds with a strong odour, called mercaptans, are added to the natural gas. This ensures that gas leaks are noticed by their strong odour.
• **Ethane** is converted into ethene, one of the most important raw materials in the chemical industry.
• **Propane and butane** are high quality fuels; the majority is bottled and sold as liquefied petroleum gas (LPG) in Europe, while in the USA the majority is used as a chemical feedstock.
• **Sulphur dioxide** is converted into sulphur and used in the manufacture of sulphuric acid.

Further Purification

Distillation alone does not produce all of the required products from crude oil. Many of the fractions may have to be adjusted in terms of quantity and quality in order to satisfy environmental and commercial requirements.

For instance, the naphtha fraction may be quite unsuitable for use as petrol and need to be further refined to create the correct blend.

Auto-Ignition (Engine 'Knocking') and Octane Number

When the petrol and oxygen mixture ignites and explodes evenly and at the correct time the engine runs smoothly. If this happens too soon in the combustion cycle, auto-ignition occurs: this is called 'pinking' or 'knocking'. A measure of how smoothly petrol burns is given by its octane number. Petrol mixtures rich in straight chain alkanes, such as hexane, ignite too quickly and explode rapidly resulting in uneven combustion. Petrol mixtures which are rich in branched chain alkanes, such as 2,2,4-trimethylheptane (isooctane), burn smoothly and efficiently. 2,2,4-trimethylheptane is given an octane number of 100 and pentane is given a rating of 0.

2,2,4-trimethylpentane: **Octane number 100**: Low tendency to auto-ignite

Heptane: **Octane number 0**: High tendency to auto-ignite

For some years, auto-ignition has been overcome by putting lead additives into the petrol to enable the fuel to burn evenly. This causes a problem because the lead compounds which pass into the air in the exhaust gases are toxic. Unleaded petrol is more expensive to produce because it contains a blend which is higher in branched chain alkanes.

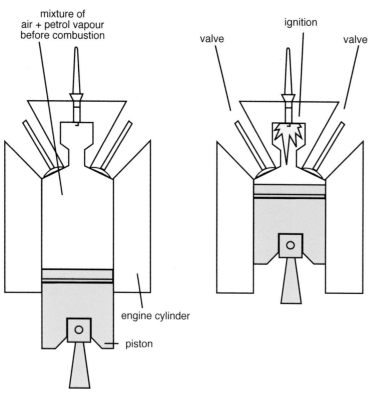

mixture of air + petrol vapour before combustion

engine cylinder

piston

Piston compresses the mixture of petrol and air

ignition

valve valve

Maximum compression of the petrol and air mixture coincides with ignition

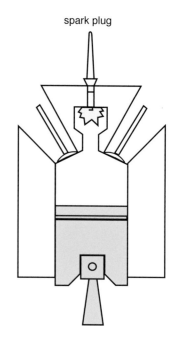

spark plug

Combustion occurs before maximum compression: this is called 'knocking'

Alternatives to Lead Additives in Petrol

Petrol companies increase the octane rating of petrol by blending different hydrocarbons together to give the best octane rating. In general, the shorter the chain of the alkane, the higher the octane number. However, this presents a problem as too many short chain alkanes, such as butane, would make the petrol too volatile. Octane number is also increased by branched chain alkanes, such as 2,2,4-trimethylheptane.

Octane number is increased by adding:
- short chain alkanes
- branched chain alkanes
- ring compounds.

The tendency of a fuel to auto-ignite is measured by its **octane number**.

Petrol is refined to increase its octane rating by:
- catalytic cracking
- reforming
- isomerisation
- addition of oxygenates.

Cracking

The amount of volatile components in crude oil is increased by breaking the large molecules up into smaller molecules. This process, where large molecules are rearranged into smaller ones, is called cracking. Cracking converts large hydrocarbons into more useful molecules such as methane, ethene, propene and buta-1,3-diene. Many of the shorter chain alkanes tend to be highly branched, which is useful in blending petrol.

The two methods of cracking are described below.

(a) Steam Cracking

In steam cracking, the naphtha or gas oil is chemically converted by a supply of heat energy alone. Steam is added to dilute the mixture. The molecules are 'cracked' or broken up into the required products.

The naphtha or gas oil feedstock is fed into vertical tubes in large ovens; the oven temperature is 1050 °C which heats the gaseous mixture in the tubes above the cracking temperature of 800–850 °C. The heat breaks the large molecules up into molecular fragments which

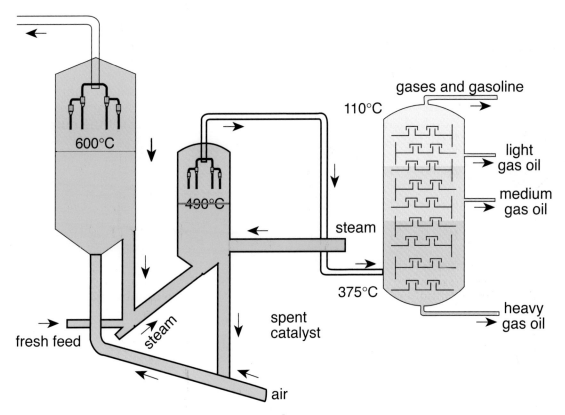

are extremely reactive radicals. These radicals not only can react with each other but also can attack all the original molecules in the feedstock. This means that several possible unwanted products could be formed. The addition of steam dilutes the reactants and minimises collisions between the reactive fragments. The fragments are also fed through the tubes in the oven very quickly; this also prevents formation of possible unwanted products.

Consider a fairly large hydrocarbon molecule such as decane:

$$
\begin{array}{cccccccccc}
H & H & H & H & H & H & H & H & H & H \\
| & | & | & | & | & | & | & | & | & | \\
H-C & -C & -C & -C & -C & -C & -C & -C & -C & -C-H \\
| & | & | & | & | & | & | & | & | & | \\
H & H & H & H & H & H & H & H & H & H
\end{array}
$$

decane

$$
\begin{array}{ccccccc}
& H & H & H & H & H & H & H \\
& | & | & | & | & | & | & | \\
\longrightarrow H-C & -C & -C & -C & -C & -C & -C-H \\
& | & | & | & | & | & | & | \\
& H & H & H & H & H & H & H
\end{array}
$$

heptane

$$
\begin{array}{cc}
H & H \\
\diagdown & \diagup \\
+ \quad C = C \\
\diagup & \diagdown \\
H & CH_3
\end{array}
$$

propene

Here, the main products are heptane and propene. Heptane is suitable for use in petrol because it has better anti-knock properties than those hydrocarbons produced by distillation. Propene is an extremely important chemical feedstock.

(b) Catalytic Cracking

Catalytic cracking requires a temperature of about 500 °C and a silica/alumina catalyst. The catalyst allows the feedstock to be cracked at a lower temperature. During the cracking reaction the catalyst becomes coated with carbon which deactivates it. The catalyst is drawn off from the bottom of the reactor and fed into a large burner where the carbon is burnt off and the catalyst is reactivated.

Reforming and Ring Formation

Reforming increases the anti-knock value of petrol components by altering the shape of the hydrocarbon molecules present in naptha.

In reforming, the molecules which are present in naphtha (mainly straight chain and branched chain hydrocarbons) are rearranged and reformed using hydrogen to produce a mixture of aromatic compounds and branched chain hydrocarbons. These have higher octane numbers and are more suitable chemical feedstocks.

hexane
Octane number 25 — reformed → cyclohexane **Octane number 83** → benzene **o.n. > 100**

heptane
0 — ring formation → methylbenzene **o.n. > 100**

Isomerisation

Isomerisation involves breaking molecules, usually straight chain alkanes, up into pieces and rejoining them again to produce new molecules, which are usually branched chain alkanes. This increases their octane rating.

pentane
o.n. 62 — isomerisation → 2-methylbutane **o.n. 93**

hexane
o.n. 25 — isomerisation → 2-methylpentane **o.n. 73**

Adding Oxygenates

Straight run gasoline has an octane number of approximately 70. As we have seen this can be increased by cracking, reforming and isomerisation. Petrol blenders can also increase the octane number of petrol by adding oxygen compounds (oxygenates) to the petrol. Two types of compounds are commonly used: alcohols and ethers. The most common alcohol used is methanol (o.n. 114) and the most common ether used is MTBE (methyl tertiary butyl ether (o.n. 118)). The amount of oxygenates added to the petrol depends on whether it is cheaper to raise the octane number by cracking, reforming and isomerisation or by adding oxygenates. It should be remembered that oxygenates cause less pollution by reducing the level of carbon monoxide in the exhaust fumes. However, a perfect blend is difficult to achieve.

Exercise 5.5

•

(1) Crude oil is a complex mixture of hydrocarbon molecules.
(a) Describe clearly how it is broken up into smaller fractions.
(b) Name the major fractions. State the approximate number of carbon atoms in each fraction and give one use for each fraction.

(2) (a) What is a petrochemical? Name some organic and some inorganic chemicals which are obtained from petrochemical sources.
(b) What is a chemical feedstock? Name the main chemical feedstocks.

(3) (a) Natural gas, like petroleum, is a mixture. Name the main components of the mixture and state the principal uses of each component.
(b) Why are mercaptans added to natural gas?

(4) (a) How is crude oil purified and processed?
(b) Why do the fractions need to be purified and processed after fractional distillation?
(c) Describe clearly the main methods used to 'crack' a large hydrocarbon molecule.
(d) Name the principal products formed when decane is cracked.
(e) How are unwanted products kept to a minimum during cracking?

(5) (a) What is meant by 'auto-ignition'?
(b) How is auto-ignition prevented in an internal combustion engine?
(c) How are octane numbers compared?

(6) The octane number of petrol can be increased by four main methods.
(a) Name each method.
(b) Write a brief note on each method.
(c) How does cost influence the type of blend used?

(7) (a) Why is lead used in petrol?
(b) Why is the use of leaded petrol prohibited in some countries?
(c) How is lead-free petrol blended?

(8) (a) What are oxygenates?
(b) Why are they added to petrol?
(c) Sometimes, oxygenates are used in the blending process. Why?

(9) (a) Why is petrol blended?
(b) How does reforming increase the anti-knock value of petrol?
(c) How does isomerisation increase the octane number of petrol?

(10) (a) What is fuel? Name three fossil fuels in common use.
(b) Name two examples of (i) a solid fuel, (ii) a liquid fuel and (iii) a gaseous fuel. Compare the advantages and disadvantages of each type. In your answer refer to their calorific values and the effect on the environment.
(c) Nuclear fuel, hydro-electric power, wind power, solar power and geothermal energy are alternative energy sources to fossil fuels. Discuss the limitations of each energy source.

(11) Natural gas from Texas differs in composition from Kinsale gas. The percentage composition by volume of natural gas from Texas is given below:
Methane 81%, Ethane 6.7%, Propane 2.7%, Butane and higher alkanes 1.5%, Nitrogen 8%, Carbon dioxide 0.1%
(a) In what way does the natural gas from Texas differ from Kinsale gas?
(b) How are ethane and the higher alkanes separated from the mixture?
(c) How is carbon dioxide removed?
(d) What are the major uses of each hydrocarbon in the gas from Texas?

155

5.6 OTHER CHEMICAL FUELS

There is an ever-increasing demand for energy in the world. The type of fuel we use is largely determined by economic and environmental factors.

Solid fuels such as coal are excellent fuels and are plentiful at present, and relatively cheap. Incomplete combustion of coal, however, does cause quite serious air pollution in some areas. Today, many coal producers are manufacturing 'smokeless' fuels such as 'Homefire', 'Coalite' and 'Phurnacite' in addition to the naturally occurring smokeless fuel 'anthracite'. Some local authorities are insisting, by law, that householders should burn only 'smokeless' fuels.

Methanol has been used as a fuel in motor racing cars. It burns more cleanly than petrol and it does not release carcinogenic vapours such as benzenze. However, it is corrosive and eats through rubber tubing and aluminium fuel tanks. It absorbs water quite easily, which means that it conducts electricity. This means that all electrical components must be well insulated or else explosions may occur.

Chemicals, like ethyne, C_2H_2, and hydrogen, H_2, are used in specialist areas as fuels.

Ethyne (Acetylene)

Ethyne (common name acetylene) is used in oxy-acetylene burners for welding and cutting.

Preparation of Ethyne

Ethyne is formed by reacting calcium carbide with water. The calcium carbide is obtained by heating calcium oxide with carbon at temperatures above 2000 °C.

$$CaC_2(s) + 2H_2O(l) = C_2H_2(g) + Ca(OH)_2(aq)$$

Combustion of Ethyne

Ethyne, like other hydrocarbons, is easily combusted. The combustion of ethyne is used to generate high temperatures for cutting and welding metals, using the oxy-acetylene torch.

$$C_2H_2(g) + \tfrac{5}{2}O_2(g) = 2CO_2(g) + H_2O(g) + heat$$

Mandatory experiment 14

The preparation of ethyne and examination of some of its properties.

Introduction

Ethyne is made in the laboratory by reacting calcium carbide with water. The calcium carbide contains several impurities (phosphine, hydrogen sulphide, water and ammonia) which must be removed. Ethyne, like other hydrocarbons, is easily combusted. Because ethyne contains a triple bond, it is an unsaturated compound. Like all other compounds with either double or triple bonds, it decolorises a solution of bromine.

Requirements

Safety glasses, test-tubes, two holed stoppers, conical flask, trough, dropping funnel, gas jars, copper sulphate solution, acidified potassium manganate(VII), solution of bromine in trichloroethane, calcium carbide and water.

Procedure

(1) Place some calcium carbide in a test-tube and arrange the apparatus as in the diagram.

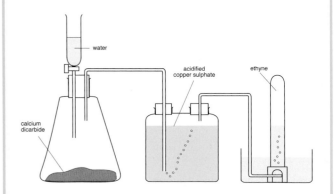

(2) Add water dropwise and collect three gas jars of gas.
(3) Carry out the following tests on the ethyne.
(a) Ignite the ethyne with a lighted splint.
(b) Add a few drops of the bromine solution to a gas jar of ethyne.
(c) Add about 1 cm³ of acidified potassium manganate(VII) to a gas jar of ethyne.

Results

Copy and complete the table below.

Combustion	Reaction with bromine	Reaction with potassium manganate(VII)

Hydrogen

Hydrogen was discovered in 1766 by John Dalton. It is very abundant on Earth: it is found combined with oxygen in water and with carbon in all organic matter. It is present as a gas in the Sun and the stars. It is a colourless, odourless gas.

Manufacture of Hydrogen

Hydrogen can be manufactured using several methods.
• By electrolysis:
In the laboratory, hydrogen is prepared by electrolysis of acidulated water.

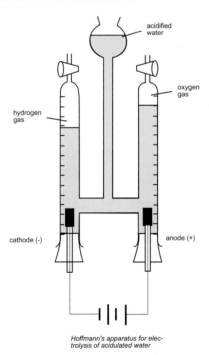

Hoffmann's apparatus for electrolysis of acidulated water

This method has been used to produce hydrogen on a large scale in areas where electrical power is cheap.

• From natural gas:
Hydrogen is produced economically from natural gas by steam reforming using a nickel catalyst.

$$CH_4(g) + H_2O(g) \xrightarrow[\text{Pressure}]{\text{Ni, heat}} CO(g) + 3H_2(g)$$

The carbon monoxide is removed by reacting it with steam.

$$CO(g) + H_2O(g) \longrightarrow CO_2(g) + H_2(g)$$

The carbon dioxide is then removed by absorbing it in potassium carbonate solution.
• Hydrogen can also be manufactured from coal and petroleum.

Industrial Uses

Approximately 20 million tonnes of hydrogen are produced each year.
(1) Approximately 50% of hydrogen produced is used in the synthesis of ammonia using the Haber process.

$$N_2(g) + 3H_2(g) \rightarrow 2NH_3(g)$$

(2) It is used to produce methanol and other alcohols.

$$CO(g) + H_2(g) \rightarrow CH_3OH$$
$$\text{methanol}$$

(3) Hydrogen is used in the food industry. Unsaturated oils are converted to saturated fats by hydrogenation.

$$\begin{array}{ccc} R & R & R \quad R \\ \backslash & / & | \quad | \\ C = C + H_2 \longrightarrow R - C - C - R \\ / & \backslash & | \quad | \\ R & R & H \quad H \end{array}$$

Unsaturated oil   Saturated fat

(4) Hydrogen is used in fuel cells. The space shuttle *Orbiter* used hydrogen as a fuel; the water produced was used as drinking water by the astronauts.

(5) It is used to make hydrochloric acid. Hydrogen reacts easily with chlorine to form hydrogen chloride gas.

$$H_2(g) + Cl_2(g) \rightarrow 2HCl(g)$$

The hydrogen chloride evolved is sprayed with water and the resulting solution of hydrochloric acid is collected.

157

(6) It is used in welding. When hydrogen gas is passed through an electric arc, the hydrogen molecules are split up into hydrogen atoms. When these hydrogen atoms recombine a great amount of energy is released, which is used to weld metal surfaces together.

Potential as a Fuel

Hydrogen has been used as a fuel in space craft as it releases large amounts of energy for a small molecule. It is a clean fuel, but it must be carried in large bulky containers.

Exercise 5.6

•

(1) (a) Name ten chemical fuels.
(b) Which of these fuels release harmful substances into the environment?
(c) In each case, name the pollutants released.
(d) Name a clean fuel.
(e) What are the disadvantages of using this fuel?

(2) (a) How is ethyne prepared? Write a balanced equation for the reaction.
(b) Draw a labelled diagram of the apparatus used.
(c) How would you test that the product formed was ethyne?
(d) What impurities might be present in the ethyne?
(e) How are the impurities removed?
(f) What is the main use of ethyne?

(3) Describe two ways in which hydrogen is manufactured.

(4) What are the main industrial uses of hydrogen?

(5) Discuss the potential use of hydrogen as a fuel.

Alternative sources of energy

(a) **Nuclear fuel** is a fuel that does not burn. Some uranium nuclei are unstable; when they are bombarded by neutrons they split up and release large amounts of energy. The main problem associated with fuels of this type is that they release radioactive material, which cannot be disposed of easily. This can cause horrific environmental problems, such as those experienced near Chernobyl.

(b) **Hydro-electric power** is a clean efficient form of energy, but relies on the availability of suitable water courses. Several hydro-electric power stations are in use in Ireland and they are also used in many other countries. Some electricity boards are developing hydro-electric power using wave power and tidal power.

(c) **Wind power** uses the kinetic energy from the wind to turn generators which produce electricity.

(d) **Solar power** The Sun's energy is absorbed by a blackened collector and then transferred to where it is required.

(e) **Geothermal energy** uses heat extracted from hot rocks from far below the ground level.

Environmental aspects of the oil industry

Crude oil and its petrochemical products are responsible for various types of pollution. Some of the major pollutants are listed below:
• oil spillages at sea cause damage to marine life
• unburned hydrocarbons cause photochemical smog
• chlorine compounds deplete the ozone layer
• oxides of nitrogen and sulphur cause acid rain
• lead compounds are neuro-toxins
• misuse of agricultural chemicals pollutes water, killing plants and animals.

Major oil, petrochemical and chemical companies have improved almost every aspect of our lives while at the same time being responsible for some environmental catastrophes. The balance between man and his environment has always been a delicate one. Industry and agriculture have played and should play a major role in the protection of the environment.

Some of the more important key terms are listed below. Other terms not included may be located by means of the index.

1. Hydrocarbons: Hydrocarbons are compounds formed by the combination of carbon and hydrogen.

2. Fuels: A fuel is a substance that can be used as an energy source.

3. Aliphatic and aromatic hydrocarbons: The alkanes, the alkenes and the alkynes are called aliphatic hydrocarbons, and their carbon atoms are joined together in chains, whereas the arenes are called aromatic hydrocarbons and they contain a special ring structure, called a benzene ring.

4. Saturated hydrocarbons: Alkanes are saturated hydrocarbons because all the chemical bonds are single bonds.

5. Alkanes: Alkanes are saturated hydrocarbons because each molecule contains carbon atoms bonded to four other atoms by single covalent bonds. The general formula is C_nH_{2n+2} where $n = 1, 2, 3 ...$

6. Systematic naming of alkanes: Today, most compounds are named in accordance with the IUPAC system. The simplest member of the alkanes is called methane, CH_4. The next three members are called ethane, propane and butane. Each subsequent alkane is named by adding the suffix -ane to a prefix indicating the number of carbon atoms present in the chain.

7. Isomerism: Isomers are substances which have the same molecular formula but different structural formulae.

8. Physical properties of alkanes: Because of the weak non-polar van der Waal's forces between them, alkanes not only have very low boiling points but also are practically insoluble in water and other polar solvents. They are, however, soluble in non-polar solvents such as trichloroethane.

9. Unsaturated hydrocarbons: Unsaturated hydrocarbons are hydrocarbons which have double or triple bonds.

10. Alkenes: Alkenes are unsaturated hydrocarbons containing a carbon–carbon double bond. The general formula is C_nH_{2n}, where $n = 2, 3, 4 ...$

11. Systematic naming of alkenes: Alkenes end with the suffix -ene: the position of the double bond is indicated by the lowest possible number.

12. Physical properties of alkenes: Alkenes have similar physical properties to alkanes. However, some alkenes, such as propene, are very slightly polar because they are not symmetrical like ethene. The slight polarity of some alkenes does not dramatically alter their physical properties.

13. Alkynes: Alkynes are unsaturated hydrocarbons which are even more unsaturated than alkenes as they contain a carbon–carbon triple bond. The general formula is C_nH_{2n-2}.

14. Systematic naming of alkynes: The alkynes are named in a similar manner to the alkenes, using the suffix -yne.

15. Preparation of ethyne: Ethyne can be prepared from calcium carbide.

16. Physical properties of ethyne: The physical properties of ethyne are similar in many ways to those of alkenes.

17. Aromatic compounds: Aromatic compounds are cyclic unsaturated hydrocarbons. Benzene, C_6H_6, has a symmetrical planar ring containing six carbon–carbon bonds of equal length.

18. Exothermic and endothermic reactions:
  Exothermic reactions are reactions in which energy is released as heat to the surroundings.
  Endothermic reactions are reactions in which heat energy is taken in from the surroundings.

19. Heat of reaction: The energy change which is used to measure the amount of the heat change in a chemical reaction is called the heat of reaction. This change is written as ΔH, called 'delta H'.

20. Standard conditions: The standard conditions for measuring heat changes are a temperature of 298 K and a pressure of 10^5 Pa.

21. Standard heat of combustion of a substance: This is the amount of heat in joules liberated when one mole of a substance is burned in excess oxygen, under standard conditions.

22. Standard heat change for a reaction: This refers to the amounts of the substances shown in the particular equation, at a temperature of 298 K and a pressure of 10^5 Pa, with the substances in the physical states normal under these conditions.

23. Bond energies: Bond energy is the energy required to break one mole of gaseous molecules into its constituent gaseous atoms.

24. Calorimeter: A calorimeter is an instrument which is used to measure the heat changes which accompany chemical reactions.

25. Units of energy: The unit of energy is the joule (J). It is a rather small amount of energy. For this reason, the energy changes of most chemical reactions are measured in kilojoules (kJ).

26. Heat changes: Heat changes are often calculated by means of the formula:

$$\Delta H = -mc\Delta T$$

27. Specific heat capacity: The specific heat capacity, c, of a substance is the amount of energy needed to raise the temperature of 1 kg of the substance by 1 degree Kelvin.

28. Heat change for one mole of reactants: The heat change for 1 mol of reactants,

$$\Delta H_m = \frac{\Delta H(\text{reaction})}{n}$$

29. Standard heat of formation of a substance: This is the amount of heat in joules evolved or absorbed when one mole of a substance is formed in its standard state from its elements in their standard states and under standard conditions.

30. Heat of formation of elements: The heat of formation of an element is zero.

31. Heat of reaction:
Heat of reaction = heat of formation of products – heat of formation of reactants

$$\Delta H = \Delta H_f(\text{products}) - \Delta H_f(\text{reactants})$$

32. Hess's law of constant heat summation: The total energy change accompanying a chemical change is independent of the route by which the chemical change takes place, provided that the initial and final conditions are the same.

33. Crude oil: Crude oil is a complex mixture of hydrocarbon molecules, whose composition varies depending on where in the world the oil is found.

34. Purification of crude oil: The components of crude oil are separated by distillation and processed further by cracking and reforming.

35. Octane number: The tendency of a fuel to auto-ignite is measured by its octane number.

RATES OF REACTION

6.1 RATES OF REACTION

Nuclear explosion.

There is a wide variation in the rates or speeds at which different chemical reactions proceed. Some reactions are slow, while other reactions are fast. The slow oxidation of iron during rusting is in marked contrast to the rapid oxidation of magnesium in photo-flash bulbs. In many industrial processes it is essential to investigate how various factors affect the rate of a chemical reaction.

Corroding diesel engine – a slow reaction.

Fireworks – a fast reaction.

However, before this can be done, it must be decided what is meant by 'rate of a chemical reaction', and how reaction rates are measured.

Measurement of Reaction Rates

The **rate of a reaction** is the change in concentration of any species involved in a reaction (either reactant or product) divided by the time taken for the change to occur.

Suppose the reaction is a simple one where substance A reacts to form substance B.

$$A \rightarrow B$$

The concentrations of the reactant A and product B are written as [A] and [B], respectively.

The rate of reaction can be written in either of the following ways:

As formation of product

$$\text{Rate of formation of product} = \frac{\Delta[B]}{\Delta t}$$

Δ means 'change in'

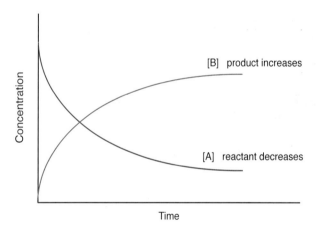

As loss or destruction of reactant

$$\text{Rate of destruction of reactant} = \frac{-\Delta[A]}{\Delta t}$$

$$\frac{\Delta[B]}{\Delta t} \text{ and } \frac{\Delta[A]}{\Delta t}$$

are the changes in concentration of the product and reactant, respectively, divided by the time taken for the changes to occur, and each rate refers to the reaction indicated above.

$$\text{Rate of a chemical reaction} = \frac{\text{change in concentration}}{\text{time taken for change}}$$

Mandatory experiment 15

To measure the rate of production of oxygen from hydrogen peroxide using manganese dioxide as a catalyst.

Introduction

Hydrogen peroxide solution, $H_2O_2(aq)$, decomposes slowly to produce oxygen. The rate of the reaction is speeded up by using a catalyst, manganese(IV) oxide, MnO_2.

In the reaction between hydrogen peroxide and manganese dioxide, oxygen gas evolves.

$$2H_2O_2(aq) \xrightarrow{MnO_2} 2H_2O(l) + O_2(g)$$

The volume of oxygen evolved can be easily measured using a gas syringe as shown. If a gas syringe is not available, an inverted burette can be used to measure the volume of water displaced by the oxygen evolved.

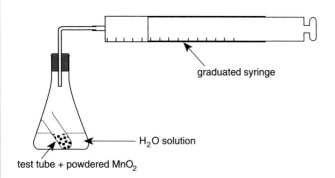

graduated syringe

H₂O solution

test tube + powdered MnO₂

Requirements

Safety glasses, conical flask, small test-tube, rubber tubing, graduated glass syringe (or burette), stop clock, hydrogen peroxide solution (30%w/v), manganese(IV) oxide.

Procedure

(1) Place 1.5 g of manganese(IV) oxide in a small test-tube.
(2) Place the test-tube in a conical flask in an upright position.
(3) Transfer 50 cm³ of the hydrogen peroxide solution into the conical flask, taking care

not to allow the solution to come in contact with the manganese(IV) oxide.
(4) Connect the conical flask to the empty glass syringe as shown above.
(5) Shake the flask gently to knock over the test-tube of manganese(IV) oxide in order to start the reaction. At the same time start the stop-clock.
(6) Record the volume of oxygen produced every 30 s until the reaction is completed.
(7) Plot a graph of volume against time.

Results

Make a table like the one at the foot of the page.

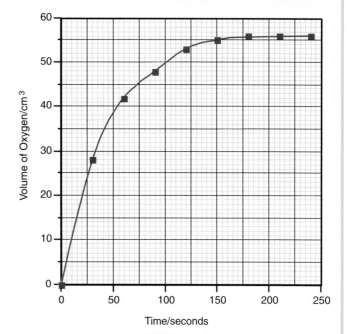

Conclusions

The rate of reaction is fast at first and then begins to slow down as the hydrogen peroxide is used up. Finally, the reaction ceases when it is all used up. This is seen from the shape of the rate curve: initially it is steep and then it flattens out.

Results

Time/s	0	30	60	90	120	150	180	210	240
Volume of oxygen/cm³	0	28	42	48	53	55	56	56	56

Average and Instantaneous Rates

The rate of a chemical reaction can be compared to the speed of a motor car on a journey. Suppose the car travels a distance of 180 km in 3 hours. This means that the average speed of the car is 60 km per hour. This does not mean that the car travelled all the time at this speed. The speedometer, which registers instantaneous speed, may read more or less than 60 km per hour at any one moment. The car may have accelerated, decelerated or even stopped during the journey.

Chemical reactions are similar, in that the rate of the reaction is constantly changing.

The rate of a chemical reaction changes as the reaction proceeds. In the beginning the rate is fast; this is called the initial rate. Later the reaction slows down; the rate is changing rapidly with time. The rate must be stated for a given time; this is called the rate at time t or the instantaneous rate.

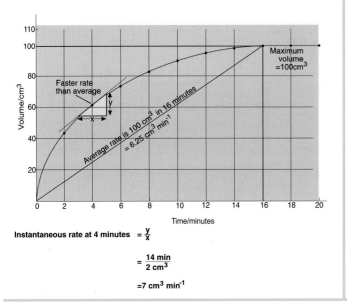

Instantaneous rate at 4 minutes $= \dfrac{y}{x}$

$$= \dfrac{14 \text{ min}}{2 \text{ cm}^3}$$

$$= 7 \text{ cm}^3 \text{ min}^{-1}$$

In order to calculate either the initial reaction rate or the rate at time t, an accurate graph must be drawn on graph paper. The reaction rate is determined by measuring the slope of the graph. This is done by drawing a tangent to the curve at the required time, t, and determining the slope of the target as shown.

1. Construct a tangent to the curve through the point (1.0, 56).
2. Construct a right-angled triangle on the tangent.
3. Measure the distances y and x.

4. $\quad \text{Slope} = \dfrac{y}{x} = \dfrac{70 \text{ cm}^3}{1.75 \text{ min}} = 40 \text{ cm}^3 \text{ min}^{-1}$

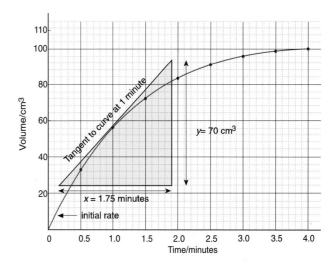

Note: the slope at zero time is called the initial reaction rate. It is used to compare reaction rates before the reactant concentrations have decreased during a reaction.

●●●●● **Example 6.1** ●●●●●

Hydrogen peroxide solution decomposes as follows:

$$\overset{\text{MnO}_2}{2\text{H}_2\text{O}_2(\text{aq}) \Rightarrow 2\text{H}_2\text{O}(\text{l}) + \text{O}_2(\text{g})}$$

The total volume of oxygen at STP liberated from 50 cm^3 of hydrogen peroxide solution is given in the table below.

(a) Using graph paper, plot a graph of volume of oxygen produced against time.
(b) From the graph on page 164 determine:
(i) when the reaction was completed
(ii) the average rate of the reaction
(iii) the instantaneous rate at 60 s
(iv) the total amount of oxygen produced at STP
(v) the concentration of the hydrogen peroxide solution.

Time/s	0	30	60	90	120	150	180	210	240
Volume of oxygen/cm^3	0	24	42	49	54	55	57	57	57

163

(a)

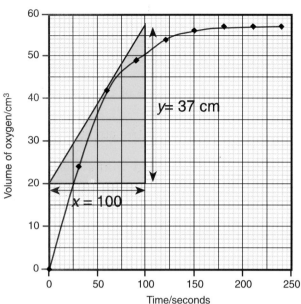

(b)

(i) The reaction is complete at 180 s because no more oxygen is produced after this time.

(ii)
$$\text{Average rate} = \frac{\text{total volume produced}}{\text{time taken}}$$

$$= \frac{57 \text{ cm}^3}{180 \text{ s}} = 0.32 \text{ cm}^3\text{s}^{-1}$$

(iii) From the graph

$$\text{Slope} = \frac{y}{x} = \frac{37 \text{ cm}^3}{100 \text{ s}} = 0.37 \text{ cm}^3\text{s}^{-1}$$

(iv) Total amount of oxygen produced at STP

$$n(O_2) = \frac{V}{V_m} = \frac{0.057 \text{ L}}{22.4 \text{ L mol}^{-1}} = 0.0025 \text{ mol}$$

(v) Concentration of the hydrogen peroxide solution

$$2H_2O_2(aq) \rightarrow 2H_2O(l) + O_2(g)$$

$$n(H_2O_2) = 2 \times n(O_2) = 2 \times 0.0025 \text{ mol}$$

$$= 0.005 \text{ mol}$$

$$c(H_2O_2) = \frac{n}{V} = \frac{0.005 \text{ mol}}{0.05 \text{ L}} = 0.1 \text{ mol L}^{-1}$$

• • • • • • **Example 6.2** • • • • • •

A mass of 1 g of manganese(IV) oxide was weighed in a test-tube. 100 cm³ of a hydrogen peroxide solution was measured into a conical flask which was then fitted with a loose plug of cotton wool. The flask was placed on the pan of a direct reading balance and, after adding the manganese(IV) oxide, the mass was recorded at 1 min intervals. The results are shown below.

(a) What is the function of the cotton wool plug?
(b) How would you find the mass at 0 min, i.e. before the reaction had started?
(c) Using graph paper, plot a graph of loss of mass against time.

Time/min	Mass/g
0	176.58
1	176.34
2	176.22
3	176.16
4	176.12
5	176.11
6	176.10
7	176.10
8	176.10

(a) The cotton wool will absorb any water vapour produced during the reaction.
(b) Weigh both reactants separately beforehand.
(c)

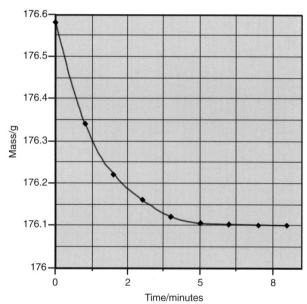

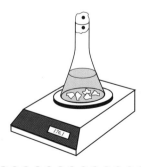

Methods for Measurement of Reaction Rates

The rate of a chemical reaction can be measured in many ways. The main methods include:

1. Volumetric or gravimetric analysis of samples removed from suitable reactions at set time intervals.

The hydrolysis of ethyl ethanoate can be monitored by titrating samples of the ethanoic acid formed with sodium hydroxide.

$$CH_3COOC_2H_5(aq) + H_2O(l) \rightleftharpoons CH_3COOH(aq) + C_2H_5OH(aq)$$

$$CH_3COOH(aq) + NaOH(aq) \rightarrow CH_3COONa(aq) + H_2O(l)$$

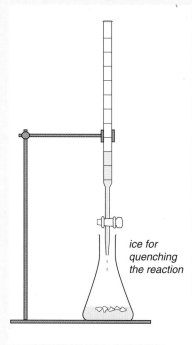

ice for quenching the reaction

This method involves 'quenching' or 'freezing' the reaction when samples are being titrated. Consequently, this method is time consuming.

2. Dilatometry – measurement of volume changes.
In reactions where the total volume of the solution changes, the progress of the reaction can be monitored by enclosing the reaction vessel in a vessel linked to a thin capillary tube. A small volume change can then be detected.

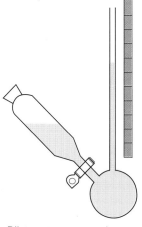

Dilatometry - measuring a small volume change

3. Measurement of gas evolved during a reaction.

Many reactions produce a gas as a product, the volume of which can be measured easily.
 The reaction of magnesium and dilute hydrochloric acid produces hydrogen gas, which can be collected in a graduated gas syringe.

$$Mg(s) + 2HCl(aq) \rightarrow MgCl_2(aq) + H_2(g)$$

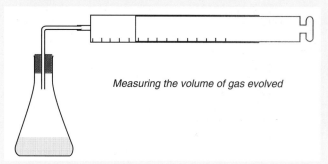

Measuring the volume of gas evolved

4. Measurement of a change in mass during a reaction.

During many chemical reactions a gas is produced as a product. As the reaction proceeds the mass of the reactants decreases. In some cases, the decrease in mass can be measured easily.
 The reaction between marble chips and dilute hydrochloric acid can be measured in this way. As the reaction proceeds, carbon dioxide gas escapes from the flask and the mass decreases.

$$CaCO_3(s) + 2HCl(aq) \rightarrow CaCl_2(aq) + H_2O(l) + CO_2(g)$$

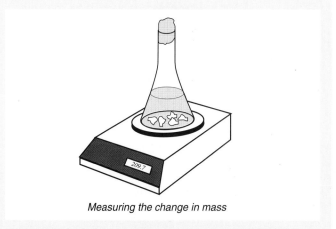

Measuring the change in mass

5. Manometry – measurement of pressure changes.

In gaseous systems the reaction often leads to a change in the number of molecules. This causes a change in the pressure of the system, which can be measured using a manometer.

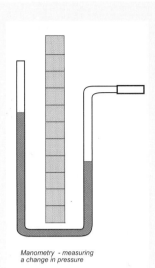

Manometry - measuring a change in pressure

In the decomposition of dinitrogen pentoxide gas, two molecules of the gas produce four molecules of nitrogen dioxide gas and one molecule of oxygen gas. The increase in volume causes a change in pressure, which can be measured easily.

$$2N_2O_5(g) \rightarrow 4NO_2(g) + O_2(g)$$

6. Colorimetry – measurement of the intensity of colour during a reaction.

Some reacting species are highly coloured. The colour intensity changes as the concentration of the coloured species decreases during the course of the reaction. The colour intensity can be measured using an instrument called a colorimeter.

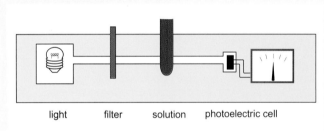

| light | filter | solution | photoelectric cell |

Colorimetry - measuring intensity of colour

When warm acidified solutions of manganate(VII) ions and ethanedioate ions are mixed a reaction occurs. The deep intense purple colour of the manganate(VII) ions changes gradually to the colourless manganese(II) ion.

$$2MnO_4^-(aq) + 16H^+(aq) + 5C_2O_4^{2-} = 2Mn^{2+}(aq) + 10CO_2(g) + 8H_2O(l)$$

7. Measurement of the electrical resistance during a reaction.

Reactions involving a change in the number or type of ions cause a change in the electrical conductivity of the solution. The conductivity change can be measured using a conductivity meter.

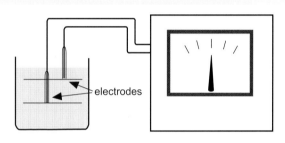

electrodes

Measuring a change in conductivity

Exercise 6.1
•

(1) (a) What is meant by the rate of a chemical reaction?

(b) Describe, in detail, a method used to measure the reaction rate, in a reaction where a gas is given off as a product.

(2) In the reaction between magnesium and dilute hydrochloric acid, hydrogen gas is produced. The volume of gas produced is measured at regular time intervals. Draw a sketch of the rate curve for this reaction and comment on the features of the curve.

(3) In the reaction of hydrogen peroxide with manganese(IV) oxide to produce oxygen, the concentration of hydrogen peroxide used was different for three experiments, A, B and C. If the time taken to collect 250 cm^3 of oxygen was 25 s, 15 s and 30 s, respectively, for experiments A, B and C, which has the fastest average rate?

(4) Calcium carbonate reacts with dilute hydrochloric acid according to the equation:

$$CaCO_3(s) + 2HCl(aq) = CaCl_2(s) + H_2O(l) + CO_2(g)$$

A student measured the volume of carbon dioxide produced at 1 min intervals until the evolution of carbon dioxide ceased.

(a) Draw a labelled diagram of the apparatus used to show how you would measure the volume of carbon dioxide produced.

(b) Draw a sketch of the graph you would

expect to obtain when the volume of carbon dioxide is plotted against time.

(5) An excess of calcium carbonate was added to dilute hydrochloric acid in a conical flask. A cotton wool plug was placed in the neck of the flask and the flask was placed on a balance and weighed. The flask was weighed again at regular time intervals. The results are given below.

Time/min	Mass/g
0	150.0
0.5	149.84
1.0	149.72
1.5	149.64
2.0	149.58
3.0	149.51
4.0	149.48
5.0	149.46
6.0	149.45
7.0	149.44
8.0	149.44
9.0	149.44

(a) Write an equation for the reaction.
(b) Why did the mass of the conical flask and its contents change?
(c) What was the purpose of the cotton wool?
(d) Plot a graph of mass against time.
(e) After what time was the reaction completed?
(f) After what time was the reaction half-way to completion?
(g) Determine the instantaneous rate at 4 min.

(6) In an experiment, 50 cm^3 of hydrogen peroxide solution and 2.5 g of manganese(IV) oxide were reacted with each other. A gas was produced and its volume was measured at regular time intervals. The results are given below.

(a) Write an equation for the reaction.
(b) What was the purpose of the manganese(IV) oxide? What mass of it would be left at the end of the reaction?
(c) Draw a labelled diagram of the apparatus you would use to do this experiment.
(d) Plot a graph of the rate curve.

Vol/cm^3	Time/min
75	1
125	2
158	3
187	4
208	5
220	6
230	7
234	8
240	9
242	10
244	11
245	12
248	13
249	14
249	15
250	16
250	18
250	20

(e) When was the maximum volume of oxygen obtained?
(f) When was the reaction completed?
(g) Determine the number of moles of oxygen gas that would be obtained if the reaction were carried out at standard temperature and pressure.
(h) Calculate the concentration of the hydrogen peroxide used.
(i) Calculate the instantaneous rate at 6 min. How does this rate compare with the initial rate?
(j) Draw simple sketches showing how the rate curve obtained would differ if:
(i) the initial concentration of the hydrogen peroxide were halved
(ii) the initial concentration was the same but the temperature were about 10 °C higher.

(7) Hydrogen peroxide decomposes to produce oxygen:

$$2H_2O_2(l) = 2H_2O(l) + O_2(g)$$

The rate of the reaction is monitored by measuring the volume of oxygen produced at regular time intervals.
(a) Sketch the apparatus used for this experiment.
(b) What catalyst is used?
(c) Why is the hydrogen peroxide stored in a dark bottle?
(d) Why does the bottle top have a small slit in it?
(e) Plot a graph of volume against time for the following data.
(f) When was the reaction most rapid?
(g) At what time was the reaction over?
(h) What volume of oxygen was produced when all of the hydrogen peroxide had decomposed?
(i) What is the average rate for the reaction?
(j) What is the instantaneous rate at 60 s?

Vol/cm^3	Time/s
50	10
80	20
93	30
105	40
115	50
120	60
125	70
126	80
126	90
126	100

(8) Oxygen gas was produced from hydrogen peroxide according to the equation:

$$2H_2O_2(l) = 2H_2O(l) + O_2(g)$$

The volume of gas produced at regular time intervals is given below.

(a) What is the initial rate of formation of oxygen?
(b) What is the average rate of formation of oxygen?
(c) What is the instantaneous rate at 75 s?

Vol/cm³	Time/s
16	25
27	50
34	75
37	100
39	125
39.5	150
40	175
40	200
40	225

6.2 FACTORS AFFECTING REACTION RATES

The main factors affecting reaction rates are:
1. The concentration of the reactants
2. The temperature of the system
3. The presence of catalysts
4. The surface area of solid reactants
5. The presence of light.
The nature of the reactants also determines whether a reaction is a slow reaction or a fast reaction.

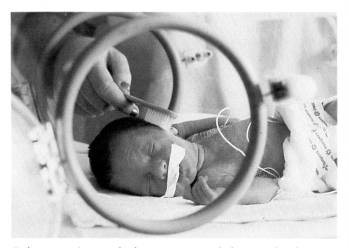

Baby in incubator – higher temperature helps it to develop more quickly. A cold storage cabinet, in contrast, will slow down the chemical reactions which cause food to deteriorate.

1. The Concentration of the Reactants

For reactions in solution, the rate depends on the concentration of one or more of the reactants.

If the reaction between magnesium and hydrochloric acid is studied, the hydrogen gas evolved can be collected in a gas syringe and its volume recorded at regular time intervals.

If the magnesium is used in excess, the reaction comes to completion when all the dilute hydrochloric acid has been used up.

A rate curve for the reaction between dilute hydrochloric acid and magnesium is illustrated.

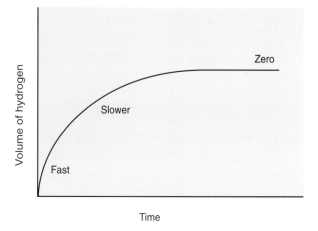

Three distinct features are noticeable.

1. The initial rate is fast.

2. As the reaction proceeds the rate becomes slower and slower.

3. The rate finally approaches zero, as is seen when the curve flattens. At this point the supply of acid has been used up.

Obviously, the rate becomes slower and slower as the concentration of the acid decreases. The amount of magnesium also decreases, but since the magnesium is in excess, it would seem that the concentration of acid is the critical factor.

This can be easily verified by doing a series of experiments, each experiment using the same quantity of magnesium, but using different concentrations of dilute hydrochloric acid for each experiment.

Some typical results are illustrated below.

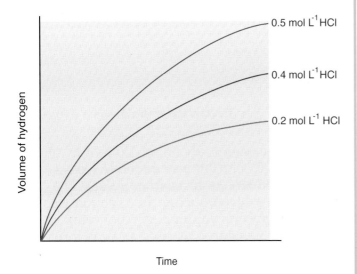

The steeper curves for the more concentrated acid solutions indicate that the reaction rate depends on the concentration of the hydrochloric acid.

Why Reaction Rates Depend on Concentration

The primary requirement for a chemical reaction to occur is that the reacting species must come in contact with each other, i.e. they must collide. Obviously, the more particles there are, the more collisions can occur.

In chemical reactions, the particles move about continuously, bumping into each other. However, not all the collisions lead to an effective or successful reaction, because some of the particles may not have enough energy to collide successfully.

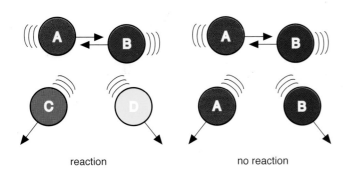

Some, but not all, of the particles have enough energy to collide and form an effective collision, resulting in the product being formed.

When the concentration of the reacting species is increased, the number of reactants with sufficient energy also increases and as a result more product forms.

> Increase in concentration
> Increases the number of collisions
> Increases the number of successful collisions

2. The Temperature of the System

The rate of a chemical reaction increases as temperature increases regardless of whether the reaction is exothermic or endothermic.

Equal volumes of hydrochloric acid solution of the same concentration are allowed to react with equal quantities of magnesium in separate reaction vessels and the temperature of each reaction is then varied.

The rate of the reactions increases as the temperature increases. The reaction rate curves are steeper at the higher temperatures.

For some reactions an increase in temperature of 10 °C roughly doubles or trebles the reaction rate. This is not a very accurate rule and must be used with extreme caution.

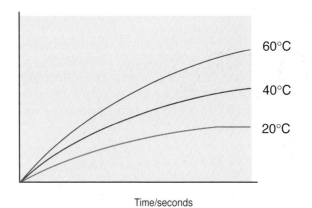

Why Increase in Temperature Increases the Rate of a Reaction

The energy of all the reacting species in a reaction is not the same. In comparison with the overall energy, some species have high energy, some have low energy, and the majority have moderate energy. The distribution of the energy of all the reacting species can be illustrated graphically.

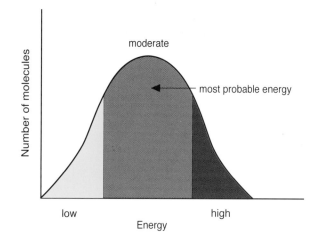

A chemical reaction occurs when the reactants have enough energy to overcome the repulsive forces of their respective electron clouds which push them apart.

If the reactants are slow moving (low kinetic energy), the collision between the colliding particles will be so gentle that they will move apart again.

If the reactants are fast moving (high kinetic energy), the collisions will be vigorous enough to overcome the repulsive forces and allow a new species (product) to form.

The minimum amount of energy that reactants need to form products is called the **activation energy**.

When the temperature of a reaction is increased the reactants move about much faster and collide more often. However, it can be shown that this increase in the number of collisions is not sufficient to account for the large increase in reaction rate as the temperature increases.

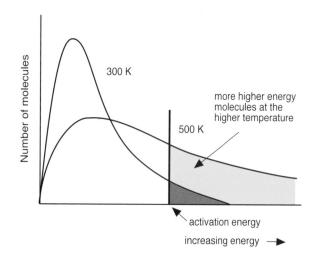

The increase in temperature increases the number of higher energy reactants, which then form new products.

Increase in temperature
Increases the number of high energy reactants
Increases the number of successful collisions

•••••• **Example 6.3** ••••••

In an experiment, 25 cm^3 of 1.0 mol L^{-1} sulphuric acid solution was added to excess granulated zinc in a reaction vessel. The hydrogen evolved during the reaction was collected and its volume (at room temperature) was measured at regular time intervals. The results are shown in the table below.

Time/min	Volume hydrogen/cm^3
0	0
1	30
2	45
3	52.5
4	56.3
5	58.2
6	60
7	61
8	61.5
9	62
10	62
11	62

1. Using a balanced equation for the reaction, calculate the maximum volume of hydrogen gas at standard temperature and pressure (STP) which would be evolved in the above experiment.

$$Zn(s) + H_2SO_4(aq) \rightarrow ZnSO_4(aq) + H_2(g)$$

The balanced chemical equation tells us that

1 mol H$_2$SO$_4$ produces 1 mol H$_2$ at STP
The amount of H$_2$SO$_4$ used,

$$n(H_2SO_4) = cV$$
$$= 0.1 \text{ mol L}^{-1} \times 0.025 \text{ L}$$
$$= 0.0025 \text{ mol H}_2SO_4$$

Therefore, 0.0025 mol H$_2$SO$_4$ produces 0.0025 mol H$_2$ at STP.

Volume of H_2 produced at STP

$$V = nV_m$$
$$= 0.0025 \text{ mol} \times 22.4 \text{ L mol}^{-1}$$
$$= 0.056 \text{ L } H_2 \text{ at STP}$$

2. Plot, on graph paper, the results given in the table. Comment on the shape of the graph.

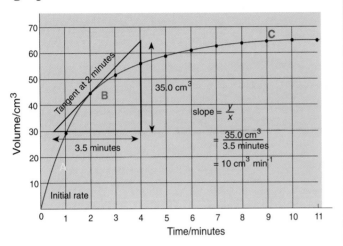

Comments
• In region A, (0–2 min) the reaction rate is rapid. The slope of the curve (y/x) is rather steep.
• In region B, (2–5 min) the reaction rate is slowing down because the concentration of sulphuric acid is decreasing. The slope of the curve is flattening out.
• In region C, (6–11 min) the reaction is nearing completion (6–9 mins) and is finally over at 9 min when all the acid has been used up. Here, the rate = 0.

3. Find, using the graph, the rate of the reaction after 2 min, i.e. the instantaneous rate, expressing the rate in terms of cubic centimetres of hydrogen liberated per minute ($cm^3 \text{ min}^{-1}$).

The instantaneous rate at 2 min is calculated by drawing a tangent to the curve and finding the value of y/x at this point as shown.

$$\text{Rate at 2 min} = \frac{y}{x} = \frac{35 \text{ cm}^3}{3.5 \text{ min}}$$
$$= 10 \text{ cm}^3 \text{ min}^{-1}.$$

4. What changes would you expect in the graph if:
(a) 25 cm^3 of 0.5 mol L^{-1} H_2SO_4 were used in the experiment, and

(b) the same quantity of 0.1 mol L^{-1} H_2SO_4 were used, but the temperature was higher.

(a) The rate of reaction is proportional to the concentrations of the reactants. Therefore, increasing the concentration of H_2SO_4 from 0.1 mol L^{-1} to 0.5 mol L^{-1} will increase the reaction rate. As a result, the slope of the graph in region A will be steeper.

(b) The rate of a reaction increases with temperature. As a result, the slope of the graph will be steeper as before.

• •

Mandatory experiment 16

A study of the reaction between sodium thiosulphate solution and hydrochloric acid to determine the effect of (a) concentration and (b) temperature on the rate of a chemical reaction.

$$Na_2S_2O_3(aq) + 2HCl(aq) = S(s) + 2NaCl(aq) + H_2O(l) + SO_2(g)$$

Introduction

The rate of a reaction is usually monitored by recording the amount of reactant or product at regular time intervals. In some reactions, this is not the most convenient method of measuring the rate. In this reaction, a solid, sulphur, is formed which makes the reaction mixture cloudy. The time taken for the reaction to reach this stage is taken as a measure of the reaction rate.

Requirements

Safety glasses, Bunsen burner or heating mantle, thermometer, conical flasks, graduated cylinders, beaker, stop-clock, 0.2 mol L^{-1} sodium thiosulphate solution, 2.0 mol L^{-1} hydrochloric acid solution.

Procedures

[a] Effect of concentration

(1) Measure 100 cm^3 of 0.2 mol L^{-1} sodium thiosulphate solution into a conical flask.
(2) Draw a distinct cross on a piece of white paper. Place the conical flask on the cross.

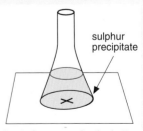

sulphur precipitate

sheet of paper under the bottom of a large beaker

(3) Measure out 10 cm^3 of 2.0 mol L^{-1} hydrochloric acid solution in a small graduated cylinder.
(4) Add the acid to the sodium thiosulphate solution. Immediately start the stop-clock.
(5) Swirl the conical flask continuously. The cross grows fainter as a yellow precipitate of sulphur forms.

(6) Stop the stop-clock when the cross just disappears.
(7) Note the time taken.
(8) Repeat the procedure at room temperature, using the quantities shown in the table below.
(9) Record the time taken in each case and plot a graph of percentage concentration against rate (1/time). (See Example 6.4).

Experiment [a]	Volume of sodium thiosulphate solution/cm^3	Volume of water/cm^3	Volume of hydrochloric acid/cm^3	Time/s	Rate 1/time
1	100	0	10		
2	80	20	10		
3	60	40	10		
4	40	60	10		
5	20	80	10		

[b] Effect of Temperature

(1) Draw a distinct cross on a piece of white paper.
(2) Measure 20 cm^3 of 0.2 mol L^{-1} sodium thiosulphate solution into a clean conical flask. Add 80 cm^3 of deionised water to the conical flask.
(3) Heat the mixture using a Bunsen burner or a heating mantle until the temperature is just above 20 °C.
(4) Add 10 cm^3 of 2.0 mol L^{-1} hydrochloric acid solution to the conical flask.
(5) Immediately start the stop-clock. Place the flask on the cross.

(6) Swirl the conical flask continuously. The cross grows fainter as a yellow precipitate of sulphur forms.
(7) Stop the stop-clock when the cross just disappears.
(8) Note the time taken.
(9) Repeat the procedure, heating the sodium thiosulphate solution, before adding the acid, to temperatures just above 30 °C, 40 °C, 50 °C and 60 °C.
(10) Record the time taken in each case and plot a graph of temperature against rate (1/time).

● ● ● ● ● ● **Example 6.4** ● ● ● ● ● ●

A series of experiments was carried out to examine the effect of changing the concentration of hydrochloric acid used in the following reaction:

$$Na_2S_2O_3 + 2HCl = S + 2NaCl + H_2O + SO_2$$

Measured volumes of each solution were poured into a conical flask and constantly swirled. The volume of sodium thiosulphate used in each experiment was the same, while the volume of hydrochloric acid used was changed each time by adding deionised water. The time taken for the solution to turn so cloudy that a cross underneath the conical flask just disappeared was recorded in the table below.

Experiment	Volume of sodium thiosulphate solution/cm^3	Volume of water/cm^3	Volume of hydrochloric acid/cm^3	Time/s	Rate (1/time)/s^{-1}
1	75	0	25	33	30.3
2	75	5	20	41	24.4
3	75	10	15	55	18.2
4	75	15	10	82	12.1
5	75	20	5	164	6.1

(a) In which experiment is the concentration of hydrochloric acid the strongest?

(b) Express each concentration of hydrochloric acid as a percentage of the concentration of the acid used in experiment 1.

(c) Plot a graph of percentage concentration of hydrochloric acid used against rate (1/time).

(d) From your graph, determine the rate of reaction if 17 cm^3 of hydrochloric acid were used.

(e) Explain why identical conical flasks must be used for experiments 1–5.

(a) Experiment 1: the hydrochloric acid was not diluted.

(b)
Experiment 2   80%
Experiment 3   60%
Experiment 4   40%
Experiment 5   20%

(c)

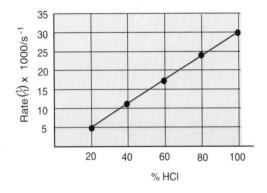

(d)
Rate of reaction using 17 cm^3 of HCl = 0.02 s^{-1}

(e) Different sizes of flasks would give different levels of cloudiness.

● ●

3. The Presence of Catalysts

The rate of a chemical reaction is increased by the presence of catalyst.

The decomposition of hydrogen peroxide to form water and oxygen gas is a slow process even when a dilute aqueous solution of hydrogen peroxide is heated. The reaction occurs at a much faster rate in the presence of manganese(IV) oxide.

$$2H_2O_2(aq) \rightarrow 2H_2O(l) + O_2(g) \quad \text{Slow reaction}$$

$$2H_2O_2(aq) \xrightarrow{\text{MnO}_2} 2H_2O(l) + O_2(g) \quad \text{Fast reaction}$$

The manganese(IV) oxide speeds up the rate of the reaction by helping the hydrogen peroxide to decompose, but is not itself used up during the reaction.

Catalysts are substances which alter the rate of a chemical reaction and which are chemically unchanged after the reaction has ended.

Catalyst in action. A small amount of potassium iodide accelerates the decomposition of hydrogen peroxide to water and oxygen. This is shown by the rapid inflation of the balloon by the oxygen produced.

Friedrich Wilhelm Ostwald 1853–1932
Wilhelm Ostwald, the German chemist and philosopher, became interested in chemistry at an early age; as an eleven year old he made his own fireworks. He developed the Ostwald process for the synthesis of nitric acid using a platinum–rhodium catalyst. He proposed that catalysts speed up chemical reactions by lowering the energy of activation. He is one of the founders of physical chemistry, the theoretical branch of chemistry which deals with the properties and reactions of ions, atoms and molecules. Curiously, although he was one of the most eminent chemists of his time, he did not accept the development of atomic theory until 1906. He was awarded the 1909 Nobel Prize for Chemistry for his work on catalysis and on the conditions of chemical equilibrium and the velocities of chemical reactions.

Most catalysts speed up chemical reactions, but a few catalysts slow them down. Catalysts which slow down reactions are called **negative catalysts** or **inhibitors**.

Inhibitors are used to slow down reactions that may occur during storage or transit of chemicals. For example, 1,2,3-propane-triol is used to stop hydrogen peroxide from decomposing while it is being stored.

Many chemical reactions are speeded up by catalysts

The synthesis of ammonia by the Haber process uses an iron catalyst, while many hydrogenation reactions, such as those used in the manufacture of margarine, use a nickel catalyst. Many industrial processes use either transition metals or their compounds as catalysts.

Some chemical reactions are catalysed by one of the products formed: this type of catalysis is called **autocatalysis**. A common example of autocatalysis is the reaction of ethane-dioic acid with manganate(VII) ions, where the Mn^{2+} ions produced during the reaction speed up the rate of the reaction.

$$2MnO_4^- + 5H_2C_2O_4 + 6H^+ \rightarrow 2Mn^{2+} + 10CO_2 + 8H_2O$$

Types of Catalyst

1. Heterogeneous catalysts

In the reaction between sulphur dioxide and oxygen to produce sulphur trioxide, the reactants and products are gases, while the catalyst used is a solid, platinum.

$$2SO_2(g) + O_2(g) \xrightarrow{Pt} 2SO_3(g)$$

Here, there are two distinct phases, a gas phase and a solid phase. Reactions which occur in different phases are called **heterogeneous reactions**. The catalyst in this reaction, platinum, is a **heterogeneous catalyst**.

Catalytic converters, which are fitted to motor car exhausts in order to remove harmful waste gases, such as carbon monoxide and the oxides of nitrogen, make up a heterogeneous catalytic system. The system consists of a mixture of the catalysts, platinum, palladium and rhodium. The carbon monoxide and oxides of nitrogen are converted into the harmless gases carbon dioxide and nitrogen. However, the catalysts are inactivated or poisoned if leaded petrol is used. Cars which are fitted with catalytic converters must run on unleaded petrol.

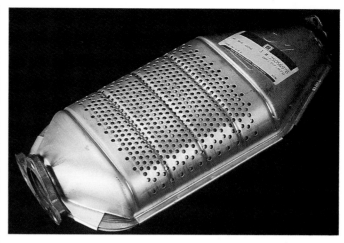

Catalytic converter, a device which reduces the toxic emissions from an internal combustion engine into relatively harmless ones. A 90% reduction of pollution emission has been achieved without loss of engine performance or fuel economy.

2. Homogeneous catalysts

In the reaction between ethanoic acid and ethanol to produce ethyl ethanoate and water, hydrogen ions are used to catalyse the reaction.

$$CH_3COOH(aq) + C_2H_5OH(aq) \overset{H^+}{\rightleftharpoons} CH_3COOC_2H_5(aq) + H_2O(l)$$

In this case, the catalyst and the reactants are all liquids. This is an example of **homogeneous catalysis**, where the catalyst and the reactants are all in the same phase. The reactants and catalyst may be either all liquid or all gaseous.

Biological catalysts

All chemical reactions occurring in all living things use catalysts called enzymes to speed up biological reactions. Enzymes are usually very specific in their action; lipase reacts with fats, but does not react with carbohydrates or proteins. Enzymes are used to make beer, whiskey, pharmaceuticals, biological washing powders and many more products.

Catalysts and Energy

For a reaction to occur an energy barrier must be overcome in order for products to form. The presence of a catalyst lowers that energy barrier. This, in effect, increases the number of reactants with sufficient energy necessary to form products.

The energy barrier which must be overcome so that compounds may react is called the

174

activation energy. A catalyst lowers the activation energy and so helps chemical bonds to break more easily.

The progress of a reaction in relation to the reduction of the energy barrier by the presence of a catalyst is illustrated below.

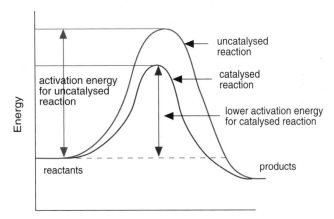

The catalyst provides a faster and easier pathway for the reaction to proceed and this pathway requires less energy.

The effect of a catalyst can also be described in terms of the increase in the number of reactants possessing the minimum energy (activation energy) for effective collisions to occur to form products.

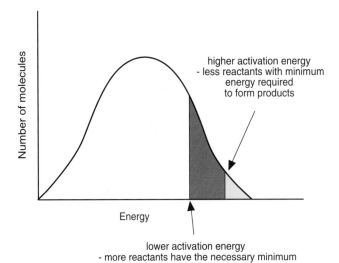

Activation energy is the minimum energy necessary for the reactants to acquire in order to overcome the energy barrier for a successful collision to occur.

Theories of Catalytic Action

The theory of how a catalyst works is often complex: for example, enzyme action in the body. However, there are two main theories which account for catalytic action in a simple manner.

[1] Surface Adsorption

In heterogeneous catalysis (two distinct phases) a surface is provided by a catalyst on which the reactants can react with each other. The process involves a physical or often a chemical interaction between the reactants and the surface of the catalyst. The physical closeness of one reactant to another allows them to react more quickly with each other.

When ethene gas reacts with hydrogen gas in the presence of a solid catalyst, nickel, ethane forms relatively easily. Both reacting gases, ethene and hydrogen, adsorb (bond) onto the nickel surface and thereby come in very close contact with each other. The product, ethane gas, formed as a result of the interaction then moves away (desorbs) from the nickel surface.

$$\overset{Ni}{C_2H_4(g) \ + \ H_2(g) \ \to \ C_2H_6(g)}$$
$$\text{ethene} \quad \text{hydrogen} \qquad \text{ethane}$$

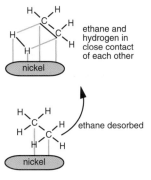

The greater the surface area of the catalyst, the easier it is for the ethene and the hydrogen to react with each other.

[2] Formation of an Intermediate Compound

This theory involves the formation of an intermediate compound which splits up to form the product more quickly than the uncatalysed reactants can.

Suppose a catalyst, C, is used to catalyse a slow reaction:

$$A + B \to AB \quad \text{Slow reaction}$$

175

The catalyst forms an intermediate in a fast reaction.

$$A + C \rightarrow A \sim C \qquad \text{Fast reaction}$$
$$\text{intermediate}$$

This intermediate reacts with B, decomposing rapidly to form the product AB and regenerate the catalyst.

$$A \sim C + B \rightarrow AB + C \quad \text{Fast reaction}$$

Because the reaction is so fast, it is usually not possible to isolate the intermediate compound formed. However, in some cases, such as the reaction of potassium sodium tartrate with hydrogen peroxide using cobalt(II) ions as a catalyst, the intermediate compound can be isolated.

Activity 6.1

Demonstration of the catalysed oxidation of methanol to methanal using a platinum or nichrome catalyst

Introduction

A catalyst speeds up a chemical reaction, in this case by providing a surface for the reactant to react on.

When methanol is heated no reaction occurs. In the presence of a hot platinum catalyst methanol (an alcohol) is oxidised to methanal (an aldehyde). The platinum catalyst speeds up the reaction.

$$\overset{\text{Pt}}{CH_3OH(l) \rightarrow} HCHO(g) + H_2(g)$$
$$\text{heat}$$

Requirements

Safety glasses, conical flask, glass rod, cardboard, Bunsen burner, methanol and platinum or nichrome wire.

Procedure

Your teacher will carry out this demonstration. Observe what happens carefully.
(1) Place about 25 cm³ of methanol in a conical flask.

(2) Heat the flask. No reaction occurs.
(3) Arrange the platinum wire into a small spiral around the glass rod and heat it until it is red hot.
(4) Hang the hot platinum spiral above the methanol as shown.
(5) Describe what happens.

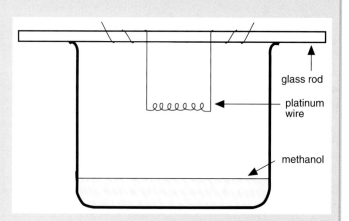

Caution

(a) Methanal is poisonous and lachrymatory (produces tears).
(b) Heat generated during the reaction often sets fire to the gases. If this happens, place a piece of cardboard over the flask for a moment. This extinguishes the flames and allows the reaction to proceed.

Activity 6.2

Demonstration of the catalysis by cobalt(II) ions of the reaction of potassium sodium tartrate and hydrogen peroxide

Introduction

Cobalt(II) ions have a distinct pink colour. When cobalt(II) ions are mixed with potassium sodium tartrate and hydrogen peroxide the colour of the mixture is pink. When the mixture is heated to about 70 °C the hydrogen peroxide starts to oxidise the potassium sodium tartrate.

As the reaction proceeds the pink colour changes to a blue-green colour because the cobalt ions have changed oxidation state. The cobalt ions are now part of an intermediate compound, which is blue-green in colour. The reaction ceases and the pink colour of the cobalt(II) ions can be seen again, indicating that the catalyst has been regenerated.

The reactions involved are quite complex and can be written simply as:

Potassium sodium tartrate +
Hydrogen peroxide + Cobalt(II) ions
Pink

↓

Intermediate complex
Blue-green

↓

Carbon dioxide + Water + Cobalt(II) ions
Pink

Requirements

Safety glasses, thermometer, stirrer, graduated cylinder, test-tube, teat pipette, 2 beakers (one containing iced water), Bunsen burner, tripod and gauze, deionised water, hydrogen peroxide solution (20 volume), potassium sodium tartrate (Rochelle salt), cobalt chloride.

Procedure

Your teacher will carry out this demonstration. Observe what happens carefully.

(1) Dissolve about 3 g of potassium sodium tartrate in about 50 cm^3 of water in a beaker. Heat the solution to about 70 °C.

(2) Add about 20 cm^3 of the hydrogen peroxide solution, and heat the solution to about 70 °C again.

(3) Dissolve about 0.3 g of cobalt chloride in about 5 cm^3 of water and add this to the hot solution (pink colour). Stir the contents.

(4) As soon as a blue-green colour appears, remove some of the solution, using a teat pipette. Place the solution in a test-tube in the beaker of iced water.

In this way, the intermediate complex is isolated for a very short time.

(5) As the reaction in the beaker ceases, the colour of the solution turns back to its original pink colour, indicating that the cobalt(II) catalyst has been regenerated.

4. The Surface Area of Solid Reactants

Most people realise that it is easier to start a fire using small sticks rather than logs. The reason for this is that the small sticks have a larger surface area. Dust explosions are another common example where finely divided particles increase the speed of a reaction. Many explosions have been caused by dust: fine particles of coal have resulted in explosions in mines, and even grain stores have exploded owing to the dust particles.

The rate of a chemical reaction involving a solid reactant is increased if the solid is subdivided.

Many chemical reactions involve a solid reacting with an aqueous solution of another substance. The reaction of calcium carbonate and dilute hydrochloric acid to produce carbon dioxide is a familiar example.

Three different samples of the calcium carbonate (marble) are allowed to react with equal volumes of hydrochloric acid of the same concentration, at the same temperature, and the reaction rates are monitored. The first sample is a large piece of marble, the second sample is marble chips, while the third sample is powdered marble.

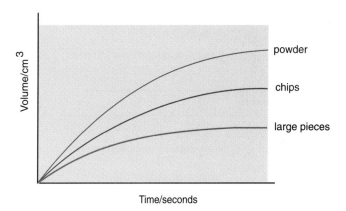

The reaction rate curves show that carbon dioxide is produced at a faster rate when the marble is in powdered form.

The reaction rate is much faster when the solid reactant is more finely subdivided. The more the solid is subdivided the greater is its surface area and consequently it is easier for the hydrochloric acid to come in contact with it.

177

This increase in the reaction surface area as the solid is subdivided is illustrated below.

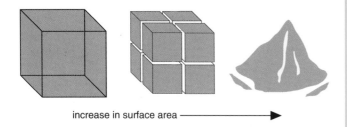

increase in surface area ⟶

The surface area of the large cube is

$$6 \times 2 \times 2 \text{ cm}^2 = 24 \text{ cm}^2$$

If the cube is divided to make eight smaller equally sized cubes the surface area becomes

$$8 \times 6 \times 1 \times 1 \text{ cm}^2 = 48 \text{ cm}^2$$

The surface area of the eight small cubes is greater than that of the first cube. If the cube is further subdivided into fine powder the surface area increases much more dramatically.

Activity 6.3

Does Surface Area Affect Reaction Rate?

Calcium carbonate reacts with dilute hydrochloric acid according to the equation:

$$CaCO_3(s) + 2HCl(aq) = CaCl_2(s) + H_2O(l) + CO_2(g)$$

The reaction can be carried out using the same amount of hydrochloric acid, but different sizes of limestone pieces in each case.

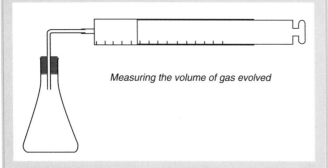

Measuring the volume of gas evolved

• Using the apparatus as shown, react 50 cm³ of 1.0 mol L⁻¹ hydrochloric acid with
(a) 20 g of powdered calcium carbonate,
(b) 20 g of small pieces of limestone, and
(c) 20 g of large pieces of limestone.
• Note the time taken for each experiment to finish.
• What do you notice about the results?

5. The Presence of Light – Photochemical Reactions

The presence of light energy speeds up the rate of some chemical reactions.

Silver salts are sensitive to light: thus silver nitrate is kept in a dark bottle. If silver salts are not shielded from the light they are converted into metallic silver. This photochemical reaction is the basis of photography, whereby a silver salt, silver bromide, is converted to metallic silver in the presence of light.

The most common photochemical reaction of all is photosynthesis, where carbohydrate is formed in plants from carbon dioxide and water in the presence of light. This reaction also uses chlorophyll as a catalyst.

The Nature of the Reactants

In general, ionic reactions are faster than covalent reactions.

There are some ionic reactions, such as those occurring in acid–base reactions, which occur almost instantaneously (10^{-9} s). There are also reactions that are so slow that they can take millions of years to complete (e.g. many geological processes). Reaction rates such as these are difficult to measure, while the rates of many other reactions are relatively easy to measure.

In the familiar reduction reactions of manganate(VII) ions using either iron(II) ions or ethanedioate ions as reducing agents, two completely different reaction rates are observed.

On the one hand, the iron(II) ions reduce the manganate(VII) ions rapidly to manganese(II) ions; this is seen by the fast discoloration of the manganate(VII) ions. On the other hand, the reduction of manganate(VII) ions by ethanedioate ions is slow; in this case the discoloration is slow and the reactants need to be heated above 60 °C to speed up the discoloration.

$$MnO_4^-(aq) + Fe^{2+}(aq) \rightarrow Mn^{2+}(aq) + Fe^{3+}(aq)$$
Fast reaction

$$MnO_4^-(aq) + H_2C_2O_4(aq) \rightarrow Mn^{2+}(aq) + CO_2(aq) + H_2O(aq)$$
Slow reaction

It seems to be in the nature of iron(II) ions to be fast and in the nature of ethanediote ions to be slow when they react with manganate(VII) ions.

Exercise 6.2

•

(1) List the factors which affect the rate of a chemical reaction.

(2) At a given temperature, dilute hydrochloric acid reacts with marble chips at a slower rate than with powdered marble.
(a) Explain why the reaction rates differ.
(b) Describe two other ways in which the speed of this reaction can be altered.

(3) For reactions in solution, the rate of reaction depends on the concentration of one or more reactants. In reactions between marble chips and dilute hydrochloric acid, where the concentration of the acid is varied and the marble chips are used in excess, it is found that the reaction rate increases with increasing hydrochloric acid concentration.
(a) Draw a simple sketch of the graph you would expect to obtain if the concentrations of the hydrochloric acid used were 0.5, 1.0 and 2.0 mol L^{-1}.
(ii) Explain why the reaction rate increases as the concentration of hydrochloric acid increases.

(4) (a) How does temperature affect the rate of a chemical reaction?
(b) Draw a profile of the energy distribution of the reacting particles in a chemical reaction.
(c) Explain how an increase in temperature changes the energy distribution of the reacting particles.
(d) What is the minimum amount of energy that reactants need to form products?

(5) (a) What is a catalyst?
(b) Name the catalysts which are used in:
(i) the Haber process, (ii) the Contact process and (iii) the manufacture of margarine.
(c) Draw a reaction profile showing how a catalyst lowers the energy of activation of a chemical reaction.

(6) Describe in terms of collision theory how the rate of a reaction in the gas phase is altered by each of the following:
(a) increasing the pressure (b) increasing the concentration (c) increasing the temperature (d) using a catalyst.

(7) Explain what is meant by each of the following:

(a) catalyst (b) inhibitor (c) catalyst poison (d) homogenous catalysis (e) heterogeneous catalysis (f) autocatalysis (g) enzyme.
In each case, give a relevant example to illustrate your answer.

(8) (a) The catalysed reaction between ethene and hydrogen occurs on a nickel surface. Explain and illustrate how this heterogenous reaction occurs by a surface adsorption mechanism.
(b) The reaction between cobalt(II) ions and sodium potassium tartrate occurs in solution by a different mechanism to the one above. Illustrate, simply, the mechanism by which this reaction proceeds.

(9) (a) What effect does a catalyst have on the activation energy of a reaction?
(b) What effect does temperature have on the activation energy of a reaction?
(c) What effect does pressure have on a reaction in:
(i) the gas phase and (ii) in solution?
(d) Explain why a small increase in temperature often results in a very large increase in reaction rate.

(10) Square-shaped pieces of aluminium foil, of different surface areas, were reacted in different experiments, each with the same amount of hydrochloric acid. Hydrogen gas was evolved in each case. The rate of reaction for each piece of aluminium is given below.

Area of aluminium/cm²	Rate of reaction/cm³ s⁻¹
1.00	120
1.69	204
3.25	390
4.41	528
6.25	750
8.41	1008

(a) Write an equation for the reaction.
(b) Draw suitable apparatus for the experiment.
(c) Plot a graph illustrating how the rate of reaction is related to surface area.
(d) Determine the rate of reaction using a piece of foil of surface area 5 cm², using the same amount of acid again.

(11) The reaction between sodium thiosulphate and hydrochloric acid is:

$$Na_2S_2O_3 + 2HCl = S + 2NaCl + H_2O + SO_2$$

In a series of experiments, the effects of changing the concentration of hydrochloric acid on the reaction rate were examined. Measured volumes of water, sodium thiosulphate solution and hydrochloric acid solution were poured into a 200 cm³ beaker. A cross was marked on a piece of filter paper and placed underneath the beaker. The reaction was started and the time taken for a yellow precipitate of sulphur to completely block out the cross was noted.
The results, taken at the same temperature, are given below.

Volume of water added /cm³	Volume of thiosulphate solution added/cm³	Volume of hydrochloric acid solution added/cm³	Time/s
40	100	10	167
30	100	20	83
20	100	30	55
10	100	40	42

(a) Describe, briefly, how the experiments were carried out.
(b) Plot a graph of 1/time against the volume of hydrochloric acid used.
(c) Why is this method called the 'initial rate method'?
(d) From your graph, determine how long it would take the cross to disappear if 25 cm³ of acid were added to 25 cm³ of water and reacted with 100 cm³ of thiosulphate solution.

(12) The reaction between sodium thiosulphate and hydrochloric acid is:

$$Na_2S_2O_3 + 2HCl = S + 2NaCl + H_2O + SO_2$$

In a series of experiments, the effects of changing the temperature on the reaction rate were examined. 20 cm³ of sodium thiosulphate solution and an excess of hydrochloric acid solution were poured into a 100 cm³ beaker. A cross was marked on a piece of filter paper and placed underneath the beaker. The reaction was started and the time taken for a yellow precipitate of sulphur to completely block out the cross was noted at different temperatures.

The results are given below.

Starting temperature/ °C	Time/s
60	6
50	10
40	16
30	27

(a) Plot a graph of temperature against the reaction rate.
(b) From your graph, determine the reaction rate at:
(i) 25 °C and (ii) 70 °C.
(c) Calculate the mass of sulphur obtained at the end of the reaction.
(d) State *three* other factors which would alter the rate of the reaction.

(13) Solutions of propane-1,2,3-triol (glycerol) in water of different concentrations were prepared. 20 cm³ of one of the propane-1,2,3 solutions was placed in a conical flask at room temperature. 10 cm³ of potassium manganate(VII) solution was then added to the flask. A stop-clock was started and the time taken for the purple colour of the manganate(VII) to disappear was measured. The procedure was repeated at room temperature for the remaining solutions. The results are given below.

Concentration of propane-1,2,3-triol /% w/w	Time for reaction/s
5	600
10	300
15	200
20	150
25	120

(a) Why did the purple colour of the potassium manganate(VII) disappear?
(b) Why was each experiment carried out at room temperature?
(c) Plot a graph of the results obtained.
(d) Using your graph, determine the time required for the reaction if a 30% w/w solution of propane-1,2,3-triol were used.
(e) Describe how the reactions would be affected by (i) raising the temperature by 10 °C and (ii) adding a few drops of sulphuric acid to the conical flask.

(f) Explain your observations in (e) in terms of the activation energy for the reaction.

(14) Oxygen gas was produced from hydrogen peroxide according to the equation:

$$2H_2O_2(l) = 2H_2O(l) + O_2(g)$$

The volume of gas produced at regular time intervals is given below.

Volume/cm³	Time/min
0	0
36	2
60	4
64	6
84	8
86	10
86	12

(a) Draw a sketch of the apparatus you would use.
(b) How would you ensure that the reaction was started at a precisely known time?
(c) Plot a graph of the results given.
(d) Which reading appears incorrect? What value would you expect?
(e) What was the average rate of the reaction?
(f) What was the instantaneous rate at 4 minutes? (taking the correct value in (d))
(g) Give *two* conditions which would slow down the speed of the reaction.

(15) The reaction between sodium thiosulphate and hydrochloric acid is:

$$Na_2S_2O_3 + 2HCl = S + 2NaCl + H_2O + SO_2$$

In a series of experiments, the effects of changing the temperature on the reaction rate were examined. 50 cm³ of sodium thiosulphate solution was placed in a conical flask. An X was drawn on a piece of paper underneath the flask. An excess of hydrochloric acid solution was added to the flask and the temperature was taken. The time taken for the X to disappear (= reaction time) was noted. The experiment was repeated at different temperatures and the following results were obtained:

Temperature/°C	20	30	40	50	60
Reaction time/s	330	210	135	90	55

(a) Plot a graph of temperature against the initial reaction rate.
(b) From your graph, determine the 'reaction time' at 35 °C.
(c) From the graph, determine the temperature, if the 'reaction time' was 2 min.
(d) What conclusion do you draw about the relationship between temperature and reaction rate?
(e) What reaction time would you expect at 30 °C if the concentration of thiosulphate was doubled?
(f) Sketch the graph you would expect to obtain when the concentration of the thiosulphate solution is plotted against reaction rate.

(16) Platinum, a solid catalyst, is used to catalyse the reaction:

$$H_2(g) + \tfrac{1}{2}O_2(g) = H_2O(l)$$

(a) Draw an energy profile diagram, showing the effect of a catalyst on the activation energy for the reaction.
(b) Is the catalysed reaction an example of homogeneous or heterogeneous catalysis? Give a reason for your answer.
(c) Give a brief outline, using diagrams, of the theory to explain how the platinum functions as a catalyst in the reaction.

Summary

Concentration	*Temperature*	*Catalyst*	*Surface area*
More particles come in contact More collisions More effective collisions	Faster moving particles More energetic collisions More high energy collisions with energy greater than the activation energy	Speeds up reaction by lowering energy of activation	Large surface area of solid More contact More collisions

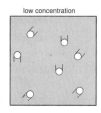
low concentration

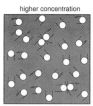
higher concentration

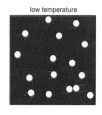
low temperature

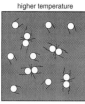
higher temperature

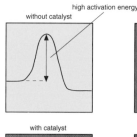
without catalyst — high activation energy — with catalyst — lower activation energy

small surface area

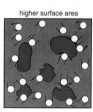
higher surface area

▼▼▼▼ Key Terms ▼▼▼▼

Some of the more important terms are listed below. Other terms not listed may be found by means of the index.

1. Rate of a reaction: The rate of a reaction is the change in concentration of any species involved in a reaction (either reactant or product) divided by the time taken for the change to occur.

2. Factors affecting reaction rates:
(1) The concentration of the reactants.
(2) The temperature of the system.
(3) The presence of catalysts.
(4) The surface area of solid reactants.
(5) The presence of light.
The nature of the reactants also determines whether a reaction is a slow reaction or a fast reaction.

3. Concentration of the reactants: For reactions in solution, the rate depends on the concentration of one or more of the reactants. Increase in concentration increases the number of successful collisions.

4. Temperature: The rate of a chemical reaction increases as temperature increases, regardless of whether the reaction is exothermic or endothermic. Increase in temperature increases the number of high energy reactants,

and this increases the number of successful collisions.

5. Catalysts: The rate of a chemical reaction is increased by the presence of a catalyst. Catalysts are substances which alter the rate of a chemical reaction and which are chemically unchanged themselves after the reaction has ended.

6. Catalysts and energy: For a reaction to occur an energy barrier must be overcome in order for products to form. The presence of a catalyst lowers that energy barrier. This, in effect, increases the number of reactants with sufficient energy to form products.

7. Activation energy: Activation energy is the minimum energy necessary for the reactants to acquire in order to overcome the energy barrier for a successful collision to occur.

8. Surface area: The rate of a chemical reaction involving a solid reactant is increased if the solid is subdivided so that it has a greater surface area.

9. Photochemical reactions: The presence of light energy speeds up the rate of some chemical reactions. The most well known photochemical reaction is photosynthesis.

10. Nature of the reactants: In general, ionic reactions are faster than covalent reactions.

• Unit 7 •

ORGANIC CHEMISTRY

Poly(ethene)-insulated copper conductors.

7.1 TETRAHEDRAL CARBON

Saturated organic compounds are compounds which contain single covalent bonds only.

Saturated compounds include
- alkanes
- chloroalkanes
- alcohols.

The bonding around each carbon atom in these compounds is **tetrahedral**.

Alkanes

The alkanes are saturated hydrocarbons. Each carbon has four electrons available for sharing; this means that it can form four bonds. In saturated compounds, each carbon atom has four atoms attached to it in a tetrahedral arrangement.

Methane is usually represented as:

$$H - \overset{\displaystyle H}{\underset{\displaystyle H}{\overset{|}{\underset{|}{C}}}} - H$$

where the central carbon atom is bound to each hydrogen atom in a **tetrahedral** arrangement.

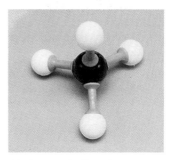

Methane – ball and spring model.

Methane – space filling model.

Other alkanes can be represented in a similar tetrahedral arrangement.

$$H - \overset{\displaystyle H}{\underset{\displaystyle H}{\overset{|}{\underset{|}{C}}}} - \overset{\displaystyle H}{\underset{\displaystyle H}{\overset{|}{\underset{|}{C}}}} - H$$
ethane

$$H - \overset{\displaystyle H}{\underset{\displaystyle H}{\overset{|}{\underset{|}{C}}}} - \overset{\displaystyle H}{\underset{\displaystyle H}{\overset{|}{\underset{|}{C}}}} - \overset{\displaystyle H}{\underset{\displaystyle H}{\overset{|}{\underset{|}{C}}}} - H$$
propane

Alkanes are excellent fuels: they burn in oxygen forming carbon dioxide and water and release large amounts of energy (see unit 5.4).

Chloroalkanes

Chloroalkanes are saturated compounds, where each carbon atom has four atoms attached to it in a **tetrahedral** arrangement. Chloroalkanes, like the fluorocarbons used in aerosol sprays, are members of the alkyl halide family. The smaller chloroalkanes, such as chloromethane and chloroethane, are gases, while the larger ones, like chlorobutane, are volatile liquids. They are non-polar compounds,

Honours • Honours • Honours • Honours • Honours • Honours • Honours • Honours • Honours • Honours • Honours • Honours • Honours • Honours • Honours •

Honours • Honours •

which means they are insoluble in water, but are soluble in non-polar solvents. They are extremely versatile reagents and are used to synthesise many other organic compounds. Chloromethane, CH_3Cl, and 1,1,1-trichloro-ethane, CCl_3CH_3, are excellent solvents.

Systematic Naming

Chloroalkanes are named in a similar way to alkanes.

(1) The parent chloroalkane is named by counting the maximum number of carbons in the longest chain containing the chlorine atom. The ending -ane is given as if the compound were an alkane.

(2) The position of the chlorine atom is indicated by the lowest possible number.

(3) Other substituting groups are indicated in the same way as in alkanes.

(4) The prefixes di-, tri- and tetra- are used to indicate two, three or four chlorine atoms present.

• • • • • Example 7.1 • • • • •

Use the IUPAC system to name the following chloroalkane.

1-chlorobutane

(1) The parent is butane: longest chain has four carbons.

(2) Chlorine is at carbon-3.

(3) The name of the compound is, therefore, 3-chlorobutane.

• • • • • Example 7.2 • • • • •

Use the IUPAC system to name the following chloroalkane.

(1) The parent is ethane: longest chain has two carbons.

(2) There are three chlorine atoms on carbon-1.

(3) The name of the compound is, therefore, 1,1,1-trichloroethane.

Alcohols

Alcohols are compounds of general formula R–OH and structural formula R–O–H, where R is any alkyl group.

We can think of an alcohol as a compound obtained by substituting a hydroxyl group, –OH, for a hydrogen atom, H, on an aliphatic hydrocarbon, e.g.

methane   methanol

Some examples include

| CH_3OH | C_2H_5OH | C_3H_7OH | C_4H_9OH |
| Methanol | Ethanol | Propanol | Butanol |

The bonding in these saturated alcohols is **tetrahedral**. An alcohol can be classified as primary, secondary or tertiary depending on the number of carbon atoms attached to the carbon atom bonded to the –OH group.

A primary alcohol has one such carbon atom, a secondary alcohol has two such carbon atoms, while a tertiary alcohol has three such carbon atoms. (Note: tertiary alcohols are outside the scope of the syllabus.)

RCH_2OH R_1R_2CHOH

ethanol   propan-2-ol
Primary – one   Secondary – two

Systematic Naming of Alcohols

(1) Alcohols are named by replacing the -e in the corresponding alkane by **-ol**.

(2) The parent alcohol is named by counting the number of carbon atoms in the longest chain.

ROH

(3) The position of the –OH group is given the lowest possible number.

(4) The position of other substituting groups is then indicated by a number.

The common names and the IUPAC names of some alcohols are given in the table below.

Formula	Common name	IUPAC name	Structural formula
CH_3OH	Methyl alcohol	Methanol	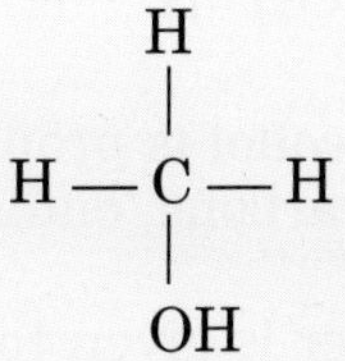
C_2H_5OH	Ethyl alcohol	Ethanol	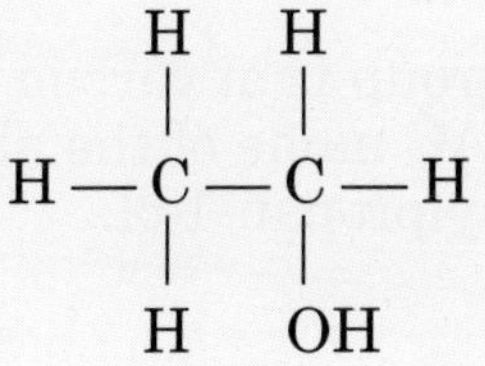
$CH_3CH_2CH_2OH$	n-Propyl alcohol	Propan-1-ol	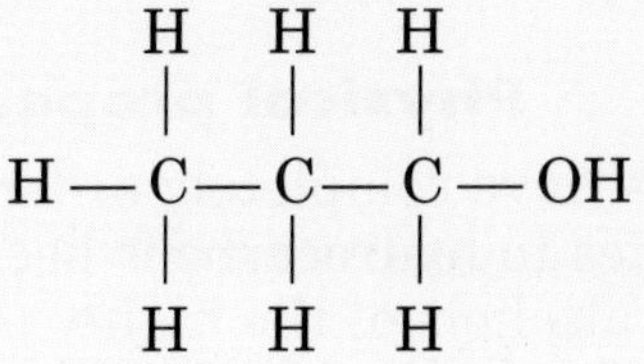
$CH_3CH(OH)CH_3$	Isopropyl alcohol	Propan-2-ol	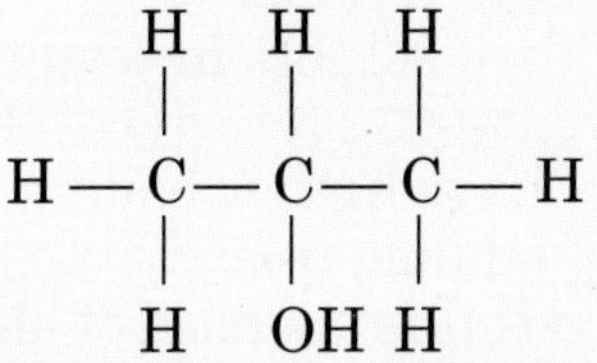
$CH_3CH_2CH_2CH_2OH$	n-Butyl alcohol	Butan-1-ol	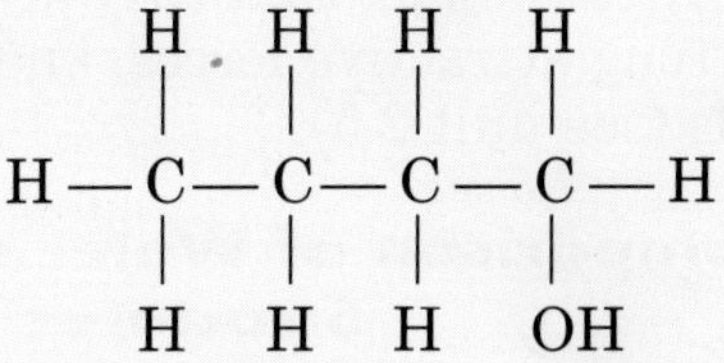
$(CH_3)_3OH$	t-Butyl alcohol	2-Methylpropan-2-ol	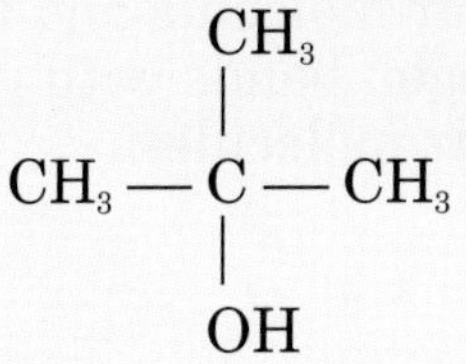

looks like $C(CH_3)_3OH$

- - - - - - **Example 7.3** - - - - - -

Give the IUPAC name for the following alcohol.

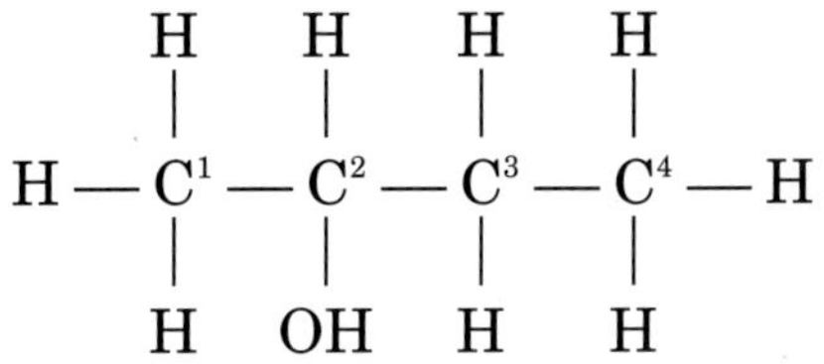

$$H—C^1—C^2—C^3—C^4—H$$
(with H, H, H, H on top and H, OH, H, H on bottom)

(1) The parent alcohol is butanol: longest chain has four carbons, ending -e becomes -ol.

(2) Lowest number is given to –OH group: carbon-2.

(3) The IUPAC name of the alcohol is, therefore, butan-2-ol.

185

•••••• **Example 7.4** ••••••

Give the IUPAC name for the following alcohol.

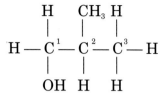

(1) The parent alcohol is propanol: longest chain has three carbons, ending -e becomes -ol.
(2) Lowest number is given to –OH group: carbon-1.
(3) Methyl group is at carbon-2.
(4) The IUPAC name of the alcohol is, therefore, 2-methylpropan-1-ol.

•••••••••••••••••••••••••••••

Physical properties

Alcohols have completely different physical properties to hydrocarbons because of the presence in alcohols of the highly polar –OH group.
• Alcohols are liquids.
• Alcohols are soluble in water and in non-polar solvents.
• Alcohols have higher boiling points than alkanes of similar relative molecular mass.

The physical properties of alcohols are due largely to the fact that they are held together by strong attractive forces, known as hydrogen bonds (see unit 2.5).

Comparison of Water and Alcohol Structures

Water is a polar molecule and is capable of forming hydrogen bonds with itself and with other non-polar molecules.

$$\overset{\delta^+}{H} - \overset{\delta^-}{O}$$
$$\qquad \diagdown \; \delta^+$$
$$\qquad\quad H$$

Alcohols such as ethanol, C_2H_5OH, are also polar, and can form hydrogen bonds. Alcohols have a small negative charge, δ^-, on the oxygen atom and a small positive charge, δ^+, on the hydrogen atom on the OH group. Alcohols are not as polar as water. The –OH group is polar, while the hydrocarbon part is non-polar.

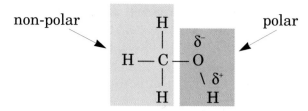

The smallest alcohols, methanol and ethanol, have a similar hydrogen-bonded structure to water. Methanol and ethanol are completely soluble in water because the hydrogen-bonded alcohol molecules can form hydrogen bonds with the water molecules, which are also linked by hydrogen bonds.

The strength of the hydrogen bonds is approximately 20 kJ mol^{-1}, which is much higher than the usual attractive forces (van der Waal's forces) between molecules, with strengths of about 1–2 kJ mol^{-1}. It should be remembered that the lower alkanes are gases because of the weak attractive forces between their molecules, whereas the lower alcohols are liquids because of the strong attractive forces between the alcohol molecules; these strong attractive forces are hydrogen bonds.

Alcohols are also soluble in non-polar solvents, such as methylbenzene. The non-polar end of the alcohol is attracted towards the non-polar methylbenzene by van der Waal's forces.

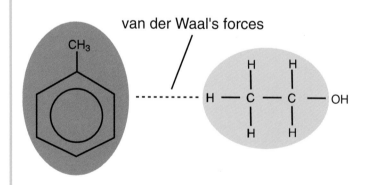

Main Source of Ethanol: Fermentation of Carbohydrates

Fermentation of sugars, from a variety of sources, such as molasses, grain, rice or potatoes, is one of the oldest chemical reactions carried out by man.

Enzymes, present in yeast, convert sucrose into a mixture of glucose and fructose and then convert the glucose into ethanol and carbon dioxide.

$$C_{12}H_{22}O_{11}(aq) + H_2O(l) \rightarrow 2C_6H_{12}O_6(aq)$$
sucrose   glucose

$$C_6H_{12}O_6(aq) \rightarrow 2C_2H_5OH(aq) + 2CO_2(g)$$
glucose   ethanol

Brewing of beers and lagers, distilling of vodkas and whiskies are all dependent on this fermentation process for the production of alcohol.

Industrial Uses of Alcohols

Methanol

• Used in the manufacture of methanal, which is used to make thermosetting plastics such as Bakelite.

• Used in the manufacture of Perspex.
• As a solvent for plastics and paints.
• As a denaturing agent.
• As a fuel when mixed with petrol.

Ethanol

• Is mainly used as a solvent.
• Is used in the manufacture of ethanal and many other chemicals.
• Is used as a fuel. Some countries, such as Brazil, grow large quantities of cane sugar which, when fermented, produces ethanol. This is a relatively cheap source of ethanol and the ethanol produced is mixed with petroleum to form a mixture called 'gasohol'.

Exercise 7.1

(1) (a) What is a saturated organic compound?
(b) Give two examples of saturated organic compounds from each of the following:
(i) alkanes (ii) chloroalkanes (c) alcohols.

(2) (a) What is meant by 'tetrahedral carbon'?
(b) Explain the concept using methane, CH_4, and ethane, C_2H_6, as examples.

(3) (a) Why are alkanes good fuels?
(b) Write chemical equations showing how methane, ethane and propane burn in oxygen.

(4) (a) How are chloroalkanes named?
(b) Name the following chloroalkanes:

(i)

```
      Cl
      |
H  —  C  —  Cl
      |
      H
```

(ii)

```
      Cl
      |
Cl —  C  —  Cl
      |
      H
```

(iii)

```
      H    H    H    H
      |    |    |    |
H  —  C  — C  — C  — C  — Cl
      |    |    |    |
      H    H    H    H
```

(iv)

```
      H    H    Cl   H
      |    |    |    |
H  —  C  — C  — C  — C  — H
      |    |    |    |
      H    H    H    H
```

(v)

```
      H    H    H    H
      |    |    |    |
H  —  C  — C  — C  — C  — H
      |    |    |    |
      H    Cl   Cl   H
```

(vi)

```
      H    Cl
      |    |
H  —  C  — C  — Cl
      |    |
      H    Cl
```

(vii)

```
      Cl   H
      |    |
H  —  C  — C  — H
      |    |
      Cl   Cl
```

(c) Draw structures for the following chloroalkanes:
(i) tetrachloromethane (ii) 1,2,3,4-tetrachlorobutane.

(5) (a) Explain why chloroalkanes are insoluble in water.
(b) Why are they soluble in non-polar solvents?
(c) Explain why chloromethane is a gas, while chlorobutane is a liquid.
(d) Show how chloroalkanes are tetrahedral in shape.

(6) (a) What are alcohols?
(b) How are alcohols similar to alkanes?
(c) How do alcohols differ from alkanes?
(d) Draw the structural formulae of the first four primary aliphatic alcohols.

(7) (a) Define (i) primary and (ii) secondary alcohols.
(b) Draw the structural formula of an example of each type of alcohol.

(8) (a) How are alcohols named using the IUPAC system?
(b) Name the following alcohol:

```
     H   H   H   H
     |   |   |   |
H —  C — C — C — C — H
     |   |   |   |
     H   Cl  OH  H
```

(9) (a) How is the structure of an alcohol similar to the structure of water?
(b) How does the structure of an alcohol affect its physical properties?

(c) The lower alkanes are gases, while the lower alcohols are liquids. Explain why they differ in this way, by referring to hydrogen bonding.
(d) Ethane is practically insoluble in water, whereas ethanol is completely miscible in all proportions. Explain why this is so.
(e) Alcohols are soluble in methylbenzene and other similar solvents. Use diagrams to explain why they are soluble in solvents of this type.

(10) Describe how you would determine whether (a) methanol, (b) ethanol and (c) butan-1-ol are soluble in (i) water and (ii) hexane.

(11) (a) What is the main industrial source of ethanol?
(b) What are the main uses of methanol and ethanol?
(c) What is 'gasohol'?

(12) Ethanol, like ethane, is tetrahedral. Explain.

7.2 PLANAR CARBON

Unsaturated Organic Compounds

Carbon compounds are not always tetrahedral: many compounds, like ethene, are **planar**.

```
H           H
 \         /
   C  =  C      ← plane
 /         \
H           H
```

Aldehydes, ketones, carboxylic acids and esters also contain planar carbon.

Alkenes

The alkenes are another hydrocarbon family. They are identified by the presence of a carbon–carbon double bond in the molecule. Because they have a double bond they are said to be **unsaturated compounds**.

The general formula is C_nH_{2n}, where $n = 2, 3, 4$ … and the structural formula is written as

```
R           H
 \         /
   C  =  C      where R = H or any alkyl group.
 /         \
H           H
```

The first three members of the family are given in the following table. They are all planar compounds.

Alkene	Formula	Structural formula	Physical state at room temperature	Molecular model
Ethene	C_2H_4	H⧵ /H C=C H/ ⧵H	Gas	

Alkene	Formula	Structural formula	Physical state at room temperature	Molecular model
Propene	C_3H_6	H—C—C=C—H (with H's)	Gas	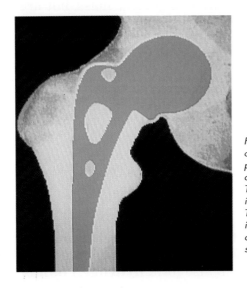
Butene	C_4H_8	H—C—C—C=C—H (with H's)	Gas	

False-colour X-ray of the human pelvis showing an artificial hip joint. The metal surface is coated with Teflon. Teflon is inert and provides a near frictionless surface.

Alkenes are used to make plastics, like poly(ethene) and poly(propene). See Option 2A.2

Systematic Naming of Alkenes

(1) The parent alkene is named by counting the maximum number of carbons in the longest chain containing the double bond. The ending -ane from the alkanes is changed to **-ene**, that is, all alkenes end in **-ene**.
(2) The position of the double bond is indicated by the lowest possible number.
(3) Other substituting groups are then indicated in the same way as in alkanes.

● ● ● ● ● **Example 7.5** ● ● ● ● ●

Name the following compound using the IUPAC system.

$$H—C^1—C^2=C^3—C^4—H$$

(1) The parent is butene: longest chain has four carbons.
(2) Double bond is at carbon-2.
(3) The name of the compound is, therefore, but-2-ene.

● ●

Carbonyl Compounds: Aldehydes and Ketones

This section is concerned with two closely related types of compound, the aldehydes and the ketones. Both of these types of compound contain the **carbonyl** or $>C=O$ group. The carbonyl group is an extremely important group; it is probably the backbone of synthetic organic chemistry. It is the most important component of some major biological compounds such as the carbohydrates and the steroids.

Aldehydes are compounds of general formula **RCHO** and structural formula

$$R—C=O$$
$$H$$

189

where R can be a hydrogen or an alkyl group such as a methyl or an ethyl group.

Examples include

$$\begin{matrix} H & & CH_3 \\ \backslash & & \backslash \\ C=O & & C=O \\ / & & / \\ H & & H \end{matrix}$$

methan**al**   ethan**al**
(formaldehyde) (acetaldehyde)

$$\begin{matrix} C_2H_5 \\ \backslash \\ C=O \\ / \\ H \end{matrix}$$

propan**al**
(propionaldehyde)

Ethanal, CH_3CHO, is the most typical member of the series. The C=O group, like the C=C bond in alkenes, is **planar**.

Systematic Naming of Aldehydes

The longest chain containing the –CHO group is the parent structure and is named by replacing the -e of the corresponding alkane by **-al**. The carbonyl carbon is always considered as carbon-1.

• • • • • **Example 7.6** • • • • •

Name the following compound using the IUPAC naming system.

$$\begin{matrix} & H & CH_3 & H \\ & | & | & | \\ H- & C^3- & C^2- & C^1=O \\ & | & | \\ & H & H \end{matrix}$$

(1) The parent aldehyde is propanal: longest chain has three carbons, ending -e becomes -al.
(2) The –CHO group is carbon-1.
(3) There is a methyl group is at carbon-2.
(4) The IUPAC name of the aldehyde is, therefore, 2-methylpropanal.

• •

Physical Properties

Aldehydes are polar compounds. The carbonyl group is permanently polarised because the electrons in the C=O double bond are unequally shared between the carbon and oxygen atoms because oxygen is more electronegative than carbon. There is a permanent δ^- charge on the oxygen and a permanent δ^+ charge on the carbon.

$$\begin{matrix} R \\ \backslash \delta^+ \quad \delta^- \\ C=O \\ / \\ H \end{matrix}$$

(1) The boiling points of aldehydes are higher than the corresponding non-polar hydrocarbons of comparable relative molecular mass, but are not as high as the corresponding alcohols because, unlike alcohols, aldehydes and ketones cannot form intermolecular hydrogen bonds.

Aldehydes can bind together by **dipole–dipole interactions**.

$$\begin{matrix} R & & R \\ \backslash \delta^+ \quad \delta^- & & \backslash \delta^+ \quad \delta^- \\ C=O & --- & C=O \\ / & & / \\ H & & H \end{matrix}$$

(2) Aldehydes of low relative molecular masses are soluble in water because not only are they polar but they can also form **hydrogen bonds** with the water molecules. The smallest, methanal and ethanal, are totally soluble in water. As the size of the aldehyde increases they become less soluble in water, but are soluble in the usual non-polar solvents such as 1,1,1-trichloroethane or methylbenzene.

$$\begin{matrix} R & & \\ \backslash \delta^+ \quad \delta^- & & \delta^+ \quad \delta^- \\ C=O & --- & H-O \\ / & & \backslash \\ H & & H \end{matrix}$$

Everyday Uses of Aldehydes

• **Methanal** (formaldehyde) is a gas and is usually handled by dissolving it in water. It is marketed as a solution called formalin (37% w/w), which is used as a disinfectant and as a preservative for many biological specimens. Its principle use is in the manufacture of various plastics and resins such as Bakelite or melamine.

• **Ethanal** (acetaldehyde) is used to make ethanoic acid, acetonitrile, polyvinyl acetate polymers used in records and to make gum resins for postage stamps.

• **Benzaldehyde** is found in almonds. It is used as a flavouring.

Ketones are compounds of general formula **RCOR′** and structural formula

$$
\begin{array}{c}
R \\
\backslash \\
C=O \\
/ \\
R'
\end{array}
$$

where R and R′ are alkyl groups which are often the same alkyl group.

Examples include

$$
\begin{array}{cc}
CH_3 & C_2H_5 \\
\backslash & \backslash \\
C=O & C=O \\
/ & / \\
CH_3 & CH_3 \\
\text{propanone} & \text{butanone} \\
\text{(acetone)} & \text{(methyl ethyl ketone)}
\end{array}
$$

Systematic Naming of Ketones

The longest chain containing the carbonyl (>C=O) group is the parent structure and is named by replacing the -e of the corresponding alkane by **-one**. The carbonyl group is given the lowest number possible.

•••••• **Example 7.7** ••••••

Draw the structure of the compound, $CH_3COC_2H_5$. Name the compound using the IUPAC naming system.

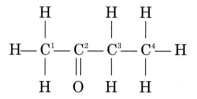

(1) The longest chain has four carbons. The parent ketone is butanone.
(2) The carbonyl group is carbon-2. There are no other substituting groups.
(3) The IUPAC name of the ketone is, therefore, butanone.

Physical Properties

Ketones, like aldehydes, are polar compounds. There is a permanent δ^- charge on the oxygen and a permanent δ^+ charge on the carbon.

$$
\begin{array}{c}
R \\
\backslash \delta+ \quad \delta^- \\
C=O \\
/ \\
H
\end{array}
$$

(1) The boiling points of ketones are high, because, like aldehydes, ketones can bind together by **dipole–dipole interactions**.

$$
\begin{array}{cc}
R & R \\
\backslash \delta^+ \quad \delta^- & \backslash \delta^+ \quad \delta^- \\
C=O - - - & C=O \\
/ & / \\
R & R
\end{array}
$$

(2) Ketones like propanone are soluble in water because not only are they polar, but they can also form **hydrogen bonds** with the water molecules. As the size of the ketone increases they become less soluble in water, but are soluble in the usual non-polar solvents such as 1,1,1-trichloroethane or methylbenzene.

$$
\begin{array}{cc}
R & \\
\backslash \delta^+ \quad \delta^- & \delta^+ \quad \delta^- \\
C=O - - - & H-O \\
/ & \backslash \\
R & H
\end{array}
$$

191

Everyday Uses of Ketones

Propanone is one of the most important organic solvents. It not only dissolves a wide variety of non-polar organic compounds but it is also miscible in water in all proportions. It is generally used to make nail-polish remover and many other solvents. It is used in the manufacture of clear sheeting (Perspex), adhesives and surface coatings.

The common names and the IUPAC names of some aldehydes and ketones are given in the table below.

Aldehyde = R(CHO) *Ketone = RCOR*

Formula	Common name	IUPAC name	Structural formula	Molecular model
HCHO	Formaldehyde	Methanal	$\begin{array}{c} H \\ | \\ H-C=O \end{array}$	
CH_3CHO	Acetaldehyde	Ethanal	$\begin{array}{c} H \quad H \\ | \quad | \\ H-C-C=O \\ | \\ H \end{array}$	
CH_3CH_2CHO	Propionaldehyde	Propanal	$\begin{array}{c} H \quad H \quad H \\ | \quad | \quad | \\ H-C-C-C=O \\ | \quad | \\ H \quad H \end{array}$	
$CH_3CH_2CH_2CHO$	Butraldehyde	Butanal	$\begin{array}{c} H \quad H \quad H \quad H \\ | \quad | \quad | \quad | \\ H-C-C-C-C=O \\ | \quad | \quad | \\ H \quad H \quad H \end{array}$	
CH_3COCH_3	Acetone	Propanone	$\begin{array}{c} H \quad O \quad H \\ | \quad || \quad | \\ H-C-C-C-H \\ | \qquad | \\ H \qquad H \end{array}$	
$CH_3COCH_2CH_3$	Methylethyl ketone	Butanone	$\begin{array}{c} H \quad O \quad H \quad H \\ | \quad || \quad | \quad | \\ H-C-C-C-C-H \\ | \qquad | \quad | \\ H \qquad H \quad H \end{array}$	

Carboxylic acids

Carboxylic acids are compounds of general formula R–COOH
and structural formula

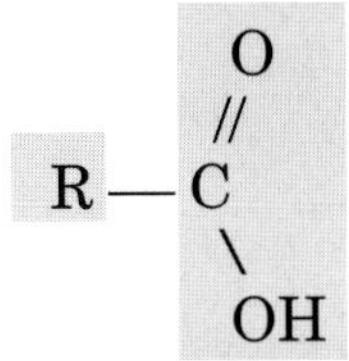

$$
\begin{array}{c}
O \\
\| \\
R - C \\
\backslash \\
OH
\end{array}
$$

where R can be a hydrogen or an alkyl group. They contain a polar carbon–oxygen double bond.

We can think of a carboxylic acid as a compound obtained by substituting a carboxyl group, –COOH, for a methyl group, CH_3, on an aliphatic hydrocarbon, e.g.

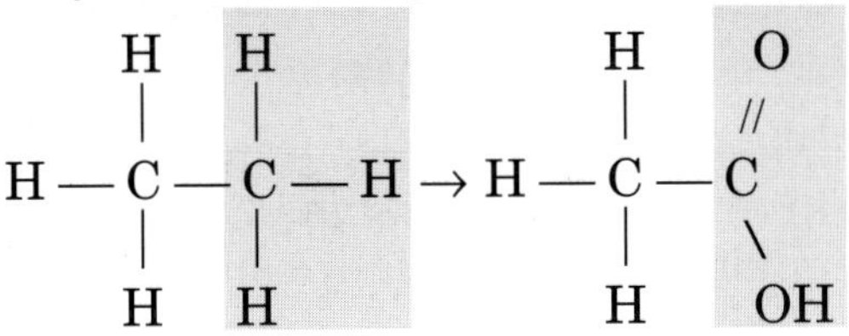

$$
\begin{array}{ccc}
H \quad H & & H \quad O \\
| \quad | & & | \quad \| \\
H - C - C - H & \rightarrow & H - C - C \\
| \quad | & & | \quad \backslash \\
H \quad H & & H \quad OH \\
\text{eth\textbf{ane}} & & \text{ethan\textbf{oic acid}}
\end{array}
$$

Some examples include:

HCOOH	CH_3COOH
methanoic acid	ethanoic acid
C_2H_5COOH	C_3H_7COOH
propanoic acid	butanoic acid

Ethanoic acid, CH_3COOH, is a typical member of the series. Carboxylic acids also contain the planar C=O group.

Systematic Naming of Carboxylic Acids

(1) Carboxylic acids are named by replacing the -e in the corresponding alkane by **-oic acid**. The longest straight chain is the parent structure.
(2) The position of the –COOH group is always carbon-1.
(3) Other substituting groups are then indicated by the appropriate number.

The common names and the IUPAC names of some carboxylic acids are given in the table below.

Formula	Common name	IUPAC name	Structural formula	Molecular model
HCOOH	Formic acid	Methanoic acid	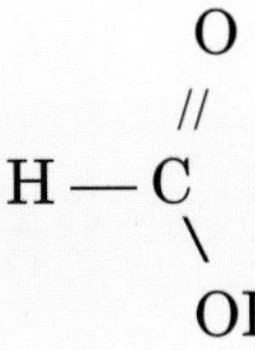 $\begin{array}{c} O \\ \| \\ H - C \\ \backslash \\ OH \end{array}$	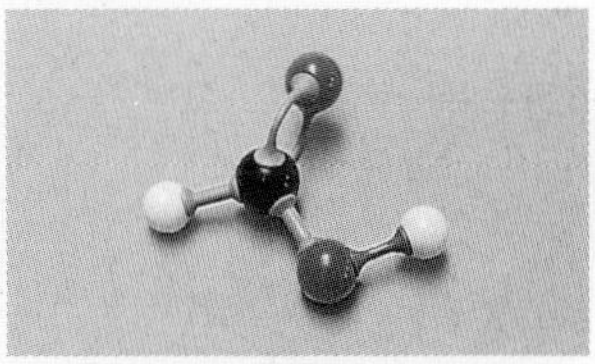
CH_3COOH	Acetic acid	Ethanoic acid	$\begin{array}{c} H \quad O \\ | \quad \| \\ H - C - C \\ | \quad \backslash \\ H \quad OH \end{array}$	
C_2H_5COOH	Propionic acid	Propanoic acid	$\begin{array}{c} H \quad H \quad O \\ | \quad | \quad \| \\ H - C - C - C \\ | \quad | \quad \backslash \\ H \quad H \quad OH \end{array}$	
C_3H_7COOH	Butyric acid	Butanoic acid	$\begin{array}{c} H \quad H \quad H \quad O \\ | \quad | \quad | \quad \| \\ H - C - C - C - C \\ | \quad | \quad | \quad \backslash \\ H \quad H \quad H \quad OH \end{array}$	

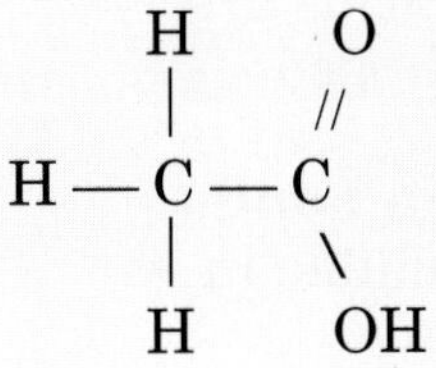

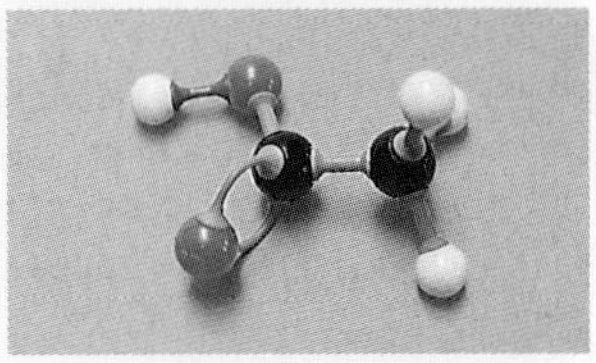

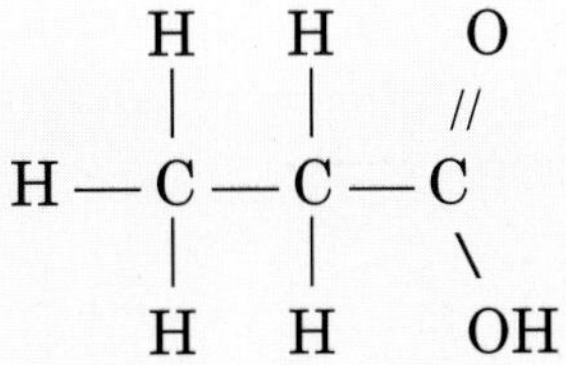

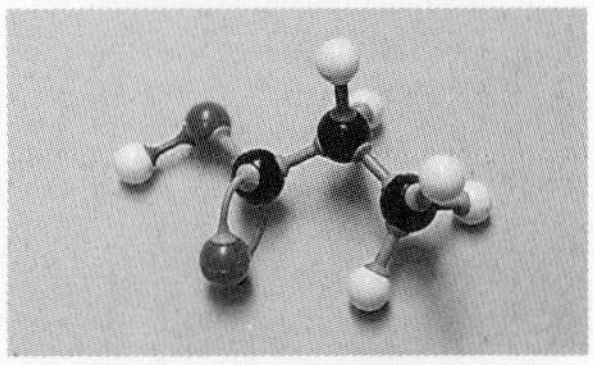

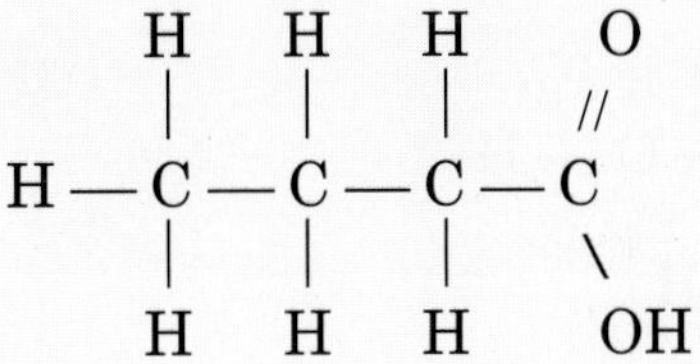

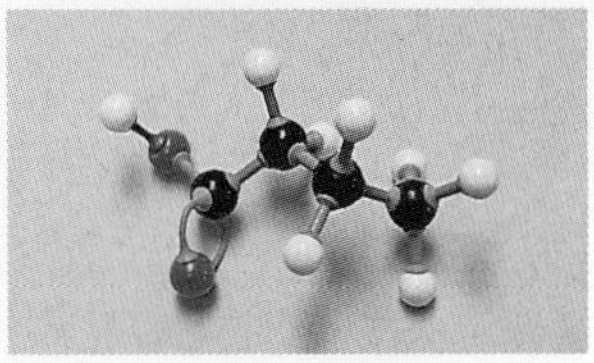

· · · · · · **Example 7.8** · · · · · ·

Give the IUPAC name for the following carboxylic acid.

$$
\begin{array}{ccc}
H & CH_3 & O \\
| & | & // \\
H-{}^3C-{}^2C-{}^1C \\
| & | & \backslash \\
H & H & OH
\end{array}
$$

(1) The parent carboxylic acid is propanoic acid: the longest straight chain contains four carbons, -e becomes -oic acid.
(2) The carboxyl group is carbon-1, therefore the methyl group is at carbon-2.
(3) The IUPAC name of the carboxylic acid is, therefore, 2-methylpropanoic acid.

· ·

Physical properties

The carboxyl group is highly polar, because it ① not only contains a hydroxyl group which can form hydrogen bonds, but it also contains a ② carbonyl group which has a high electron density. The high electron density can be redistributed throughout the carboxyl group causing the –OH bond to be more polar than normal.

$$
\begin{array}{c}
\delta^- \\
O \\
// \\
R-C^{\delta^+} \\
\backslash \delta^- \quad \delta^+ \\
O-H
\end{array}
$$

Owing to their polarity, carboxylic acids are capable of forming even stronger hydrogen bonds than alcohols.

As a result of the **strong hydrogen bonds**:
(1) the smaller carboxylic acids are soluble in water; the solubility decreases with an increase in the size of the non-polar alkyl group.

$$
\begin{array}{c}
H \\
\delta^- \quad \delta^+ \quad / \\
O---H-O \\
// \\
R-C \\
\backslash \delta^- \quad \delta^+ \quad \delta^- \\
O-H---O-H \\
/ \\
H
\end{array}
$$

(2) the boiling points are even higher than those of the corresponding alcohols, because carboxylic acids can be held together by two hydrogen bonds between the molecules.

$$
\begin{array}{c}
\delta^- \qquad \delta^+ \\
O---H-O \\
// \qquad\qquad \backslash \\
R-C \qquad\qquad C-R \\
\backslash \quad \delta^+ \quad \delta^- // \\
O-H---O
\end{array}
$$

The smaller carboxylic acids are liquids with pungent odours. Methanoic acid is the acid found in nettles. Ethanoic acid smells of vinegar, while butanoic acid smells of rancid butter (which is partly butanoic acid).

Social Impact of Carboxylic Acids

• Ethanoic acid is used on a large scale as a solvent and as a starting material for the production of many important esters, such as ethenyl ethanoate, which is used to produce cellulose acetate. Vinegar is about 5% ethanoic acid, which is often produced by fermentation using suitable enzymes.
• Lactic acid (2-hydroxypropanoic acid) is the acid found in sour milk and is also produced by the muscles during vigorous exercise often causing tightening of the muscles (cramp).
• Acetylsalicyclic acid, better known as aspirin, is one of the most common pain killers used and has been marketed since 1899 by Bayer.
• Propanoic acid, benzoic acid and their salts are often used as food preservatives.

Mandatory experiment 17

To recrystallise benzoic acid and determine its melting point.

Introduction

Recrystallisation is often used to separate and purify organic solids. The substance to be purified is dissolved in the minimum amount of hot solvent and this is filtered to remove any insoluble impurities. The filtrate is then cooled to give crystals. The solvent used depends on the particular substance to be dissolved.

The solvent used should increasingly dissolve the solute as the temperature is increased. Common solvents include water, methanol, ethanol, ethanoic acid and 1,1,1-trichloroethane.

In this case, impure benzoic acid is dissolved in hot water. The solution is filtered rapidly to remove the insoluble impurities and allowed to cool. The solid then crystallises out and is filtered under reduced pressure using a Buchner funnel and flask as shown. The crystals are dried and tested for purity by determining their melting point.

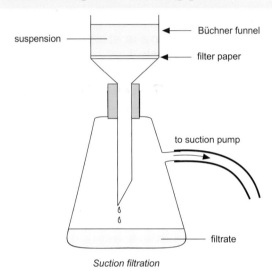

Suction filtration

Accurate melting points of organic compounds can be determined by heating a small sample of the solid in a melting point tube; the melting point can be determined to within 1 K. The melting point of benzoic acid can be determined by heating the sample in a melting point tube in a boiling tube containing liquid paraffin.

Requirements

Safety glasses, 100 cm^3 beaker, benzoic acid, Buchner funnel, vacuum pump, filter paper, melting point tubes, rubber band, thermometer, Bunsen burner, boiling tube fitted with three-holed cork to accommodate thermometer and stirrer, liquid paraffin and clock glass.

Procedure

Recrystallisation

(1) Place about 5 g benzoic acid in a clean 100 cm^3 beaker.
(2) Dissolve in the minimum amount of hot water.
(3) Set up the Buchner funnel as shown.

(4) Filter the hot solution through the filter paper in the Buchner funnel.
(5) Allow the filtrate to cool slowly.
(6) Filter off the excess solvent and allow the crystals to dry.
(7) Determine the melting point of the benzoic acid crystals.

Determination of Melting Point

(1) Take a melting point tube and push the open end into a sample of the benzoic acid on a clock glass until some crystals enter the tube.
(2) Tap the closed end of the tube vertically on the laboratory bench until the crystals fall to the bottom. Repeat the procedure until about 0.5 cm of the solid is at the bottom of the tube. Prepare two or three other samples in a similar way.
(3) Using a rubber band, attach a prepared sample to the thermometer as shown.

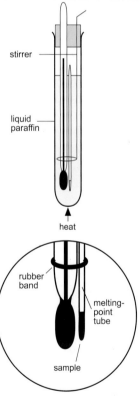

Melting point determination

(4) Half fill the boiling tube with liquid paraffin and position the thermometer, the attached melting point tube and stirrer as shown.
(5) Place the apparatus over a Bunsen burner and gently heat the apparatus, stirring the paraffin continuously with the stirrer.

(6) Keep watching the crystals until they begin to melt. Record the temperature and use it as a guide to determine a more accurate melting point.

(7) Remove the Bunsen burner and allow the temperature to drop to about 10 °C below the rough melting point. Place a fresh melting-point tube containing another sample of benzoic acid in the apparatus.

(8) Repeat the above procedure, but heat the paraffin very slowly this time (approximately 2 °C per minute).

(9) Record the temperature at which the crystals melt. Melting is indicated by the formation of a visible meniscus.

(10) Check with your teacher that you have obtained the correct value.

Esters

Esters are compounds in which the acidic hydrogen of a carboxylic acid has been replaced by an alkyl group.

Members of the ester family have strong pleasant fruity smells. Many occur naturally in fruits, such as bananas or pineapples, and flowers, and have very distinctive odours. Esters are often used to provide flavour and odour in many products such as sweets, cakes and cosmetics.

Esters contain a carbonyl group, $>C=O$, which means that they are polar compounds.

Because they are polar:

(1) the smaller esters are soluble in water.

(2) the boiling points are similar to those of aldehydes and ketones. The boiling points are lower than alcohols (and carboxylic acids) because, unlike these, they cannot form hydrogen bonds with each other.

Esters are usually liquids or solids. The smaller esters (those with up to four carbons) are soluble in water. They are all soluble in non-polar solvents such as 1,1,1-trichloroethane and methylbenzene.

Systematic Naming of Esters

Esters are named by changing the ending -oic acid from the carboxylic acid to -oate and then putting the alkyl prefix from the alcohol in front.

• • • • • • **Example 7.9** • • • • • •

Name the following ester using the IUPAC naming system.

(1) The ester is derived from ethanoic acid. Therefore, the ending is ethanoate.

(2) The prefix is methyl.

(3) The name of the ester is, therefore, methyl ethanoate. ? *looks like methyl propanoate*

Natural Esters: Fats and Oils

Animal fats and vegetable oils are esters formed between the tri-hydroxy alcohol, propane-1,2,3-triol (also called glycerol or glycerine) and long-chain carboxylic acids, such as octadecanoic acid (stearic acid), e.g.

acid	alcohol	ester
$C_{17}H_{35}COOH$	$HOCH_2$	$C_{17}H_{35}COOCH_2$
$+$		
$C_{17}H_{35}COOH$	$HOCH_2 \rightarrow$	$C_{17}H_{35}COOCH$
$+$		
$C_{17}H_{35}COOH$	$HOCH_2$	$C_{17}H_{35}COOCH_2$
octadecanoic acid	propane-1,2,3-triol	fat

Many other long-chain carboxylic acids, some of which are saturated, such as lauric acid, others which are unsaturated, such as oleic acid, can form fats or oils. Most fats and oils often contain two or three fatty acids.

Vegetable oils, which contain unsaturated fatty acids, are liquids because unsaturated fatty acids with double bonds do not pack easily together. Animal fats are usually saturated solids.

Fats are a rich source of chemical energy. Fats are used by animals as a long-term energy reservoir and also to insulate the body against cold.

The names of some esters are given in the table below.

Formula	Common name	IUPAC name
$HCOOCH_3$	Methyl formate	Methyl methanoate
$HCOOC_2H_5$	Ethyl formate	Ethyl methanoate
CH_3COOCH_3	Methyl acetate	Methyl ethanoate
$CH_3CH_2COOCH_3$	Ethyl acetate	Ethyl ethanoate

Esters in Industry

• Esters are used as flavours and odorisers in the food and cosmetics industries.
• Ethyl ethanoate and other lower esters are used as solvents for many paints, varnishes, adhesives and glues.
• Animal fats and vegetable oils are natural esters. Fats can be converted into soaps by reaction with alkalis.
• Many esters are polymerised to form polyesters which are used to make a variety of goods, such as plastic film, Perspex and Terylene.

Chemistry in Action

Soaps and Detergents

Fats can be converted into soaps by alkaline hydrolysis (saponification) with sodium hydroxide. The ester (fat) is hydrolysed by sodium hydroxide and forms the sodium salt of the carboxylic acid (soap) and an alcohol, glycerol.

$$C_{17}H_{35}COOCH_2$$
$$|$$
$$C_{17}H_{35}COOCH + 3NaOH \rightarrow 3C_{17}H_{35}COONa + HO—CH$$
$$|$$
$$C_{17}H_{35}COOCH_2 \qquad\qquad HO—CH_2$$

fat   alkali   soap   glycerol (propane-1,2,3-triol)

The structure of this particular soap, sodium stearate, or any other soap can be given simply as:

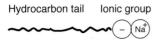

Hydrocarbon tail   Ionic group

It consists of a long non-polar hydrocarbon tail and an ionic group at the other end. The soap acts as if it were a hydrocarbon and a salt at the same time. The hydrocarbon tail is soluble in other non-polar substances, such as oil or grease, whereas the ionic end is soluble in water. Soap acts by forming an emulsion between two immiscible liquids, oil and water.

How soap cleans

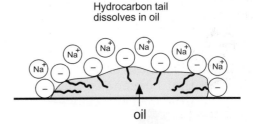

Hydrocarbon tail dissolves in oil

oil

Here the hydrocarbon tails point inwards towards each other and away from the water while the ionic ends point towards the water. The oil or grease is absorbed into the hydrophobic (water-hating) centres to form an aggregate and then is washed away.

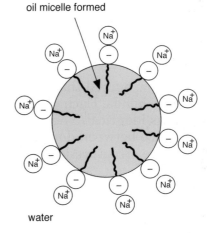

oil micelle formed

water

Soap also acts by lowering the surface tension of water, enabling the surface water molecules to move apart from each other and be absorbed onto the fabric being cleaned.

Soap was the first detergent used by man and is still the most common. However, other detergents, called soapless detergents, are now replacing soap, because they do not form a scum with hard water. Their structure is similar to soap, containing a hydrocarbon tail and an ionic head, but with some modifications ensuring that the detergent does not react with calcium ions in hard water. The alkylbenzene sulphonates are common examples of soapless detergents.

Aromatic Compounds

Benzene is an aromatic hydrocarbon: it is a cyclic unsaturated hydrocarbon. The term aromatic was first used to describe a group of compounds which had a pleasant aroma or smell. The group of compounds included benzene and its many derivatives. Today, the term aromatic is used to describe a range of compounds which have a similar structure to benzene. The aromatic compounds have many similar properties, unlike the aliphatic compounds, which, as we have seen, have many completely different properties.

The molecular formula of benzene is C_6H_6. This indicates that benzene is unsaturated. Possible structures for benzene were proposed showing an arrangement consisting of alternate and single bonds.

However, it is now agreed that benzene does not contain double bonds. Each carbon–carbon bond in benzene is identical. The bonds are intermediate in size between double and single bonds. The benzene structure is represented by

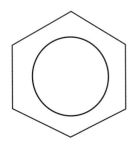

where the hexagonal corners represent H–C bonds and the inner circle represents the six equivalent carbon–carbon bonds.

The six carbon–carbon bonds in benzene are of equal length (0.139 nm). The length of a carbon–carbon single bond is 0.154 nm and that of a carbon–carbon double bond is 0.134 nm.

Sigma and Pi Bonding

The structure of benzene can be explained in terms of sigma and pi bonds (see Unit 2.3). Each carbon atom has four electrons available for bonding. Three of these electrons can form

strong sigma bonds: two with the adjacent carbon atoms and one with a hydrogen atom. The remaining available electrons on the carbon atoms form pi bonds by overlapping above and below the plane. The pi orbitals cover the entire carbon–carbon ring and the pi electrons are said to be delocalised.

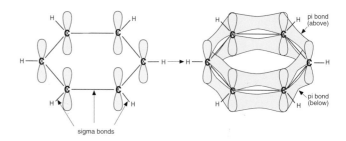

Carbon–carbon bond lengths

		Bond length/nm
Alkane	C — C	0.154
Benzene ring	C -- C	0.139
Alkene	C = C	0.134

Range of Aromatic Chemistry

Since the beginning of time man has explored the world in search of natural products. Many aromatic compounds have common names that indicate where they come from. Toluene, $C_6H_5CH_3$, is an aromatic resin coming from *Tolu balsam*, a South American tree. Many of the great explorers, such as Columbus and Magellan, traded in spices and herbs: this enabled the early chemists to try and isolate flavours and fragrances from these natural products. Many of these aromatic substances have fairly simple structures similar to benzene.

The range of aromatic chemistry is vast. We encounter aromatic compounds in almost every aspect of our lives.

- **Fuels**: Coal is composed of a large arrangement of aromatic molecules.
- **Solvents**: Toluene and benzene are common solvents.
- **Flavours**: Benzaldehyde is extracted from bitter almonds. Vanillin is extracted from vanilla beans.
- **Dyes**: Naphthalene is used in the synthesis of dyes.
- **Detergents**: Soapless detergents contain benzene rings.
- **Herbicides**: DDT was used during World War II to kill mosquitoes. Today its use is restricted, because it accumulates in fat tissues

198

with harmful effects. Modern herbicides, such as atrazine and chlorosulfuron, also contain benzene rings.
• **Hormones**: Oestrogen is the main female sex hormone.
• **Vitamins**: Many vitamins, such as vitamin K, are aromatic compounds.
• **Pharmaceuticals**: Aspirin, paracetamol and ibuprofen are aromatic pain killers.
• **Indicators**: Methyl orange and phenolphthalein are aromatic acid–base indicators.

• **Polymers**: Many polymers, such as polystyrene, Terylene and Kevlar, contain aromatic rings.

Many aromatic compounds are known carcinogens. Benzene is rarely used as a solvent because it causes cancer. Lung cancer is caused by an aromatic compound. However, most aromatic compounds in consumer use are non-carcinogenic. Aspirin, paracetemol and ibuprofen are non-carcinogenic.

Activity 7.1

Identification of Some Consumer Products with Benzene Rings

The following consumer compounds contain benzene rings. Examine the packaging of household goods and see if you can identify more products which contain benzene rings.

Structure	*Name*	*Use or Effect*
Naphthalene	Naphthalene	Mothballs and dyes
Aspirin	Aspirin	Painkiller
Terylene	Terylene	Material
DDT	DDT	Herbicide
Benzaldehyde	Benzaldehyde	Oil of almonds

Activity 7.2

Demonstration of Solubility of Organic Compounds

This activity can be carried out in parts as each type of organic compound is encountered or may be carried out as one entire activity as a revision exercise as shown.

Required

Safety glasses, water, cyclohexane, methanol, ethanol, butan-1-ol, propanone, ethanoic acid, ethyl ethanoate, methylbenzene, disposable Pasteur pipettes, test-tubes and rack.

Procedure

(1) Place about 5 cm³ of water into seven different test-tubes.
(2) Place about 5 cm³ of cyclohexane into another seven different test-tubes.
(3) In turn, place about 1 cm³ of methanol, ethanol, butan-1-ol, propanone, ethanoic acid, ethyl ethanoate, and methylbenzene into one test-tube containing water and one test-tube containing cyclohexane.
(4) Shake each test-tube and observe what happens.
(5) Copy and complete the table.
(6) Discuss your conclusions.

Results

Solute	Type	Solubility in water	Solubility in cyclohexane
Methanol	Polar		
Ethanol	Polar + non-polar		
Butan-1-ol	Polar + non-polar		
Propanone	Polar + non-polar		
Ethanoic acid	Polar + non-polar		
Ethyl ethanoate	Polar + non-polar		
Methylbenzene	Non-polar		

Exercise 7.2

(1) (a) What does the term 'planar carbon' mean?
(b) Draw the structures of some compounds which have planar carbon.

(2) (a) What are the alkenes?
(b) Name and give the formulae of the first four alkenes.

(3) (a) What is the general formula of the alkenes? Give the names and the molecular formulae of the first four alkenes.
(b) Draw the structural formulae of the first four alkenes.
(c) Name the following alkenes using the IUPAC system:

(i)

$$H-\overset{\overset{\displaystyle H}{|}}{\underset{\underset{\displaystyle H}{|}}{C}}-\overset{\overset{\displaystyle H}{|}}{C}=\overset{\overset{\displaystyle H}{|}}{C}-H$$

(ii)

$$H-\overset{\overset{\displaystyle H}{|}}{C}=\overset{\overset{\displaystyle H}{|}}{\underset{\underset{\displaystyle C}{|}}{C}}-\overset{\overset{\displaystyle H}{|}}{\underset{\underset{\displaystyle H}{|}}{C}}-H$$

(d) Draw the structures of the following alkenes:
(i) but-1-ene (ii) but-2-ene (iii) prop-2-ene.

(4) Explain why ethene has similar physical properties to ethane.

(5) What is the structural difference between an aldehyde and a ketone?

(6) Give the formula and draw the structure of each of the following aldehydes:
(a) methanal (b) ethanal (c) propanal.

(7) (a) How are aldehydes named according to the IUPAC system?
(b) Name the following aldehydes:

(i)

$$H-\overset{\overset{\displaystyle H}{|}}{C}=O$$

(ii)

$$H-\overset{\overset{\displaystyle H}{|}}{\underset{\underset{\displaystyle H}{|}}{C}}-\overset{\overset{\displaystyle H}{|}}{\underset{\underset{\displaystyle H}{|}}{C}}-\overset{\overset{\displaystyle H}{|}}{C}=O$$

(iii)

$$H-\overset{\overset{\displaystyle H}{|}}{\underset{\underset{\displaystyle H}{|}}{C}}-\overset{\overset{\displaystyle H}{|}}{\underset{\underset{\displaystyle H-C-H}{|}}{C}}-\overset{\overset{\displaystyle H}{|}}{C}=O$$
$$|$$
$$H$$

(8) (a) Why are some aldehydes soluble in water?
(b) Which alkanes correspond to the following aldehydes: (i) methanal, (ii) ethanal and (iii) propanal?
(c) Why are the boiling points of aldehydes higher than those of the corresponding alkanes?
(d) Mention some everyday uses of some named aldehydes.

(9) Describe how you would determine whether ethanal is soluble in (a) water and (b) hexane.

(10) Why are the boiling points of aldehydes higher than the corresponding hydrocarbons of comparable relative molecular mass and lower than the corresponding carboxylic acids of comparable relative molecular mass?

(11) Explain why aldehydes and ketones are described as polar carbonyl compounds.

(12) Give the formula and draw the structure of each of the following ketones:
(a) propanone (b) butanone.

(13) (a) How are ketones named according to the IUPAC system?
(b) Name the following ketones:

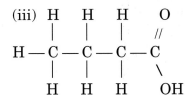

(14) (a) Name a ketone which is soluble in water.
(b) Why is it soluble in water?
(c) Why are the boiling points of ketones higher than those of the corresponding alkanes?
(d) Mention some everyday uses of some named ketones.

(15) Why are the boiling points of ketones higher than the corresponding hydrocarbons of comparable relative molecular mass and lower than the corresponding carboxylic acids of comparable relative molecular mass?

(16) Propanone is miscible in water in all proportions. Explain why this is so, using the structural formulae of water and propanone.

(17) Describe how you would determine whether propanone is soluble in (a) water and (b) cyclohexane.

(18) Draw the structural formulae of the first four carboxylic acids. Explain why only the fourth member of this homologous series has isomeric forms.

(19) (a) Describe how you would name a carboxylic acid using the IUPAC system.
(b) Name the following carboxylic acids:

(i)
$$H - \overset{\displaystyle O}{\underset{\displaystyle OH}{\overset{\displaystyle /\!/}{\underset{\displaystyle \backslash}{C}}}}$$

(ii)
$$H - \overset{\displaystyle H}{\underset{\displaystyle H}{\overset{\displaystyle |}{\underset{\displaystyle |}{C}}}} - \overset{\displaystyle H}{\underset{\displaystyle H}{\overset{\displaystyle |}{\underset{\displaystyle |}{C}}}} - \overset{\displaystyle O}{\underset{\displaystyle OH}{\overset{\displaystyle /\!/}{\underset{\displaystyle \backslash}{C}}}}$$

(c) What are the common names of these acids?

(iii)
$$H - \overset{\displaystyle H}{\underset{\displaystyle H}{\overset{\displaystyle |}{\underset{\displaystyle |}{C}}}} - \overset{\displaystyle H}{\underset{\displaystyle H}{\overset{\displaystyle |}{\underset{\displaystyle |}{C}}}} - \overset{\displaystyle H}{\underset{\displaystyle H}{\overset{\displaystyle |}{\underset{\displaystyle |}{C}}}} - \overset{\displaystyle O}{\underset{\displaystyle OH}{\overset{\displaystyle /\!/}{\underset{\displaystyle \backslash}{C}}}}$$

(d) Mention any uses of the following acids:
(i) ethanoic acid (ii) propanoic acid (iii) benzoic acid (iv) acetylsalicylic acid.

(20) The carboxylic acids contain a carboxyl group, which is composed of the carbonyl group found in aldehydes and ketones and the hydroxyl group found in alcohols. Explain why the physical properties of the carboxylic acids differ from the other types of compounds given above.

(21) Ethanoic acid is more soluble in water than butanoic acid, while butanoic acid is more soluble in non-polar solvents. Explain why this is so.

(22) Butane, propan-1-ol, propanal and ethanoic acid have nearly the same relative molecular mass but different boiling points. Place them in order of ascending boiling point and explain the order.

(23) Describe how you would determine whether ethanoic acid is soluble in (a) water and (b) hexane.

(24) A student obtained a pure sample of benzoic acid by <u>recrystallisation</u> and then

obtained an accurate melting point of the sample.
(a) What is the meaning of the underlined word?
(b) Describe the steps taken to obtain a pure sample of the benzoic acid.
(c) How was the melting point measured?

(25) (a) What are esters?
(b) How are esters named?
(c) Draw the structures of the following esters:
(i) methyl methanoate (ii) methyl ethanoate (iii) methyl propanoate.

(26) Esters are polar compounds.
(a) Explain why esters are polar.
(b) Explain why ethyl ethanoate is soluble in water.
(c) Why are esters soluble in non-polar compounds?
(d) Give some common industrial uses of esters.

(27) Explain why fats and oils are called natural esters.

(28) Describe how you would determine whether ethyl ethanoate is soluble in
(a) water and (b) hexane.

(29) (a) How is the structure of benzene represented?
(b) What are aromatic compounds?
(c) Why are they called aromatic compounds?

(30) 'The bonds in benzene are intermediate between single and double.' How does this statement help us to represent the structure of benzene?

(31) Benzene, C_6H_6, contains <u>sigma bonds</u> and <u>pi bonds</u>.
(a) Explain the meanings of the underlined terms.
(b) How do these terms help to explain the structure of benzene?

(32) (a) Name any *four* consumer products that contain benzene rings.
(b) Name (i) a dye (ii) a detergent (iii) a herbicide (iv) a pharmaceutical (v) an indicator, which contains a benzene ring.
(c) Name an aromatic command which is carcinogenic.
(d) Name an aromatic command which is not carcinogenic.

(33) (a) What are aromatic compounds?
(b) Draw the structure of benzene. What evidence is there to show that this is the correct structure?
(c) Draw the structure of any other aromatic compound. Indicate the range and scope of aromatic chemistry.

(34) Describe how you would determine whether methylbenzene is soluble in
(a) water and (b) hexane.

(35) Explain clearly, in terms of bonding, the differences in boiling points in the table below.

Compound	Formula	M_r	Boiling point/°C
Butane	$CH_3CH_2CH_2CH_3$	58	0
Propanal	CH_3CH_2CHO	58	49
Propanone	CH_3COCH_3	58	57
Propanol	$CH_3CH_2CH_2OH$	60	98
Ethanoic acid	CH_3COOH	60	118

7.3 ORGANIC CHEMICAL REACTION TYPES

(A) ADDITION REACTIONS

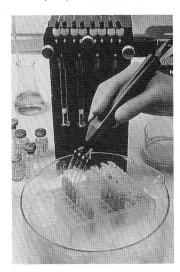

Automated mixing of chemicals.

Reactions of Alkenes

The characteristic feature of the alkene structure is the carbon–carbon double bond, $C = C$. The typical reactions of alkenes are those that take place at the double bond.

Functional Groups

The group of atoms that defines the structure and properties of a particular family of organic compounds is called the functional group.

For instance, the functional group in alkenes

is the $-\overset{|}{C} = \overset{|}{C}-$ group, while in alcohols the functional group is the $-OH$ group.

The double bond in alkenes consists of two types of bond: one strong sigma (σ) bond and one weak pi (π) bond. When another substance reacts with an alkene it breaks the weak pi bond and forms two strong sigma bonds in its place. Reactions of this type, where one substance adds on to another, are called addition reactions.

$$-\overset{|}{C} = \overset{|}{C}- + A - B \rightarrow -\overset{|}{\underset{A}{C}} - \overset{|}{\underset{B}{C}}-$$

Alkene ⇒ Alkane or substituted alkane

Many reactants, such as hydrogen, H_2, chlorine, Cl_2 and hydrogen chloride, HCl, can react with alkenes. Alkene molecules can also react with themselves and form much larger molecules called polymers.

(1) Addition of Hydrogen (Hydrogenation)

Hydrogen reacts with alkenes when heated in the presence of a catalyst to form alkanes. The catalyst, platinum, palladium or nickel, lowers the energy of activation for the reaction by providing a surface where the reactants can come in close contact with each other and then form products.

$$H - \overset{H}{\underset{}{C}} = \overset{H}{\underset{}{C}} - H + H - H \rightarrow H - \overset{H}{\underset{H}{C}} - \overset{H}{\underset{H}{C}} - H$$

ethene   hydrogen   ethane

Industrial importance

Hydrogenation is industrially important in the hardening of unsaturated vegetable oils, such as those extracted from soya beans or rape oil seed. Hydrogenation converts them into saturated compounds, which have higher melting points and hence are solids rather than liquids. Solids of this type can be used to make margarine and other fats. However, for health reasons, many people prefer to use softer margarines which are high in unsaturated oil.

(2) Addition of Halogens (Halogenation)

Alkenes react readily with fluorine, chlorine and bromine forming saturated compounds with two halogens attached to adjacent carbons. The reaction is carried out by dissolving the halogen in 1,1,1-trichloroethane and mixing it with the alkene.

$$H - \overset{H}{\underset{}{C}} = \overset{H}{\underset{}{C}} - H + Cl - Cl \rightarrow H - \overset{H}{\underset{Cl}{C}} - \overset{H}{\underset{Cl}{C}} - H$$

ethene   chlorine

1,2-dichloroethane

The reaction with bromine is similar.

$$H-C=C-H + Br-Br \rightarrow H-\overset{\displaystyle H}{\underset{\displaystyle Br}{C}}-\overset{\displaystyle H}{\underset{\displaystyle Br}{C}}-H$$

 ethene   bromine   1,2-dibromoethane

The typical dark brown colour of the bromine is decolorised as it reacts with the carbon–carbon double bond. This can be used as a simple test for unsaturation in a compound.

Industrial importance

Chloroalkanes, such as 1,1,1-trichloroalkanes, are used as solvents. Others are as used as alternatives to CFCs as aerosol propellants. Some are used in fire extinguishers and as flame retardants.

Mechanism of Addition to a Double Bond: Ionic Addition

Addition of chlorine or other reactants, such as hydrogen chloride, to a double bond involves the formation of ions. The mechanism for the addition of chlorine to an alkene, such as ethene, is as follows.

(a) The initial stage involves the chlorine molecule approaching the carbon–carbon double bond in a 'head-on' manner.

$$Cl-Cl \longrightarrow \overset{\displaystyle H \quad H}{\underset{\displaystyle H \quad H}{\overset{\diagdown\,\diagup}{\underset{\diagup\,\diagdown}{\overset{C}{\underset{C}{\|}}}}} \;\leftsquigarrow \text{High electron density}$$

Non-polar

(b) The carbon–carbon double bond is a region of high electron density because it contains two bonds, one sigma bond and one pi bond. This region of high electron density disturbs the electrons in the chlorine molecule, causing the normally non-polar chlorine molecule to become polarised. One side of the chlorine now has a small positive charge, δ^+, while the other side has a small negative charge, δ^-.

$$\overset{\delta^-}{Cl}-\overset{\delta^+}{C}---\overset{\displaystyle H \quad H}{\underset{\displaystyle H \quad H}{\overset{\diagdown\,\diagup}{\underset{\diagup\,\diagdown}{\overset{C}{\underset{C}{\|}}}}}$$

Polar

(c) The more positive chlorine is then taken up by the ethene to form a new positive ion called a carbonium ion. The negative chlorine atom which remains then reacts with this carbonium ion and forms the final product, 1,2-dichloroethane.

$$\begin{array}{ccc} Cl-\overset{\displaystyle H}{\underset{}{C}}-H & & Cl-\overset{\displaystyle H}{\underset{}{C}}-H \\ \mid & \rightarrow & \mid \\ Cl^- \rightarrow + C-H & & Cl-C-H \\ \mid & & \mid \\ H & & H \end{array}$$

carbonium ion   1,2-dichloroethane

The mechanism for the addition of hydrogen chloride to an alkene is similar, the only difference being that polarisation is not induced in the HCl molecule because it is already a polar molecule.

(a)
$$\overset{\delta^-}{Cl}-\overset{\delta^+}{H}---\overset{\displaystyle H \quad H}{\underset{\displaystyle H \quad H}{\overset{\diagdown\,\diagup}{\underset{\diagup\,\diagdown}{\overset{C}{\underset{C}{\|}}}}}$$

Polar

(b)
$$\begin{array}{ccc} H-\overset{\displaystyle H}{\underset{}{C}}-H & & H-\overset{\displaystyle H}{\underset{}{C}}-H \\ \mid & \rightarrow & \mid \\ Cl^- \rightarrow + C-H & & Cl-C-H \\ \mid & & \mid \\ H & & H \end{array}$$

carbonium ion   chloroethane

Evidence for Ionic Addition

When bromine, Br_2, reacts with ethene only one product, 1,2-dibromoethane, is formed.

$$\begin{array}{ccc} Br-\overset{\displaystyle H}{\underset{}{C}}-H & & Br-\overset{\displaystyle H}{\underset{}{C}}-H \\ \mid & \rightarrow & \mid \\ Br^- \rightarrow + C-H & & Br-C-H \\ \mid & & \mid \\ H & & H \end{array}$$

carbonium ion   1,2-dibromoethane

When bromine water containing sodium chloride, is added to ethene, three products are formed: 1,2-dibromoethane, 2-bromoethanol, and 1-bromo-2-chloroethane. This indicates that the bromine first forms a positive carbonium ion to which a Br^- ion or an OH^- or a Cl^- ion is then added.

$$
\begin{array}{ccc}
& H & & H \\
& | & & | \\
Br-C-H & & Br-C-H \\
& | & \rightarrow & | \\
OH^- \rightarrow +C-H & & HO-C-H \\
& | & & | \\
& H & & H \\
\text{carbonium ion} & & \text{2-bromoethanol}
\end{array}
$$

$$
\begin{array}{ccc}
& H & & H \\
& | & & | \\
Br-C-H & & Br-C-H \\
& | & \rightarrow & | \\
Cl^- \rightarrow +C-H & & Cl-C-H \\
& | & & | \\
& H & & H \\
\text{carbonium ion} & & \text{1-bromo-2-chloroethane}
\end{array}
$$

(3) Addition of Hydrogen Chloride

Ethene reacts with hydrogen chloride to give chloroethane.

$$
\underset{\text{ethene}}{H-C=C-H} + \underset{\substack{\text{hydrogen} \\ \text{chloride}}}{H-Cl} \rightarrow \underset{\text{chloroethane}}{H-C-C-H}
$$

Industrial importance

Ethene and hydrogen chloride can react together to form dichloroethane, CH_2ClCH_2Cl. Dichloroethane can be converted into chloroethene. Chloroethene is used to make poly(chloroethene), which is commonly called PVC (polyvinyl chloride). PVC is a plastic with many uses, e.g. hosepipes and insulation.

(4) Addition of Water (Hydration)

Water combines with the more reactive alkenes in the presence of acids to form alcohols. Ethanol is manufactured mainly by the catalytic hydration of ethene.

$$
\underset{\text{ethene}}{H-C=C-H} + \underset{\text{water}}{H_2O} \rightarrow \underset{\text{ethanol}}{H-C-C-H}
$$

Industrial importance

When ethene is isolated cheaply from petroleum, it can be used to produce ethanol.

(5) Addition Polymerisation: Formation of Plastics

Polymerisation is the vital stage in the manufacture of plastics. Addition polymerisation involves the adding together of thousands of small molecules (or monomers) by addition reactions to form a giant molecule or polymer.

Ethene is used to make polythene, more correctly called poly(ethene), by subjecting the ethene monomers to an extremely high pressure in the presence of a catalyst.

$$
n \underset{\text{ethene}}{\left(\begin{array}{c} H \quad\quad H \\ \diagdown \quad\ \diagup \\ C=C \\ \diagup \quad\ \diagdown \\ H \quad\quad H \end{array} \right)} \longrightarrow \underset{\text{Poly(ethene)}}{\left[\begin{array}{c} H \quad H \\ | \quad\ | \\ C-C \\ | \quad\ | \\ H \quad H \end{array} \right]_n}
$$

Side chains may form during the reaction; the more side chains formed, the less dense the polymer. The side chains prevent other side chains packing closely together, which results in a low density. The reaction conditions and the particular catalyst used determines whether the polythene is low-density poly(ethene) (LDPE) or high-density poly(ethene) (HDPE). (See pages 327/8)

Polymerisation of ethene

Ethene + Ethene + Ethene + – – – – – $\longrightarrow$ Polythene chain

$$
\underset{}{C=C} + C=C + C=C \longrightarrow \left[\begin{array}{c} H \ H \ H \ H \ H \\ | \ | \ | \ | \ | \\ C-C-C-C-C \\ | \ | \ | \ | \ | \\ H \ H \ H \ H \ H \end{array} \right]_n
$$

Low density poly(ethene) is used as a plastic film for packaging food and other materials, while high-density poly(ethene) is mostly used for injection moulded products.

Propene can be polymerised to form poly(propene).

$$n \begin{pmatrix} \begin{matrix} H \\ \diagdown \\ C \\ \diagup \\ H \end{matrix} = \begin{matrix} H \\ \diagup \\ C \\ \diagdown \\ CH_3 \end{matrix} \end{pmatrix} \longrightarrow \begin{bmatrix} \begin{matrix} H \\ | \\ C \\ | \\ H \end{matrix} - \begin{matrix} H \\ | \\ C \\ | \\ CH_3 \end{matrix} \end{bmatrix}_n$$

propene   Poly(propene)

Polymers can be tailor-made to suit the purposes for which they are needed. Polymers can be thought of as strands of spaghetti. If the strands are linked strongly together you get a solid mass, like Formica. If they are not strongly linked together you get a soft solid which can be bent, like polythene. Polymers can be made into fibres by drawing the polymer through a small hole. The polymer molecules line up alongside one another to form fibres, like Terylene.

Common polymers from the petrochemical industry

Common name	Chemical name	Monomer	Use
Polythene	Poly(ethene)	Ethene	Packaging
Polypropylene	Poly(propene)	Propene	Kitchenware
Polystyrene	Poly(phenylethene)	Phenylethene	Plastic cups
Polyvinylchloride	Poly(chloroethene)	Chloroethene	Insulation
Polytetrafluoroethylene	Poly(tetrafluoroethene)	Fluoroethene	Hip joints
Polyacrylonitrile	Poly(propenonitrile)	Propenonitrile	Acrylic fibres

Unreactivity of Benzene

Unlike other unsaturated hydrocarbons, such as ethene, benzene is reluctant to take part in addition reactions. Unlike ethene, benzene does not react with bromine. If benzene is shaken up with bromine rapid decoloration does not occur. This indicates that C=C bonds are not present.

The stability of benzene is due to the fact that it has neither three carbon–carbon single bonds nor three carbon–carbon double bonds, but has six equivalent carbon–carbon bonds of size intermediate between a carbon–carbon single bond and a carbon–carbon double bond. These six equivalent bonds share the available bonding electrons equally with each other; the electrons are delocalised, i.e. they are not confined to any particular carbon atom but are spread out around the benzene ring. These delocalised electrons give additional bonding strength.

Exercise 7.3a
•

Addition reactions

(1) What is an addition reaction? Explain why most of the reactions of alkenes are addition reactions.

(2) Write balanced equations, using structural formulae, for the following reactions:
(a) hydrogenation of ethene (b) chlorination of ethene (c) bromination of ethene (d) addition of water to ethene (e) addition of HCl to ethene (f) addition of H_2SO_4 to ethene.

(3) Write equations (using structural formulae) for each of the following reactions:
(a) hydrogenation of propene (b) bromination of but-1-ene (c) addition of HCl to propene.

(4) What is the industrial importance of each of the following reactions:
(a) addition of hydrogen to unsaturated compounds
(b) addition of halogens to alkenes
(c) addition of water to ethene
(d) addition of hydrogen chloride to ethene
(e) addition of ethene to ethene
(f) addition of propene to propene?

(5) Ethene can be prepared by the following methods:
(a) cracking of crude oil, (b) heating of natural gas, and (c) dehydration of ethanol.

Describe briefly how each method is used. How would you test that the product obtained is an alkene?

(6) (a) What is ionic addition?
(b) Describe the mechanism by which
(i) HCl and (ii) Cl_2 reacts with ethene.
(c) What evidence is there for this mechanism?

(7) Describe the mechanism by which
(a) Br_2 and (b) HCl reacts with propene.

(8) (a) What is a polymer? Describe how ethene can be converted into poly(ethene).
(b) Explain why polythene is referred to as an addition polymer.
(c) How does poly(propene) form?
(d) Give some of the uses of poly(ethene) and poly(propene).
(e) On the one hand, poly(ethene) can be used as a plastic film for packaging and on the other hand it can also be used for making hard plastic containers. Explain why it can be used to make such different products.

(9) Alkenes such as ethene and propene have often been described as the building blocks of the organic chemical industry. Discuss the properties of alkenes which enable them to be used in so many ways.

(10) Explain why benzene is unreactive, relative to ethene, with regard to addition reactions.

(B) SUBSTITUTION REACTIONS

A substitution reaction is a reaction where an atom, or a group of atoms, is replaced in a molecule by another atom or group of atoms.
For example,

$$CH_4 + Cl_2 \rightarrow CH_3Cl + HCl$$

The main types of substitution reaction include:
• halogenation of alkanes
• esterification
• hydrolysis of esters.

(1) Halogenation of Alkanes

Because alkanes are saturated compounds, they can react only by substitution. In a substitution reaction involving an alkane, a hydrogen atom is replaced by another atom or group of atoms. In the following example, a hydrogen atom is substituted for or swapped for a halogen such as chlorine.

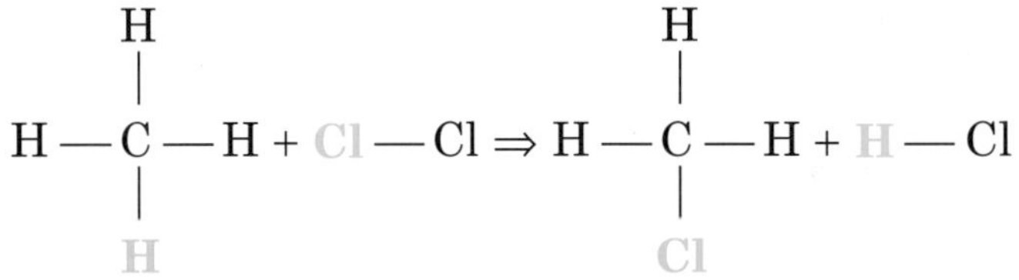

methane   chloromethane

Chlorination of methane

When chlorine and methane are mixed in the dark at room temperature no reaction occurs. However, if they are heated or allowed to react in the presence of ultraviolet light a vigorous reaction occurs yielding a mixture of products. The main products are chloromethane, dichloromethane, trichloromethane and tetrachloromethane.

The reaction mechanism depends on the formation of free radicals and consists of three main steps:

(a) Initiation

The chlorine molecule is split into chlorine atoms which have an unpaired electron. Species of this type are extremely reactive and are called free radicals. (Free radicals are usually indicated by a dot.)

$$Cl - Cl \xrightarrow{uv} Cl^{\bullet} + Cl^{\bullet}$$

molecule   free radicals

(b) Propagation

The extremely reactive chlorine free radicals, Cl, attack the methane molecules and form hydrogen chloride, HCl, and methyl free radicals, $CH_3^{\bullet}$. The newly formed methyl free radicals, $CH_3^{\bullet}$ are also extremely reactive and attack the chlorine molecules and form chloromethane, CH_3Cl, and more chlorine free radicals. A chain reaction has now started.

$$CH_4 + Cl^{\bullet} \rightarrow HCl + CH_3^{\bullet}$$
$$Cl_2 + CH_3^{\bullet} \rightarrow CH_3Cl + Cl^{\bullet}$$
chloromethane

(c) Termination

The chain reaction comes to an end when the various free radicals combine with each other to form unreactive molecules.

Some of the possible combinations are given below:

$$CH_3^{\cdot} + Cl^{\cdot} \rightarrow CH_3Cl$$

$$Cl^{\cdot} + Cl^{\cdot} \rightarrow Cl_2$$

$$CH_3^{\cdot} + CH_3^{\cdot} \rightarrow C_2H_6 \, (ethane)$$

Chlorination of ethane

The reaction mechanism is similar to the reaction of methane with chlorine.

(a) Initiation

$$Cl - Cl \xrightarrow{uv} Cl^{\cdot} + Cl^{\cdot}$$
$$molecule \qquad\qquad free\ radicals$$

(b) Propagation

$$C_2H_6 + Cl^{\cdot} \rightarrow HCl + C_2H_5^{\cdot}$$

$$Cl_2 + C_2H_5^{\cdot} \rightarrow C_2H_5Cl + Cl^{\cdot}$$
$$chloroethane$$

(c) Termination

The chain reaction comes to an end when the various free radicals combine with each other to form unreactive molecules.

Some of the possible combinations are given below:

$$C_2H_5^{\cdot} + Cl^{\cdot} \rightarrow C_2H_5Cl$$

$$Cl^{\cdot} + Cl^{\cdot} \rightarrow Cl_2$$

$$C_2H_5^{\cdot} + C_2H_5^{\cdot} \rightarrow C_4H_{10} \, (butane)$$

Evidence for Free Radical Mechanism

(1) A chain reaction occurs if ultraviolet light is used even for a very short period.

(2) Ethane is formed during the chlorination of methane when two methyl free radicals combine with each other.

$$CH_3^{\cdot} + CH_3^{\cdot} \rightarrow C_2H_6 \, (ethane)$$

Butane is formed during the chlorination of ethane when two ṃethyl free radicals combine with each other.   ethyl

$$C_2H_5^{\cdot} + C_2H_5^{\cdot} \rightarrow C_4H_{10} \, (butane)$$

(3) Tetraethyl lead is a known source of free radicals. Chlorination of methane and ethane is speeded up by the addition of tetraethyl lead.

(2) Esterification

Alcohols react with carboxylic acids to form sweet-smelling compounds called esters. The reaction is slow unless a catalyst, such as sulphuric acid or hydrochloric acid, is used.

For example, ethanol reacts with ethanoic acid to form ethyl ethanoate and water.

ethanol   ethanoic acid

ethyl ethanoate   water

This is an example of a reversible reaction, which eventually reaches equilibrium when the rate of formation of products equals the rate of decomposition of reactants. The same catalyst, the H^+ ion, catalyses both the forward and reverse reactions to the same extent. This reversibility is a disadvantage when preparing an ester in this way: the water must be removed as it is formed, thereby preventing the reverse reaction.

Other examples of esterification include:

methanol   ethanoic acid

methyl ethanoate   water

ethanol   methanoic acid   ethyl methanoate   water

208

(3) Hydrolysis of Esters

Esters are hydrolysed on heating with mineral acids or alkalis. The acid catalysed hydrolysis is reversible, and is the opposite to esterification.

Basic Hydrolysis of Esters

This reaction can go to completion if sufficient alkali is used and the carboxylic acid ends up as a salt. This process is called saponification, the type of reaction used in the manufacture of soap from fats and oils, e.g.

$$CH_3COOC_2H_5 + NaOH \rightarrow CH_3COONa + C_2H_5OH$$

| ethyl | sodium | sodium | ethanol |
| ethanoate | hydroxide | ethanoate | |

A similar reaction is used in the manufacture of soap. Fats are esters which can be converted into soaps by alkaline hydrolysis (saponification) with sodium hydroxide. The ester (fat) is hydrolysed by sodium hydroxide and forms the sodium salt of the carboxylic acid (soap) and an alcohol, glycerol.

C$_{17}$H$_{35}$COOCH$_2$ HO — CH$_2$
|
C$_{17}$H$_{35}$COOCH + 3NaOH → 3C$_{17}$H$_{35}$COONa + HO — CH
|
C$_{17}$H$_{35}$COOCH$_2$ HO — CH$_2$

  fat   alkali   soap   glycerol
                                     (propane-1,2,3-triol)

Mandatory experiment 18

To prepare soap in the laboratory.

Introduction

Soaps are the sodium or potassium salts of long chain carboxylic acids. They can be formed quite easily by boiling a fat (ester) in potassium hydroxide. Ethanol is used as a solvent to dissolve the fat. Potassium hydroxide is more soluble in ethanol than sodium hydroxide and is therefore used to make the soap.

Requirements

Safety glasses, quick-fit distillation apparatus, heating mantle (or Bunsen burner), thermometer, filter funnel, filter paper, glass rod, lard or castor oil, potassium hydroxide, ethanol, saturated sodium chloride solution and deionised water.

Procedure

(1) Set up the apparatus for reflux as shown.

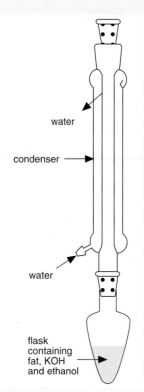

water

condenser

water

flask containing fat, KOH and ethanol

(2) Place 4 g of the fat (lard or castor oil), 4 g of potassium hydroxide pellets and approximately 50 cm^3 of ethanol in the 100 cm^3 round bottomed flask.
(3) Boil, under reflux, for 30 minutes.
(4) Rearrange the apparatus for distillation. Distill the ethanol off until the temperature of the distilling vapour is greater than 80 °C.
(5) Add the contents of the round bottomed flask to the saturated sodium chloride solution to dissolve the excess alkali and to precipitate the soap.
(6) Filter with saturated sodium chloride solution and then with deionised water.
(7) Test the soap formed by:
(a) forming a lather, making sure that you wash your hands well afterwards to remove any alkali still present
(b) making a soap solution with hot water.

Exercise 7.3b

•

Substitution reactions

(1) (a) What is a substitution reaction?
(b) Explain why alkanes can react only by substitution (excluding decomposition reactions).

(2) (a) What is a free radical substitution?
(b) Methane and the other alkanes react with halogens forming substituted alkanes. Describe the main steps in the mechanism by which methane reacts with chlorine.
(c) What evidence is there to support this mechanism?

(3) (a) Describe the mechanism of free radical chlorination of ethane.
(b) Give three pieces of evidence in support of this mechanism.

(4) Write balanced equations for the following reactions:
(a) propane + bromine
(b) methane + chlorine
(c) ethane + chlorine.

(5) (a) Describe, outlining the experimental details, how soap is prepared in the laboratory.
(b) How are the impurities removed?
(c) How would you test the soap sample?

(6) Using structural formulae, write equations for the following reactions:
(a) ethanol with ethanoic acid
(b) methanol with ethanoic acid
(c) methanol with methanoic acid
(d) methanol with propanoic acid
(e) methanol with ethanoic acid
(f) methanol with ethanoic acid
(g) propan-1-ol with methanoic acid.

(7) Using structural formulae, write equations for the following reactions:
(a) ethyl ethanoate with sodium hydroxide
(b) methyl ethanoate with sodium hydroxide.

(8) Write equations for each of the following:
(a) saponification of methyl methanoate
(b) saponification of ethyl ethanoate.

(9) (a) Write an equation, indicating the structures of the reactants and products, for the formation of soap.
(b) Describe how soap removes greasy substances from clothes.

(C) ELIMINATION REACTIONS

An **elimination reaction** is a reaction where a small molecule is removed from a larger molecule leaving an unsaturated compound.
  For example,

$$C_2H_5OH \Rightarrow C_2H_4 + H_2O$$
ethanol   ethene   water

Dehydration of Alcohols

Dehydration of alcohols involves the elimination of a water molecule, H_2O, from the alcohol. Alcohols may be dehydrated to form alkenes or in some cases to form ethers, depending on the reaction conditions.

If the alcohol has at least one hydrogen on the carbon atom next but one to the hydroxyl group, it can be dehydrated when heated using an excess of concentrated sulphuric acid. An alkene is formed as a result.

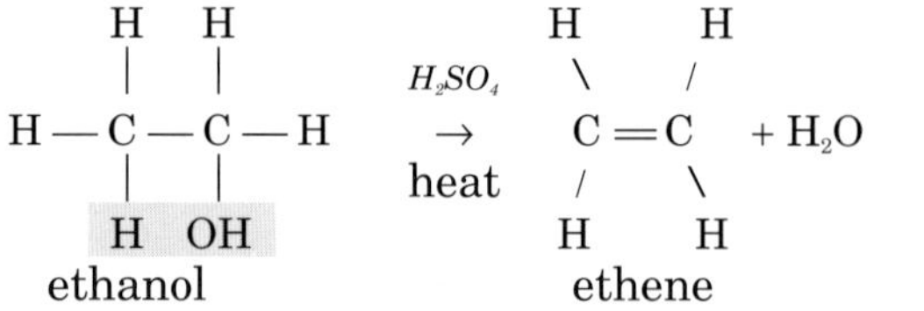

If an alcohol is heated in the presence of concentrated sulphuric acid and the alcohol is in excess, an ether is formed.

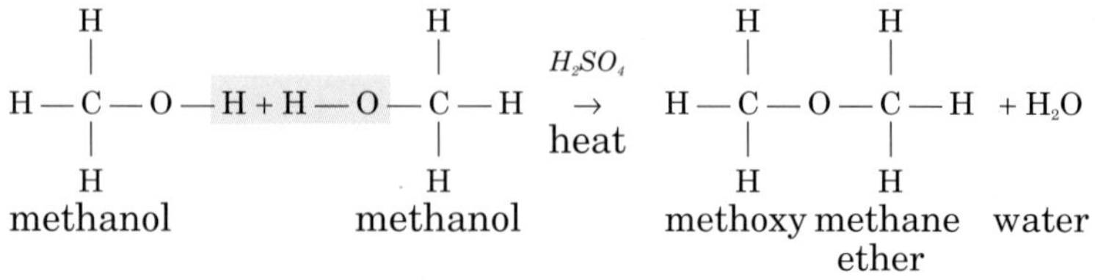

Mandatory experiment 19

The laboratory preparation and some properties of ethene.

Introduction

Ethene can be prepared by heating ethanol in the presence of an aluminium oxide catalyst.

The following tests can be performed to test the product:
(a) combustion in air
(b) reaction with Br_2 in 1,1,1-trichloroethane
(c) reaction with acidified potassium manganate(VII).

Requirements

Safety glasses, test-tubes, delivery tubing, one-holed test-tube bung, trough, glass wool, ethanol, aluminium oxide, solution of bromine in 1,1,1-trichloroethane, acidified potassium manganate(VII).

Procedure

(1) Pour some ethanol into a test-tube to a depth of about 2 cm. Push in a plug of glass wool to soak up the ethanol. Assemble the apparatus as in the diagram.

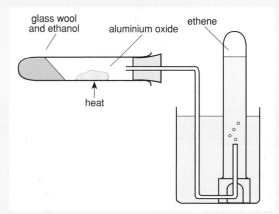

(2) Add about 1 g of aluminium oxide to the centre of the test-tube.
(3) Heat the aluminium oxide catalyst gently.
(4) Collect the ethene in a test-tube. Describe what happens when it is lighted.
(5) Bubble some ethene into a dilute solution of
(a) bromine and
(b) potassium manganate(VII).
Record your results in a table like the one below.

Combustion	Reaction with bromine solution	Reaction with potassium manganate(VII)

Exercise 7.3c
•

Elimination reactions

(1) (a) What is an elimination reaction?
(b) Write a balanced equation for the dehydration of ethanol.
(c) Can any other balanced equation be used for the dehydration?

(2) Show, giving relevant equations, how ethanol is converted into ethene.

(3) A hydrocarbon gas was prepared using the apparatus shown opposite.
The ethanol was placed in a test-tube as shown. Solid A was placed half way along the test-tube, and was heated gently. The gas produced was collected. When production was complete, the delivery tube was immediately removed. A jar of the gas was burnt.

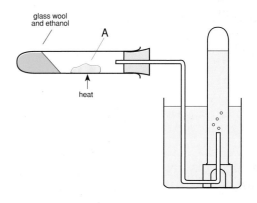

Bromine water was added to a second gas jar and a very dilute solution of potassium manganate(VII) was added to a third gas jar.
(a) Name the solid A and the gas produced.
(b) Why was the ethanol soaked in glass wool?
(c) Why was the delivery tube removed immediately the gas production was complete?

211

(d) Describe the flame produced when the gas was burnt. What conclusion do you draw?

(e) What happens when bromine water is added to the second gas jar? What conclusion do you draw? Write an equation for the reaction occurring.

(f) What happens when very dilute potassium manganate(VII) solution is added to the third gas jar? What conclusion do you draw?

(g) Why was *very* dilute potassium manganate(VII) solution used?

(D) REDOX REACTIONS

Oxidation of alcohols

Distilling whiskey.

Primary alcohols are readily oxidised to aldehydes and can be further oxidised to carboxylic acids.

Several possible oxidising agents in solution may be used, the most common being either acidified sodium or potassium dichromate(VI), or acidified potassium manganate(VII).

In general,

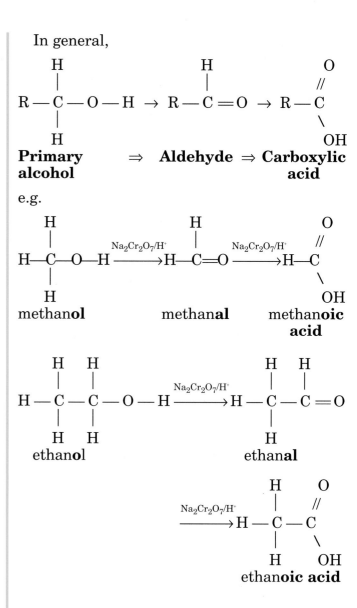

Primary alcohol ⇒ **Aldehyde** ⇒ **Carboxylic acid**

e.g.

methan**ol** methan**al** methan**oic acid**

ethan**ol** ethan**al**

ethan**oic acid**

Social Aspects

• The reaction of ethanol with potassium dichromate(VII) has been used as a breathalyser to determine the amount of alcohol present in breath. The ethanol in the drinker's breath reduces the orange dichromate(VII) ion to the green chromium(III) ion, thereby indicating the presence of alcohol.

• When ethanol is metabolised in the human body it is oxidised to ethanal.

Mandatory experiment 20

(a) Preparation of ethanal and (b) examination of some of its properties.

(a) Preparation of ethanal

Introduction

Acidified potassium or sodium dichromate solution is frequently used as an oxidising agent. It can be used to oxidise a primary alcohol to form an aldehyde, e.g.

$$C_2H_5OH \xrightarrow{\text{K}_2\text{Cr}_2\text{O}_7,\ \text{H}_2\text{SO}_4} CH_3CHO$$
$$\text{ethanol} \qquad\qquad\qquad \text{ethanal}$$

It should be remembered that care must be taken so that the ethanol is not oxidised to ethanoic acid.

Requirements

Safety glasses, quick-fit apparatus for distillation, dropping funnel, conical flask, graduated cylinders (100 cm³), rubber tubing, Bunsen burner (a heating mantle is preferable), ethanol, sodium dichromate, concentrated sulphuric acid, deionised water, anti-bumping granules.

Procedure

(1) Set up the apparatus for distillation with addition, as shown in the diagram.

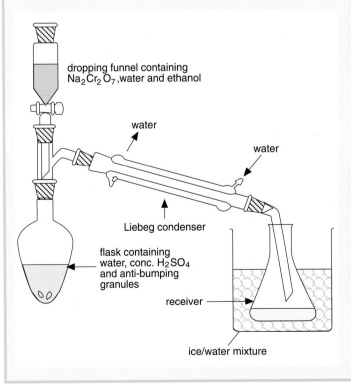

dropping funnel containing
Na₂Cr₂O₇, water and ethanol

water

water

Liebeg condenser

flask containing
water, conc. H₂SO₄
and anti-bumping
granules

receiver

ice/water mixture

(2) Place 12 cm³ of water and some anti-bumping granules in the pear-shaped flask and slowly add 4 cm³ of concentrated sulphuric acid, swirling the flask gently and cooling it under the tap.

(3) Dissolve 10 g of sodium dichromate in 10 cm³ of water in a clean beaker and then add 8 cm³ of ethanol. Place the mixture in the dropping funnel.

(4) Heat the dilute acid until it boils and then turn off the Bunsen burner (or heating mantle).

(5) Add the alcohol mixture slowly from the dropping funnel, at such a rate that the liquid is maintained at its boiling point.

(6) Collect the distillate in a receiver surrounded by ice.

(7) Replace the dropping funnel by a thermometer and redistill the distillate at a temperature of between 20 and 23 °C, keeping the receiver surrounded by ice.

Experimental Notes

• The mixing of sulphuric acid and water is an exothermic process. The sulphuric acid must be added to water (not water to acid), otherwise a violent reaction occurs with subsequent splashing of acid. The reaction flask should be cooled in iced water.

• Further heating of the dilute acid is unnecessary because the reaction between the acid and the ethanol mixture is exothermic.

• Sodium dichromate is used in preference to potassium dichromate because it is more soluble in ethanol.

• The formation of ethanal (and not ethanoic acid) is favoured by:

(a) use of an excess of ethanol over acidified sodium dichromate

(b) distilling off ethanal as soon as it is formed.

• In the reaction, ethanol is oxidised to ethanal and dichromate(VI) ions are reduced to chromium(III) ions.

This is observed by:

(a) cautious smelling of the product ethanal, which has a distinct pungent odour.

(b) a colour change from orange to green

$$Cr_2O_7{}^{2-}(aq) \longrightarrow 2Cr^{3+}(aq)$$
$$\text{Orange} \qquad\qquad \text{Green}$$

i.e. chromium +6 is reduced to chromium +3

• The distillate is colourless and contains ethanal as well as ethanol, ethanoic acid and water impurities. Ethanal boils at 21 °C; consequently the impurities can be removed by redistilling the mixture and collecting the fraction boiling between 20 and 23 °C.

(b) Properties of ethanal

Reactions with (i) acidified potassium manganate (VII) solution, (ii) Fehling's solution and (iii) Tollens reagent (ammoniacal silver nitrate solution).

Requirements

Safety glasses, test-tubes, beaker, Bunsen burner, 0.02 mol L^{-1} potassium manganate(VII), dilute sulphuric acid, Fehling's solutions A and B, 2.0 mol L^{-1} sodium hydroxide solution, 0.05 mol L^{-1} silver nitrate solution, 2 mol L^{-1} ammonia solution, ethanal.

Procedure

(i) Oxidation with acidified potassium manganate(VII)

(1) Put approximately 2 cm³ of ethanal, 1 cm³ of potassium manganate(VII) solution and 4 cm³ of dilute sulphuric acid into a test-tube.
(2) Shake the test-tube gently and warm it in a water bath for a few minutes.
(3) Record your observations in a copy of the table below.

(ii) Oxidation using Fehling's solution

(1) Put approximately 1 cm³ of Fehling's solution A into a test-tube. Slowly add Fehling's solution B dropwise until the blue precipitate just dissolves.
(2) Add about 1 cm³ of ethanal. Heat the test-tube very gently. Shake it slightly.
(3) Record your observations in a copy of the table below.

(iii) Oxidation using Tollens reagent (silver mirror test)

(1) Put about 3 cm³ of silver nitrate solution into an extremely clean test-tube. Add about 1 cm³ of sodium hydroxide solution.
(2) Add aqueous ammonia dropwise until the silver oxide precipitate just dissolves.
(3) Add two to three drops of ethanal, shake the test-tube gently and place it in a water bath.

(4) Record your observations in a copy of the table below.

Results

Reaction of ethanal with	Observation
Acidified potassium manganate(VII)	
Fehling's solution	
Tollens reagent (ammoniacal silver nitrate)	

(5) Rinse out the test-tube immediately after use.

Chemistry in Action: Ethanol, A Profile of an Everyday Chemical

Throughout the ages, man has always known how to make ethanol. Man has drunk ethanol rather than water and survived, due to the disinfectant properties of the alcohol. The common alcoholic beverages, such as beer, wine, whiskey, gin and rum are all aqueous mixtures of ethanol. They differ from one another due to flavours in the substances from which they are made.

Ethanol can be made from ethene, a major petrochemical feedstock. It can ***also be made by fermentation***, where sugars are converted, using enzymes, into ethanol and carbon dioxide.

A process was developed in Ireland to make ethanol from ***milk***. The whey left over after cheese making contains a sugar, lactose, which can be fermented using a suitable enzyme to form ethanol. The alcohol is used in the manufacture of cream liqueurs and vodka and as an industrial chemical.

Fermentation is only one of the steps involved in the making of an alcoholic drink. ***Whiskey***, for instance, is made from barley which is first ***germinated*** and then the starch in the barley is converted into ***maltose***.

The malted barley is then ***roasted and mashed*** up in water in order to extract all the flavours. Yeast is then added and the 'mash' is fermented in large vats, eventually forming ethanol.

The ethanol present after fermentation is then concentrated by distilling it in copper stills called pot stills. The illegal drink, poteen, gets its name from these *pot* stills.

Finally, the whiskey is *left in wooden barrels to mature*. Sometimes several whiskies are blended together to give a particular taste to the whiskey.

Ethanol is used for many other things, other than as a drink. It is not only the *starting point in the manufacture of a variety of chemicals*, but is also an excellent *solvent*. It is often said that if a chemist were left on earth with only six chemicals to choose from, he would pick ethanol as one of them. It can be used to purify many substances by *dissolving* them first in hot ethanol and then recrystallising the pure compound.

Ethanol is also an excellent *fuel*. When one mole of ethanol is burned in air approximately *1400 kJ of energy* is released. Some countries, such as Brazil and the United States of America, which grow large quantities of cane sugar or grain, can use these sources of starch to form ethanol at a relatively low cost. The ethanol is then mixed with petroleum to form a new fuel called *'gasohol'*. This is used in motor cars.

As you have seen, ethanol is a very versatile chemical. However, it is often *abused* by many people. It must be remembered that ethanol, like some other alcohols, is both *hypnotic and sleep inducing*. It also affects your brain quite rapidly, because being totally miscible in water it can enter the blood stream directly. At a high concentration of ethanol the body can absorb no more and the person reaches a state where he is *dead drunk* and eventually dies.

Determination of Percentage Yields

In a chemical reaction it is extremely important that as much as possible of the desired product is obtained. Often, a low yield of product is obtained owing to a variety of factors.

These factors may include:
- use of an insufficient amount of one of the reactants.
- incorrect reaction conditions, such as low reaction temperature.
- loss of product through poor distillation, inadequate recrystallation, loss by evaporation or poor separation of impurities.

The percentage yield is calculated from the actual amount obtained during the reaction and from the theoretical amount obtained in accordance with the balanced chemical equation.

$$\% \text{ yield} = \frac{\text{actual yield}}{\text{theoretical yield}} \times 100\%$$

The theoretical yield is sometimes difficult to calculate when the experimental procedure often contains the amounts of catalysts, acidifying agents, solvents and other substances used which have no direct bearing on the theoretical yield.

In all cases the theoretical yield is obtained by finding out which reactant is the limiting reactant.

• • • • • **Example 7.10** • • • • •

Calculation of percentage yield of ethanal

A sample of ethanal was prepared by oxidation of ethanol with acidified sodium dichromate(VI) solution. The half equation for the reaction is given as:

$$3C_2H_5OH + Cr_2O_7^{2-} + 8H^+ \Rightarrow 3CH_3CHO + 2Cr^{3+} + 7H_2O$$

11.5 cm^3 of ethanol, C_2H_5OH (density = 0.80 g cm^{-3}), and 14.9 g of sodium dichromate(VI) crystals, $Na_2Cr_2O_7.2H_2O$, were reacted together with about 10 cm^3 of concentrated sulphuric acid dissolved in 25 cm^3 of water. After purification, 1.6 g of ethanal was obtained. Determine (a) the limiting reactant and (b) the percentage yield of ethanal.

(Remember that only the 14.9 g of sodium dichromate(VI) and the 11.5 cm^3 of ethanol determine the theoretical yield. The sulphuric acid is a catalyst and the water is a solvent.)

(a) Determination of the limiting reactant

Amount of $Na_2Cr_2O_7.2H_2O$ in moles

$$n(Na_2Cr_2O_7.2H_2O) = \frac{m}{M} = \frac{14.9\ g}{298\ g\ mol^{-1}} = 0.05\ mol$$

Mass of ethanol = Volume used × density
$$= 11.5\ cm^3 \times 0.8\ g\ cm^{-3}$$
$$= 9.2\ g$$

Amount of C_2H_5OH in moles

$$n(C_2H_5OH) = \frac{m}{M} = \frac{9.2\ g}{46\ g\ mol^{-1}} = 0.2\ mol$$

The balanced chemical equation tells us that 3 mol of C_2H_5OH reacts with 1 mol of $Na_2Cr_2O_7.2H_2O$

Therefore in the experiment,
0.2 mol of C_2H_5OH reacts with 0.2/3 mol of $Na_2Cr_2O_7.2H_2O$

0.2 mol of C_2H_5OH reacts with 0.066 mol of $Na_2Cr_2O_7.2H_2O$

However, the amount of $Na_2Cr_2O_7.2H_2O$ used in the experiment was 0.05 mol. This amount (0.05 mol) is less than that required from the balanced equation (0.066 mol): 0.05 mol $Na_2Cr_2O_7.2H_2O$ is the limiting reactant.

(b) Calculation of % yield
The balanced equation tells us that:

1 mol $Na_2Cr_2O_7.2H_2O \Rightarrow$ 3 mol CH_3CHO

0.05 mol $Na_2Cr_2O_7.2H_2O \Rightarrow 3 \times 0.05$ mol CH_3CHO

0.05 mol $Na_2Cr_2O_7.2H_2O \Rightarrow 0.15$ mol CH_3CHO

Mass of ethanal, $m = nM$
$$= 0.15\ mol \times 44\ g\ mol^{-1}$$
$$= 6.6\ g$$
This is the maximum theoretical yield according to the balanced equation.

$$\% \text{ yield} = \frac{\text{actual yield}}{\text{theoretical yield}} \times 100\%$$

$$= \frac{1.6\ g}{6.6\ g} \times 100\% = 24.2\%$$

• • • • • • Example 7.11 • • • • • •

Calculation of % yield of ethanoic acid

A sample of ethanoic acid was prepared by oxidation of ethanol with acidified sodium dichromate(VI) solution.

The half equation for the reaction is given as:

$$3C_2H_5OH + 2Cr_2O_7^{2-} + 16H^+ \Rightarrow 3CH_3COOH + 4Cr^{3+} + 11H_2O$$

40 cm³ of water was put into a flask and 20 cm³ of concentrated sulphuric acid was added. A mass of 29.8 g of sodium dichromate(VI) crystals, $Na_2Cr_2O_7.2H_2O$, was dissolved in this solution. A solution of 6.9 cm³ of ethanol, C_2H_5OH (density = 0.80 g cm⁻³), in 20 cm³ of water was reacted with the acidified sodium dichromate(VI) solution. After purification, 5.7 g of ethanoic acid was obtained. Determine (a) the limiting reactant and (b) the percentage yield of ethanoic acid.

(Remember that only the 29.8 g of sodium dichromate(VI) and the 6.9 cm³ of ethanol determine the theoretical yield. The sulphuric acid is a catalyst and the water is a solvent.)

(a) Determination of limiting reactant
Amount of $Na_2Cr_2O_7.2H_2O$ in moles

$$n(Na_2Cr_2O_7.2H_2O) = \frac{m}{M} = \frac{29.8\ g}{298\ g\ mol^{-1}} = 0.1\ mol$$

Mass of ethanol = Volume used × density of ethanol
$$= 6.9\ cm^3 \times 0.8\ g\ cm^{-3}$$
$$= 5.52\ g$$

Amount of C_2H_5OH in moles

$$n(C_2H_5OH) = \frac{m}{M} = \frac{5.52\ g}{46\ g\ mol^{-1}} = 0.12\ mol$$

The balanced chemical equation tells us that 3 mol of C_2H_5OH reacts with 2 mol of $Na_2Cr_2O_7.2H_2O$

Therefore in the experiment,
0.12 mol of C_2H_5OH reacts with 2/3(0.12) mol of $Na_2Cr_2O_7.2H_2O$

0.12 mol of C_2H_5OH reacts with 0.08 mol of $Na_2Cr_2O_7.2H_2O$

However, the amount of $Na_2Cr_2O_7.2H_2O$ used in the experiment was 0.1 mol. This amount (0.1 mol) is greater than that required from the balanced equation (0.08 mol): sodium dichromate(VI) is in excess and 0.12 mol C_2H_5OH is the limiting reactant.

(b) Calculation of % yield

The balanced equation tells us that:

3 mol C_2H_5OH ⇒ 3 mol CH_3COOH

0.12 mol C_2H_5OH ⇒ 0.12 mol CH_3COOH

Mass of ethanoic acid, $m = nM$
= 0.12 mol × 60 g mol⁻¹ = 7.2 g. This is the maximum theoretical yield according to the balanced equation.

$$\% \text{ yield} = \frac{\text{actual yield}}{\text{theoretical yield}} \times 100\%$$

$$= \frac{5.7 \text{ g}}{7.2 \text{ g}} \times 100\% = 79.2\%$$

● ●

Oxidation of Aldehydes and Ketones

The aldehyde group is easily oxidised to the carboxylic acid group. The ketone group is very difficult to oxidise. The difference in the ease of oxidation of aldehydes and ketones helps to distinguish one from the other.

Aldehydes are oxidised to form carboxylic acids by mild oxidising agents, such as the manganate(VII) ion and dichromate(VI) ion, or even by very weak oxidising agents, such as the silver ion (see the silver mirror test) or the copper(II) ion (see Fehlings test), e.g.

$$H-\underset{\underset{H}{|}}{\overset{\overset{H}{|}}{C}}-\overset{\overset{H}{|}}{C}=O \xrightarrow{Na_2Cr_2O_7/H^+} H-\underset{\underset{H}{|}}{\overset{\overset{H}{|}}{C}}-\overset{\overset{O}{\parallel}}{C}\diagdown OH$$

ethanal   **ethanoic acid**

Ketones are oxidised to form carboxylic acids by using very vigorous oxidising agents such as concentrated nitric acid, e.g.

$$\underset{\underset{CH_3}{|}}{\overset{\overset{CH_3}{|}}{C}}=O \xrightarrow{HNO_3} CH_3-\overset{\overset{OH}{/}}{\underset{\underset{O}{\backslash\backslash}}{C}} + CO_2 + H_2O$$

propanone   ethanoic   carbon   water
                       acid   dioxide

Fehling's Solution — copper sulphate, sodium potassium tartrate,

R−CHO
Aldehydes reduce Fehling's solution on warm-
R−COOH
ing; methanal reduces it without warming. The solution is prepared fresh for use by adding Solution B (an alkaline solution of sodium potassium tartrate) to approximately 1 cm³ of Solution A (copper(II) sulphate) until the deep blue precipitate just dissolves. A few drops of the aldehyde to be tested are added and warmed gently. The reaction can be represented by: *sodium potassium tartrate + copper sulphate = Fehlings solution*

$$\underset{\text{aldehyde}}{R-\overset{\overset{H}{|}}{C}=O} + \underset{\substack{\text{Blue}\\(+2)}}{Cu^{2+}} \longrightarrow \underset{\text{carboxylic acid}}{R-\overset{\overset{OH}{/}}{\underset{O}{\backslash\backslash}}C} + \underset{\substack{\text{Red}\\(+1)}}{Cu_2O}$$

Here the aldehyde is oxidised to a carboxylic acid and at the same time the Fehling's solution, represented by Cu^{2+}, is reduced to Cu_2O.

sodium hydroxide + silver nitrate → silver oxide + silver oxide + ammonia precipitate

Tollens Reagent (Silver Mirror Test) — *Tollens reagent*

Aldehydes reduce an ammoniacal solution of

Silver mirror test for the presence of aldehydes. If aldehydes are present, they reduce the silver ions in solution to form silver metal which forms a deposit on the inside of the tube.

silver nitrate to metallic silver; ketones do not. The solution is prepared by first adding a drop of dilute sodium hydroxide to 1 cm³ of a solution of silver nitrate to form a precipitate of silver oxide. Dilute aqueous ammonia is added, drop by drop, until the brown silver oxide precipitate just dissolves. This solution is known as Tollens reagent. Pour the ammoniacal silver nitrate solution into a clean test-tube and then add a few drops of the aldehyde to be tested. Silver is deposited on the wall of the test-tube in the form of a mirror. The reaction can be represented by:

$$R\!-\!\overset{\overset{\displaystyle H}{|}}{C}\!=\!O + Ag(NH_3)_2^{2+} \longrightarrow R\!-\!\overset{\overset{\displaystyle /O^-}{}}{\underset{\displaystyle \backslash\backslash O}{C}} + 2\,Ag\downarrow$$

aldehyde   silver(1) complex   carboxylate ion   metallic silver(0)

Here the aldehyde is oxidised in alkaline solution to the carboxylate ion, while at the same time the Ag^+ ion is reduced to metallic silver.

Reduction of Carbonyl Compounds (Aldehydes and Ketones)

The carbonyl group can add hydrogen just as the alkene group can.

Aldehydes are reduced to primary alcohols, and ketones to secondary alcohols by hydrogenation using a nickel catalyst, i.e.

$$R\!-\!\overset{\overset{\displaystyle H}{|}}{C}\!=\!O + H\!-\!H \overset{Ni}{\longrightarrow} R\!-\!\overset{\overset{\displaystyle H}{|}}{\underset{\displaystyle |}{\underset{H}{C}}}\!-\!OH$$

aldehyde   hydrogen   primary alcohol

$$R\!-\!\overset{\overset{\displaystyle R}{|}}{C}\!=\!O + H\!-\!H \overset{Ni}{\longrightarrow} R\!-\!\overset{\overset{\displaystyle R}{|}}{\underset{\displaystyle |}{\underset{H}{C}}}\!-\!OH$$

ketone   hydrogen   secondary alcohol

Examples are:

$$H\!-\!\overset{\overset{\displaystyle H}{|}}{C}\!=\!O + H\!-\!H \overset{Ni}{\longrightarrow} H\!-\!\overset{\overset{\displaystyle H}{|}}{\underset{\displaystyle |}{\underset{H}{C}}}\!-\!OH$$

methanal   hydrogen   methanol

$$CH_3\!-\!\overset{\overset{\displaystyle H}{|}}{C}\!=\!O + H\!-\!H \overset{Ni}{\longrightarrow} CH_3\!-\!\overset{\overset{\displaystyle H}{|}}{\underset{\displaystyle |}{\underset{H}{C}}}\!-\!OH$$

ethanal   hydrogen   ethanol

$$CH_3\!-\!\overset{\overset{\displaystyle CH_3}{|}}{C}\!=\!O + H\!-\!H \overset{Ni}{\longrightarrow} CH_3\!-\!\overset{\overset{\displaystyle CH_3}{|}}{\underset{\displaystyle |}{\underset{H}{C}}}\!-\!OH$$

propanone   hydrogen   propan-2-ol

Mandatory experiment 21

To prepare ethanoic acid and examine some of its properties.

Introduction

Ethanoic acid is prepared by oxidation of ethanol using acidified sodium dichromate(VI).

Ethanoic acid is the final oxidation product of ethanol. The oxidation occurs in two stages: the ethanol is first oxidised to ethanal and then to ethanoic acid. To ensure that ethanoic acid and not ethanal is produced the oxidising agent, sodium dichromate(VI), is used in excess and the reaction mixture is constantly refluxed = *to heat so that the vapours formed condense and return to be heated again*

Requirements

Safety glasses, quick-fit apparatus (suitable for reflux and also for distillation), dropping funnel, water bath, graduated cylinder ($10\ cm^3$), thermometer, anti-bumping granules, ethanol, sodium dichromate(VI), dilute ($1\ mol\ L^{-1}$) and concentrated sulphuric acid, universal indicator paper, magnesium, sodium carbonate.

Procedure

(1) Pour $10\ cm^3$ of the dilute sulphuric acid into a pear-shaped flask.
(2) Add some anti-bumping granules and then add 10 g of sodium dichromate(VI). Swirl the flask to mix the contents.

(3) Carefully add 2 cm³ of concentrated sulphuric acid. Cool the flask under a running tap.

(4) Arrange the apparatus for reflux with addition, as shown in the diagram.

REFLUX = BACKFLOW/return

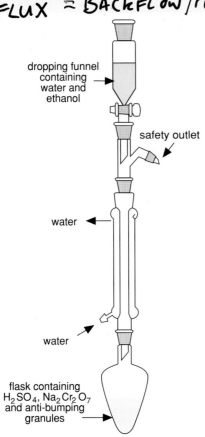

dropping funnel containing water and ethanol

safety outlet

water

water

flask containing H₂SO₄, Na₂Cr₂O₇ and anti-bumping granules

(5) Mix 5 cm³ of ethanol and 10 cm³ of deionised water in the dropping funnel. Add this, a drop at a time, down the neck of the condenser, to the oxidising mixture.

(6) Swirl the flask after each addition to mix the contents thoroughly. Cool the flask, if necessary, after each addition.

(7) When all the ethanol has been added, boil the mixture gently on a water bath for about 25 minutes.

(8) Rearrange the apparatus for distillation. Distill off about 20 cm³ of distillate.

DISTILLATION

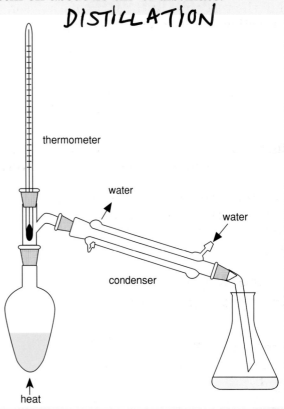

thermometer

water

water

condenser

heat

(9) Divide the distillate into five portions and test as follows:

(a) Smell cautiously and compare the smell with that of ethanol.

(b) Add some universal indicator paper to the distillate. Observe what happens.

(c) Add a small piece of magnesium to the distillate. Observe what happens.

(d) Add some sodium carbonate to the distillate. Observe what happens.

(e) Add 1 cm³ of ethanol and 2–3 drops of concentrated sulphuric acid to 2 cm³ of the distillate. Warm gently. Note the odour.

Results

Copy the table below and record your results in it.

Test	Effect/reaction	Inference
Odour of distillate		
Effect on universal indicator		
Reaction with magnesium		
Effect on sodium carbonate		
Reaction with ethanol + sulphuric acid		

Destructive Oxidation: Combustion of Organic Compounds

In Unit 5.4 the combustion of a fuel was described as an exothermic process, a reaction where heat was generated. Fossil fuels are organic compounds, composed mainly of carbon, hydrogen and oxygen. The combustion of an organic compound is another example of a redox reaction, where the particular compound is oxidised by oxygen to produce heat energy.

For example, when methane is oxidised to carbon dioxide and water, 890 kJ mol^{-1} of energy is released.

$$CH_4(g) + 2O_2(g) \rightarrow CO_2(g) + 2H_2O(l);$$
$$\Delta H = -890 \text{ kJ mol}^{-1}$$

Other organic compounds also release energy when they are burnt in oxygen. Alcohols release considerable energy when they undergo combustion.

Methanol, CH_3OH, releases 751 kJ mol^{-1} of energy when it is burnt in oxygen.

$$CH_3OH(l) + \tfrac{3}{2}O_2(g) = CO_2(g) + 2H_2O(l);$$
$$\Delta H = -751 \text{ kJ mol}^{-1}$$

Ethanol, C_2H_5OH, releases 1370 kJ mol^{-1} of energy when it is burnt in oxygen.

$$C_2H_5OH(l) + 3O_2(g) = 2CO_2(g) + 3H_2O(l);$$
$$\Delta H = -1370 \text{ kJ mol}^{-1}$$

Alcohols as Motor Fuels

Formula 1 car.

Methanol and ethanol burn cleanly in a motor car engine. The combustion is almost complete, so carbon monoxide emissions are reduced. Petrol blenders use organic compounds containing oxygen and blend them with petrol in order to increase the octane number of the petrol. Alcohols and ethers are the most common 'oxygenates' used by petrol blenders. Methanol has been used by Formula One racing drivers for some years, because it is safer than petrol as it is less volatile and less likely to explode. Methanol and ethanol have been used as part mixtures with petrol in some countries where alcohol can be produced cheaply. However, some difficulties arise because methanol and ethanol absorb water easily and because they do not mix very well with the non-polar petrol.

Non-Flammability of Halogenated Alkanes

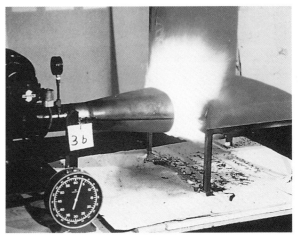

Testing polyurethane aeroplane seats for flammability.

Hydrocarbons and alcohols are excellent fuels because they release energy when they combine with oxygen. They release energy depending on the number of C–O and O–H bonds made during the particular reaction.

Some compounds, such as the haloalkanes, are quite unreactive. They do not burn in oxygen because they do not easily form chemical bonds with oxygen. This is because halogens such as fluorine and chlorine form very strong bonds with carbon and these bonds are difficult to break. Brominated hydrocarbons, such as 1,2-dibromoethane, are used in fire extinguishers. Fully halogenated alkanes, such as the chlorofluorocarbons (CFCs) which are used in refrigerators and as aerosol propellants, are non-flammable. Fully halogenated alkanes are used as flame retardants. However, care must be taken in choosing a particular haloalkane as some are toxic compounds.

Non-flammable halogenated alkanes

Compound	Formula	Flammability	Toxicity
Tetrachloromethane	CCl_4	Non-flammable	Toxic
Tetrafluororomethane	CF_4	Non-flammable	Non-toxic
Hexachloroethane	C_2Cl_6	Non-flammable	Toxic
Hexafluoroethane	C_2F_6	Non-flammable	Non-toxic

Exercise 7.3d

•

Redox reactions

(1) (a) Write an equation showing the final products, when ethanol is oxidised by sodium dichromate(VI).
(b) How is this reaction used in the breathalyser?

(2) (a) Write an equation showing the final products when ethanol is oxidised by potassium manganate(VII).
(b) What is the main product formed in the body when ethanol is metabolised?

(3) (a) Write an equation for the reaction of methanol with potassium manganate(VIII).
(b) Write an equation for the reaction of propan-1-ol with sodium dichromate(VI).

(4) Ethanal was prepared in the laboratory by the oxidation of ethanol using the apparatus shown. The flask was heated until boiling occurred and then heating was stopped. The mixture in the dropping funnel was then slowly added.
(a) What oxidising agent would you use?
(b) What colour change would you expect to observe?
(c) Why was the alcohol added slowly?
(d) Why was heating stopped before the mixture was added?
(e) How would you test that you had obtained ethanal?
(f) Draw a diagram of the apparatus you would use to prepare ethanoic acid.

(5) (a) Draw a labelled diagram of the apparatus you would use to prepare ethanal.
(b) Why is it important to remove the ethanal as soon as it is formed?
(c) What colour is the ethanol/dichromate solution? What would you notice as this solution was added to the hot sulphuric acid solution? Explain your answer.
(d) One of the physical properties of ethanal makes it difficult to collect under normal laboratory conditions. What is this property and how is the difficulty overcome?
(e) What would you observe if ethanal were warmed gently (i) with Fehling's solution and (ii) with ammoniacal silver nitrate solution (Tollens reagent)? In each case, explain the changes observed.

(6) Ethanol can be oxidised quite readily, first to ethanal and then to ethanoic acid.
(a) Outline the principle steps involved in both reactions.
(b) Describe the apparatus used to prepare both products.

(7) Ethanol has a wide range of applications and some very definite reactions which have widespread uses. Outline the properties of ethanol which account for the following:
(a) Breathalysers containing potassium dichromate(VI) can be used to test for alcohol.

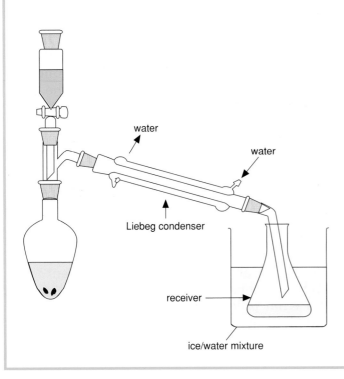

water

water

Liebeg condenser

receiver

ice/water mixture

221

(b) Vinegar can be made from ethanol.
(c) Ethanol is used in perfumes.
(d) Ethanol is often used as a cleaning agent.
(e) Ethanol is used as a motor fuel.

(8) Ethanoic acid can be prepared in the laboratory by oxidation of ethanol, first to ethanal and then to ethanoic acid.
(a) Describe and explain the colour change associated with the oxidation of the ethanol by acidified sodium dichromate(VI).
(b) Discuss the effect (if any) of each of the following on the ethanal formed initially and then on the final product, ethanoic acid:
(i) universal indicator (ii) sodium carbonate
(iii) sodium metal (iv) magnesium metal.

(9) (a) Describe and explain the changes in colour during the oxidation of ethanal using:
(i) acidified potassium manganate(VII)
(ii) acidified sodium chromate(VI)
(iii) ammoniacal silver nitrate
(iv) Fehling's solution.
(b) Explain why:
(i) acid is used with potassium manganate(VII)
(ii) ammonia is required in Tollens reagent.

(10) Ethanoic acid may be prepared in the laboratory by the oxidation of ethanol. The procedure used involves refluxing the ethanol with an excess of acidified sodium dichromate(VI) solution and then distilling the resulting mixture to obtain a pure sample of ethanoic acid.
The reaction is represented as:
$3C_2H_5OH + 2Cr_2O_7^{2-} + 16H^+ \Rightarrow 3CH_3COOH + 4Cr^{3+} + 11H_2O$
(a) Explain the meaning of each of the underlined terms.
(b) Draw a labelled diagram of the apparatus used during refluxing.
(c) What material, other than ethanol and acidified sodium dichromate(VI), should be put into the reaction vessel? Why is this material used?
(d) Why is an excess of acidified sodium dichromate(VI) solution used?
(e) What impurities are removed by distillation?
(f) Describe what would happen if a piece of sodium metal were added to the ethanoic acid.
(g) If 10 cm³ of ethanol (density, 0.8 g cm⁻³) were used in the experiment, calculate the

maximum mass of ethanoic acid which could be obtained.
(h) If 5.0 g of ethanoic acid was the actual mass obtained, calculate the percentage yield of ethanoic acid.

(11) To prepare a sample of ethanal, 30 cm³ of sulphuric acid was heated with some anti-bumping granules. The heating was stopped and a mixture of 15 g of sodium dichromate, $Na_2Cr_7.2H_2O$, 25 cm³ of water and 15 cm³ of ethanol was added slowly, from a dropping funnel. The half equation for the reaction is given as:
$3C_2H_5OH + Cr_2O_7^{2-} + 8H^+ \rightarrow 3CH_3CHO + 2Cr^{3+} + 7H_2O$
(a) Draw a labelled diagram of the apparatus used.
(b) Why was the heating stopped before the ethanol was added ?
(c) How was the ethanal collected ?
(d) Explain the change of colour during the reaction.
(e) If the density of ethanol is 0.8 g cm⁻³ and the yield of ethanal was 3.5 g, (i) determine the limiting reactant and (ii) calculate the percentage yield of ethanal.
(f) Suggest reasons why the yield was low.

(12) To prepare a sample of ethanoic acid, 10 cm³ of water was measured into a flask and 5 cm³ of concentrated sulphuric acid was added. 12.0 g of sodium dichromate(VI), $Na_2Cr_2O_7.2H_2O$, was dissolved in the acid solution and the flask was fitted with a reflux condenser. A solution of 3.0 cm³ of ethanol (density, 0.8 g cm⁻³) and 12.0 cm³ of water was added to the flask through the condenser. The apparatus was assembled and the reaction mixture was heated under reflux for about 30 minutes. Finally, the ethanoic acid was recovered from the reaction mixture by distillation. The yield of ethanol was 2.6 cm³.
The reaction is represented as:
$3C_2H_5OH + 2Cr_2O_7^{2-} + 16H^+ \Rightarrow 3CH_3COOH + 4Cr^{3+} + 11H_2O$
(a) Explain, using diagrams where necessary, the meaning of each of the underlined terms.
(b) What precautions should be taken when adding the concentrated sulphuric acid to the water at the start of the experiment? Why are these precautions necessary?
(c) What procedure should be followed when adding the ethanol solution to the flask? Why is this procedure used?

(d) Describe and explain the colour change taking place during the reaction.

(e) Describe the changes in odour which take place during the reaction.

(f) Determine the limiting reactant.

(g) Calculate the percentage yield of ethanoic acid (density, 1.06 g cm^{-3}).

(13) An ester, methyl ethanoate, was prepared by reacting 5 cm³ of methanol and 5 cm³ of ethanoic acid together using sulphuric acid as a catalyst, in suitable apparatus.

The reaction is as follows:

$$CH_3OH + CH_3COOH = CH_3COOCH_3 + H_2O$$

(a) If the densities of the methanol, ethanoic acid and the methyl ethanoate were 0.8, 1.0 and 0.9 g cm⁻³, respectively, calculate which of the reactants is the limiting reactant.

(b) If 4.8 cm³ of the ester was obtained at the end, calculate its percentage yield.

(14) Ethanoic acid can be prepared in the laboratory using the same reactants as in the preparation of ethanal. Describe, in detail, how you would would ensure that ethanal and not ethanoic acid was the principal product in the preparation.

(15) Ethanal is easily oxidised, while propanone is not easily oxidised. Explain why ethanal is more reactive than propanone.

(16) Describe how you would distinguish between propanal and propanone.

(17) Give some everyday uses of methanal, ethanal, propanone and describe how Tollens test is used in commercial applications.

(18) An unknown compound, C_3H_6O, reacted with acidified potassium manganate(VII) to form a compound, $C_3H_6O_2$. The unknown compound also reacted with Tollens reagent and gave a positive result. Write the structures of C_3H_6O and $C_3H_6O_2$.

(19) (a) The C=C and the C=O groups both have a high electron density about the double bond. Describe a reaction which shows how they are alike. In your answer, refer to the distribution of electrons about each double bond.

(b) Write equations for the reactions of (i) methanal, (ii) ethanal and (iii) propanone with hydrogen using a nickel catalyst.

(20) Using structural formulae, write equations for the following reactions. Name the compound formed in each case.

(a)
methanal + acidified potassium manganate(VII) ⇒
methanal + ammoniacal silver nitrate ⇒
methanal + Fehling's solution ⇒
methanal + hydrogen/nickel ⇒

(b)
ethanal + acidified potassium manganate(VII) ⇒
ethanal + ammoniacal silver nitrate ⇒
ethanal + Fehling's solution ⇒
ethanal + hydrogen/nickel ⇒

(c)
propanone + acidified potassium manganate(VII) ⇒
propanone + ammoniacal silver nitrate ⇒
propanone + concentrated nitric acid ⇒
propanone + Fehling's solution ⇒
propanone + hydrogen/nickel ⇒

(21) Write a brief note on each of the following topics:
(a) Combustion of organic compounds
(b) Alcohols as fuels
(c) Non-flammability of halogenated alkanes.

(E) REACTIONS AS ACIDS

Reaction of Alcohols with Sodium

Sodium reacting with alcohol.

Alcohols are extremely weak acids. Ethanol ($K_a = 10^{-16}$) is in fact a weaker acid than water ($K_a = 10^{-14}$), but is a stronger acid than either ethyne or ammonia.

An alcohol, ROH, is acidic, because, like water, it has a hydrogen atom bonded to a strong electronegative atom, oxygen.

Like water it dissociates into ions; in this case into an alkoxide ion, RO⁻ and a hydrogen ion, H⁺.

$$ROH \rightleftharpoons RO^- + H^+$$

This enables alcohols to react with active metals, such as sodium. The alcohol reacts forming a salt and hydrogen gas is released, e.g.

$$\text{H}-\overset{\overset{\text{H}}{|}}{\underset{\underset{\text{H}}{|}}{\text{C}}}-\text{OH} + \text{Na} \rightarrow \text{H}-\overset{\overset{\text{H}}{|}}{\underset{\underset{\text{H}}{|}}{\text{C}}}-\text{ONa} + \tfrac{1}{2}\text{H}_2$$

methanol   sodium methoxide

$$\text{H}-\text{C}-\text{C}-\text{OH} + \text{Na} \rightarrow \text{H}-\text{C}-\text{C}-\text{ONa} + \tfrac{1}{2}\text{H}_2$$

ethanol   sodium ethoxide

Acidic Nature of the Carboxylic Acid Group

Carboxylic acids are weak acids which dissociate to a small extent (approximately 1–2%) in aqueous solution. The value of K_a, the acidity constant, is about 10^{-4} to 10^{-5} for most carboxylic acids. Although they are weak acids, they are stronger acids than either water or alcohols. The value of K_a for ethanoic acid is 1.8×10^{-5} at 298 K.

A carboxylic acid dissociates into ions, the carboxylate ion, RCOO⁻, and the hydrogen ion, H⁺.

$$RCOOH \rightleftharpoons RCOO^- + H^+$$

This enables carboxylic acids to react with sodium and magnesium to form salts, with sodium hydroxide to form a salt and water and with sodium hydrogencarbonate to form a salt, water and carbon dioxide.

Reactions with sodium and magnesium

ethanoic acid   sodium ethanoate

Magnesium reacts with ethanoic acid forming magnesium ethanoate and hydrogen gas.

$$2\text{CH}_3\text{COOH} + \text{Mg} \rightarrow (\text{CH}_3\text{COO})_2\text{Mg} + \text{H}_2$$

methanoic acid   sodium methanoate

Magnesium reacts with methanoic acid forming magnesium methanoate and hydrogen gas.

$$2\text{HCOOH} + \text{Mg} \rightarrow (\text{HCOO})_2\text{Mg} + \text{H}_2$$

Reaction with sodium hydroxide

ethanoic acid   sodium ethanoate

propanoic acid   sodium propanoate

Reaction with sodium carbonate

ethanoic acid   sodium ethanoate

methanoic acid   sodium methanoate

The release of carbon dioxide indicates that carboxylic acids are stronger acids than other organic acids.

Acidity of Carboxylic Acids is Related to Structure

The hydroxyl group, –OH, of a carboxylic acid is more polar than the –OH group of an alcohol and consequently can donate a proton more readily. When a carboxylic acid such as ethanoic acid dissociates into a proton and an ethanoate ion, the negative ethanoate ion stabilises itself by a process called resonance.

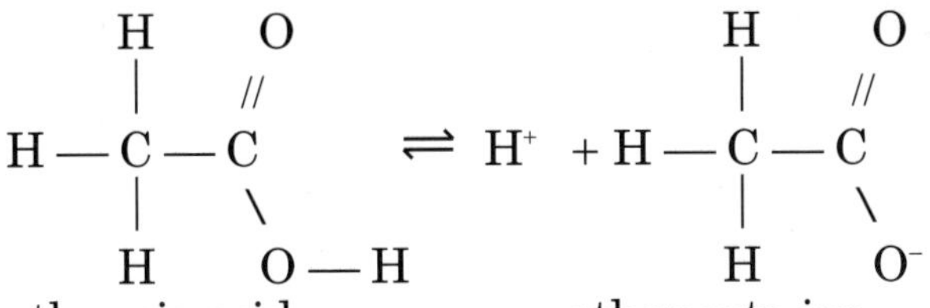

```
    H   O                      H   O
    |   //                     |   //
H — C — C      ⇌  H⁺ + H — C — C
    |   \                      |   \
    H   O — H                  H   O⁻
ethanoic acid               ethanoate ion
```

The ethanoate ion can stabilise itself by redistribution of the negative charge by resonance.

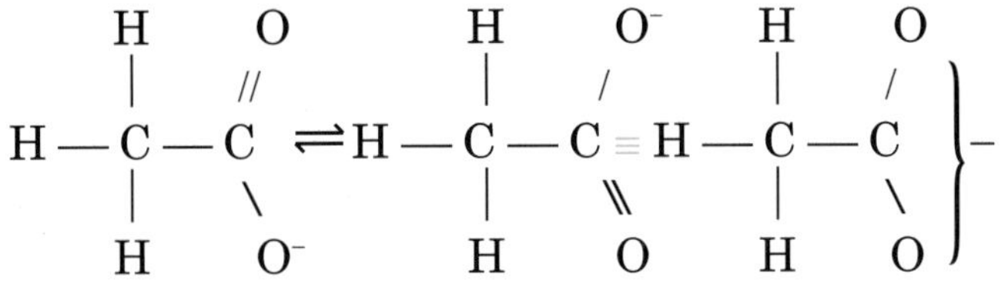

```
    H   O            H   O⁻           H   O        ⎫
    |   //           |   /            |   /        ⎪
H — C — C  ⇌  H — C — C ≡ H — C — C        ⎬ —
    |   \            |   \\           |   \        ⎪
    H   O⁻           H   O            H   O        ⎭
       I                II
```

The negative charge on the ion shifts from one oxygen to the other, the structure of the ion being neither one structure nor the other but a combination or resonance hybrid of both structures I and II. X-ray diffraction studies show that the two carbon–oxygen bonds are of equal length.

This resonance hybridisation makes the ethanoate ion more stable than the ethanoic acid; this means that the hydrogen ion, H⁺, is also stable. The ethanoic acid can therefore dissociate more easily to form a hydrogen ion and an ethanoate ion.

Exercise 7.3e

•

Reactions as acids and further reactions

(1) (a) Explain why alcohols are weak acids.
(b) How does the fact that alcohols are acids enable them to react with reactive metals like sodium?

(2) Write equations, using structural formulae, for the following reactions:
(a) methanol + sodium ⇒
(b) ethanol + sodium ⇒
(c) propan-1-ol + sodium ⇒
(d) propan-1-ol + sodium ⇒

(3) (a) Why are carboxylic acids weak acids?
(b) Write equations, using structural formulae, for the following reactions:
(i) ethanoic acid and sodium
(ii) ethanoic acid and magnesium
(iii) ethanoic acid and sodium carbonate.

(4) Write equations, using structural formulae for the following reactions:
(a) methanoic acid and sodium
(b) methanoic acid and magnesium
(c) methanoic acid and sodium carbonate
(d) propanoic acid and sodium
(e) propanoic acid and magnesium
(f) propanoic acid and sodium carbonate.

(5) Explain why the acidity of carboxylic acids is related to the structure of the carboxylic acid.

(6) Using structural formulae, write equations for the following reactions:
(a) the reaction of methanol with sodium
(b) the reaction of ethanol with calcium
(c) the oxidation of methanol using potassium manganate(VII)
(d) the oxidation of ethanol using sodium dichromate(VI)
(e) the combustion of propanol
(f) the dehydration of methanol using sulphuric acid
(g) the two possible dehydration reactions of ethanol with sulphuric acid.

(7) Complete the following:
(a) $CH_3CHO + KMnO_4 \Rightarrow$
(b) $HCHO + Ag^+ \Rightarrow$
(c) $CH_3CHO + Cu^{2+} \Rightarrow$
(d) $CH_3CHO + H_2 \Rightarrow$
(e) $HCHO + H_2 \Rightarrow$
(f) $CH_3COCH_3 + H_2 \Rightarrow$

(8) Give equations for the reactions which follow:
(a) oxidation of a primary alcohol
(b) oxidation of a secondary alcohol
(c) reduction of a ketone
(d) oxidation of an aldehyde
(e) oxidation of a ketone.

(9) Using structural formulae, write equations for the following reactions:
(a) methanoic acid and sodium hydroxide
(b) ethanoic acid and sodium carbonate
(c) propanoic acid and sodium
(d) propanoic acid with ethanol
(e) ethanoic acid with methanol
(f) ethanoic acid with ethanol
(g) ethyl ethanoate with sodium hydroxide.

(10) Describe chemical tests which would differentiate between the following pairs of compounds:
(a) propanal and propanoic acid
(b) butanone and ethyl ethanoate
(c) ethanol and methanoic acid.

(F) ORGANIC SYNTHESIS
Principles and Examples

Synthesis of a new product at BASF.

Thousands of new organic compounds are synthesised each year and enhance the quality of our lives. These include: new medicines and drugs, new agri-chemicals, new plastics and polymers and many other compounds. Many older traditional organic compounds are also synthesised by newer methods which may be more economical and more environmentally friendly.

The synthesis of an organic compound, whether in a school or in an industrial laboratory, involves similar techniques. The process involves reacting the starting compound with a chemical reagent to produce the final product.

$$\text{Starting material} \xrightarrow{\text{Chemical reagent}} \text{Final product}$$

For example, ethanol may be synthesised by reacting the starting compound, ethene, with a chemical reagent, water, using a catalyst.

$$\text{Ethene} \xrightarrow{\text{Water}} \text{Ethanol}$$

Here, the structure of ethene is changed and converted by the chemical reagent, water, into a new compound, ethanol. Care must be taken to ensure that the desired product is obtained by correct choice of reagent and the optimum use of catalysts, solvents, heat and pressure.

There are usually **three stages** in the production of an organic compound:

1. Planning the synthesis: Careful choice of starting compounds and chemical reagents and proper use of conditions is necessary to ensure that the maximum yield of desired product is obtained. It should also be remembered that many organic compounds are volatile and inflammable, while some are toxic. Therefore, the method used and the solvents involved must be environmentally acceptable.

There are three main processes involved in any synthesis:
• heating and cooling
• mixing of reactants
• stirring the mixture.

2. Separation and purification of crude product: The product obtained at the end of the chemical reaction may not be sufficiently pure owing to unused reactants and catalysts or unwanted by-products. The pure product is isolated by choosing the most suitable separation technique.

Some of the separation techniques include:
• distillation
• filtration
• recrystallisation
• chromatography.

3. Analysis of the pure product: The pure product is tested using various analytical techniques to determine its purity.

Some of the techniques include:
• melting point and boiling point determination
• chromatography
• mass spectrometry
• infrared absorption
• nuclear magnetic resonance.

Bond Breaking and Bond Forming in Chemical Synthesis

Bond Breaking

During a chemical reaction, chemical bonds are broken and new bonds are formed. The breaking of a chemical bond usually occurs in one of two ways.

1. The bond breaks in such a way that **both electrons remain with one part.**

$$X—Y \rightarrow X^+ + Y^-$$

This results in the formation of positive and negative ions.

For example, when chlorine reacts with ethene, a positive ion, $C_2H_5^+$, called a carbonium ion, is formed as well as a negative ion, Cl^-. These ions combine with each other to form the final product.

2. The bond breaks in such a way that **both parts take an equal share of electrons.**

$$X—Y \rightarrow X^• + Y^•$$

This results in the formation of very reactive particles called **free radicals**. Free radicals are atoms or groups of atoms with unpaired electrons.

For example, in the reaction of chlorine with methane, the chlorine molecule is broken up into two chlorine free radicals

$$Cl : Cl \rightarrow Cl^• + Cl^•$$

Bond Forming

[1] Bringing Positive and Negative Centres Together

In the addition reaction of chlorine to ethene, the non-polar chlorine molecule becomes polarised as it approaches the double bond in ethene. The positive end of the chlorine attaches itself to the double bond. This generates a positive ion called a carbocation and at the same time generates a negative chloride ion.

The negative ion is attracted towards the positive ion, thereby completing the reaction.

carbonium ion   1,2-dichloroethane

[2] Bringing Reactive Radical Centres Together

In the reaction of chlorine with methane, the chlorine–chlorine bond is broken into two parts, each with one unpaired electron. These chlorine atoms with an unpaired electron are called free radicals and are extremely reactive. The chlorine free radicals react easily with methane and other free radicals.

(i) $Cl—Cl \rightarrow Cl^• + Cl^•$

(ii) $CH_4 + Cl^• \rightarrow HCl + CH_3^•$

(iii) $CH_3^• + Cl^• \rightarrow CH_3Cl$ Here two reactive radicals are brought together.

HONOURS • HONOURS • HONOURS • HONOURS • HONOURS • HONOURS • HONOURS • HONOURS • HONOURS • HONOURS •

PVC from ethene

Poly(chloroethene), PVC, is a polymer made from chloroethene (vinyl chloride).

• **Synthesis of chloro(ethene):** Chloroethene is not easily made from ethene in the laboratory. In industry it is synthesised from ethene by reaction with hydrogen chloride and oxygen using a catalyst.

$$
\begin{array}{ccc}
\text{H} & \quad & \text{H} \\
\diagdown & \diagup & \\
\text{C} = \text{C} \quad +\text{H—Cl} + \text{O} = \text{O} \Rightarrow \\
\diagup & \diagdown & \\
\text{H} & \quad & \text{H}
\end{array}
\qquad
\begin{array}{cc}
\text{H} & \text{H} \\
| & | \\
\text{H—C—C—H} \\
| & | \\
\text{Cl} & \text{Cl}
\end{array}
$$

$$
\begin{array}{cc}
\text{H} & \text{H} \\
| & | \\
\text{H—C—C—H} \\
| & | \\
\text{Cl} & \text{Cl}
\end{array}
\quad \overset{heat}{\Rightarrow} \quad
\begin{array}{cc}
\text{H} & \text{H} \\
\diagdown & \diagup \\
\text{C} = \text{C} \\
\diagup & \diagdown \\
\text{H} & \text{Cl}
\end{array}
$$

Step 1.

$$CH_2{=}CH_2 + HCl + O_2 \Rightarrow CH_2ClCH_2Cl$$

Step 2.

$$CH_2ClCH_2Cl \overset{heat}{\Rightarrow} CH_2{=}CHCl$$

• **Polymerisation of chloroethene:**
Poly(chloroethene) is synthesised from chloroethene using a catalyst, heat and pressure.

Step 3.

$$CH_2{=}CHCl \Rightarrow \left[\begin{array}{cc} \text{H} & \text{H} \\ | & | \\ \text{C—C} \\ | & | \\ \text{H} & \text{H} \end{array}\right]_n$$

chloroethene   poly(chloroethene),
                            PVC

Socially Acceptable Solvents

Many chemical reactions are carried out in solution. The solvent used must be carefully chosen. Some of the following factors may affect the choice of solvent.

• **Polar and non-polar solvents:** Some reactants and products dissolve in polar solvents, while others dissolve in non-polar solvents. Sometimes it may be possible to dissolve the required compounds in a mixed solvent, such as alcohol and water.

• **Cost:** Chemists must not only consider the cost of a particular solvent but also must con-sider how difficult it is to separate the product from the solvent.

• **Safety:** Many solvents are flammable, others may be carcinogenic. Proper care must be taken in choosing the correct solvent.

• **Environmental considerations:** Most solvents are volatile and may be released into the atmosphere. Some, such as CFCs, cause damage to the ozone layer. Others are often released into water and may be harmful to fish or may enter the food chain.

• **Acceptable and non-acceptable solvents:** Water is perhaps the only environmentally acceptable solvent. Oxygenated solvents, like ethanol and propanone, do not cause any environmental damage as they are biodegradable. Chlorinated solvents, in particular PCBs (polychlorinated biphenyls), and those containing benzene rings, in particular the dioxins, can become established in food chains and build up in body fat. The build up of these substances often causes cancer. Benzene is a carcinogen liked to leukaemia.

Exercise 7.3f

•

(1) How do the products of organic synthesis enhance the quality of our lives? In your answer name some familiar products from which we benefit.

(2) (a) What type of process is involved in a synthesis?
(b) Using ethene as a starting compound, explain how ethanol is formed.
(c) Write a brief note on the three main stages involved in a synthesis.

(3) A chemical synthesis involves (a) <u>bond breaking</u> and (b) <u>bond forming</u>. Explain the meanings of the underlined terms.

(4) Describe, using structural formulae, the stages involved in the synthesis of poly(chloroethene), PVC, from ethene.

(5) Discuss the need for acceptable solvents in organic synthesis.

(6) The formulae of *four* organic compounds are given in the table at the top of p. 229.
(a) Copy and complete the table.

HONOURS • HONOURS •

Compound	Name	Homologous series	Functional group	Structural formula
C_2H_4 (A)	Ethene			
C_2H_5OH (B)		Alcohols		
CH_3CHO (C)			—CHO	
CH_3COOH (D)				(structure shown)

The structural formula shown:

$$H-\underset{\underset{H}{|}}{\overset{\overset{H}{|}}{C}}-\overset{O}{\underset{OH}{C\!\!\nearrow\!\!\searrow}}$$

(b) Using structural formulae, describe how you would synthesise B from A.

(c) Using structural formulae, describe how you would synthesise A from B.

(d) Using structural formulae, describe how you would synthesise C from B.

(e) Using structural formulae, describe how you would synthesise D from B.

(7) The formulae of *four* organic compounds are given below.

(A)	(B)
$CH_3CH = CHCH_3$	$CH_3CH_2CH_2CH_3$

(C)	(D)
$CH_3COCH_2CH_3$	$CH_3CH_2CH(OH)CH_3$

(a) Name the homologous series to which A, B, C and D belong.

(b) Using structural formulae and the reagents necessary, write equations for the conversion of

(i) B to A and (ii) A to D.

(8) Study the reaction scheme below and answer the questions which follow.

Ethene ⇒ Ethanol ⇒ Ethanoic acid

(a) Name the homologous series to which each compound belongs.

(b) Draw the structural formula of each compound.

(c) Outline, using structural formulae and the reagents required, how you would synthesise (i) ethanol from ethene and (ii) ethanoic acid from ethanol.

(d) How would you convert ethanol back to ethene?

(9) Answer the following questions in relation to the following functional groups:

(A)	(B)	(C)	(D)
—O—H	$-\overset{|}{\underset{H}{C}}=O$	$-C=O$	$-\overset{|}{\underset{OH}{C}}=O$

(a) Name the homologous series to which A, B, C and D belong.

(b) Using structural formulae and the reagents necessary, outline the conversion of the C-3 compound of (i) A to B, (ii) A to C and (iii) A to D.

(10) The formulae of *four* organic compounds are given below:

A	B
C_2H_5OH	CH_3CHO

C	D
CH_3COOH	CH_3COCH_3

(a) Name the compounds A, B, C and D.

(b) Name the homologous series to which A, B, C and D belong.

(c) Using structural formulae and the reagents necessary, outline the conversion of (i) A to B and (ii) A to C.

(d) Show, using an equation, how can D be converted into an alcohol.

7.4 ORGANIC NATURAL PRODUCTS

Herbal medicine, aromatherapy and alternative medicines have become increasingly popular in recent years. Most of the chemicals used in these areas have been extracted from plants. The native Indians in the rain forests and other areas have in fact practised herbal medicine for thousands of years. It should be remembered that natural organic products are the same as those chemicals which are made in the laboratory.

Chemicals extracted from natural products have widespread uses.

• Citral extracted from lemon grass has been used to synthesise vitamin A.

• Clove oil, eucalyptus oil, citrus oils and other natural oils are used as aromas and flavours in

229

the food and cosmetics industries.
• Acetylsalicylic acid (aspirin) was isolated from willow trees (Latin name: *salix*).

Natural products – aspirin from willow trees

Extraction Techniques

When organic chemicals are prepared in the laboratory they may contain impurities, such as unused reactants, by-products of side reactions and the solvent used. Isolation of the pure product from the impurities may require various separation techniques depending on the substances involved.

Distillation, crystallisation and chromatography are the main methods used. The extraction of organic natural products such as citral, oil of wintergreen, oil of cloves, eucalyptus oil and other essential oils usually involves solvent extraction and steam distillation.

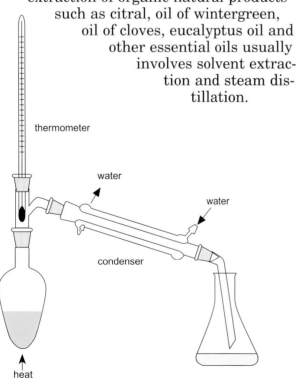
Separation by distillation

Distillation

When the product required is the only volatile (low boiling point) component in the reaction mixture it can be separated from the other components by distillation. When the boiling point is lower than 140 °C, a water condenser is used. If the boiling point is higher than 140 °C, an air condenser is used because a water condenser would crack at the higher temperatures.

The liquid is heated to its boiling point; its vapour distils over, condenses on the cold condenser surface and is collected. Porcelain chips are usually used to prevent bumping (uneven boiling).

When the product required is part of a mixture containing more than one volatile component, fractional distillation is used. Fractional distillation is used in many large industrial processes, such as the separation of crude oil into its main components or the separation of water and alcohol in the brewing and distilling industries. In fractional distillation the most volatile component in the mixture rises to the top of the column while the least volatile component is drawn off at the bottom.

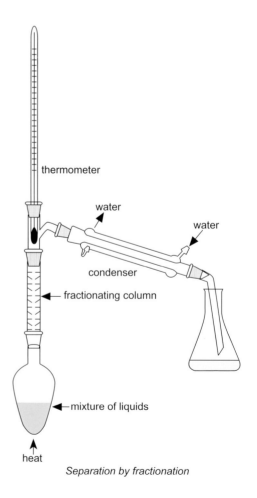
Separation by fractionation

Solvent Extraction

Solvent extraction is used to separate organic materials from an aqueous reaction mixture. The organic material is usually dissolved in a non-polar solvent. The mixture is then shaken in a separating funnel and allowed to separate into two layers: the aqueous layer on top, the organic layer on the bottom. The organic layer is then allowed to run off. The required organic product is obtained by allowing the solvent to evaporate.

Steam Distillation

When two immiscible liquids are combined, their boiling point is lower than that of either of the two pure components. Steam is used to distill many organic compounds at lower temperatures than their boiling points. This technique, called 'steam distillation', is very useful because it allows substances to be heated which might undergo decomposition at their boiling points.

Mandatory experiment 22

Extraction of citrus oils from fruit peel (or clove oil from cloves) by steam distillation.

Introduction

Many natural products are isolated by steam distillation for use as perfumes, flavours and fragrances. Fruit peel contains citrus oils. When fruit peel is steam distilled two layers, consisting of citrus oils and water, separate out. The two layers can be separated using a separating funnel. Clove oil can be isolated from cloves in a similar manner.

Requirements

Safety glasses, heating mantle (or Bunsen burner), steam generator, three-holed round-bottomed flask, distillation apparatus, water, fruit peel (or cloves), separating flask.

Procedure

(1) Arrange the steam distillation apparatus as shown.
(2) Place some fruit peel (or cloves) and water in the distillation flask. Seville oranges/marmalade are most suitable.
(3) Allow the steam to pass continuously through the distillation flask.

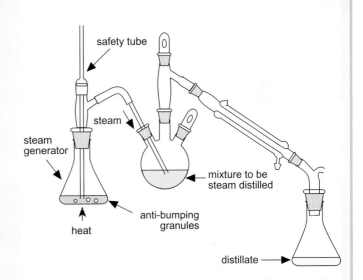

(4) Collect the distillate.
(5) Add about 15 cm³ of 1,1,1-trichloroethane to the distillate. The citrus (or clove) oils dissolve in the 1,1,1-trichloroethane.
(6) Place the distillate in a separating funnel. Allow the lower organic layer to run off into a small beaker.
(7) Allow the 1,1,1-trichloroethane to evaporate off leaving the citrus (or clove) oils behind.

231

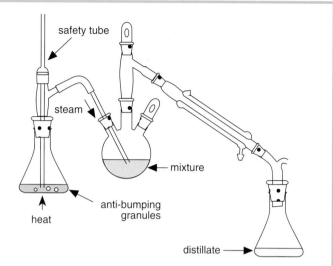
7.5 CHROMATOGRAPHY AND INSTRUMENTATION IN ORGANIC CHEMISTRY

Chromatography

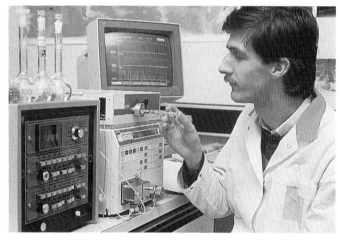

Traces of substances can be detected using this gas chromatograph at the Bayer Agrochemicals Centre at Monheim.

Chromatography is a technique which is used to separate and identify the components of a mixture. There are a number of different types of chromatography: column chromatography, thin-layer chromatography, paper chromatography, gas chromatography, high-performance liquid chromatography (HPLC) and many other forms.

All forms of chromatography can be divided into two types: adsorption chromatography and partition chromatography. Both depend on the separation of the components of the mixture between two phases, a stationary phase and a mobile phase.

(1) Adsorption chromatography

In adsorption chromatography the stationary phase is a solid, while the mobile phase can be a liquid or a gas. The components of the mixture are adsorbed onto a solid at different rates as the mobile phase passes over the solid.

Column chromatography and thin-layer chromatography are forms of adsorption chromatography.

Column chromatography

This is the simplest form of chromatography. The stationary phase is a solid, usually aluminium oxide, Al_2O_3, while the mobile phase is a solvent. A large glass burette (column) is packed with a slurry made from aluminium oxide and the solvent and allowed to settle. The mixture to be separated is dissolved in the solvent

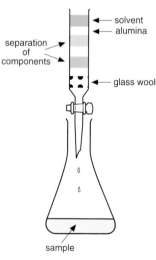

Column chromatography

232

and allowed to run through the column. The components of the mixture separate as they move gradually down the column. If the components are coloured, such as the pigments chlorophyll or carotene, they can be separated visually. The component which moves through the column fastest is collected and then distilled to remove the solvent, leaving the pure component.

Thin-layer chromatography (TLC)

TLC is used to separate dyes from fibres in forensic work.

In thin-layer chromatography, the solid phase is usually silica gel, aluminium oxide or cellulose containing a binder such as starch. This phase, the adsorbent, is mixed into a slurry and coated onto a thin plate of glass. The mixture to be separated is applied as a spot on a line near one edge of the plate. Another line is drawn on the opposite end of the plate and the spot is allowed to dry. The plate is dipped in a small amount of solvent in a small tank or beaker and covered to prevent evaporation. As the solvent rises past the spot components begin to move and are separated. When the solvent front nears the top of the plate, it is removed and allowed to dry. If the components of the mixture are coloured they can be visually identified. If they are colourless they may be identified by exposure to ultraviolet light, by iodine vapour or by heat.

Each component of the mixture is identified by means of its R_f value, where the R_f value is given by

$$R_f = \frac{\text{Distance travelled by the component from the original spot}}{\text{Distance travelled by the solvent from the original spot}}$$

(2) Partition chromatography

In partition chromatography, the stationary phase is a liquid. The components of the mixture are separated between the liquid phase and the mobile phase.

Paper chromatography, gas–liquid chromatography and high-performance liquid chromatography (HPLC) are all forms of partition chromatography.

Paper chromatography

In paper chromatography, the stationary phase is a liquid: water adsorbed onto a sheet of paper. This process is similar in many ways to thin-layer chromatography (TLC). A mixture is dissolved in a solvent; a spot of the sample solution is placed on a line on the moist chromatography paper. The paper is dipped into the solvent and the mixture separates in the same way as in TLC. The R_f values are calculated in a similar way.

There are two types of paper chromatography: ascending chromatography, where the paper strip dips into the solvent at the bottom of the tank, and descending chromatography, where the solvent is contained in a trough at the top of the tank.

Paper chromatograms are particularly useful in that they can be stored easily for future reference.

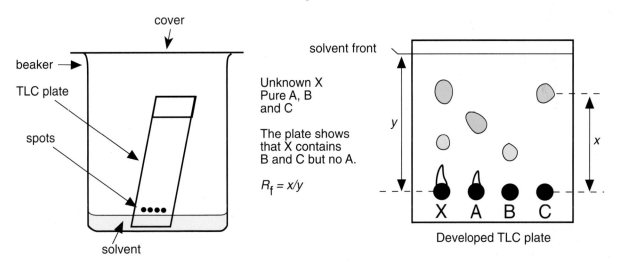

Thin layer chromatography

233

Mandatory experiment 23

To separate some indicator dyes using paper chromatography

Introduction

A mixture of three indicators is separated by means of paper chromatography. The R_f values are calculated and compared with the values obtained from the pure indicators. The experiment can also be done using thin-layer or column chromatography.

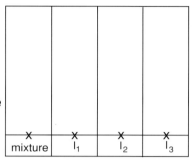

Requirements

Safety glasses, fume hood, chromatography tank, solvent trough for descending chromatography, chromatography paper, deionised water, ethanol, ammonia solution (0.880 ammonia), melting point tubes, methyl red, methyl orange and bromothymol blue solutions, a mixture of the three indicator solutions, pencil, ruler, hair drier.

Procedure

(1) In a fume hood, prepare the solvent mixture by pouring 25 cm³ of water, 10 cm³ of ethanol and 5 cm³ of ammonia solution into a beaker. Pour about 10 cm³ of the solvent into the bottom of the trough and the remainder into the chromatography tank.
Place the trough in position as shown. Cover the tank and allow to stand to saturate the column with solvent vapour.

(2) Using a pencil, not a pen, draw lines on the chromatography paper as shown.
(3) Dip a clean melting point tube into the mixture of indicators and touch the appropriate spot on the chromatography paper for a brief moment.

I_1 = methyl red
I_2 = methyl orange
I_3 = bromothymol blue

(4) Using a fresh tube each time, repeat the procedure for each of the three pure indicators.
(5) Attach the chromatography paper to the solvent trough in the tank, taking care not to handle the paper. Do not allow the solvent in the trough to touch the samples or allow the paper to touch the small amount of solvent in the bottom of the tank.
(6) Remove the chromatogram from the trough just before the solvent front reaches the bottom of the paper.
(7) Attach the chromatogram between two retort stands and dry it using a hair drier.
(8) Calculate the R_f value of each indicator and compare each value with the values obtained for the components of the mixture. Record your results in a table like the one below.

$$R_f = \frac{\text{distance travelled by the component from the original spot}}{\text{distance travelled by the solvent from the original spot}}$$

Indicator	Distance travelled by indicator	Distance travelled by solvent	R_f value
Pure methyl red			
Methyl red in mixture			
Pure methyl orange			
Methyl orange in mixture			
Pure bromothymol blue			
Bromothymol blue in mixture			

Gas–liquid chromatography

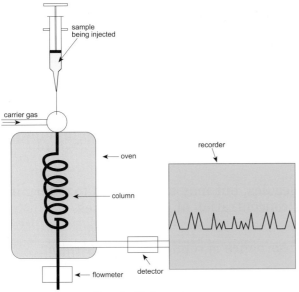

Gas - liquid chromatogram

In gas–liquid chromatography the stationary phase is a liquid supported on an inert porous solid such as charcoal. This is packed into a coiled column through which the sample to be determined is injected by means of a hypodermic syringe. The column is heated to vaporise the sample and it is carried through the column by the mobile phase, which is a gas such as nitrogen or helium. The different components of the mixture separate out at different rates as they are carried through the column. The composition of the pure carrier gas and the carrier gas containing the particular component is determined using a detector and a gas chromatogram is produced on a recorder.

Gas chromatography has many uses. It is used to identify sewage effluents, to test athletes for drugs and to determine the amount of alcohol in blood.

High-performance liquid chromatography (HPLC)

HPLC is used in the determination of the non-volatile components of a mixture. It is used to a large extent in the food industry to determine the amount of impurities or additives in foodstuffs, such as growth promoters in meat or vitamins in food. High pressure is used to force the mobile liquid phase through a column containing minute quantities of the food material.

Other Instrumental Methods Used in Separation and Analysis

Mass Spectrometry

A quick and accurate method of determining the relative molecular mass of an organic compound is mass spectrometry.

When an organic compound is passed into a mass spectrometer it is bombarded with high energy electrons. The compound is broken up into fragments,

$$M \rightarrow M^+$$

For example, a molecule of ethanol, C_2H_5OH, can break up into several fragments such as $C_2H_5O^+$, CH_3^+ and so on. The mass spectrum of the compound is detected and analysed by computer. Many peaks are noticed in the mass spectrum, but the one with the heaviest mass corresponds to the ethanol, in this case $M_r = 58$.

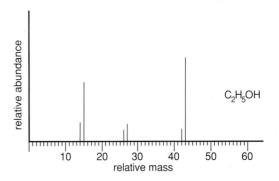

Mass spectrometry can be used to analyse gases from waste dumps, to identify small amounts of organic pollutants in water and to separate and identify metal ions in blood samples.

Spectroscopy

The main method of determining the molecular structure of a compound is spectroscopy. The most widely used spectroscopic techniques used are: infra-red (IR) spectroscopy, ultra-violet–visible (UV-VIS) spectroscopy and nuclear magnetic resonance (NMR) spectroscopy. The details of the methods are beyond the scope of this text. However, a brief outline of infra-red spectroscopy and ultraviolet spectroscopy is given.

In each spectroscopic method a sample is exposed to radiation such as infra-red or ultraviolet. Different substances absorb radiation to different extents depending on their structures.

Honours • Honours • Honours • Honours • Honours • Honours • Honours • Honours • Honours • Honours • Honours • Honours • Honours • Honours •

Honours • Honours • Honours • Honours • Honours • Honours • Honours • Honours • Honours • Honours • Honours • Honours • Honours •

Infra-red (IR) Absorption Spectroscopy

In infra-red spectroscopy, a sample of the material to be tested is irradiated with infra-red light (2500–25000 nm). Radiation of a specific frequency is absorbed causing the molecule to rotate and the bonds to begin to vibrate. A knowledge of the frequency (or wavelength) of the radiation absorbed by a particular functional group in the molecule helps to determine the structure of the compound. From this information a 'fingerprint' pattern of the compound is built up.

For example, propanone, CH_3COCH_3, and butanone, $CH_3COCH_2CH_3$ have IR spectra which are similar but differ slightly; the absorption spectra of their carbonyl groups differ sufficiently to distinguish between them.

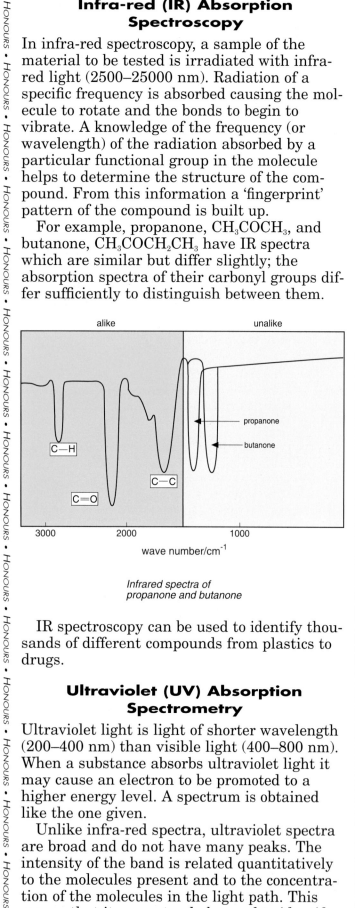

Infrared spectra of propanone and butanone

IR spectroscopy can be used to identify thousands of different compounds from plastics to drugs.

Ultraviolet (UV) Absorption Spectrometry

Ultraviolet light is light of shorter wavelength (200–400 nm) than visible light (400–800 nm). When a substance absorbs ultraviolet light it may cause an electron to be promoted to a higher energy level. A spectrum is obtained like the one given.

Unlike infra-red spectra, ultraviolet spectra are broad and do not have many peaks. The intensity of the band is related quantitatively to the molecules present and to the concentration of the molecules in the light path. This means that it can not only be used to identify

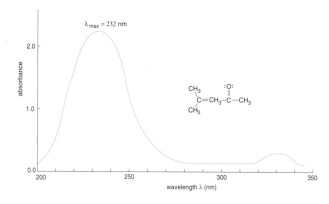

molecules, but may also be used to determine the concentration of the molecules.

Absorbance is related to concentration.

Applications of UV spectroscopy

• Detection of drug metabolites: Steroids, such as clenbuterol, taken by athletes are broken down into other compounds in the body. Even very small traces of these metabolites can be determined using ultraviolet spectroscopy.
• Analysis of plant pigments: Plant pigments like chlorophyll can be analysed using UV spectroscopy.

Exercise 7.5
•

(1) Chromatography is a technique which is used to <u>separate</u> and <u>identify</u> the components of a mixture. Explain, using diagrams, how the technique carries out the underlined functions.

(2) (a) Write notes on the following types of chromatography:
(i) column chromatography, (ii) thin-layer chromatography and (iii) paper chromatography.
(b) What is the difference between thin-layer chromatography and paper chromatography, in terms of the mobile phase involved?
(c) Draw a diagram of a column chromatogram.

(3) (a) Describe how you would separate some indicator dyes using chromatography.
(b) What is meant by an R_f value?

(4) Write a brief account of the following instrumental methods of separation and analysis. In each of your answers, give some everyday uses of the particular technique.
(a) thin-layer chromatography (TLC)
(b) gas chromatography (c) high-performance liquid chromatography (HPLC) (d) mass spectrometry.

(5) (a) What is absorption spectrometry?
(b) Write a brief account of the principles involved in (i) infra-red spectrometry and (ii) ultraviolet spectrometry.
(c) How does an infra-red spectrum differ from an ultraviolet one?
(d) Mention some uses of each method.

▼▼▼▼ Key Terms ▼▼▼▼

Some of the more important key terms are listed below. Other terms not included may be found by means of the index.

1. Saturated organic compounds: Saturated compounds are compounds which contain only single covalent bonds.

2. Systematic naming of chloroalkanes: Chloroalkanes are named in a similar way to alkanes. The position of the chlorine is indicated by the lowest possible number.

3. Alcohols: Alcohols are compounds of general formula R–OH and structural formula R–O–H, where R is any alkyl group.

4. Systematic naming of alcohols: Alcohols are named by replacing the -e in the corresponding alkane by -ol.

5. Physical properties of alcohols: The highly polar –OH group is capable of forming hydrogen bonds and influences the boiling points and solubilities of alcohols.

6. Unsaturated organic compounds: Unsaturated organic compounds are those which contain double or triple bonds.

7. Alkenes: The alkenes are hydrocarbons which have a carbon–carbon double bond. The general formula of alkenes is C_nH_{2n}, where $n = 2, 3, 4 \ldots$

8. Systematic naming of alkenes: Alkenes are named in a similar way to alkanes. The ending -ane from the alkanes is replaced by -ene.

9. Physical properties of alkenes: The boiling points of the lower alkenes are similar to those of the alkanes. Alkenes are insoluble in water, but soluble in non-polar solvents such as 1,1,1-trichloroethane or methylbenzene.

10. Carbonyl compounds:
Aldehydes and ketones contain the carbonyl >C═O group.
Aldehydes are compounds of general formula RCHO.
Ketones are compounds of general formula RCOR.

11. Systematic naming of aldehydes: The longest chain containing the –CHO group is the parent structure and is named by replacing the -e of the corresponding alkane by -al. The carbonyl carbon is always considered as carbon-1.

12. Physical properties of aldehydes: Aldehydes are polar compounds. The boiling points of aldehydes are higher than the corresponding non-polar hydrocarbons and are lower than the corresponding alcohols. The lower aldehydes are soluble in water.

13. Systematic naming of ketones: The longest chain containing the carbonyl (>C═O) group is the parent structure and is named by replacing the -e of the corresponding alkane by -one. The carbonyl group is given the lowest number possible.

14. Physical properties of ketones: Ketones are polar compounds. The boiling points of ketones are high. Ketones like propanone are soluble in water. Other ketones are less soluble in water, but are soluble in 1,1,1-trichloroethane or methylbenzene.

15. Carboxylic acids: Carboxylic acids are compounds of general formula R–COOH.

16. Systematic naming of carboxylic acids: Carboxylic acids are named by replacing the -e in the corresponding alkane by -oic acid.

17. Physical properties of carboxylic acids: The carboxyl group is highly polar. Carboxylic acids are soluble in water and have higher boiling points than the corresponding alcohols.

18. Esters: Esters contain a carbonyl group, >C═O, which means that they are polar compounds. The smaller esters are soluble in water. Their boiling points are similar to those of aldehydes and ketones.

19. Systematic naming of esters: Esters are named by changing the ending -oic acid from the carboxylic acid to -oate and then putting the alkyl prefix from the alcohol in front.

20. Range of aromatic chemistry: The range of aromatic chemistry is vast. It includes fuels, flavours, dyes, detergents, pharmaceuticals, polymers and many more compounds.

21. Functional groups: The group of atoms that defines the structure and properties of a particular family of organic compounds is called the functional group.

22. Addition reactions: Reactions where one substance adds on to another are called addition reactions, e.g.

$$\begin{array}{ccc} | & | & \\ -C = C- & + \ A-B \ \rightarrow & -C-C- \\ & & | \ \ | \\ & & A \ \ B \end{array}$$

23. Reactions of alkenes: The reactions of the alkenes are addition reactions. Reactants such as hydrogen, H_2, chlorine, Cl_2, hydrogen chloride, HCl, bromine, Br_2 and water, H_2O, can react with alkenes. Alkene molecules can also react with themselves to form much larger molecules called polymers.

24. Unreactivity of benzene: Unlike other unsaturated hydrocarbons, such as ethene, benzene is reluctant to take part in addition reactions. Unlike ethene, benzene does not react with bromine.

25. Substitution reactions: Substitution reactions are reactions where an atom, or a group of atoms, is replaced in a molecule by another atom or group of atoms. The main types include halogenation of alkanes, esterification and hydrolysis of esters.

26. Elimination reactions: Elimination reactions are reactions where a small molecule is removed from a larger molecule leaving an unsaturated compound.

27. Dehydration of alcohols: Dehydration of alcohols involves the elimination of a water molecule, H_2O, from the alcohol.

28. Redox reactions:
(a) Oxidation of alcohols: Primary alcohols are readily oxidised to aldehydes and can be further oxidised to carboxylic acids.
(b) Oxidation of aldehydes and ketones: The aldehyde group is easily oxidised to the carboxylic acid group. The ketone group is very difficult to oxidise.
(c) Reduction of aldehydes and ketones: Aldehydes are reduced to primary alcohols, and ketones are reduced to secondary alcohols by adding hydrogen.
(d) Combustion of organic compounds: The combustion of an organic compound is a redox reaction, where the particular compound is oxidised by oxygen to produce heat energy.

29. Reactions as acids:
(a) Alcohols are extremely weak acids. Alcohols react with active metals, such as sodium.
(b) Carboxylic acids are weak acids but are stronger acids than either water or alcohols. Carboxylic acids react with sodium and magnesium to form salts, with sodium hydroxide to form a salt and water and with sodium hydrogencarbonate to form a salt, water and carbon dioxide.

30. Organic synthesis: Synthesis involves bond making and bond breaking. The process involves reacting the starting compound with a chemical reagent to produce the final product.

31. Stages in production of an organic compound: There are usually three stages in the production of an organic compound:
(a) planning the synthesis
(b) separation and purification of the pure product
(c) analysis of the pure product.

32. Extraction techniques: Isolation of pure products from impurities may require various separation techniques depending on the substances involved. Solvent extraction, steam distillation, fractional distillation, crystallisation and chromatography are the main methods used.

33. Instrumental methods: Some of the methods used in separation and analysis include thin-layer chromatography (TLC), gas chromatography, high-performance liquid chromatography (HPLC), mass spectrometry, infra-red (IR) absorption spectroscopy and ultraviolet (UV) absorption spectroscopy.

SUMMARY OF REACTIONS

(a) Addition Reactions of Alkenes

Addition of hydrogen (hydrogenation) (reduction)

$$RCH = CH_2 \xrightarrow[\text{Catalyst = Ni, Pd or Pt}]{H_2, \text{ heat}} RCH_2CH_3$$

alkene → alkane

Example: $C_2H_4 + H_2 \longrightarrow C_2H_6$

Addition of halogens (halogenation)

$$RCH = CH_2 \xrightarrow[\text{(X = Cl or Br)}]{X_2} RCHXCH_2X$$

alkene → chloroalkane

Example: $C_2H_4 + Br_2 \longrightarrow C_2H_4Br_2$

Addition of water (hydration)

$$RCH = CH_2 \xrightarrow[\text{or (ii) H}_2\text{O, cat = H}_3\text{PO}_4/\text{SiO}_2]{\text{(i) Conc. H}_2\text{SO}_4 \text{ followed by H}_2\text{O}} RCH(OH)CH_3$$

alkene → alcohol

Examples: $C_2H_4 + H_2O \longrightarrow C_2H_5OH$

Addition of hydrogen chloride

$$RCH = CH_2 \xrightarrow{HCl} RCHClCH_3$$

alkene → chloroalkane

Example: $C_2H_4 + HCl \longrightarrow C_2H_5Cl$

Self addition (polymerisation)

$$n(RCH = CH_2) \longrightarrow [RCHCH_2]\ n$$

Example: $C_2H_4 \longrightarrow \left(CH_2\right)_n$

(b) Substitution Reactions

(i) Halogenation of alkanes

$$R\text{-}CH_2CH_3 \xrightarrow{Cl_2} R\text{-}CH_2CH_2Cl$$

alkane → chloroalkane

Example: $CH_4 + Cl_2 \longrightarrow CH_3Cl + HCl$

(ii) Esterification of alcohols

$$ROH \xrightarrow[\text{Cat = H}_2\text{SO}_4]{R^1COOH} R^1COOR$$

alcohol → ester

Example: $C_2H_5OH + CH_3COOH \rightleftharpoons CH_3COOC_2H_5 + H_2O$

(iii) Hydrolysis (basic) of esters

$$R^1COOR \xrightarrow{NaOH} R^1COONa + ROH$$

ester → salt + alcohol

Example: $CH_3COOC_2H_5 + NaOH \rightarrow CH_3COONa + C_2H_5OH$

(c) Elimination Reactions

Dehydration of alcohols

(i)

$$-\underset{\underset{H}{|}}{C} - \underset{\underset{OH}{|}}{C} - \xrightarrow[\text{heat}]{H_2SO_4} C = C$$

alcohol → alkene

Example: $C_2H_5OH \longrightarrow C_2H_4 + H_2O$

(ii)

$$-C-C-OH + HO-C-C- \rightarrow -C-C-O-C-C-$$

alcohols → ether

Example: $2\ C_2H_5OH \longrightarrow C_2H_5OC_2H_5 + H_2O$

(d) Redox Reactions

Combustion (oxidation)

alkanes
$$R - CH_3CH_2 \xrightarrow[\text{heat}]{O_2} CO_2 + H_2O$$

Example: $CH_4 + O_2 \longrightarrow CO_2 + 2H_2O$

alkenes
$$RCH = CH_2 \xrightarrow[\text{heat}]{O_2} CO_2 + H_2O$$

Example: $C_2H_4 + O_2 \longrightarrow 2CO_2 + 2H_2O$

alkynes
$$RC \equiv CH \xrightarrow[\text{heat}]{O_2} CO_2 + H_2O$$

Example: $C_2H_4 + \frac{5}{2}O_2 \longrightarrow 2CO_2 + H_2O$

alcohols
$$R\text{–}OH \xrightarrow[\text{heat}]{O_2} CO_2 + H_2O$$

Examples: $CH_3OH + \frac{3}{2}O_2 \longrightarrow CO_2 + 2H_2O$

aldehydes
$$R\text{–}CHO \xrightarrow[\text{heat}]{O_2} CO_2 + H_2O$$

Examples: $HCHO + O_2 \longrightarrow CO_2 + H_2O$

ketones
$$R\text{–}COR \xrightarrow[\text{heat}]{O_2} CO_2 + H_2O$$

Examples: $CH_3COCH_3 + 4O_2 \longrightarrow 3CO_2 + 3H_2O$

carboxylic acids

$$\text{R–COOH} \xrightarrow[\text{heat}]{O_2} CO_2 + H_2O$$

Example: $CH_3COOH + 2O_2 \longrightarrow 2CO_2 + 2H_2O$

esters

$$\text{R–COOR} \xrightarrow[\text{heat}]{O_2} CO_2 + H_2O$$

Example: $CH_3COOC_2H_5 + 5O_2 \longrightarrow 4CO_2 + 4H_2O$

Oxidation of alcohols and aldehydes

$$\text{R–CH}_2\text{OH} \xrightarrow[\text{or } Na_2Cr_2O_7/H^+]{KMnO_4/H^+} \text{RCHO} \xrightarrow[\text{or } Na_2Cr_2O_7/H^+]{KMnO_4/H^+} \text{RCOOH}$$

alcohol | aldehyde | carboxylic acid

Examples: $C_2H_5OH \xrightarrow{O} CH_3CHO \xrightarrow{O} CH_3COOH$

Reduction

$$\text{R–CHO} \xrightarrow[\text{cat. = Ni}]{H_2} RCH_2OH$$

aldehyde | primary alcohol

Example: $CH_3CHO + H_2 \longrightarrow C_2H_5OH$

$$\text{R–COR} \xrightarrow[\text{cat. = Ni}]{H_2} RCH(OH)R$$

ketone | secondary alcohol

Example: $CH_3COCH_3 + H_2 \longrightarrow CH_3CH(OH)CH_3$

$$\text{R–COOH} \xrightarrow[\text{cat. = Ni}]{H_2} RCH_2OH$$

carboxylic acid | alcohol

Example: $CH_3COOH + H_2 \rightarrow CH_3CH_2OH$

(e) Reactions as Acids

Reaction of alcohols with sodium

$$RCH_2OH \xrightarrow{Na} RCH_2ONa$$

alcohol | sodium salt

Example: $C_2H_5OH + Na \rightarrow C_2H_5ONa + \frac{1}{2}H_2$

Acidic reactions of carboxylic acids

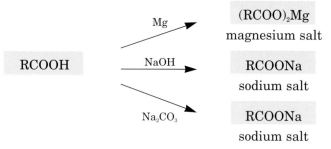

$$RCOOH \xrightarrow{Mg} (RCOO)_2Mg$$ magnesium salt

$$RCOOH \xrightarrow{NaOH} RCOONa$$ sodium salt

$$RCOOH \xrightarrow{Na_2CO_3} RCOONa$$ sodium salt

Examples:

$2CH_3COOH + Mg \rightarrow (CH_3COO)_2Mg + H_2$

$CH_3COOH + NaOH \rightarrow CH_3COONa + H_2O$

$2CH_3COOH + Na_2CO_3 \rightarrow 2CH_3COONa + CO_2 + H_2O$

CHEMICAL EQUILIBRIUM

Spider's web with dewdrops. When the atmosphere is saturated with water vapour, liquid water is in equilibrium with gaseous water.

8.1 CHEMICAL EQUILIBRIUM

Children balanced on a see-saw.

Equilibrium is defined as 'a state of balance'. Two children balancing on a see-saw are in a state of balance when the see-saw does not move. When no movement of any kind takes place the system is in a state of **static equilibrium**.

A person running on a running machine in gymnasium seems to be static. The person and the machine are both moving at the same speed but against each other in opposite directions. They are in a state of **dynamic equilibrium**.

In **chemical equilibrium**, changes take place between molecules in such a way that no overall change is noticed. Since movement is involved the system is said to be in a state of dynamic equilibrium.

Walking machine in a gym. The person seems to be at rest because the rate of movement of the machine equals the rate of walking of the person.

When the atmosphere is saturated with water vapour, liquid water is in equilibrium with gaseous water.

The water at the top of the waterfall remains at the same level when the flow rate of water in equals the flow rate of water out, i.e. it is at equilibrium.

Reversible reactions

Many chemical reactions go to completion, where the reaction continues until one of the reactants is completely used up. For example, when sodium reacts with water, sodium hydroxide and hydrogen gas are formed. Usually hydrogen gas escapes into the atmosphere and cannot recombine with the sodium hydroxide to reform the sodium and water

$$Na(s) + H_2O(l) \rightarrow NaOH(aq) + \tfrac{1}{2}H_2(g)$$

Other chemical reactions often seem to stop before they are completed: these are reversible reactions. In reversible reactions, the original reactants form products, and then the products react with each other to reform the reactants. There are two reactions occurring, a forward reaction and a reverse reaction, which eventually form a mixture of reactants and products.

Water, if left in an open vessel, has a tendency to evaporate, even at room temperature. However, if the water is kept in a closed container no evaporation occurs because the vapour is prevented from escaping into the atmosphere. At the same time, liquid water molecules change into gaseous water molecules, and gaseous water molecules change into liquid water molecules. The system is in a state of dynamic equilibrium.

$$H_2O(l) \rightleftharpoons H_2O(g)$$

The **reversible** processes of evaporation and condensation can occur under certain conditions of temperature and pressure; when gaseous water molecules and liquid water molecules are both present, the system is in dynamic equilibrium.

Equilibrium reactions are described using reversible arrows ($\rightleftharpoons$), where the reversible arrows indicate that the reaction can go either way.

In **reversible reactions**, when substances react, they eventually form a mixture of reactants and products in dynamic equilibrium. A reaction in a state of dynamic equilibrium consists of a **forward reaction** where the reactants react to form products and a **reverse reaction** where the **products react to form the original reactants**. Both the forward and the reverse reactions occur at the same rate (speed).

Consider the reaction of iron and steam in a **closed container** (no product can escape). If the iron is heated it is oxidised to iron oxide while the steam is reduced to hydrogen gas. After a while a state of equilibrium is reached where all four substances, iron, steam, iron oxide and hydrogen, are present and where there is no net change in either the forward or reverse directions. The system is in a state of dynamic equilibrium, because both the forward and the reverse reactions are still proceeding, but **at equal rates**.

$$3Fe(s) + 4H_2O(g) \rightleftharpoons Fe_3O_4(s) + 4H_2(g)$$

It should be noted that it is important industrially for reactions to go to completion; this is usually accomplished by removing one of the reactants as it is formed.

Consider a general reaction where, for simplicity, the reactants and products are gases.

$$A(g) \rightleftharpoons B(g)$$

Here, one mole of A reacts to form one mole of B in a closed container. As the reaction proceeds the concentration of A initially decreases quite rapidly, while the concentration of B increases quite rapidly. After a while the concentration of A changes very little and settles down and becomes constant. The concentration of B, which in the beginning was zero, increases as time goes by and eventually settles down and becomes constant. Thus, after a certain time, the concentrations of both substances A and B become constant.

At this stage the system is said to be at equilibrium and shows no tendency to change its composition with time.

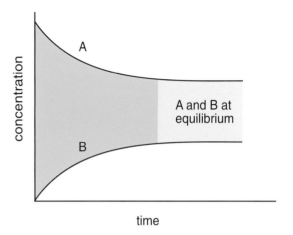

The Equilibrium Constant

Experimental evidence has shown that there is a simple mathematical relationship between the amounts of the reactants and the amounts of products in an equilibrium mixture, provided that the temperature remains constant. This relationship is called the **Law of Chemical Equilibrium**.

For a general reaction of the form

$$aA + bB \rightleftharpoons cC + dD$$

where a, b, c and d are the numbers of moles of A, B, C and D, respectively, the **equilibrium constant expression** is given as

$$K_c = \frac{[C]^c \, [D]^d}{[A]^a \, [B]^b}$$

The **equilibrium constant**, K_c, is the value obtained when the equilibrium concentration (in mol L^{-1}) of each species is substituted in the equilibrium constant expression.

Here, the molar concentrations are denoted by brackets [ ] and the subscript $_c$ refers to the fact that the equilibrium constant is expressed in terms of molar concentrations.

The value of K, the equilibrium constant, at a given temperature, indicates the amount of reactants and products in the equilibrium mixture.

Some equilibrium constant expressions and the values for the equilibrium constant at stated temperatures are given for the following reactions.

It should be noted from the examples shown that the equilibrium constant can only be defined if the balanced chemical equation is known.

①

$$CH_3COOH(l) + C_2H_5OH(l) \rightleftharpoons CH_3COOC_2H_5(l) + H_2O(l)$$

$$K_c = \frac{[CH_3COOC_2H_5] \, [H_2O]}{[CH_3COOH] \, [C_2H_5OH]} = 4 \text{ at } 298 \text{ K}$$

Because the value of K_c is large (>1), the concentration of the products in the equilibrium mixture is greater than that of the reactants. This means that the equilibrium lies to the right-hand side.

②

$$N_2O_4(g) \rightleftharpoons 2NO_2(g)$$

$$K_c = \frac{[NO_2]^2}{[N_2O_4]} = 4.8 \times 10^{-3} \text{ at } 298 \text{ K}$$

Because the value of K_c is small (<1), the concentration of the reactant, N_2O_4, in the equilibrium mixture is greater than that of the product, NO_2. This means that the equilibrium lies to the left-hand side.

If K is large, then the amount of products is large.
If K is small, then the amount of products is small.

Calculations using the Equilibrium Constant, K_c

• • • • • **Example 8.1** • • • • •

Determination of the Equilibrium Constant, K_c

One mole of ethanoic acid and one mole of ethanol were mixed in a 1 L vessel and allowed to come to equilibrium at 298 K. Analysis of the equilibrium mixture showed that 0.33 mol of ethanoic acid remained. Calculate the value of the equilibrium constant at 298 K.

$$CH_3COOH(l) + C_2H_5OH(l) \rightleftharpoons CH_3COOC_2H_5(l) + H_2O(l)$$

Required

$$K_c = \frac{[CH_3COOC_2H_5] \, [H_2O]}{[CH_3COOH] \, [C_2H_5OH]} = ? \text{ at } 298 \text{ K}$$

• The balanced chemical equation tells us that one mole of ethanoic acid and one mole of ethanol react to form one mole of ethyl ethanoate and one mole of water.
• This tells us that if 0.33 mol of CH_3COOH remains at equilibrium, then 0.33 mol of C_2H_5OH also remains.
• If 0.33 mol of CH_3COOH and 0.33 mol of C_2H_5OH remain at equilibrium, then 0.67 mol of each must have been used to form 0.67 mol of $CH_3COOC_2H_5$ and 0.67 mol of H_2O.
• Because the reaction was carried out in a 1 L vessel, the concentrations are easily expressed in mol L^{-1}.

It is useful to set up a table showing the initial concentrations and the equilibrium concentrations of the reactants and products.

Concentration /mol L^{-1}	CH_3COOH	+ C_2H_5OH	$\rightleftharpoons CH_3COOC_2H_5$	+ H_2O
Initial	1.0	1.0	0	0
Equilibrium	0.33	0.33	0.67	0.67

Substituting the equilibrium concentrations into the equilibrium constant expression

$$K_c = \frac{[CH_3COOC_2H_5] \, [H_2O]}{[CH_3COOH] \, [C_2H_5OH]}$$

$$= \frac{0.67 \text{ mol } L^{-1} \times 0.67 \text{ mol } L^{-1}}{0.33 \text{ mol } L^{-1} \times 0.67 \text{ mol } L^{-1}}$$

$$= 4 \text{ at } 298 \text{ K.}$$

Units: It should be noted that in this particular case K_c **has no units**, because when the total number of moles of reactants equals the total number of moles of products, the units cancel each other out in the equilibrium constant expression.

• •

• • • • • • **Example 8.2** • • • • • •

Determination of the Equilibrium Constant, K_c

When 0.2 mol of H_2 and 0.3 mol of CO_2 were mixed in a 1 L vessel and allowed to come to equilibrium at a certain temperature, the equilibrium mixture contained 0.04 mol of water. Calculate the value of the equilibrium constant at that temperature for the reaction:

$H_2(g) + CO_2(g) \rightleftharpoons H_2O(g) + CO(g)$

$$\text{Required} \quad K_c = \frac{[H_2O]\,[CO]}{[H_2]\,[CO_2]} = ?$$

• The balanced chemical equation tells us that one mole of hydrogen and one mole of carbon dioxide react to form one mole of water and one mole of carbon monoxide.

• This means that if the equilibrium mixture contains 0.04 mol of H_2O, then it also contains 0.04 mol of CO.

• If there are 0.04 mol of H_2O and 0.04 mol of CO are at equilibrium, then there are 0.16 mol of H_2 and 0.26 moles of CO_2 remaining at equilibrium.

• As the reaction was carried out in a 1 L vessel, the concentrations are easily expressed in $mol\,L^{-1}$.

Draw up a table showing the initial concentrations and the equilibrium concentrations of the reactants and products.

Concentration /mol L^{-1}	H_2	+ CO_2	$\rightleftharpoons H_2O$	+ CO
Initial	0.20	0.30	0	0
Change	–	–	+	+
Equilibrium	0.16	0.26	0.04	0.04

Substituting the equilibrium concentrations in the equilibrium constant expression

$$K_c = \frac{[H_2O]\,[CO]}{[H_2]\,[CO_2]}$$

$$= \frac{0.04\ mol\,L^{-1} \times 0.04\ mol\,L^{-1}}{0.16\ mol\,L^{-1} \times 0.26\ mol\,L^{-1}}$$

$$= 0.0385$$

Units: In this reaction, K_c **has no units**, because when the total number of moles of reactants equals the total number of moles of products, the units cancel each other out in the equilibrium constant expression.

• •

• • • • • **Example 8.3** • • • • •

Determination of the Equilibrium Constant, K_c, from the Equilibrium Concentrations

Phosphorus pentachloride was heated in a 10 L flask at 525 K and was allowed to come to equilibrium. Analysis of the equilibrium mixture showed that it contained 0.02 mol PCl_5, 0.2 mol PCl_3 and 0.2 mol Cl_2. Calculate the value of K_c at 525 K.

$PCl_5(g) \rightleftharpoons PCl_3(g) + Cl_2(g)$

$$\text{Required} \quad K_c = \frac{[PCl_3]\,[Cl_2]}{[PCl_5]} = ?\ \text{at 525 K}$$

• The balanced chemical equation tells us that one mole of PCl_5 reacts to form one mole of PCl_3 and one mole of Cl_2.

• The equilibrium concentration is obtained by dividing the amount by the volume (10 L).

$$c = \frac{n}{V}$$

• The equilibrium concentrations are

$$[PCl_5] = \frac{0.02\ mol}{10\ L} = 0.002\ mol\,L^{-1}$$

$$[PCl_3] = \frac{0.2\ mol}{10\ L} = 0.02\ mol\,L^{-1}$$

$$[Cl_2] = \frac{0.2\ mol}{10\ L} = 0.02\ mol\,L^{-1}$$

$$K_c = \frac{[PCl_3]\,[Cl_2]}{[PCl_5]}$$

$$= \frac{0.02\ mol\,L^{-1} \times 0.02\ mol\,L^{-1}}{0.002\ mol\,L^{-1}}$$

$$= 0.2\ mol\,L^{-1} \leftarrow \text{Units}$$

Two points should be noted in this example.

(i) When there is a change in the total number of moles from reactants to products then K_c **has units**.

(ii) Concentrations are expressed in $mol\,L^{-1}$.

• •

HONOURS • HONOURS • HONOURS • HONOURS • HONOURS • HONOURS • HONOURS • HONOURS • HONOURS • HONOURS • HONOURS • HONOURS • HONOURS •

HONOURS • HONOURS • HONOURS • HONOURS • HONOURS • HONOURS • HONOURS • HONOURS • HONOURS • HONOURS • HONOURS • HONOURS • HONOURS •

• • • • • • **Example 8.4** • • • • • •

Determination of Concentration using K_c

Two moles of ethanoic acid and three moles of propanol were mixed in a 1 L flask and allowed to equilibrate. Calculate the equilibrium concentrations of each of the reacting species, given that the equilibrium constant K_c is 6.25 at 298 K.

$$CH_3COOH(l) + C_3H_7OH(l) \rightleftharpoons CH_3COOC_3H_7(l) + H_2O(l)$$

$$K_c = \frac{[CH_3COOC_2H_5]\,[H_2O]}{[CH_3COOH]\,[C_2H_5OH]} = 6.25 \text{ at } 298 \text{ K}$$

In many equilibrium problems initial concentrations are given which change during the reaction to give the equilibrium concentrations. In this problem and in many others it is useful to set up a table showing the values of the initial concentration, the change and the equilibrium concentration.
• The balanced chemical equation tells us that one mole of ethanoic acid reacts with one mole of propanol to form one mole of propyl ethanoate and one mole of water.
• Let x be the molar change in concentration of each species.
• Because we have one mole of each reacting species, each species changes by x; the reactants decrease by x, while the products increase by x.
• Because the volume of the flask is 1 L, the initial concentrations of the reactants are 2.0 mol L^{-1} CH_3COOH and 3.0 mol L^{-1} C_3H_7OH.

Concentration / mol L^{-1}	CH_3COOH +	C_3H_7OH $\rightleftharpoons$	$CH_3COOC_3H_7$ +	H_2O
Initial	2.0	3.0	0	0
Change	$-x$	$-x$	$+x$	$+x$
Equilibrium	$2.0 - x$	$3.0 - x$	$+x$	$+x$

Now,

$$K_c = \frac{[CH_3COOC_2H_5]\,[H_2O]}{[CH_3COOH]\,[C_2H_5OH]} = 6.25 \text{ at } 298 \text{ K}$$

then,

$$K_c = \frac{(x)\,(x)}{(2-x)\,(3-x)} = 6.25 \text{ at } 298 \text{ K}$$

$$6.25 = \frac{x^2}{6 - 5x + x^2}$$

Solving the quadratic equation,

$$5.25x^2 - 31.25x + 37.5 = 0$$

$x = 1.67$ (or 4.29, which is not a possible value, as x must be less than the smallest initial concentration; in this case, 2 mol L^{-1})

The equilibrium concentrations are as follows:

$[CH_3COOH] = (2.0 - 1.67)$ mol L^{-1} = 0.33 mol L^{-1}

$[C_3H_7OH] = (3.0 - 1.67)$ mol L^{-1} = 1.33 mol L^{-1}

$[CH_3COOC_3H_7] = 1.67$ mol L^{-1}

$[H_2O] = 1.67$ mol L^{-1}

• •

• • • • • • **Example 8.5** • • • • •

Determination of Concentration using K_c

0.02 mol hydrogen and 0.02 mol iodine were mixed in a 1 L flask and allowed to form hydrogen iodide at 720 K. Calculate the equilibrium concentrations of each of the reacting species, given that the equilibrium constant K_c is 50 at 720 K.

$$H_2(g) + I_2(g) \rightleftharpoons 2HI(g)$$

$$K_c = \frac{[HI]^2}{[H_2]\,[I_2]} = 50 \text{ at } 720 \text{ K}$$

• The balanced chemical equation tells us that one mole of hydrogen reacts with one mole of iodine to form two moles of hydrogen iodide.
• Let x be the molar change in concentration of each species.
• Because we have one mole of each reactant, each reactant decreases by x. Because we have two moles of product, the product increases by $2x$.
• Because the volume of the flask is 1 L, the initial concentrations of the reactants are 0.02 mol L^{-1} H_2 and 0.02 mol L^{-1} I_2.

Concentration /mol L^{-1}	H_2 +	I_2 $\rightleftharpoons$	$2HI$
Initial	0.02	0.02	0
Change	$-x$	$-x$	$+2x$
Equilibrium	$0.02 - x$	$0.02 - x$	$+2x$

Now

$$K_c = \frac{[\text{HI}]^2}{[\text{H}_2]\,[\text{I}_2]} = 50 \text{ at } 720 \text{ K}$$

then,

$$K_c = \frac{(2x)^2}{(0.02 - x)(0.02 - x)} = 50 \text{ at } 720 \text{ K}$$

$$50 = \frac{4x^2}{0.0004 - 0.04x + x^2}$$

Solving the quadratic equation,

$$46x^2 - 2x + 0.02 = 0$$

$x = 0.0156$ (or 0.0279, which is not possible as it is greater than the initial concentration)

The equilibrium concentrations are as follows:

$[\text{H}_2] = (0.02 - 0.0156) \text{ mol L}^{-1} = 0.0044 \text{ mol L}^{-1}$

$[\text{I}_2] = (0.02 - 0.0156) \text{ mol L}^{-1} = 0.0044 \text{ mol L}^{-1}$

$[\text{HI}] = 2 \times 0.0156 \text{ mol L}^{-1} = 0.0312 \text{ mol L}^{-1}$

• •

• • • • • **Example 8.6** • • • • •

Determination of Concentration using K_c

Carbon monoxide and chlorine were allowed to equilibrate in a 10 L flask and form an equilibrium mixture containing phosgene, COCl_2. If the initial mixture contained 0.2 mol CO and 0.1 mol Cl_2, calculate the equilibrium concentrations of each of the reacting species, given that the equilibrium constant K_c is 0.42 L mol^{-1}.

$$\text{CO(g)} + \text{Cl}_2\text{g} \rightleftharpoons \text{COCl}_2\text{(g)}$$

$$K_c = \frac{[\text{COCl}_2]}{[\text{CO}]\,[\text{Cl}_2]} = 0.42 \text{ L mol}^{-1} \leftarrow \text{Units}$$

• The balanced chemical equation tells us that one mole of carbon monoxide reacts with one mole of chlorine to form one mole of phosgene.
• Let x be the molar change in concentration of each species.
• Because we have one mole of each reacting species, each species changes by x; the reactants decrease by x, while the product increases by x.
• Because the volume of the flask is 10 L, the initial concentrations of the reactants are 0.2/10 mol L^{-1} CO and 0.1/10 mol L^{-1} Cl_2.

Concentration /mol L^{-1}	CO	+	Cl_2	$\rightleftharpoons$	$COCl_2$
Initial	0.2/10 = 0.02		0.1/10 = 0.01		0
Change	$-x$		$-x$		$+x$
Equilibrium	$0.02 - x$		$0.01 - x$		$+x$

Now,

$$K_c = \frac{[\text{COCl}_2]}{[\text{CO}]\,[\text{Cl}_2]} = 0.42 \text{ L mol}^{-1}$$

$$K_c = \frac{x}{(0.02 - x)(0.01 - x)} \text{ L mol}^{-1} = 0.42 \text{ L mol}^{-1}$$

$$\frac{x}{0.0002 - 0.03x + x^2} = 0.42$$

Solving the quadratic equation,

$$0.42x^2 - 1.0126x + 0.000084 = 0$$

$x = 0.000083$ (or 2.41, which is not possible)

The equilibrium concentrations are as follows:

$[\text{COCl}_2] = 8.3 \times 10^{-5} \text{ mol L}^{-1}$
$[\text{CO}] = (0.02 - 8.3 \times 10^{-5}) \text{ mol L}^{-1} = 0.019 \text{ mol L}^{-1}$

$[\text{Cl}_2] = (0.01 - 8.3 \times 10^{-5}) \text{ mol L}^{-1} = 0.0099 \text{ mol L}^{-1}$

When there is a change in the total number of moles from reactants to products, then amounts (moles) must be changed to concentrations (mol L^{-1}).

• •

••••• Example 8.7 •••••

Determination of Equilibrium Concentrations Starting from a Mixture

Phosphorus pentachloride forms an equilibrium mixture with phosphorus trichloride and chlorine at 435 K. If this equilibrium mixture contained 0.05 mol L^{-1} of each species, calculate the new equilibrium concentrations at 525 K, given that $K_c = 0.2$ mol L^{-1} at 525 K.

$$PCl_5(g) \rightleftharpoons PCl_3(g) + Cl_2(g)$$

$$K_c = \frac{[PCl_3][Cl_2]}{[PCl_5]} = 0.2 \text{ mol L}^{-1} \text{ at } 525 \text{ K}$$

• Construct a table making sure that the initial concentrations of PCl_5, PCl_3 and Cl_2 are each 0.05 mol L^{-1}.

Concentration /mol L^{-1}	PCl$_5$	$\rightleftharpoons$ PCl$_3$	+ Cl$_2$
Initial	0.05	0.05	0.05
Change	$-x$	$+x$	$+x$
Equilibrium	$0.05 - x$	$0.05 + x$	$0.05 + x$

Now, $K_c = \dfrac{[PCl_3][Cl_2]}{[PCl_5]} = 0.2$ mol L^{-1}

$$K_c = \frac{(0.05 + x)(0.05 + x)}{(0.05 - x)} \text{ mol L}^{-1} = 0.2 \text{ mol L}^{-1}$$

$$\frac{(0.05 + x)(0.05 + x)}{(0.05 - x)} = 0.2$$

$$0.0025 + 0.1x + x^2 = 0.01 - 0.2x$$

$$x^2 + 0.3x - 0.0075 = 0$$

Solving the quadratic equation,
$x = -0.0275$ (or -0.2725, which is not possible)
The equilibrium concentrations are as follows:

$[PCl_5] = [0.05 - (-0.00275)]$ mol L^{-1}
$\qquad = 0.05275$ mol L^{-1}

$[PCl_3] = [0.05 + (-0.00275)]$ mol L^{-1}
$\qquad = 0.04725$ mol L^{-1}

$[Cl_2] = [0.05 + (-0.00275)]$ mol L^{-1}
$\qquad = 0.04725$ mol L^{-1}

•••••••••••••••••••••••••••••

Exercise 8.1

•

(1) Explain the meaning of each of the following terms:
(a) reversible reaction (b) equilibrium mixture (c) dynamic equilibrium (d) equilibrium constant (e) equilibrium constant expression.

(2) For each of the following reactions write the equilibrium constant expression:

(a) $CH_3COOH(l) + C_2H_5OH(l) \rightleftharpoons CH_3COOC_2H_5(l) + H_2O(l)$

(b) $H_2(g) + I_2(g) \rightleftharpoons 2HI(g)$

(c) $2CO(g) + O_2(g) \rightleftharpoons 2CO_2(g)$

(d) $PCl_5(g) \rightleftharpoons PCl_3(g) + Cl_2(g)$

(e) $N_2(g) + 3H_2(g) \rightleftharpoons 2NH_3(g)$

(f) $2SO_2(g) + O_2(g) \rightleftharpoons 2SO_3(g)$

(g) $N_2O_4(g) \rightleftharpoons 2NO_2(g)$

(3) The values for the equilibrium constants are given for the following reactions. Explain clearly in which direction the equilibrium lies by referring to the size of K_c.

(a) $CH_3COOH(l) + C_2H_5OH(l) \rightleftharpoons CH_3COOC_2H_5(l) + H_2O(l)$

$K_c = 4$ at 298 K

(b) $CO_2(g) + H_2(g) \rightleftharpoons CO(g) + H_2O(g)$

$K_c = 0.08$ at 400 K

(4) Ethene reacts with water to form ethanol as follows:

$$C_2H_4(g) + H_2O(g) \rightleftharpoons C_2H_5OH(g)$$

If an equilibrium mixture, at a certain temperature, contained 0.15 mol L^{-1} C_2H_4, 0.36 mol L^{-1} H_2O and 1.8 mol L^{-1} C_2H_5OH, determine the value of the equilibrium constant, K_c.

(5) In the equilibrium system:

$$CH_3COOH(l) + C_2H_5OH(l) \rightleftharpoons CH_3COOC_2H_5(l) + H_2O(l)$$

where $K_c = 4.0$ at 298 K, analysis showed that there were 20 g of CH_3COOH, 58 g of $CH_3COOC_2H_5$ and 12 g of H_2O at equilibrium. Calculate the amount of C_2H_5OH at equilibrium.

(6) Methanoic acid, HCOOH, decomposes as follows:

$$HCOOH(g) \rightleftharpoons CO(g) + H_2O(g);$$
$$K_c = 2.9 \text{ at } 400 \text{ °C}$$

If a mixture was prepared containing 0.2 mol L^{-1} CO and 0.3 mol L^{-1} H$_2$O, determine the concentration of HCOOH at equilibrium.

(7) Explain why some values of K_c have units, while others have no units. In your answer refer to each of the examples in question (2), and assume that all the concentrations are expressed in mol L^{-1}.

(8) 0.4 mol of methanoic acid was placed in a 2 L reaction vessel and allowed to come to equilibrium:

$$HCOOH(g) \rightleftharpoons CO(g) + H_2O(g); K_c = 4.3 \times 10^4 \text{ at } 1000 \text{ °C}$$

Calculate the concentration of each species at equilibrium. Explain your answer.

(9) Thionyl chloride, SO$_2$Cl$_2$, decomposes on heating to form sulphur dioxide, SO$_2$, and chlorine, Cl$_2$. When 13.5 g of SO$_2$Cl$_2$ was placed in a 2 L flask and heated to 380 K and was allowed to equilibrate, the amount of chlorine at equilibrium in the flask was 0.07 mol. Calculate the equilibrium constant at 380 K for the reaction:

$$SO_2Cl_2(g) \rightleftharpoons SO_2(g) + Cl_2(g)$$

(10) Nitric oxide, NO, one of the main constituents of acid rain, is produced during combustion in motor cars by the reaction:

$$N_2(g) + O_2(g) \rightleftharpoons 2NO(g);$$
$$K_c = 2.5 \times 10^3 \text{ at } 2400 \text{ K}$$

If an equilibrium mixture of these gases contained 0.016 mol L^{-1} N$_2$ and 0.036 mol L^{-1} O$_2$, calculate the concentration of NO in the equilibrium mixture.

(11) 1.00 mol L^{-1} of phosphorus pentachloride, PCl$_5$, dissociates at 433 K to give 0.135 mol L^{-1} of phosphorus trichloride, PCl$_3$, at equilibrium.

$$PCl_5(g) \rightleftharpoons PCl_3(g) + Cl_2(g)$$

Calculate the equilibrium constant at 433 K.

(12) Methanol, CH$_3$OH, can be manufactured from synthesis gas by the following equilibrium reaction:

$$CO(g) + 2H_2(g) \rightleftharpoons CH_3OH(l)$$

When 0.1 mol L^{-1} of CO and 0.2 mol L^{-1} of H$_2$ were mixed at 225 °C and allowed to equilibrate, the mixture contained 0.0791 mol L^{-1} of CO. Calculate the value of the equilibrium constant, K_c, for the reaction at 225 °C.

(13) In the equilibrium system:

$$H_2(g) + I_2(g) \rightleftharpoons 2HI(g); K_c = 49.0 \text{ at } 717 \text{ K}$$

If 3.00 mol of hydrogen and 2.00 mol of iodine were heated in a 5 L flask at 717 K and allowed to come to equilibrium, calculate the equilibrium composition in mol L^{-1}.

(14) In the equilibrium system:

$$Br_2(g) + Cl_2(g) \rightleftharpoons 2BrCl(g); K_c = 7.0 \text{ at } 25 \text{ °C}$$

If 0.6 mol Br$_2$ and 0.6 mol Cl$_2$ are heated in a 10 L flask and allowed to come to equilibrium, calculate the concentration of BrCl when equilibrium is established.

(15) In the equilibrium system:

$$Br_2(g) + Cl_2(g) \rightleftharpoons 2BrCl(g); K_c = 7.0 \text{ at } 25 \text{ °C}$$

If 0.03 mol L^{-1} Br$_2$, 0.02 mol L^{-1} Cl$_2$ and 0.04 mol L^{-1} BrCl are mixed and allowed to come to equilibrium, calculate the concentrations when equilibrium is established.

(16) Explain, using K_c, which of the following reactions you would expect to go almost to completion:

(a) $N_2(g) + 2O_2(g) \rightleftharpoons 2NO(g); K_c = 3 \times 10^{-17}$

(b) $2SO_2(g) + O_2(g) \rightleftharpoons 2SO_3(g); K_c = 8 \times 10^{25}$

(17) At 670 K, hydrogen iodide is 20% dissociated. Calculate the value of the equilibrium constant, K_c, at this temperature.

(18) When 0.5 mol of COCl$_2$ is placed in a 2 L flask at a certain temperature and allowed to come to equilibrium. If 84% of the COCl$_2$ remains at equilibrium, calculate the value of K_c at the same temperature for the equilibrium:

$$COCl_2(g) \rightleftharpoons CO(g) + Cl_2(g)$$

(19) 2.0 mol of $CO(g)$ and 2.0 mol $H_2O(g)$ were mixed in a 5 L container at 800 K and allowed to come to equilibrium.

$$CO(g) + H_2O(g) \rightleftharpoons CO_2(g) + H_2(g)$$

If the equilibrium mixture contained 0.665 mol of CO_2 and 0.665 mol of H_2, determine the value of K_c at 800 K.

(20) For the reaction:

$$CO(g) + H_2O(g) \rightleftharpoons CO_2(g) + H_2(g);$$
$$K_c = 4.06 \text{ at } 500\,°C$$

If 0.1 mol of CO and 0.2 mol of H_2O are mixed in a 5 L reaction vessel at this temperature, calculate the concentrations of the components of the equilibrium mixture.

(21) Ammonia is synthesised by the Haber process:

$$N_2(g) + 3H_2(g) \rightleftharpoons 2NH_3(g); K_c = 0.135 \text{ at } 100\,°C$$

If the equilibrium mixture contained 0.03 mol L^{-1} NH_3 and 0.05 mol L^{-1} N_2, calculate the concentration of H_2 at equilibrium.

(22) In the equilibrium system:

$$H_2(g) + I_2(g) \rightleftharpoons 2HI(g)$$

(a) If 3.00 mol of hydrogen and 3.00 mol of iodine were heated in a 2 L flask at 773 K and allowed to come to equilibrium, it was found that 0.35 mol L^{-1} of iodine was present in the equilibrium mixture. Calculate the value of the equilibrium constant at this temperature.
(b) A similar experiment, using the same concentrations of hydrogen and iodine, was carried out at 623 K. It was found that the value of K_c had increased to 64. Calculate (i) the number of moles and (ii) the mass of each species at equilibrium.

(23) A mixture of 27 g of ethanoic acid and 20.7 g of ethanol was allowed to come to equilibrium at a certain temperature. It was found by experiment that 9 g of ethanoic acid was present in the equilibrium mixture.
(a) Calculate the value of the equilibrium constant for the reaction:

$$CH_3COOH(l) + C_2H_5OH(l) \rightleftharpoons CH_3COOC_2H_5(l) + H_2O(l)$$

(b) If 13.8 g of ethanol was added to the above equilibrium mixture, determine the mass of each species when equilibrium has been re-established.

(24) A mixture of 6 moles of methane gas and 6 moles of steam was allowed to come to equilibrium in a closed 60 L reaction vessel at a temperature of 800 °C. If at equilibrium 50% of the methane had reacted, calculate the value of K_c at 800 °C for the reaction:

$$CH_4(g) + H_2O(g) \rightleftharpoons CO(g) + 3H_2(g)$$

(25) Ten moles of nitrogen and thirty moles of hydrogen were mixed and allowed to come to equilibrium in a 7.5 L container at a certain temperature. The equilibrium was found to contain 15 moles of ammonia.
(a) Calculate the number of moles of nitrogen and hydrogen and also the total number of moles of gas present in the equilibrium mixture. (b) Calculate the value of the equilibrium constant, K_c, at that temperature.

8.2 LE CHATELIER'S PRINCIPLE

Equilibrium Changes and Le Chatelier's Principle

When chemical products are manufactured, it is important to control the reaction conditions in order to obtain the maximum yield of product in the quickest possible time.

The yield in a chemical reaction can be altered by:
1. changing the concentrations: products may be removed or reactants may be added
2. changing the pressure of the reactants or changing the volume in which they react
3. changing the temperature.
  • A **catalyst** cannot alter the equilibrium composition, but it can change the speed at which a product is formed.

Henri Le Chatelier studied the factors which affect equilibrium reactions and summarised his results as a general principle.

Le Chatelier's Principle

If an equilibrium system is subjected to a stress, such as a change in concentration, temperature or pressure, then the system tries, if possible, to reduce that stress by adjusting the amounts of the substances present in the equilibrium mixture.

He observed that when the concentration of any of the reacting species is changed in an equilibrium system, then the system readjusts by **changing the position of equilibrium** to counteract that change.

The factors which affect an equilibrium system are:
(1) change in concentration
(2) change in pressure
(3) change in temperature
(4) use of catalysts.

(1) Change in Concentration

If the concentration of one of the reactants or of one of the products in an equilibrium system is changed, the effect of the change is counteracted by the establishment of a new equilibrium system. The value of the equilibrium constant, K_c, does not change.

Consider the equilibrium system

$$H_2(g) + I_2(g) \rightleftharpoons 2HI(g)$$

$$K_c = \frac{[HI]^2}{[H_2][I_2]} = 50 \text{ at } 720 \text{ K}$$

Suppose more H_2 was added:
The system will readjust to this stress by forming more HI; thus using up some H_2 (and some I_2 as well). When equilibrium is re-established, the concentration of HI will have increased and the position of equilibrium will have shifted to the right-hand side. The increase in HI is compensated for by a decrease in H_2 and I_2. This means that K_c does not change in value; it is still 50 at 720 K.
Suppose some HI was removed:
The system readjusts by using up more H_2 and I_2 to form more HI. The position of equilibrium shifts to the right-hand side and K_c does not change.
Removal or addition of any of the other reacting species affects the equilibrium system in a similar way.

(2) Change in Pressure

If the pressure of an equilibrium system involving gases is increased, the system readjusts in such a way as to establish a new equilibrium so that the volume is decreased. Consider the equilibrium system:

$$2CO_2(g) \rightleftharpoons 2CO(g) + O_2(g)$$
$$2 \text{ mol} \qquad 2 \text{ mol} + 1 \text{ mol}$$

When the reaction moves in a forward direction, two moles of reactants ($2CO_2$) form a total of three moles of products ($2CO + O_2$)

In accordance with Boyle's law, $PV = k$, an increase in pressure causes the volume of the system to decrease, while a decrease in pressure causes the volume of the system to increase.

Therefore, an increase in pressure in the reaction above will cause the equilibrium to shift to the position of least volume; that is the equilibrium will tend to the left-hand side. Thus more CO_2 will be formed.

In general for reactions involving a change in the total number of moles of gas

Pressure increase
$$\xrightarrow{\hspace{3cm}}$$
$$2A(g) + B(g) \rightleftharpoons 2C(g)$$
$$\xleftarrow{\hspace{3cm}}$$
Pressure decrease

For reactions where there is no change in the total number of moles of gas, a change in pressure has no effect on the position of equilibrium.

(3) Change in Temperature

A change in temperature has a large effect on most chemical reactions. Some reactions, which are normally slow, are speeded up if the temperature is raised. An increase in temperature allows a chemical reaction to reach equilibrium sooner and **alters the value of the equilibrium constant K_c.**

In order to predict the effect of a temperature change on an equilibrium system it is necessary to know the type of the heat change accompanying the particular reaction, that is, it must be known whether the reaction is **exothermic** or **endothermic**.

Consider the reaction

$$N_2(g) + O_2(g) \rightleftharpoons 2NO(g); \Delta H = +184 \text{ kJ mol}^{-1}$$

Here, the forward reaction is endothermic, and consequently the reverse reaction is exothermic.

If heat is added (increase in temperature) the position of equilibrium will shift to the right-hand side, the direction in which heat is absorbed.

If the reaction mixture is cooled (decrease in temperature) the equilibrium will shift to the left-hand side, the direction in which heat is evolved.

In general for an endothermic reaction
It should be noted that an increase in the temperature of an equilibrium system results in an increase in the value of K_c, if the reaction involved is endothermic, while a decrease in temperature results in an decrease in K_c, if the reaction involved is endothermic.

$$\begin{array}{c} \xrightarrow{\text{Temperature increase}} \\ A(g) + B(g) \rightleftharpoons C(g) + D(g);\ \Delta H = + \\ \xleftarrow{} \\ \text{Temperature decrease} \end{array}$$

(4) Use of Catalysts

A catalyst cannot alter the composition of an equilibrium mixture, but it can change the speed at which the product is formed.

In accordance with chemical kinetics, a catalyst speeds up the rate of the forward reaction and the rate of the reverse reaction to an **equal extent**. A catalyst, does, however, help a system **attain equilibrium more quickly**.

Factors Affecting Equilibrium

Factor	Equilibrium constant	Position of equilibrium	Rate at which equilibrium is attained
Change in concentration	No change	Changes	Changes
Change in pressure	No change	Changes	Changes
Change in temperature	Changes	Changes	Changes
Use of catalysts	No change	No change	Changes

Mandatory Experiment 24

To illustrate Le Chatelier's principle.

Introduction

Some reactions which occur in aqueous solution are easy to observe because in some cases a clear colour change is noticed.
Reactions of these types include:

1.
$$CoCl_4^{2-}(aq) + 6H_2O(l) \rightleftharpoons Co(H_2O)_6^{2+}(aq) + 4Cl^-(aq)$$
blue   pink

Cobalt(II) chloride ions + water $\rightleftharpoons$ hydrated cobalt(II) ions + chloride ions

2.
$$2CrO_4^{2-}(aq) + 2H^+(aq) \rightleftharpoons Cr_2O_7^{2-}(aq) + H_2O(l)$$
yellow   orange

Chromate(VI) ions + hydrogen ions $\rightleftharpoons$ dichromate(VI) ions + water

3.
$$Fe^{3+}(aq) + SCN^-(aq) \rightleftharpoons Fe(SCN)^{2+}(aq)$$
pale yellow   colourless   blood-red

Iron(III) ions + thiocyanate ions $\rightleftharpoons$ iron thiocyanate ions

The colour observed in each of the reactions 1–3 depends on how far the reaction has gone to the right-hand side or to the left-hand side.
If the reaction goes to the right-hand side, the position of equilibrium lies to the right-hand side. If the reaction goes to the left-hand side, the position of equilibrium lies to the left-hand side.

Requirements

Safety glasses, test-tubes and test-tube rack, teat pipettes, beakers, glass rod, white cardboard, deionised water, cobalt(II) chloride, concentrated hydrochloric acid, 0.1 mol L^{-1} sodium dichromate solution, 2 mol L^{-1} sodium hydroxide solution, dilute hydrochloric acid, 0.5 mol L^{-1} potassium thiocyanate solution, 0.5 mol L^{-1} iron(III) chloride solution, ammonium chloride.

Procedure

Investigation 1. To demonstrate the effects of both temperature changes and concentration changes.

$Co(H_2O)_6^{2+}(aq) + 4Cl^-(aq) \rightleftharpoons CoCl_4^{2-}(aq) + 6\ H_2O$; ΔH is positive

(1) Dissolve 5 g of cobalt(II) chloride in 100 cm³ of water.
(2) Divide this solution equally into three beakers. Place one beaker to one side as a reference colour.
(3) Using a teat pipette, carefully add some concentrated hydrochloric acid dropwise to the second beaker. Observe what happens.
(4) Add some water dropwise to the same beaker. Observe what happens.
(5) Heat the solution in the third beaker. Observe what happens.
(6) Place the same beaker in an ice bath. Observe what happens.

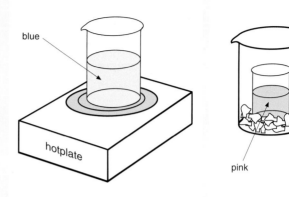

$CoCl_4^{2-}(aq) + 6H_2O \rightleftharpoons Co(H_2O)_6^{2+} + 4Cl^-(aq)$
Blue   Pink
Cooling in ice shifts the equilibrium to the right, the pink complex. Heating shifts the equilibrium to the left, the blue complex.

(7) Copy and complete the table below. For each observation, state which side the equilibrium has shifted to.

Test performed	*Observation*	*Reason*
Addition of HCl		
Addition of water		
Heating the solution		
Cooling the solution		

Investigation 2. To demonstrate the effect of concentration changes.

$2CrO_4^{2-}(aq) + 2H^+(aq) \rightleftharpoons Cr_2O_7^{2-}(aq) + H_2O(l)$

(1) Make up dilute solutions of potassium chromate and potassium dichromate. Note the colour of each solution.
(2) Place approximately 5 cm³ of each solution in separate test-tubes.
(3) Using a teat pipette, slowly add dilute hydrochloric acid to each test-tube. Observe what happens.

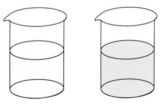

Change in position of equilibrium
$2CrO_4^{2-}(aq) \rightleftharpoons Cr_2O_7^{2-}(aq)$

(4) Using a teat pipette, slowly add dilute sodium hydroxide to each test-tube. Observe what happens.
(5) Copy and complete the table below. For each observation, state which side the equilibrium has shifted to.

Solution	*Colour*	*Effect of addition of HCl*	*Effect of addition of NaOH*	*Position of equilibrium*
Potassium chromate				
Potassium dichromate				

252

Investigation 3. To demonstrate the effect of concentration changes.

$$Fe^{3+}(aq) + SCN^-(aq) \rightleftharpoons Fe(SCN)^{2+}(aq)$$

(1) Mix together equal volumes of the iron(III) chloride and the potassium thiocyanate solutions. Add deionised water until the solution changes from a blood-red colour to a light orange-brown colour.
(2) Divide this solution into four equal parts in four test-tubes. Put one test-tube aside as a colour reference.
(3) Add some iron(III) chloride solution dropwise to the first test-tube.
(4) Add some potassium thiocyanate solution dropwise to the second test-tube.

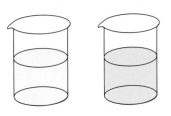

$Fe^{3+}(aq) \rightarrow Fe(SCN)^{2+}(aq)$

(5) Add some ammonium chloride to the third test-tube. The ammonium chloride removes iron(III) ions from solution by forming an iron–chloride complex.
(6) Compare the colours in each test-tube to the reference colour.
(7) Copy and complete the table below. For each observation, state which side the equilibrium has shifted to.

Addition of	Observation	Reason	Shift in equilibrium
Iron(III) chloride			
Potassium thiocyanate			
Ammonium chloride			

Industrial Applications: Controlling Reaction Conditions

Obtaining the maximum amount of product from a chemical reaction depends on the proper selection of reaction conditions. In an industrial process, it is important that the maximum yield is obtained at the lowest cost.

An equilibrium mixture can be affected in various ways; some will favour the formation of products, others will favour the formation of reactants. A change in concentration, pressure, or temperature will alter the yield of product, while the use of a catalyst will change the speed at which the product is formed.

The work of the chemist and the chemical engineer ensures that factors which may compete with each other during a chemical reaction are minimised and that the optimum use is made of the reaction conditions and of the use of catalysts.

The Haber–Bosch Process

Before World War I, fertilisers and explosives were made from various nitrate deposits. In 1909, a German chemist, Fritz Haber, discovered the conditions under which atmospheric nitrogen could be fixed to hydrogen to give ammonia. The process was further developed by a chemical engineer, Karl Bosch. Both men were later awarded the Nobel Prize for Chemistry.

Fritz Haber (1868–1934)
Fritz Haber was born in Breslau, Germany. In 1906 he began to investigate the possibility of synthesising ammonia from its elements. In 1909, in the presence of Karl Bosch, he made the first 100 grams of synthetic ammonia. The Haber–Bosch process for the large scale synthesis of ammonia was vital to the Germans during World War I as it enabled them to make nitric acid from which explosives were made. He developed the first use of a gas, chlorine, as a weapon in 1915. France reciprocated by using phosgene the following year. He received the Nobel Prize for Chemistry in 1918, the year the war ended, and was denounced by many scientists for his patriotic assistance to the German army. As an act of reparation he tried unsuccessfully to extract gold from the sea to help pay for the war damage. He was director of the Kaiser Wilhelm Institute for Chemistry from its founding in 1911 to the year of his death in 1934. When the Nazis rose to power in 1933 they demanded the dismissal of all Jewish workers. Haber, himself a Jew, refused to do this. He died, broken in spirit and in health, of a heart attack in exile in Switzerland.

The equilibrium reaction involved in the production of ammonia is:

$$N_2(g) + 3H_2(g) \rightleftharpoons 2NH_3(g); \Delta H = -92 \text{ kJ mol}^{-1}$$

The conditions which would yield the maximum amount of product in accordance with Le Chatelier's principle are:

(i) high pressure favours the position of least volume.

In the forward direction, four volumes of reactants ($N_2 + 3H_2$) form two volumes of product ($2NH_3$). Therefore, the equilibrium will lie to the right-hand side if the pressure is increased, favouring the formation of NH_3.

(ii) for an exothermic reaction the amount of product increases if the temperature is lowered.

In this reaction, a decrease in temperature creates a problem, because the reaction rate would be far too slow and the reaction would take too much time to come to equilibrium. The use of low temperature is impractical, even in the presence of catalysts.

The optimum choice of temperature, pressure and the use of a catalyst enabled Haber and Bosch to develop the process which is responsible for the manufacture of about 80% of ammonia today.

> Today, the reaction is carried out at 450 °C and 200 atm, using an iron catalyst containing potassium hydroxide as promoter.

The Contact Process

The most important stage in the manufacture of sulphuric acid involves the oxidation of sulphur dioxide to sulphur trioxide.

The reaction takes place in the gas phase and is exothermic:

$$2SO_2(g) + O_2(g) \rightleftharpoons 2SO_3(g); \Delta H = -98 \text{ kJ mol}^{-1}$$

(i) high pressure favours the position of least volume.

In the forward direction, three volumes of reactants ($2SO_2 + O_2$) form two volumes of products ($2SO_3$). Therefore, the equilibrium will lie to the right-hand side if the pressure is increased, favouring the formation of SO_3.

(ii) for an exothermic reaction the amount of product increases if the temperature is lowered.

The use of a catalyst, vanadium(V) oxide, V_2O_5, with potassium sulphate promoter supported on silica helps the reaction attain equilibrium sooner.

In this particular case, the catalyst enables a high yield (99.5%) of product to be formed without the use of high pressure. The catalyst will not work at a temperature below 400 °C and decomposes at a temperature higher than 620 °C.

> A high yield of SO_3 is obtained at atmospheric pressure and at a low operating temperature of 450 °C using a vanadium(V) oxide catalyst.

Exercise 8.2

•

(1) Complete each of the following equilibrium reactions. State the colours (if any) of each reactant and product.

(a) $Co(H_2O)_6^{2+}$ (aq) + 4Cl⁻ (aq) $\rightleftharpoons$

(b) $2CrO_4^{2-}$ (aq) + $2H^+$(aq) $\rightleftharpoons$

(c) Fe^{3+} (aq) + SCN⁻ (aq) $\rightleftharpoons$

(2) (a) State Le Chatelier's principle.
(b) Explain the effects on each of the following equilibrium systems of:
(i) increasing the concentration of reactants
(ii) increasing the total pressure
(iii) increasing the temperature
(iv) using a catalyst.

(a) $H_2(g) + I_2(g) \rightleftharpoons 2HI(g); \Delta H = -10 \text{ kJ mol}^{-1}$

(b) $2CO(g) + O_2(g) \rightleftharpoons 2CO_2(g) \Delta H = -566 \text{ kJ mol}^{-1}$

(c) $PCl_5(g) \rightleftharpoons PCl_3(g) + Cl_2(g); \Delta H = +124 \text{ kJ mol}^{-1}$

(d) $N_2(g) + 3H_2(g) \rightleftharpoons 2NH_3(g); \Delta H = -92 \text{ kJ mol}^{-1}$

(e) $2SO_2(g) + O_2(g) \rightleftharpoons 2SO_3(g); \Delta H = -197 \text{ kJ mol}^{-1}$

(f) $N_2O_4(g) \rightleftharpoons 2NO_2(g); \Delta H = + 57 \text{ kJ mol}^{-1}$

(3) Using Le Chatelier's principle, predict any equilibrium shifts in the reaction:

$$CO(g) + H_2O(g) \rightleftharpoons CO_2(g) + H_2(g);$$
$$\Delta H \text{ is negative}$$

when the following changes occur:
(a) the temperature of the equilibrium mixture is decreased
(b) the concentration of H_2O is increased
(c) a catalyst is added to the system
(d) the pressure of the system is increased.

Change	Concentration of N_2	Concentration of H_2	Concentration of NH_3
Increase in pressure			
Increase in temperature			
Increase in amount of H_2			
Remove NH_3			
Use a catalyst			

(4) Ammonia is made by the Haber process as follows:

$N_2(g) + 3H_2(g) \rightleftharpoons 2NH_3(g); \Delta H = -92$ kJ mol^{-1}

Copy and complete the table above, by applying Le Chatelier's principle.

(5) Hydrogen gas for use in ammonia production is produced as follows:

$$CH_4(g) + H_2O(g) \rightleftharpoons CO(g) + 3H_2(g);$$
$$\Delta H = 205 \text{ kJ mol}^{-1}$$

Describe what happens to the reaction mixture at equilibrium if:
(a) steam is removed (b) the temperature is increased (c) carbon monoxide is removed (d) the pressure is increased (e) a catalyst is used.

(6) Hydrogen gas can be prepared for commercial use by the following reaction:

$$CO(g) + H_2O(g) \rightleftharpoons H_2(g) + CO_2(g)$$

Describe the shift in equilibrium for each of the following:
(a) carbon dioxide is removed (b) carbon monoxide is added (c) the pressure is decreased.

(7) Describe the changes which take place when (a) the temperature is changed and (b) the concentrations are changed in the following reaction:

$CoCl_4^{2-}(aq) + 6H_2O \rightleftharpoons Co(H_2O)_6^{2+}(aq) + 4Cl^-(aq)$

(8) When concentrated hydrochloric acid is added to a solution of cobalt(II) chloride, the following equilibrium system is observed:

$CoCl_4^{2-}(aq) + 6H_2O \rightleftharpoons Co(H_2O)_6^{2+}(aq) + 4Cl^-(aq)$

(a) What colour change would you notice as the concentrated hydrochloric acid was added? Explain your answer.
(b) If sufficient water was now added, what would happen?

(c) What would happen if the original solution was (i) heated and (ii) cooled in iced water?
(d) Is the reaction exothermic or endothermic?

(9) Describe the changes which take place when the concentrations are changed in the following reactions:

(a)
$2CrO_4^{2-}(aq) + 2H^+(aq) \rightleftharpoons Cr_2O_7^{2-}(aq) + H_2O(l)$

(b) $Fe^{3+}(aq) + SCN^-(aq) \rightleftharpoons Fe(SCN)^{2+}(aq)$

(10) (a) How will the number of moles of SO_3 change in the equilibrium reaction,

$$2SO_2(g) + O_2(g) \rightleftharpoons 2SO_3(g);$$
$$\Delta H = -197 \text{ kJ mol}^{-1}$$

in each of the following cases?
(i) the pressure is increased (ii) the temperature is decreased (iii) oxygen is added (iv) a catalyst is used (v) sulphur dioxide is removed.
(b) What is the significance of these changes in terms of the commercial production of SO_3?

(11) In which direction will the equilibrium shift for the reaction,

$$4NH_3(g) + 5O_2(g) \rightleftharpoons 4NO(g) + 6H_2O(g);$$
$$\Delta H = -905 \text{ kJ mol}^{-1}$$

if the following changes are made:
(a) the concentration of O_2 is increased
(b) the concentration of H_2O is decreased
(c) the concentration of NO is increased
(d) the pressure is increased
(e) the temperature is decreased?

(12) The combustion of petrol in motor engines produces oxides of nitrogen. Nitrogen dioxide is produced as follows:

$$N_2(g) + 2O_2(g) \rightleftharpoons 2NO_2(g); \Delta H = 66 \text{ kJ mol}^{-1}$$

Explain how the concentration of nitrogen dioxide changes if the following changes occur:
(a) the combustion chamber is made smaller
(b) the engine overheats
(c) less air is drawn into the combustion chamber.

(13) What would you expect to be the general pressure and temperature conditions for optimum yields in each of the following industrial processes?

(a) $4 NH_3(g) + 5O_2(g) \rightleftharpoons 4NO(g) + 6H_2O(g)$;

$\Delta H = -905 \text{ kJ mol}^{-1}$

(b) $2SO_2(g) + O_2(g) \rightleftharpoons 2SO_3(g)$;

$\Delta H = -197 \text{ kJ mol}^{-1}$

(c) $CH_4(g) + H_2O(g) \rightleftharpoons CO(g) + 3H_2(g)$;

$\Delta H = 205 \text{ kJ mol}^{-1}$

(d) $N_2(g) + 3H_2(g) \rightleftharpoons 2NH_3(g)$;

$\Delta H = -92 \text{ kJ mol}^{-1}$

▼▼▼ Key Terms ▼▼▼

Some of the more important key terms are listed below. Other terms not listed may be located by means of the index.

1. Equilibrium reactions: Equilibrium reactions are reversible reactions. Such reactions are described using reversible arrows, which indicate that the reaction can go in the forward direction or in the reverse direction.

2. Dynamic equilibrium: A reaction in a state of dynamic equilibrium consists of a forward reaction and a reverse reaction. Both the forward reaction and the reverse reaction occur at the same rate (speed).

3. The equilibrium constant, K_c: The equilibrium constant expression is given as

$$K_c = \frac{[C]^c [D]^d}{[A]^a [B]^b}$$

4. Values of the equilibrium constant, K: The value of K, at a given temperature, indicates the amount of reactants and products in the equilibrium mixture.

If K is large, the amount of products is large.

If K is small, the amount of reactants is large.

5. Le Chatelier's principle: If an equilibrium system is subjected to a stress, such as a change in concentration, temperature or pressure, then the system tries, if possible, to reduce that stress by adjusting the amounts of the substances present in the equilibrium mixture.

6. Factors which affect an equilibrium system:
(i) change in concentration
(ii) change in pressure
(iii) change in temperature
(iv) use of catalysts.

7. Change in concentration: If the concentration of one of the reactants or of one of the products in an equilibrium system is changed, the effect of the change is counteracted by the establishment of a new equilibrium system. The value of the equilibrium constant, K, does not change.

8. Change in pressure: If the pressure of an equilibrium system involving gases is increased, the system re-adjusts in such a way as to establish a new equilibrium so that the volume is decreased.

9. Change in temperature: An increase in temperature allows a chemical reaction to reach equilibrium sooner and alters the value of the equilibrium constant K.

10. Use of catalysts: A catalyst cannot alter the composition of an equilibrium mixture, but it can change the speed at which the product is formed.

ENVIRONMENTAL CHEMISTRY: WATER

INTRODUCTION

Water – a wonderful amenity.

Water is a unique substance. It is the only chemical compound that occurs naturally in all three physical states: as a solid, ice, as a liquid, water, and as a gas, steam.

Approximately four-fifths of the Earth's surface is covered with water as sea, rivers or lakes. The atmosphere contains about 1.38×10^{10} gallons of water. Water existed on this planet before any form of life evolved. As life developed in water it is not unexpected that nearly 70% of the human body is composed of water. Nearly all of the processes which take place in animals and plants depend on water.

Water has several properties which make it a very unique substance. It has a maximum density at 4 °C, it has an extremely high heat capacity, it has a high melting point and boiling point and many other unusual properties. Most of these properties of water are due to the vast network of hydrogen bonds throughout the liquid.

Water is the universal solvent. There is hardly anything on Earth which does not dissolve in water to some extent. It is an excellent solvent for ionic substances and polar covalent substances. It is unusual to find really pure water on this planet, because water is such an excellent solvent that it dissolves solids, such as sodium chloride, or gases, such as oxygen or sulphur dioxide. These dissolved solutes sometimes enable the water to become a better conductor of electricity, or allow fish to breathe, but at other times solutes kill the fish by pollution.

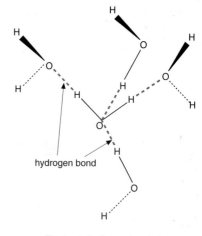

The basic hydrogen-bonded tetrahedral structure of water

Water is a national asset. In Ireland we often take water for granted. Its many benefits include recreational activities, like sailing, fishing and swimming, which are not only beneficial to us individually, but also encourage tourists whose economic contribution to our country is vast.

9.1 HARDNESS IN WATER

The water you drink contains some dissolved gases, such as oxygen, nitrogen and carbon dioxide, and may also contain some dissolved solids. This dissolved material usually is not harmful; often it makes the water taste more pleasant. Indeed, some Irish companies sell large quantities of bottled water containing dissolved gases and solids.

As water flows through soil and rocks it dissolves many ions. Some ions, such as Ca^{2+} ions and Mg^{2+}, ions cause hardness in water. These ions are present in rocks such as limestone (calcium carbonate), gypsum (calcium sulphate) and dolomite (magnesium carbonate).

Water which does not form a lather easily with soap is called hard water. This water, although pleasant to taste, causes difficulties not only in washing but also in many industrial processes by causing deposits to build up in pipes and tanks.

Scanning electron micrograph of 'fur' from a domestic kettle, showing the crystals that form in a hard water area. The fur consists of crystals of calcium carbonate and crystals of calcium sulphate.

How scum is formed

Soap is made of the sodium and potassium salts of organic acids such as octadecanoic acid (stearic acid). The sodium salt, for simplicity, may be represented as Na^+St^- and reacts with the calcium or magnesium ions as follows:

$$2Na^+St^- + Ca^{2+} \Rightarrow Ca(St)_2 + 2Na^+$$
$$\text{scum}$$

The calcium stearate, unlike the sodium stearate, is insoluble in water and forms a curdy precipitate or scum which is difficult to remove from clothes. Modern detergents do not form a scum with hard water.

Water Softening

The removal of Ca^{2+} and Mg^{2+} ions from hard water is called water softening. The softening process used depends on the type of hardness in the water, i.e. whether it is temporary hardness or permanent hardness.

Temporary Hardness

Temporary hardness is hardness caused by the presence of dissolved calcium (or magnesium) hydrogencarbonate. This form of hardness is easily removed because the hydrogen carbonates decompose on heating.

$$Ca(HCO_3)_2(aq) \Rightarrow CaCO_3(s) + H_2O(l) + CO_2(g)$$
$$\text{soluble} \qquad \text{insoluble}$$

The insoluble $CaCO_3$ falls to the bottom as a white solid, called fur or scale. This fur blocks kettles and pipes causing poor thermal conduction. In large industrial tanks and vats it is not only difficult to remove scale but it can also involve shutting down the plant to scrape off the scale. This is often an expensive process. Recently large magnets have been successfully used to remove the scale.

Scale in pipes.

Permanent Hardness

Permanent hardness is caused by calcium and magnesium ions combined with either sulphate, SO_4^{2-}, or chloride ions, Cl^-.

Permanent hardness cannot be removed by boiling the water.

Permanent hardness (and temporary hardness) can be removed by distillation and the use of ion-exchange resins.

(a) Distillation

This removes all the dissolved solids from the water and produces very pure water. However, due to the energy required to heat the water, it is too expensive and is generally only used in analytical laboratories where extremely pure water is required.

(b) Use of Ion-Exchange Resins

Ion-exchange is the process by which a water solution is passed through a column of a material which replaces ions of one kind with ions of another kind. The process can be used for domestic, laboratory and industrial applications. Domestic and commercial ion-exchange resins contain cation exchange resins which consist of sodium ions and negatively charged ions bound to a macromolecular surface. The calcium and magnesium ions, which cause water hardness, are exchanged for a sodium ion which is present on the resin.

If the ion-exchange resin is represented as Na—Resin, the exchange can be written as

$$CaSO_4(aq) + 2Na\text{—Resin} \Rightarrow Ca\text{—Resin} + Na_2SO_4(aq)$$

The calcium ions swop with the sodium ions on the surface of the ion-exchange resin.

When all the sodium ions on the resin have been exchanged for calcium (or magnesium) ions the resin can be regenerated by passing a solution of sodium chloride through it, where the Ca^{2+} ions are again replaced by Na^+ ions.

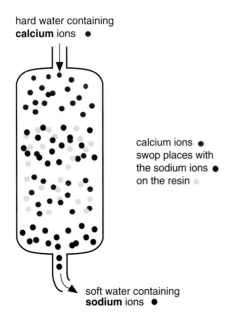

hard water containing **calcium** ions ●

calcium ions ●
swop places with
the sodium ions ●
on the resin ●

soft water containing
sodium ions ●

Ion-exchange works like this

It is possible to remove all the ions from a water sample by ion-exchange, if the water is passed through a resin which contains a mixture of cation and anion exchange resins. In this case the cation which is exchanged is an H^+ ion and the anion which is exchanged is an OH^- ion. Suppose, for instance, that the substance to be removed from the water is sodium chloride. The Na^+ ions are exchanged for H^+ ions and the Cl^- ions are exchanged for OH^- ions. The H^+ ions react with the OH^- ions to form water and the NaCl is removed.

High-purity Water

Many industries require water which is particularly pure, for example, the electronics industry. Water of extremely high purity is achieved using a process called reverse osmosis, where the water is purified by forcing it through a membrane under high pressure. This process removes not only ions dissolved in the water but also the organic compounds in the water.

Mandatory experiment 25

(a) To estimate the total suspended solids in a water sample.
(b) To estimate the total dissolved solids in a water sample.
(c) To determine the pH of a water sample.

Introduction

Water often contains suspended solids which give it an unsightly appearance and may also cause damage to fish gills. When polluted water is treated and then discharged into rivers, lakes or the sea the amount of suspended solids should have fallen to approximately 20 ppm. Water suitable for drinking should have no suspended solids.

Water also contains dissolved ions which may alter the pH of the water. Before water leaves the treatment plant its pH must be in the range 6.5–7.0.

Requirements

(a) Filter paper and funnel (Buchner funnel and suction pump if possible), beakers, 1 L volumetric flask, water samples from a river or a lake.
(b) Filtered water samples, 250 cm³ beaker, burette and an oven.
(c) Water samples, pH paper (preferably a pH meter), 50 cm³ beakers, deionised water, filter paper.

Procedure

(a) Total suspended solids
(1) Weigh a sheet of filter paper.
(2) Filter 1 L of the water to be tested through the filter paper. Put the water aside for part (b).
(3) When all the water sample has been filtered allow the filter paper to dry completely and then reweigh the filter paper.

(4) Calculate the amount of total dissolved solids in mg L^{-1} (ppm).
(5) Record your results in a table like the one below and repeat the procedure using another sample of filtered water.

(b) Total dissolved solids
(1) Weigh a clean dry beaker.
(2) Using a burette put exactly 100 cm^3 of the filtered water from part (a) into the clean dry beaker.
(3) Weigh the beaker + filtered water.
(4) Place the beaker in an oven and evaporate to dryness.
(5) Cool the beaker in a dessicator and reweigh it.
(6) Calculate the amount of total dissolved solids in mg L^{-1} (ppm).
(7) Record your results in a table like the one below and repeat the procedure using another sample of filtered water.

(c) Determination of pH
(1) Place some different water samples in some clean 50 cm^3 beakers.
(2) Using a pH meter (or pH paper) determine the pH of each water sample.
(3) Record your results in a table like the one below.

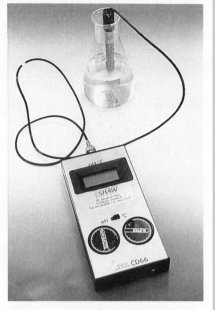

Measuring changes in pH using a pH meter.

Results

(a) Suspended solids

Water sample	Mass of filter paper/g	Mass of filter paper after filtration and drying/g	Mass of suspended solids/g	Total suspended solids/ppm
Tap				
River				
Stream				
Lake				

(b) Dissolved solids

Water sample	Mass of beaker/g	Mass of beaker + filtered water/g	Mass of beaker + dried solids/g	Total dissolved solids/ppm
Tap				
River				
Stream				
Lake				

(c) pH determination

	Tap	River	Stream	Lake
pH				

Mandatory experiment 26

To determine the total hardness of a water sample using EDTA.

Introduction

Hard water contains calcium ions, Ca^{2+}, and magnesium ions, Mg^{2+}. If water containing Ca^{2+} and Mg^{2+} ions is titrated with ethylenedi-aminetetraacetic acid (EDTA) a complex is formed.

EDTA has the ability to complex with many metallic ions, such as Ca^{2+} and Mg^{2+} ions.

The acid is not very soluble in water and so its disodium salt is used in titrations because it is more soluble in water.

The structures of the acid and its disodium salt are as follows:

CH_2COOH CH_2COONa

CH_2—N CH_2—N

CH_2COOH CH_2COOH

CH_2COOH CH_2COOH

CH_2—N CH_2—N

CH_2COOH CH_2COONa

EDTA   disodium salt of EDTA

For simplicity we can represent the acid as H_4Y and the disodium salt as Na_2H_2Y. Calcium (or magnesium) ions can react with either the acid or its disodium salt and form a complex as shown below:

$Ca^{2+} + Na_2H_2Y \Rightarrow CaH_2Y + 2Na^+$

disodium salt   calcium complex
of EDTA   of EDTA

It is important to remember that Ca^{2+} ions or Mg^{2+} ions react with EDTA in a 1:1 ratio.

Eriochrome Black T (often called Solochrome Black T) is used as an acid–base indicator during the titration. This indicator is extremely pH sensitive. It forms a wine-red complex with the Mg^{2+} ions in water at pH 10.

When EDTA is added to the hard water, the indicator prefers to form a complex with the EDTA rather than with the Mg^{2+} ions or the Ca^{2+} ions in the hard water. Thus, as the EDTA is added the Ca^{2+} ions are first removed from the water and then the Mg^{2+} ions are removed from the wine-red indicator and Mg^{2+} complex. The wine-red colour changes to a navy blue colour at the end-point.

The pH is maintained at pH 10 by using an ammonium chloride–ammonia buffer.

If the water sample does not contain any Mg^{2+} ions, some should be added, otherwise the initial wine-red colour will not be present.

Requirements

Safety glasses, burette, pipette, conical flask, beaker, funnel, Eriochrome Black T, 0.01 mol L^{-1} EDTA solution, buffer solution (made by dissolving 6.75 g of ammonium chloride in 57 cm^3 of 0.88 ammonia and making the solution up to 100 cm^3 with deionised water), hard water sample.

Procedure

(1) Pipette out 100 cm^3 of the hard water sample into a conical flask.

(2) Add about 2 cm^3 of the buffer solution, and then add a tiny pinch of the powdered Eriochrome Black T indicator. The indicator and magnesium complex is now wine-red in colour.

(3) Fill the burette with the EDTA solution and start the titration. Swirl the conical flask constantly towards the end of the titration.

(4) When the end-point is reached the colour changes from wine-red to navy blue. Record your result and repeat the titration at least twice.

(5) Study Example 9.1 and then use your results to calculate the total hardness of the water sample.

Results

	Titre 1	Titre 2	Titre 3	Titre 4
Volume of EDTA used				

Mean titre =

Reminder: 1 mol $Ca^{2+} \equiv$ 1 mol EDTA

Total hardness is usually expressed as ppm $CaCO_3$, even though other compounds such as $MgCO_3$ are dissolved in the water.

• • • • • • **Example 9.1** • • • • • •

In a titration to measure the total hardness of a water sample it was found that 30 cm^3 of a 0.01 mol L^{-1} EDTA solution reacted with 50 cm^3 of the water sample to reach the end-point. Calculate the total hardness of the water sample, expressed as ppm CaCO$_3$.

$$1 \text{ mol Ca}^{2+} \equiv 1 \text{ mol EDTA}$$

$$\frac{c_A V_A}{a} = \frac{c_B V_B}{b}$$

$$\frac{c_A \times 50 \text{ cm}^3}{1} = \frac{0.01 \text{ mol L}^{-1} \times 30 \text{ cm}^3}{1}$$

$$c_A = \frac{0.01 \text{ mol L}^{-1} \times 30 \text{ cm}^3}{50 \text{ cm}^3}$$

$$= 0.006 \text{ mol L}^{-1} \text{ CaCO}_3$$

change moles to grams using $m = nM$

$$= 0.006 \text{ mol L}^{-1} \times 100 \text{ g mol}^{-1} \text{ CaCO}_3$$

$$M(\text{CaCO}_3) = 100 \text{ g mol}^{-1}$$

$$= 0.6 \text{ g L}^{-1} \text{ CaCO}_3$$

change g to mg by multiplying by 1000

$$= 600 \text{ mg L}^{-1} \text{ CaCO}_3$$

$$= 600 \text{ ppm CaCO}_3$$

• •

Exercise 9.1

•

(1) (a) How do calcium ions, Ca^{2+} and magnesium ions, Mg^{2+}, get into water?
(b) What is hard water?
(c) Describe, using simple equations, how scum is formed with soap.
(d) What is water softening?

(2) (a) Explain, using equations, the difference between temporary hardness and permanent hardness.
(b) Describe clearly how (i) temporary hardness and (ii) permanent hardness are removed from water.

(3) Explain using a labelled diagram how an ion-exchange resin works.

(4) (a) What is the difference between distilled and deionised water?
(b) Explain which is the purer type of water.

(5) Three water samples, X , Y and Z, were tested for hardness by shaking them with soap solution. 250 cm^3 of sample X required 1 cm^3 of soap solution to produce a lather, while 250 cm^3 of sample Y required 20 cm^3 of soap solution to produce a lather. After boiling, 250 cm^3 of sample Y required 1 cm^3 to produce a lather. 250 cm^3 of sample Z required 15 cm^3 to produce a lather both before and after boiling.
(a) Which of the samples contained soft water? Explain your answer.

(b) Which of the samples contained temporary hardness? What was the cause of the temporary hardness?
(c) Which of the samples contained permanent hardness? What was the cause of the permanent hardness?
(d) 250 cm^3 of sample Y was evaporated to dryness, leaving 0.06 g of solid. Calculate the total solids in mg L^{-1} and in ppm.

(6) A student analysed a water sample and found that one litre contained:
(i) 0.2 g of suspended solids
(ii) 1.6 g of dissolved solids
(a) Express the amount of (i) suspended solids, (ii) dissolved solids present in mg L^{-1} (ppm).
(b) Describe, briefly, how the student determined the mass of suspended solids and the mass of the dissolved solids.

(7) (a) What is meant by hardness in water?
(b) Distinguish between temporary and permanent hardness of water and state how each type is caused. How may temporary hardness be removed?
(c) Describe, briefly, how an ion exchange resin may be used to remove permanent hardness.
(d) 1 L of a water sample was filtered and it was found that the mass of the filter paper, after drying, had increased by 0.25 g. Calculate the total suspended solids in parts per million (ppm). How might the total dissolved solids content of the water sample be obtained?

(8) The following statement ended a student's account of an experiment.
'The total suspended solids in a litre of water was determined to be 150 ppm.'
(a) What is meant by suspended solids?
(b) Describe the apparatus used and the procedures followed which led to the above conclusion.

(9) (a) Which calcium compound is a cause of temporary hardness? How does this compound get into water?
(b) A white fur or scale is often found in kettles in hard water districts. Write an equation for the reaction involved in producing this product.
(c) How would you test the inside of a kettle for deposits of scale?

9.2 WATER TREATMENT

Water Resources and Uses

There are approximately 10^8 tonnes of water on Earth, of which only 20% is available to man through the water cycle. Most of the water on Earth is locked up as ice in the ice caps. Water evaporates from the ice caps and from rivers, lakes and the sea and accumulates in the atmosphere. The water vapour can return to the land when it precipitates from the clouds as rain or snow.

Water uses

• Domestic uses: about 10% of the water withdrawn from the water cycle each year is used in the home for washing clothes, food, the body, the car, for watering the garden and for flushing the toilet.

• Industrial uses: most of the water required by industry is used to cool down machinery and in various chemical reactions. The remainder is used for washing and as a solvent.

• Agricultural uses: agriculture is the main user of water withdrawn (82%) from the water cycle. Water is used to irrigate crops from carrots to cotton.

• Power: large quantities of water are used to generate electrical power. This water is not lost as it is returned to its original source.

Water quality is usually determined by its:

 turbidity (cloudiness)
 odour
 colour
 taste
 pH
 dissolved organic material
 dissolved inorganic ions
 micro-organisms and algae
 temperature.

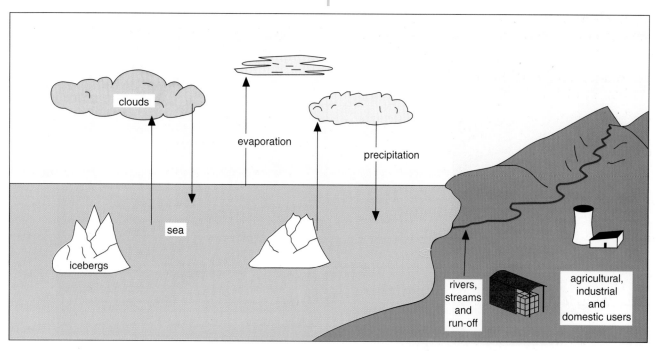

Water Pollution

Water pollution is the introduction by man, either directly or indirectly, of matter or energy into the water which alters the suitability of water for any of its beneficial uses.

Pollution is caused by people. Thousands of years ago, when man lived in small communities scattered in various parts of the world there was no real pollution. As man became more inventive he built more villages, farms and factories and because the population was not scattered any more, the amount of waste from the towns, factories and farms accumulated. The waste produced was usually disposed of by dumping it into rivers and other water sources.

Water becomes polluted when too much waste enters it from domestic, industrial and agricultural sources.

The principal ways in which water is polluted include:
1. Deoxygenation
2. Eutrophication
3. Oil Pollution
4. Thermal Pollution
5. Chemical Pollution.

1. Deoxygenation

Water normally contains about 10 ppm of dissolved oxygen. Fish and other aquatic organisms need the oxygen dissolved in water to survive. If this oxygen supply decreases below a certain level these creatures die. Trout and salmon, for example, need a concentration of approximately 5 ppm of dissolved oxygen to survive, while other aquatic species need less.

Dead fish.

The release of organic material (from domestic, industrial or agricultural waste) and of inorganic material (from factories, chemicals or fertilisers) into the water is a rich source of food for the micro-organisms (mainly bacteria and fungi) in the water. These tiny micro-organisms use oxygen to break up the waste into simpler substances, causing a decrease in the level of dissolved oxygen (DO).

The amount of dissolved oxygen (DO) can be determined by a method known as the Winkler titration. The amount of oxygen consumed by the micro-organisms is based on the Biological Oxygen Demand (BOD) of the waste.

Biological Oxygen Demand (BOD) Determination

1. An aqueous sample of the waste or water to be tested is taken and kept at less than 4 °C before testing.
2. The amount of dissolved oxygen (DO) is determined by titration.
3. A well stoppered, perfectly full (no air bubbles) bottle of the sample is placed in an incubator at 20 °C in the dark for 5 days.
4. After 5 days the dissolved oxygen (DO) is measured again.
5. The difference between the two values for dissolved oxygen is the Biological Oxygen Demand.

Typical BOD values are given in the table below.

Clean water	< 2 ppm (mg $O_2 L^{-1}$)
Sewage effluent	300–400 ppm
Dairy washings	2000 ppm
Pig slurry	up to 35 000 ppm
Silage effluent	up to 67 000 ppm

The higher the BOD value, the more environmentally dangerous is the material.

These values clearly indicate that agricultural wastes are the major possible cause of pollution in rivers and lakes.

Slurry spraying.

2. Eutrophication

Eutrophication is the stimulation of plant growth in water by nutrients, such as nitrates and phosphates. These nutrients find their way into rivers and lakes owing to the uncontrolled use of nitrate fertilisers and detergents. Natural processes which may have taken thousands of years occur in as little as fifty years owing to over-fertilisation. The effect is seen as dense algal blooms, which sink to the bottom of the water and die, using up the dissolved oxygen.

An algal bloom in a small pond caused by fertiliser run off from surrounding farmland.

3. Thermal pollution

Thermal pollution is the addition of excessive quantities of heat to water as a result of industrial activities. Many biological organisms are seriously affected by temperature changes. For instance, trout and salmon will not survive at temperatures above 25 °C, while their eggs will only develop below 15 °C.

4. Chemical pollution

Various chemical compounds, including pesticides and heavy metal compounds, are directly poisonous to aquatic organisms. Some may kill the aquatic organisms directly, while others may slow down their growth rate or reproduction.

Crop spraying.

Most chemical pollutants come from industrial wastes and include heavy metals, cyanides, acids and alkalis. Everyday sources of heavy metals include lead from car batteries and platinum, palladium and rhodium from catalytic converters. If these metals are not recycled, they can find their way into the food chain. Some metal ions, such as lead, mercury and cadmium, are known to cause brain damage, while aluminium is associated with Alzheimer's disease.

5. Oil pollution

Oil pollution is mainly associated with oil spillages at sea or from large industrial users such as the petrochemical industry. Severe damage is often caused to a variety of marine animals and plants by various means, such as coating of fish gills and bird feathers, poisoning and deoxygenation.

1989 – Oil spillage in Alaska. Cleaning up consisted of washing, manual removal, mechanical treatment and bioremediation. Bioremediation occurs when bacteria feed on oil and convert it to carbon dioxide and water.

1992 – Three years on, widespread shoreline recovery. In 1990 and 1991 pink salmon harvests were the largest and second largest on record.

265

Water Treatment

Water is treated for the following reasons:
• to provide clean water for domestic use (drinking, washing, cooking) and industrial use (cooling, cleaning, dissolving)
• to control agricultural (silage, slurry, fertilisers) and industrial wastes (chemicals, heavy metals)
• to treat domestic sewage.

Water for public use (drinking water) is treated in a different way to sewage. Water is treated in three stages depending on the level of pollution of the water and on where the water is being discharged.

The three stages are: (a) primary treatment, (b) secondary treatment and (c) tertiary treatment.

(a) Primary Treatment

This involves passing the water through screens which remove large solids. Smaller solids are allowed to settle and fall to the bottom by gravity settlement in large tanks, pits or lagoons. The water is then passed through sand filters which remove very fine solids.

(b) Secondary Treatment

Secondary treatment involves various techniques which depend on the quality of the water being treated. The order in which treatment is carried out depends on the particular effluent.

The techniques include: (i) pH adjustment, (ii) chemical oxidation of toxins, (iii) flocculation and solid separation, (iv) biological oxidation and (v) chemical precipitation.

(i) pH Adjustment

Water may be acidic or basic depending on the particular substances dissolved in it. For example, the water may be too acidic as a result of chlorination. Lime is usually added to neutralise acidic water. In fact, some countries, such as Norway and Sweden, add large quantities of lime to rivers and lakes in order to neutralise the effects of acid rain.

(ii) Chemical Oxidation

Many toxic substances, such as cyanides, are removed from water by oxidation with chlorine gas or sodium hypochlorite (bleach).

(iii) Flocculation or Coagulation

Tiny suspended solids in water are removed by first making the tiny particles stick together or coagulate and form a larger solid. The coagulated material or flocculate (floc) then sinks to the bottom where the solid is separated from the water by gravity settlement or by filtration. The particles are usually organic substances which are negatively charged at their surface. This causes the particles to repel each other. When chemicals containing large positive ions, such as Al^{3+} or Fe^{3+}, are placed into the water they make the tiny particles coagu-

---

Dublin corporation: Providing enough clean water until the end of the 21st century.

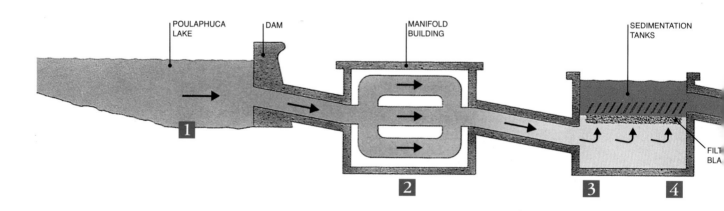

POULAPHUCA LAKE　　DAM　　MANIFOLD BUILDING　　SEDIMENTATION TANKS　　FILT BLA

1. Most of the water supply for Dublin comes from Poulaphuca lake. It is sent via pipelines into the treatment plant at Ballymore Eustace.

2. The water is distributed into the manifold building where flocculating agents are added to maximise the elimination of suspected solid impurities.

3. The chemically treated water flows upwards into the sedimentation tanks where impurities are removed.

late or flocculate into larger particles. Aluminium sulphate, $Al_2(SO_4)_3.18H_2O$, and iron(III) chloride, $FeCl_3$, are typical coagulants. Other materials called polyelectrolytes are added which help the colloidal material to settle more quickly.

(iv) Biological Oxidation

Waste organic material in water is removed mainly by biological oxidation using bacteria which convert the organic material into carbon dioxide and inorganic ions. The BOD is reduced to a large extent as a result.

The process requires a dense bacterial culture and a good supply of oxygen. The waste water is continually aerated by stirring or by air injection. The aerobic bacteria feed on the waste material and convert it into harmless substances like CO_2 and H_2O. The water is then filtered and disinfected with chlorine. The remaining material, called sludge, is compressed in large presses to remove excess water and is then spread out on land or dumped at sea.

(v) Chemical Precipitation

Many heavy metals are removed by treating the water with lime. A precipitate of the metal hydroxide is formed, settles on the bottom, and is removed.

Phosphates can also be removed from water by precipitation with lime.

(c) Tertiary Treatment

The effluent resulting from secondary treatment will usually have a BOD level of approximately 20 ppm. The water may require further treatment before it is discharged into a water course because it may still have too high a BOD level, it may contain excessive amounts of phosphates or nitrates or it may contain undesirable ions.

Tertiary treatment usually involves:
(i) filtration through fine sand
(ii) filtration through activated charcoal to remove offensive odours such as H_2S
(iii) removal of ions using ion exchange resins.

Treatment of Drinking Water

In Ireland, drinking water is taken from rivers and lakes where the initial level of pollution is low (low BOD).
• Before it is collected in a reservoir, the water passes through screens which filter out any floating debris. It is then passed into the treatment process.
• Coagulation (flocculation) agents are added which make the fine organic particles stick together to form larger particles. These larger particles are allowed to settle in sedimentation or settling tanks and are then thickened and pressed into cakes by large presses and removed.

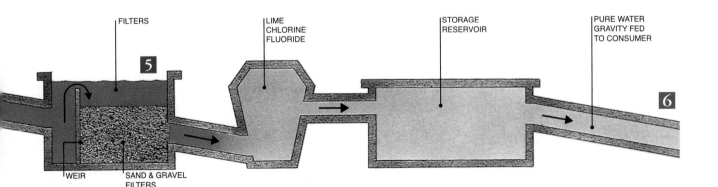

4. The water is removed through a series of decanting channels.

5. The water is filtered using rapid-gravity filters. Final chemical treatment including disinfection and fluoridation takes place.

6. The water is stored and fed by gravity feed to the consumer.

- Further impurities are removed by passing the water through specially graded sand beds. The sand beds are washed clean regularly by back washing with air.
- The pH is adjusted by adding lime to neutralise the water if it is too acidic.
- Chlorine or ozone is added to disinfect the water.

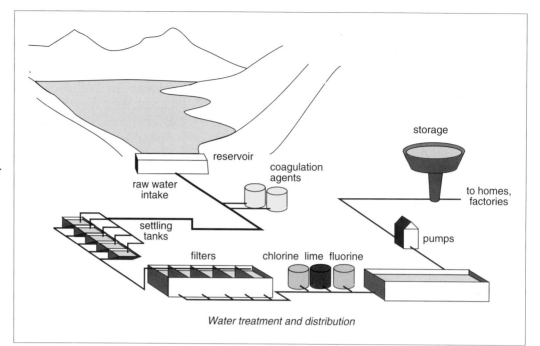

Water treatment and distribution

- Fluoride (hexafluorosilicic acid) is added to prevent tooth decay.
- The water is now fit for human consumption (in accordance with a European Union directive of 1980) and is fed by pumps or by gravity feed to various reservoirs before it is distributed for general domestic and industrial use.

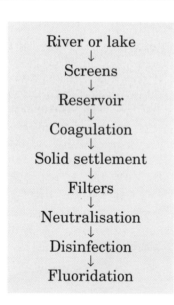

River or lake
↓
Screens
↓
Reservoir
↓
Coagulation
↓
Solid settlement
↓
Filters
↓
Neutralisation
↓
Disinfection
↓
Fluoridation

Nature's Monitors

Fish, such as rainbow trout, are used as biological monitors to sample the water. The water is passed through the fish tanks, where electrodes pick up impulses from the fish if they become agitated by any trace impurities.

Fish being used as biological monitors.

Treatment of Sewage

Sewage contains our general domestic waste products, such as washing water, human faeces and urine as well as anything else that happens to go down the toilet or drain. It has a high BOD level (300–400 ppm) and contains nitrates and phosphates which must be treated before it is passed out into the sea.

Sewage can be treated in three stages: primary, secondary and tertiary. The treatment used depends on how much the treated water is diluted when it is discharged into an inland waterway or into the sea. Inland towns usually require secondary treatment and sometimes tertiary treatment. Coastal towns which can discharge the treated water into the sea often need only primary treatment.

268

Sludge press

Primary Treatment

Large particles are removed from the sewage by passing it through screens. The waste water is allowed to settle in large tanks, where the suspended solids sink to the bottom as a sludge.

Secondary Treatment (Biological Oxidation)

The organic material in sewage is oxidised biologically by bacteria in activated sludge tanks or lagoons. Millions of bacteria feed on the organic waste suspended in the water and convert it into carbon dioxide and at the same time produce new cell material by rapidly reproducing themselves. This new mass of bacteria is called 'activated sludge' and is rapidly produced in the presence of a constant supply of oxygen. The activated sludge is allowed to settle in large sedimentation tanks or may be reintroduced into the effluent at the primary treatment stage. The settled solids are then concentrated and compressed to remove excess liquid and are either dumped at sea or spread out on land.

Most new sewage schemes which discharge water to inland waterways use secondary (biological) treatment.

Tertiary Treatment

The biologically treated water may still be relatively impure. In Ireland, sewage is rarely purified at the tertiary stage, because of the high costs involved. The water can be further purified by passing it through sand and graphite filters before it is discharged into the sea.

High levels of nutrients, like nitrates and phosphates, may need to be reduced. Phosphates are removed by adding aluminium or iron salts, which cause the phosphorus to settle out as a sludge.

An effluent quality of 20 ppm BOD and 30 ppm suspended solids is normally aimed for in sewage discharge from large urban areas into rivers.

European Union (EU) Limits

There are EU limits in relation to many substances which are discharged into water. The main limits concern heavy metals, phosphates and nitrates.

• **Heavy metals:** Some heavy metals are toxic to humans. Even at very low levels of concentration they may be hazardous to fish and may accumulate and pass along the food chain. Cadmium, chromium, copper, mercury, tin and many other metals are toxic to either man or fish or both. Pollution by heavy metals is caused by localised industries, such as electroplating and electronics. These industries and others require a licence before they can discharge waste into a waterway. Since 1994, new industries must apply to the Environmental Protection Agency for a licence. Existing industries are monitored by the relevant corporation or county council. The EU limits for metals vary, for example, the EU limit for cadmium is 0.005 ppm.

• **Nitrates:** Nitrates in water come mainly from artificial fertilisers. High nitrate concentrations in water are hazardous to infants, often causing the 'blue baby' syndrome. Surface water should contain less than 50 ppm nitrate.

• **Phosphates:** Phosphates are used widely in fertilisers and in detergents. Phosphates have no effect on human health but promote the growth of algal blooms. Values of more than 0.01 ppm of phosphates in water are thought to promote excessive algal growth.

Mandatory experiment 27

To determine the amount of dissolved oxygen (DO) in a water sample using the Winkler method.

The value obtained can be used to calculate the biological oxygen demand (BOD) of the sample.

Introduction

Winkler's reagent consists primarily of a mixture of concentrated solutions of manganese(II) sulphate and alkaline potassium iodide. In alkaline conditions the manganese(II) ions, Mn^{2+}, react with the hydroxide ions, OH^-, and form a white precipitate of manganese hydroxide, $Mn(OH)_2$.

$$Mn^{2+} + 2OH^- \Rightarrow Mn(OH)_2$$
white precipitate

If there is enough dissolved oxygen in the water, the white $Mn(OH)_2$ reacts with the oxygen in the water sample and forms a brown precipitate of manganese(III) oxide, Mn_2O_3.

$$4Mn(OH)_2 + O_2 \Rightarrow 2Mn_2O_3 + 4H_2O$$
brown precipitate

This brown precipitate is then acidified with concentrated sulphuric acid, where under these acidic conditions the Mn(III) ions are reduced to Mn(II) ions. At the same time the iodide ions, I^-, in the water are oxidised to free iodine, I_2.

$$Mn_2O_3 + 2I^- + 6H^+ \Rightarrow 2Mn^{2+} + I_2 + 3H_2O$$

The liberated iodine, I_2, can be determined in the usual way with sodium thiosulphate using starch as an indicator.

$$I_2 + 2S_2O_3^{2-} \Rightarrow 2I^- + S_4O_6^{2-}$$

From the above equations it can be seen that:

$$1 \text{ mol } O_2 \equiv 2 \text{ mol } Mn_2O_3 \equiv 2 \text{ mol } I_2$$
$$\equiv 4 \text{ mol } Na_2S_2O_3$$

Requirements

Safety glasses, two 250 cm³ stoppered bottles, pipettes, burette, white cardboard, water sample, manganese(II) sulphate solution (120 g of $MnSO_4.4H_2O$ in 250 cm³ of solution), alkaline potassium iodide (dissolve 125 g of NaOH and 37.5 g of KI in 250 cm³ solution), concentrated sulphuric acid, 0.02 mol L⁻¹ sodium thiosulphate solution, starch solution.

Procedure

(1) Rinse out two 250 cm³ bottles, first with deionised water and then with the water sample to be tested.
(2) Fill the bottles completely with the water sample. Ensure that there are no air bubbles. Stopper the bottles well.
(3) Using pipettes, transfer 1 cm³ of manganese sulphate solution and then 1 cm³ of alkaline potassium iodide solution to each water sample well below the level of liquid in each bottle.
(4) Restopper the bottles and mix well by inverting several times. Allow to stand and mix again by inversion. Allow to stand until the brown precipitate of Mn_2O_3 settles out.
(5) Carefully add 2 cm³ of concentrated sulphuric acid down the neck of each bottle. Stopper and mix well by inverting until all the solid dissolves. The yellow-brown colour of iodine should be noticed at this point.
(6) Titrate 100 cm³ samples of this solution with the 0.02 mol L⁻¹ sodium thiosulphate in the usual way using starch as an indicator near the end-point.
(7) Record your results using a table like the one below and calculate the amount of dissolved oxygen in ppm.

Results

	Titre 1	Titre 2	Titre 3	Titre 4
Volume Na₂S₂O₃				

Calculations

The concentration of iodine used is calculated in the usual way (see Example 9.2). The concentration of oxygen used up is exactly half the amount of iodine obtained,

i.e. $1 \text{ mol } O_2 \equiv 2 \text{ mol } I_2 \equiv 4 \text{ mol } Na_2S_2O_3$

The amount of dissolved oxygen (DO) is determined first in mol L⁻¹ and then converted to mg L⁻¹ and finally to ppm.

Determination of BOD – some comments

(1) The biological oxygen demand is a test carried over five days by first determining the amount of DO of the water sample.
(2) The sample is then kept in the dark to prevent photosynthesis and is incubated at

20 °C for five days, after which the amount of DO is redetermined.

(3) The test is not particularly suitable for rivers and streams because the laboratory conditions cannot mimic the actual river or stream conditions. The main use of the test is in measuring treatment of sewage and other effluents with high BOD.

(4) Samples to be tested may undergo changes during transit between the sampling stage and the laboratory. In order to minimise changes during transit and testing the sample should be kept at 4 °C.

(5) Samples with little or no dissolved oxygen should be diluted with deionised water and aerated, to increase the initial DO above the actual BOD.

(6) If the waste water does not contain any micro-organisms, some should be introduced to oxidise the waste. This is called seeding.

• • • • • • Example 9.2 • • • • • •

A sample of polluted water was analysed by Winkler titration. 300 cm^3 of the water liberated sufficient iodine to react with 9.0 cm^3 of 0.02 mol L^{-1} sodium thiosulphate solution. Calculate the amount of dissolved oxygen in the water in ppm.

$$1 \text{ mol } O_2 \equiv 2 \text{ mol } I_2 \equiv 4 \text{ mol } Na_2S_2O_3$$

$$\text{i.e. } 1 \text{ mol } O_2 \equiv 4 \text{ mol } Na_2S_2O_3$$

$$\frac{c_A V_A}{a} = \frac{c_B V_B}{b}$$

$$\frac{c_A \times 300 \text{ cm}^3}{1} = \frac{0.02 \text{ mol L}^{-1} \times 9.0 \text{ cm}^3}{4}$$

$$c_A = \frac{0.02 \text{ mol L}^{-1} \times 9.0 \text{ cm}^3}{4 \times 300 \text{ cm}^3}$$

$= 0.00015 \text{ mol L}^{-1} O_2 \longleftarrow$ change moles to grams using $m = nM$

$= 0.00015 \times 32 \text{ g L}^{-1} O_2 \longleftarrow M(O_2) = 32 \text{ g mol}^{-1}$

$= 0.048 \text{ g L}^{-1} O_2 \longleftarrow$ change g to mg by multiplying by 1000

$= 4.8 \text{ mg L}^{-1} O_2$

$= 4.8 \text{ ppm dissolved } O_2.$

• •

Exercise 9.2
•

(1) What are the main uses of water?

(2) Define 'water pollution'. Write a brief note on the main causes of water pollution in Ireland.

(3) Describe the main stages in the treatment of drinking water.

(4) 'Water in rivers and lakes is mainly polluted by deoxygenation and by eutrophication.'
(a) Explain clearly what is meant by
(i) deoxygenation and (ii) eutrophication.
(b) What are the main sources of the materials which cause deoxygenation and eutrophication? (c) Describe possible ways in which each of these can be controlled.

(5) (a) Why does sewage cause fresh water to become smelly and murky?
(b) Why are bacteria used to treat sewage and other forms of organic pollution?
(c) Why is water containing sewage continuously aerated at the treatment plant?

(6) (a) Describe what happens to a lake when large amounts of phosphate and nitrate ions are added to it from polluted waters. What are the EU limits for nitrates and phosphates in rivers and lakes?
(b) How does thermal pollution affect fish and other aquatic species?
(c) Why is it necessary to remove organic materials from water before it is returned to rivers and lakes?

(7) 'Water is treated in three main stages, primary, secondary and tertiary treatment.'
(a) Describe, briefly, each treatment stage.
(b) Describe in detail how water is purified by: (i) coagulation and (ii) biological oxidation.

(8) Describe the use of each of the following in relation to the treatment of drinking water:
(a) activated sludge (b) charcoal filters
(c) pH adjustment (d) aluminium sulphate
(e) settling tanks (f) lime.

(9) 'Chemical pollution is mainly caused by small localised industries.'
(a) Name the main pollutants from some industries.
(b) How are these industries controlled?
(c) State the value of an EU limit for a chemical species in water.

(10) Explain any possible effects on fish and aquatic life if each of the following is changed in a river or lake:
(a) pH (b) total suspended solids (c) amount of dissolved oxygen.

(11) (a) What is BOD? What are the main causes of BOD?
(b) List the main precautions taken when testing for BOD.

(12) The amount of dissolved oxygen in a water sample is usually determined by the Winkler titration. Describe:
(a) the colour changes that take place during each stage of the determination
(b) the main precautions taken during the determination.

(13) A water sample was taken from a canal and was analysed for the amount of dissolved oxygen using Winkler's method. The main reactions involved are:

$$Mn^{2+} + 2OH^- \Rightarrow Mn(OH)_2$$

$$4Mn(OH)_2 + O_2 \Rightarrow 2Mn_2O_3 + 4H_2O$$

$$Mn_2O_3 + 2I^- + 6H^+ \Rightarrow 2Mn^{2+} + I_2 + 3H_2O$$

(a) Give reasons for the precautions which should be taken when collecting the water sample.
(b) Describe how the Mn^{2+} and the OH^- solutions should be added to the water sample. Describe how the colour changes as each is added to the water sample.
(c) If 100 cm^3 of the treated sample was titrated against 0.02 mol L^{-1} sodium thiosulphate solution and reacted completely with 4.5 cm^3 of the latter, calculate the amount of dissolved oxygen in the water sample.

(14) A water sample was taken from a river and was analysed for the amount of dissolved oxygen using Winkler's method. The main reactions involved are:

$$Mn^{2+} + 2OH^- \Rightarrow Mn(OH)_2$$

$$4Mn(OH)_2 + O_2 \Rightarrow 2Mn_2O_3 + 4H_2O$$

$$Mn_2O_3 + 2I^- + 6H^+ \Rightarrow 2Mn^{2+} + I_2 + 3H_2O$$

A bottle was filled to the brim with some of the water sample; two concentrated solutions were added well below the surface of the water in the bottle. The bottle was stoppered and the contents mixed well by inverting several times. A brown precipitate formed when the mixture settled. Some concentrated sulphuric acid was then added down the neck of the bottle. The brown precipitate dissolved and a yellow-brown element was released into the solution.
100 cm^3 samples of the solution was titrated against 0.02 mol L^{-1} sodium thiosulphate solution; the mean titre was 4.0 cm^3.
(a) Why was the bottle filled right up to the brim with the water sample?
(b) Name the two concentrated solutions which were added at the start of the experiment.
(c) How were the solutions added below the surface of the water? Why were they added below the surface of the water?
(d) Sometimes a white precipitate is formed instead of a brown precipitate. Explain, using equations, how each precipitate may form.
(e) Why was concentrated sulphuric acid added?
(f) Calculate the amount of dissolved oxygen in the water sample both in g L^{-1} and in ppm.
(g) Explain why you think that the river from which the sample was taken could support fish such as trout or salmon.

(15) (a) Explain the terms:
(a) effluent (b) sewage (c) eutrophication.
(b) 1 L of water was found to contain 0.126 g of total suspended solids. Express this result in ppm.
(c) State one effect of eutrophication.
(d) What principles are involved in the primary treatment of sewage?

(16) (a) Write a brief note on the primary, secondary and tertiary treatment of sewage.
(b) A town of 1000 people discharges raw sewage into a river containing 10 ppm of oxygen. The average daily sewage from one person requires 60 g of oxygen from the water.
(i) How many milligrams of oxygen are present in each litre of water of river water?
(ii) What is the total demand for oxygen of the town's sewage in kilograms?
(iii) How many litres of oxygen are completely deoxygenated by the town's sewage output?

(17) (a) Flocculation, chlorination, pH adjustment and chemical precipitation are all involved in secondary treatment of water. Write a brief note on each process.
(b) Sewage treatment involves three stages, (i) primary, (ii) secondary and (iii) tertiary treatment. Explain, briefly, the procedures involved in each stage.

(18) (a) Explain why most coastal towns and cities in Ireland require only primary sewage treatment, while inland towns require further treatment.
(b) Why is tertiary treatment of sewage costly?

(19) (a) What use is made of each of the following substances in the production of water for drinking?
(i) aluminium sulphate (ii) chlorine (iii) dilute sulphuric acid (iv) hexafluorisilicic acid.
(b) Outline the processes which take place in the secondary (biological) treatment of sewage.
(c) Some sewage treatment plants carry out a tertiary process which further reduces the levels of certain substances. What are these substances and why is the process necessary?

(20) A 100 cm³ sample of polluted water was diluted to 2 L using freshly oxygenated water. 300 cm³ of this water liberated sufficient iodine to react with 18.0 cm³ of 0.02 mol L⁻¹ sodium thiosulphate. The remaining water was immediately incubated at 20 °C for five days, the usual precautions having been taken. The amount of dissolved oxygen was redetermined and the mean thiosulphate titre was 6.0 cm³.
(a) Describe some possible sources of error (other than titration errors) during the determination of the BOD.
(b) Why was the water first diluted and then aerated?
(c) Calculate the BOD of the original water sample.

(21) To find the biochemical oxygen demand (BOD) of a sample of polluted river water, 25 cm³ of the water was diluted to 1 litre with well-oxygenated water. Two bottles, A and B, were filled with the diluted water and their dissolved oxygen concentrations were determined. The analysis was carried out immediately for bottle A and five days

later for bottle B. The results obtained were 12.8 ppm and 8.2 ppm, respectively.
(a) Why was it necessary to dilute the polluted river water?
(b) What was the advantage of using well-oxygenated water for this purpose?
(c) Under what conditions should bottle B have been kept for the five days before it was analysed? In the case of one of the conditions you have stated, explain why it is necessary.
(d) What is the BOD of the polluted river water?

9.3 pH SCALE

Many substances dissolve in water, and this can make the water either acidic or basic. The pH scale tells us whether an aqueous solution is acidic or basic.

Self-Ionisation of Water

Water is an amphoteric (amphiprotic) substance that can react with itself.

$$H_2O(l) + H_2O(l) \rightleftharpoons H_3O^+(aq) + OH^-(aq)$$

Even the purest water reacts with itself to produce a small number of ions which can conduct an electrical current.

The equation for the ionisation of water given above can be simplified to:

$$H_2O(l) \rightleftharpoons H^+(aq) + OH^-(aq)$$

If the equilibrium law is now applied to this expression:

$$K_c = \frac{[H^+][OH^-]}{[H_2O]}$$

As very few water molecules are dissociated into ions, the concentration of the water molecules can be taken as a constant, that is $[H_2O]$ = constant.

The equilibrium constant, K_c, can now be rewritten as another constant, K_w,

$$K_c[H_2O] = K_w = [H^+][OH^-]$$

where K_w is called the ion product of water.

K_w has been determined by experiment and is found to be 1.0×10^{-14} mol² L⁻² at 298 K.

It should be noted that K_w is an equilibrium constant and therefore changes as temperature changes.

273

In pure water all the H^+ and OH^- ions come from the dissociation of water molecules. The balanced chemical equation tells us that if x mol of H^+ ions are produced then x mol of OH^- ions are also produced; therefore since

$$K_w = [H^+][OH^-] = 1.0 \times 10^{-14} \, mol^2 \, L^{-2}$$

and $[H^+] = [OH^-]$

then $[H^+]^2 = 1.0 \times 10^{-14} \, mol^2 \, L^{-2}$

Therefore, $[H^+] = 1.0 \times 10^{-7} \, mol \, L^{-1}$

and $[OH^-] = 1.0 \times 10^{-7} \, mol \, L^{-1}$

If an acid is added to pure water the concentration of H^+ ions increases above 1.0×10^{-7}. However, the ion product must remain at 1.0×10^{-14}; consequently the OH^- concentration must decrease. If a base is added to pure water the opposite happens; the H^+ concentration decreases and the OH^- concentration increases.

pH and the pH Scale

The pH of a substance tells us how acidic or how basic that substance is.

The product of the concentrations of the hydrogen ions and the hydroxide ions is always constant, for dilute solutions at a stated temperature.

We can calculate the concentration of hydrogen ions in a solution if we know the concentration of hydroxide ions. We can also measure the concentration of hydroxide ions if we know the concentration of hydrogen ions.

It is easier to express small numbers as logarithms because the logarithms of small numbers are positive. The use of logarithms in relation to the concentration of acids is the basis of the pH scale, which is a measure of the acidity or basicity of a substance.

pH is defined as

$$pH = -\log [H^+]$$

A similar expression exists which measures the OH^- concentration and is defined as

$$pOH = -\log [OH^-]$$

The relationship between pH and pOH can be derived from the ion product of water, K_w.

$$K_w = [H^+][OH^-] = 10^{-14}$$

By taking the log of each term

$$pH + pOH = 14$$

Some pH values are given below.

H^+	10^0	10^{-3}	10^{-7}	10^{-10}	10^{-14}
pH	0	3	7	10	14

The pH scale is used to record the strength of an acid in solution. The scale ranges from 0 to 14; an acid would have a pH below 7, while a base would have a pH above 7. Substances which are neither acidic or basic are neutral and have a pH equal to 7.

The pH scale is summarised below.

pH 1 2 3 4 5 6   7   8 9 10 11 12 14
Acidic solutions   N   Alkaline solutions
                    e
                    u
                    t
                    r
                    a
                    l
Strong acid Weak acid   Weak alkali Strong alkali
   HCl CH_3COOH H_2O NH_3 NaOH

pH and Indicators

Acids and bases have an effect on a group of coloured substances called indicators. If a solution is acidic (pH < 7) it turns an indicator a particular colour, whereas if the solution is basic (pH > 7) it turns the indicator a different colour.

Indicators can be extracted from plants such as lichens, blackberries and red cabbage. Litmus is an indicator which is extracted from certain kinds of lichen. Blue litmus turns red when acid is added to it and turns back to blue when the effect of the acid is neutralised by a base.

Indicator	Colour in acid solution	Colour in basic solution
Litmus	Red	Blue
Red cabbage extract	Pink	Green

Universal indicator is made from a mixture of indicators and is used to estimate pH readings. It can be used as a solution or in paper form. The pH of the sample to be tested is

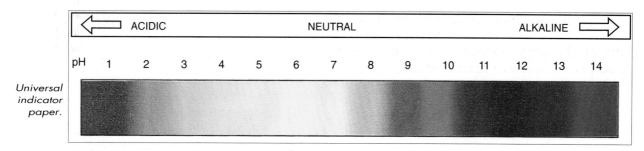

	ACIDIC					NEUTRAL						ALKALINE	

pH 1   2   3   4   5   6   7   8   9   10   11   12   13   14

Universal indicator paper.

found by dipping the universal indicator paper into the sample. The colour obtained is then compared with a scale of colours supplied with the indicator paper. The particular colour obtained will correspond to a particular pH.

Limitations of the pH scale

The pH scale goes from 0 to 14. Values outside this range are those of very strong acids and strong bases.

The highest hydrogen ion concentration on the pH scale is 1. This concentration gives a pH of 0. Higher concentrations than 1 give a negative pH reading which is off the pH scale.

For example, the pH of a 10 mol L^{-1} HCl solution is equal to –1. Hydrogen ion concentrations greater than 1 always give negative pH values. Consequently, the pH scale is limited to dilute aqueous solutions.

· · · · · Example 9.3 · · · · · ·

Calculate the pH of a 0.1 mol L^{-1} solution of hydrochloric acid.

Hydrochloric acid is a strong acid and is almost fully dissociated in aqueous solution.

$$HCl(aq) \Rightarrow H^+(aq) + Cl^-(aq)$$
$$1 \text{ mol} \qquad 1 \text{ mol}$$

Therefore

$$[H^+] = [HCl] = 0.1 \text{ mol L}^{-1}$$

$$pH = -\log [H^+]$$

$$= -\log 0.1$$

Using an electronic calculator,

$$pH = - (-1) = +1$$

· ·

· · · · · Example 9.4 · · · · · ·

Calculate the pH of a 0.01 mol^{-1} solution of sulphuric acid.

Sulphuric acid is a strong acid and is almost 100% dissociated in aqueous solution.

$$H_2SO_4(aq) \Rightarrow 2H^+(aq) + SO_4^{2-}(aq)$$
$$1 \text{ mol} \qquad 2 \text{ mol}$$

Therefore,

$$[H^+] = 2 \times [H_2SO_4] = 2 \times 0.01 \text{ mol L}^{-1}$$

$$pH = -\log [H^+]$$

$$= -\log (2 \times 0.01)$$

$$= -\log (0.02)$$

Using an electronic calculator,

$$pH = -(-1.669) = +1.699$$

· ·

Hydrochloric acid, HCl, because it contains one replaceable H$^+$ ion, is called a monobasic (monoprotic) acid.

Sulphuric acid, H$_2$SO$_4$, because it contains two replaceable H$^+$ ions, is called a dibasic (diprotic) acid.

····· **Example 9.5** ····· ·

Calculate the pH of a solution containing 0.4 g of sodium hydroxide in 1 L of aqueous solution.

$$n(\text{NaOH}) = \frac{m}{M} = \frac{0.4 \text{ g}}{40 \text{ g mol}^{-1}} = 0.01 \text{ mol}$$

$$[\text{NaOH}] = \frac{n}{V} = 0.01 \text{ mol L}^{-1}$$

Now, sodium hydroxide is a strong electrolyte and is almost totally dissociated into ions in aqueous solution.

$$\text{NaOH(aq)} \Rightarrow \text{Na}^+\text{(aq)} + \text{OH}^-\text{(aq)}$$
1 mol   1 mol

Therefore,
$$[\text{OH}^-] = [\text{NaOH}] = 0.01 \text{ mol L}^{-1}$$

$$\text{pOH} = -\log [\text{OH}^-]$$

$$= -\log 0.01$$

Using an electronic calculator,

$$\text{pOH} = -(-2) = +2$$

As pH + pOH = 14,

then pH = 14 − 2 = 12

····················

····· **Example 9.6** ····· ·

Determine the concentration of hydrogen ions in a solution of hydrochloric acid with a pH of 10.6.

As $\text{pH} = -\log [\text{H}^+]$

$$10.6 = -\log [\text{H}^+]$$

Using an electronic calculator,

$$[\text{H}^+] = \text{antilog } (-10.6)$$

$$= 2.5 \times 10^{-11} \text{ mol L}^{-1}$$

····················

Weak Acids and Weak Bases

In everyday terms, we usually compare acidity and basicity with water. An acid has more hydrogen ions than water, while a base has fewer. Strong acids such as hydrochloric acid, HCl, have more hydrogen ions than weak acids such as ethanoic acid, CH_3COOH.

When acids are dissolved in water they dissociate into ions. A strong acid such as HCl is completely ionised (dissociated) in water.

$$\text{HCl(aq)} + \text{H}_2\text{O(l)} \Rightarrow \text{H}_3\text{O}^+\text{(aq)} + \text{Cl}^-\text{(aq)}$$

A weak acid is an acid which does not fully dissociate into ions in water. For example, ethanoic acid dissociates only very slightly in water.

$$\text{CH}_3\text{COOH(aq)} + \text{H}_2\text{O(l)} \rightleftharpoons \text{H}_3\text{O}^+\text{(aq)} + \text{CH}_3\text{COO}^-\text{(aq)}$$

Weak acids are easily recognised by the values of their dissociation constants, K_d, which are very small. The value for ethanoic acid is 1.75×10^{-5} at 298 K.

pH of Weak Acids and Weak Bases

In order to calculate the pH of a weak acid or a weak base, the value of the dissociation constant, K_d, for the particular acid or base must be known.

····· **Example 9.7** ····· ·

Calculate the approximate pH of a 0.1 M solution of ethanoic acid at 298 K, given that the value of the dissociation constant, K_d, is 1.75×10^{-5} mol L^{-1} at 298 K.

Ethanoic acid dissociates in aqueous solution as follows:

$$\text{CH}_3\text{COOH(aq)} + \text{H}_2\text{O(l)} \rightleftharpoons \text{H}_3\text{O}^+\text{(aq)} + \text{CH}_3\text{COO}^-\text{(aq)}$$

This is usually simplified to:

$$\text{CH}_3\text{COOH} \rightleftharpoons \text{H}^+ + \text{CH}_3\text{COO}^-$$

$$K_d = \frac{[\text{CH}_3\text{COO}^-][\text{H}^+]}{[\text{CH}_3\text{COOH}]} = 1.75 \times 10^{-5} \text{ at 298 K}$$

The concentration of the hydrogen ions, H^+, is calculated using the equilibrium law.

Concentration/ mol L^{-1}	CH_3COOH	$\rightleftharpoons$ H^+	$+$ CH_3COO^-
Initial	0.1	0	0
Change	$-x$	$+x$	$+x$
Equilibrium	$0.1 - x$	$+x$	$+x$

Using the equilibrium expression,

$$K_d = \frac{x.x}{0.1 - x} = 1.75 \times 10^{-5}$$

For dilute concentrations x is small and can be neglected compared with 0.1,

i.e. $0.1 - x \approx 0.1$

then

$$\frac{x.x}{0.1} = 1.75 \times 10^{-5}$$

$$x^2 = 0.1 \times 1.75 \times 10^{-5}$$

Using an electronic calculator,

$$x = 1.323 \times 10^{-3}$$

As $x = [H^+]$, then

$$pH = -\log [H^+]$$
$$= -\log (1.323 \times 10^{-3})$$
$$= 2.8785$$

● ●

The calculations in 9.7 and 9.8 can also be solved using the formula

$$pH = -\log [\sqrt{(c \times K_d)}]$$

where c is the concentration of the acid or base and K_d is the dissociation constant of the acid or base.

This formula should only be used when the student understands why an approximation is allowed.

● ● ● ● ● **Example 9.8** ● ● ● ● ●

Calculate the approximate pH of an aqueous 0.20 mol L^{-1} solution of ammonia, given that the value of the dissociation constant, K_d, is 1.8×10^{-5} mol L^{-1} at 298 K.

Ammonia dissociates in aqueous solution as follows:

$$NH_3(aq) + H_2O(l) \rightleftharpoons NH_4^+(aq) + OH^-(aq)$$

This is usually simplified to

$$NH_3 \rightleftharpoons NH_4^+ + OH^-$$

$$K_d = \frac{[NH_4^+][OH^-]}{[NH_3]} = 1.8 \times 10^{-5} \text{ at 298 K}$$

The concentration of the hydroxide ions, OH^-, is calculated using the equilibrium law.

Concentration/ mol L^{-1}	NH_3	$\rightleftharpoons$ NH_4^+	$+$ OH^-
Initial	0.2	0	0
Change	$-x$	$+x$	$+x$
Equilibrium	$0.2 - x$	$+x$	$+x$

Inserting into the equilibrium expression

$$K_d = \frac{x.x}{0.2 - x} = 1.8 \times 10^{-5}$$

For dilute concentrations x is small and can be neglected compared with 0.2,

i.e. $0.2 - x \approx 0.2$

then

$$\frac{x.x}{0.2} = 1.8 \times 10^{-5}$$

$$x^2 = 0.2 \times 1.8 \times 10^{-5}$$

Using an electronic calculator,

$$x = 1.897 \times 10^{-3}$$

As $x = [OH^-]$, then

$$pOH = -\log [OH^-]$$
$$= -\log (1.897 \times 10^{-3})$$
$$= 2.7218$$

As $pH + pOH = 14$,

then

$$pH = 14 - 2.7218$$
$$= 11.2782$$

The dissociation constant, K_d, is often written as K_a, the dissociation constant of an acid, and as K_b, the dissociation constant of a base.

● ●

• • • • • • **Example 9.9** • • • • •

At 283 K, the ion product of water, K_w, is 0.30×10^{-14} mol^2 L^{-2}. Calculate the pH of pure water at 283 K.

$$H_2O(l) \rightleftharpoons H^+(aq) + OH^-(aq)$$

$$K_w = [H^+][OH^-] = 0.3 \times 10^{-14} \text{ mol}^2 \text{L}^{-2} \text{ at } 283 \text{ K}$$

For pure water $[H^+] = [OH^-]$

then

$$K_w = [H^+][H^+] = 0.3 \times 10^{-14}$$

$$[H^+]^2 = 0.3 \times 10^{-14}$$

$$[H^+] = 5.48 \times 10^{-6}$$

$$pH = -\log [H^+]$$

$$= -\log (5.48 \times 10^{-6})$$

$$= 5.2612$$

It should be noted that while the pH of pure water does not equal 7 at this particular temperature, the water is still neutral because $[H^+] = [OH^-]$.

• •

Theory of Acid–Base Indicators

A distinctive property of acids and bases is their ability to change the colour of indicators. Indicators are compounds that change colour in solution as the pH changes. Methyl orange, for example, is red in colour in solutions of pH below 3.1 and yellow in solutions of pH above 4.5. The colour of methyl orange varies from pink to orange between the values 3.1 and 4.5.

Indicators are usually weak acids or weak bases which are highly coloured in aqueous solution. The colour of the indicator depends on whether it is in its dissociated or in its undissociated form. This, in turn, depends on the pH of the solution.

An acid–base indicator is regarded as a weak acid, represented by the general formula HIn. If its conjugate base In$^-$ has a different colour to the acid HIn in solution, it can then be used as an indicator.

$$HIn(aq) \rightleftharpoons H^+(aq) + In^-(aq)$$
$$\text{colour A} \qquad\qquad \text{colour B}$$

If, for example, HIn is the litmus indicator, which is red, and In$^-$ is its conjugate base, which is blue, the equilibrium can be given as:

$$HIn(aq) \rightleftharpoons H^+(aq) + In^-(aq)$$
$$\text{red} \qquad\qquad\qquad \text{blue}$$

In accordance with Le Chatelier's principle, an increase in H$^+$ ions shifts the equilibrium to the left-hand side; consequently the red (acid) colour of the HIn predominates. On the other hand, if OH$^-$ ions are added the equilibrium will shift to the right-hand side; consequently the blue (alkaline) colour of the In$^-$ predominates.

The dissociation constant of an indicator, K_i, is written as

$$K_i = \frac{[H^+][In^-] \longleftarrow \boxed{\text{Blue}}}{[HIn] \longleftarrow \boxed{\text{Red}}}$$

It can be seen from this expression that the predominant colour clearly depends on the concentration of H$^+$ ions, i.e. if H$^+$ is in excess then the colour in this case of litmus is red, if it is not in excess then the colour is blue.

The acid–base indicator, litmus, can be extracted from lichens.

Indicator	Colour in acid	pH range	Colour in alkali
Methyl orange	Red	3.2–4.4	Yellow
Methyl red	Red	4.2–6.3	Yellow
Litmus	Red	5.0–8.0	Blue
Bromothymol blue	Yellow	8.0–9.6	Blue
Phenolphthalein	Colourless	8.2–10.0	Red

Choosing an Indicator

The pH range over which a given indicator works depends on the value of K_i, the ionisation constant of the indicator. For most indicators the range is approximately ±1 either side of the value of $-\log K_i$.

Phenolphthalein, for example, has a value of K_i equal to 1.0×10^{-9}. The mid-point of the colour change occurs when $[HIn] = [In^-]$.

As

$$K_i = \frac{[H^+][In^-]}{[HIn]} = 1.0 \times 10^{-9},$$

then

$$pH = -\log (1.0 \times 10^{-9})$$
$$= 9$$

Therefore, the indicator range for phenolphthalein is approximately 8–10, i.e one unit either side of 9.

Titration Curves

The progress of an acid–base titration can be observed by plotting the pH changes which occur against the volume of either acid or base added. The curve obtained is called a titration curve.

During an acid–base titration, the concentration of an acid or a base is determined by the volume of an acid or base which neutralises it. An indicator must be chosen which changes colour close to the end-point of the titration. The end-point may or may not be at pH 7. A suitable indicator must be chosen with a colour range which coincides with the end-point of the titration. Titration curves help us to select a suitable indicator for a particular titration.

When selecting an indicator for a titration we must know:
(i) the pH of the solution at the equivalence point
(ii) the pH range at which the indicator changes colour sharply.

Titration of a Strong Acid with a Strong Base

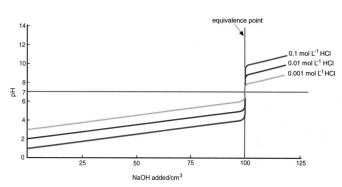

*pH changes during the titration of a **strong acid with a strong base***

When a strong acid, like hydrochloric acid, HCl, is titrated with a strong base, like sodium hydroxide, NaOH, the pH changes slowly at first as the base is added to the acid and then it changes very quickly between pH 4 and pH 10. An indicator must be chosen which changes colour within this range. Methyl orange (3.2–4.4) is just within this range, while phenolphthalein (8.2–10.0) and methyl red (4.2–6.3) are well within this range. These and many other indicators are suitable for strong acid–strong base titrations.

Titration of a Strong Acid with a Weak Base

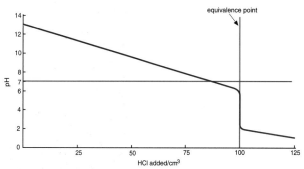

*pH changes during the titration of a **strong acid with a weak base**;*
0.1 mol L^{-1} HCl with 100 cm³ 0.1 mol L^{-1} NH₃ (aq)

When a strong acid, such as hydrochloric acid, HCl, is titrated with a weak base, such as aqueous ammonia, $NH_3(aq)$, the pH changes very quickly between pH 6.5 and pH 3.5. Methyl orange (3.2–4.4) and methyl red (4.2–6.3) would be suitable indicators for this type of titration.

Titration of a Weak Acid with a Strong Base

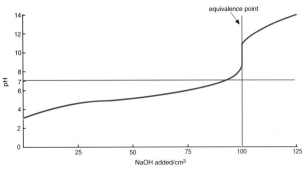

*pH changes during the titration of a **weak acid with a strong base**; 100 cm³ of 0.1 mol L⁻¹ CH₃COOH with 0.1 mol L⁻¹ NaOH*

When a weak acid, such as ethanoic acid, CH_3COOH, is titrated against a strong base, such as sodium hydroxide, NaOH, the pH changes very quickly between approximately 7.5 and 10.5. Phenolphthalein (8.2–10.0) is a suitable indicator for this type of titration.

Titration of a Weak Acid with a Weak Base

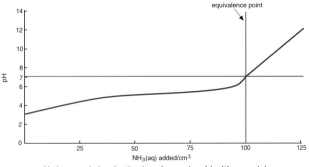

*pH changes during the titration of a **weak acid with a weak base**; 100 cm³ of 0.1 mol L⁻¹ CH₃COOH with 0.1 mol L⁻¹ NH₃ (aq)*

When a weak acid, such as ethanoic acid, CH_3COOH, is titrated against a weak base, such as aqueous ammonia, $NH_3(aq)$, there is no sharp change in pH. It is very difficult to detect the end-point using an indicator. The end-point is usually detected by using a pH meter.

Exercise 9.3

•

(1) (a) Define the terms pH and pOH.
(b) What is the normal range of the pH scale?
(c) Draw a clearly labelled diagram illustrating the pH scale. Indicate the position on the scale of one of each of the following: a strong acid, a weak acid, a strong base, a weak base.

(2) Explain each of the following terms:
(a) dissociate (b) strong acid (c) weak acid.

(3) Calculate the pH of each of the following solutions:
(a) 0.01 mol L^{-1} HCl (b) 0.01 mol L^{-1} H_2SO_4
(c) 0.01 mol L^{-1} NaOH

(4) Calculate the pH of the following solutions:
(a) 250 cm³ of a solution containing 0.4 g of sodium hydroxide
(b) 500 cm³ of a solution containing 3.65 g of hydrochloric acid.

(5) (a) What are indicators?
(b) What effect do indicators have on acids and bases?

(c) Name an indicator which can be extracted from a plant source. What is the colour of that indicator in acidic and in basic solution?
(d) What is universal indicator? How is it used?

(6) (a) What value must the hydrogen ion concentration, [H^+], be to have negative pH?
(b) What are the limitations of the pH scale?

(7) Define the ion product of water. If the value of the ion product of water is 1.0×10^{-14} at 298 K, calculate the fraction of molecules in pure water which are ionised, assuming that the concentration of un-ionised water molecules is 55.6 mol L^{-1} at 298 K.

(8) Calculate the pH of pure water at 323 K, given that the ion product of water, K_w, is 5.47×10^{-14} mol² L^{-2} at 323 K.

(9) Determine whether an aqueous solution which has a pH of 6.9 at 333 K is acidic or basic, given that the value of the ion product of water, K_w, is 9.55×10^{-14} mol² L^{-2} at 333 K.

(10) Calculate the hydrogen ion concentration in solutions which have the following pH values at 298 K:
(a) 10 (b) 9.40 (c) 1.40 (d) 4.82 (e) 0.05 (f) 7.7.

(11) Given the equilibrium system:

$$C_2H_5COOH + H_2O(l) \rightleftharpoons H_3O^+ + C_2H_5COO^-$$

(a) Write down the simple form for the dissociation constant of the acid, K_a.
(b) Calculate the pH of a 0.1 mol L^{-1} solution of the acid, given that the value of K_a is 1.35×10^{-5} at 298 K.

(12) Methanoic acid (formic acid) is the acid found in ants. Calculate the pH of a 0.02 mol L^{-1} solution of methanoic acid, HCOOH, if the value of the dissociation constant, K_a, is 1.8×10^{-4} at 298 K.

$$HCOOH(aq) \rightleftharpoons H^+(aq) + HCOO^-(aq)$$

(13) Calculate the pH of a 0.01 mol L^{-1} solution of iodic(V) acid, HIO$_3$, given that K_a is 0.17 mol L^{-1} at 298 K.

$$HIO_3 \rightleftharpoons H^+ + IO_3^-$$

(14) Calculate the pH of 0.02 mol L^{-1} nitrous acid, HNO$_2$, given that K_a is 4.7×10^{-4} mol L^{-1} at 298 K.

$$HNO_2 \rightleftharpoons H^+ + NO_2^-$$

(15) Given the following equilibrium:

$$NH_3(aq) + H_2O(aq) \rightleftharpoons NH_4^+(aq) + OH^-(aq)$$

(a) Write down the simple form for the dissociation of the base, K_b.
(b) Calculate the pH of a 0.4 mol L^{-1} solution of ammonia, NH$_3$, given that the value of K_b is 1.75×10^{-5} at 298 K.

(16) Morphine is a naturally occurring base, or alkaloid, which is used to relieve pain. Its formula may be represented as Mor and it ionises as follows:

$$Mor\ (aq) + H_2O(l) \rightleftharpoons Mor^-H^+(aq) + OH^-(aq)$$

Determine the pH of a 0.01 mol L^{-1} solution of morphine at 298 K, given that the value of K_b is 1.60×10^{-6} at 298 K.

(17) The degree of dissociation of ethanoic acid in a 0.1 mol L^{-1} solution is 0.013 at a set temperature. Calculate the value of the dissociation constant, K_a, at that temperature.

(18) If the pH of a 0.10 mol L^{-1} solution of ethanoic acid is 3.40 at a certain temperature, calculate the value of the dissociation constant, K_a, at that temperature.

(19) The dissociation constant of ethanoic acid is 1.75×10^{-5} mol L^{-1} at 298 K. Calculate the concentration of an aqueous solution of this acid if its pH is 2.00.

(20) Lactic acid, HC$_3$H$_5$O$_3$, is a weak monobasic acid found in sour milk. Calculate the value of the ionisation constant, K_a, for the acid, given that a 0.025 mol L^{-1} solution of the acid has a pH of 2.75.

$$HC_3H_5O_3 \rightleftharpoons H^+ + C_3H_5O_3^-$$

(21) The indicator phenolphthalein is a weak monobasic acid represented as HIn. Explain how it works in a titration of ethanoic acid against sodium hydroxide.

(22) When a weak acid was titrated against sodium hydroxide solution the equivalence occurred at a pH of 8.0. Illustrate the changes which occurred during the titration by drawing a titration curve. Explain which indicator is suitable for the titration.

(23) The pH of a 1.0 M solution of a weak monobasic acid is 3.7 at 293 K. Determine the value of the acid dissociation constant, K_a, at this temperature.

(24) (a) What is meant by the ion product of water?
(b) If the ion product of water at 330 K is 9×10^{-14}, calculate the hydrogen ion concentration in mol L^{-1}. Is the water neutral? Explain your answer.
(c) Explain why the pH concept is of little value outside pH 0–14.
(d) A solution contains 0.23 g of methanoic acid in 250 cm^3 of solution. Calculate its pH, given that the value of its dissociation constant, K_d, is 2.4×10^{-4}.

9.4 WATER ANALYSIS

Water can be analysed using various techniques. Some of the techniques listed below have already been examined in the text.

The main techniques include:
• tests for anions: Unit 2.2, pages 44, 45
• determination of total suspended and dissolved solids: Unit 9.1, page 259
• determination of pH: Unit 9.1, page 259
• estimation of dissolved oxygen: Unit 9.2, page 270
• atomic absorption spectra: Unit 1.4, pages 18, 19
• colorimetry.

Instrumental Methods of Analysis

pH Meter

A pH meter is an electrochemical cell. One electrode is a reference electrode and the other is the working electrode. The working electrode changes its potential as the pH of the electrolytic solution changes. The change in electrode potential is related to the change in pH.

pH meters are used to analyse many different types of water samples.
• Inland waters: pH must be measured regularly, as fish and their eggs will die if the water is too acidic.
• Drinking water: the pH of drinking water must be in the range 6–9 to comply with EU limits.
• Swimming pools: the pH of the water must be maintained in the range 7.4–7.8.

Atomic Absorption Spectrometry (AAS)

When a sodium salt is heated the flame appears yellow-orange. When sodium atoms are heated to a high temperature they become excited by absorbing energy. Afterwards, this energy is emitted as light of a definite wavelength in the visible region. This emission of light radiation can be precisely measured in an atomic absorption spectroscope. The orange colour emitted by sodium is due to an intense emission at a wavelength of 589 nm. Different metals have characteristic emissions at different wavelengths.

Some metal ions, such as cadmium, lead and mercury, are harmful, even in very low concentrations. Small concentrations of metals are usually measured using AAS.

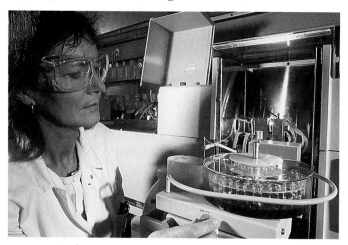
Atomic absorption spectrometer in use.

Colorimetry

Some reacting species are highly coloured. The colour intensity changes as the concentration of the coloured species decreases during the course of the reaction. The colour intensity can be measured using an instrument called a colorimeter. The level of chlorine in swimming pools is monitored using colorimetic methods. Chlorine reacts with certain chemicals producing a definite colour. The intensity of the colour can be measured against a known standard in a colorimeter.

Mandatory experiment 28

To estimate the amount of chlorine in a swimming pool using a colorimetric method.

Introduction

Swimming pool water contains bacteria and organic material which must be removed for health and sanitary reasons. These substances are usually removed by chlorination. The chlorine reacts with the water and forms an acid, hypochlorous acid, HOCl, which oxidises the bacteria and converts them into carbon dioxide and other harmless substances.

Many swimming pools are chlorinated using sodium hypochlorite, NaOCl, or calcium hypochlorite, Ca(OCl)$_2$, both of which are solids, while others use chlorine gas, Cl$_2$. Chlorine gas, Cl$_2$, is perhaps easier to use as the gas can be bubbled slowly into the water before it is pumped into the swimming pool. Solids, like sodium hypochlorite, are quite bulky and often messy to work with.

Too much chlorine in the water can cause itchiness, sore eyes, and on a large scale, can cause poisoning. The level of chlorine in a swimming pool must be constantly monitored: it is usually measured every hour, using a colorimeter or a comparator. In Ireland the health boards make regular random visits to swimming pools to check that the levels of chlorine are correct.

Swimming pool

Requirements

Safety glasses, Lovibond 2000 Comparator, DPD No. 1 and DPD No. 3 tablets to test for free and total chlorine, swimming pool water samples.

Procedure

(1) Place approximately 10 cm^3 of swimming pool water in one optical prism of the comparator.
(2) Place approximately 10 cm^3 of the DPD No. 1 solution in the second optical prism.

(3) Rotate the comparator disc and compare the colours.
(4) Read off the level of chlorine in the sample.

Results

If a reading of less than 1.5 ppm is obtained, the pool needs to be chlorinated, while if the level exceeds 2.0 ppm, the amount of chlorine needs to be reduced.

Exercise 9.4

•

(Note: most analysis questions are included in exercises 9.1, 9.2 and 9.3.)

(1) Describe, briefly, how each of the following instruments works:
(a) pH meter (b) atomic absorption spectrometer (c) colorimeter.

(2) 'pH meters can be used to test different types of water samples.' Name three different types of water samples and discuss why their pH needs to be monitored.

(3) Describe, briefly, how heavy metals such as lead and cadmium are analysed in water samples.

(4) 'Fertilisers contain high levels of nitrates and phosphates.'
(a) Why are there European Union (EU) limits for nitrates and phosphates in water?
(b) How do nitrates and phosphates find their way into water?
(c) How would you test a water sample for (i) nitrates and (ii) phosphates?
(d) How are high levels of nitrates and phosphates removed from water?

(5) 'Swimming pool water is chlorinated on a regular basis.'
(a) Why is the water chlorinated?
(b) Some pools are chlorinated with sodium hypochlorite, while others are chlorinated with chlorine gas. Discuss the advantages and disadvantages of both types of chlorination.
(c) Describe, briefly, how the level of chlorine in swimming pools is measured.

(6) 'Chlorine in swimming pools is often monitored using a comparator.'
(a) What is a comparator? On what principle does it work?
(b) Describe how you would analyse chlorine in water using a comparator.

(c) Are there limits for the amount of chlorine in water?

(d) Who monitors the amount of chlorine in swimming pools?

(7) How are simple experiments using flame tests similar to atomic absorption spectra?

▼▼▼ Key Terms ▼▼▼

Some of the more important terms are listed below. Other terms not listed may be found by means of the index.

1. Some properties of water: Water has several unique properties, such as a maximum in density at 4 °C, its extremely high heat capacity, its high melting point and boiling point, its high viscosity and many other eccentric properties.

Water is the universal solvent. There is hardly anything on Earth which does not dissolve in water to some extent. It is an excellent solvent for ionic substances and polar covalent substances.

Pure water is a very poor conductor of electricity.

2. pH: pH is defined as
$$pH = -\log [H^+]$$

3. pH of weak acids and weak bases:
$$pH = -\log[\sqrt{(c \times K_d)}]$$

where c is the concentration of the acid or base and K_d is the dissociation constant of the acid or base.

4. Total hardness of water: Hard water contains calcium ions, Ca^{2+}, and magnesium ions, Mg^{2+}. If water containing Ca^{2+} and Mg^{2+} ions is titrated with ethylenediaminetetraacetic acid (EDTA) a complex is formed.

$$1 \text{ mol } Ca^{2+} \equiv 1 \text{ mol EDTA}$$

Total hardness is usually expressed as ppm $CaCO_3$, even though other compounds, such as $MgCO_3$, may be dissolved in the water.

5. Water uses: These include domestic uses, industrial uses, agricultural uses and as a source of power.

6. Water pollution: Water pollution is the introduction by man, either directly or indirectly, of matter or energy into the water which alters the suitability of water for any of its beneficial uses.

7. Principal ways in which water is polluted: These are deoxygenation, eutrophication, oil pollution, thermal pollution and chemical pollution.

8. Typical BOD values: BOD, the biological oxygen demand, is the difference in the amount of dissolved oxygen in a water sample after 5 days. Typical values are:

Clean water	< 2 ppm (mg O_2 L^{-1})
Sewage effluent	300–400 ppm
Dairy washings	2000 ppm
Pig slurry	up to 35 000 ppm
Silage effluent	up to 67 000 ppm

9. Water treatment:
(a) Primary: screening, settling and filtering.
(b) Secondary: pH adjustment, chemical oxidation, flocculation, biological oxidation and chemical precipitation.
(c) Tertiary: sand and charcoal filtration and ion exchange.

10. Treatment of drinking water: The processes involved are: screening, filtration, coagulation, sedimentation, sand filtration, addition of lime, disinfection and fluoridation.

11. Treatment of sewage:
Primary: screening and solid settlement.
Secondary: biological oxidation and solid settlement.
Tertiary: filtration and ion exchange.

12. European Union (EU) limits: There are EU limits in relation to many substances which are discharged into water. The main limits concern heavy metals, phosphates and nitrates.

13. Winkler titration:

1 mol O_2 ≡ 2 mol Mn_2O_3 ≡ 2 mol I_2
≡ 4 mol $Na_2S_2O_3$

14. Water analysis: Water can be analysed using various techniques. The main techniques include:
(a) Tests for anions
(b) Determination of total suspended and dissolved solids
(c) Determination of pH
(d) Estimation of dissolved oxygen
(e) Atomic absorption spectra
(f) Colorimetry.

• Option 1 •

ADDITIONAL INDUSTRIAL CHEMISTRY AND ATMOSPHERIC CHEMISTRY

1A.1 GENERAL PRINCIPLES

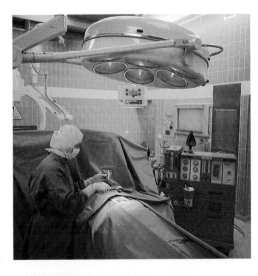

PVP–iodine being used in preparation for an operation.

Dyes to colour any occasion.

PVC for air inflated halls.

Plastic dispenser for detergents.

The chemical industry impinges on every aspect of our lives. Without the chemical industry many of us would not be alive today. Medicines and drugs have allowed us to live longer. New materials have enabled engineers and scientists to build and equip better hospitals and have allowed us to travel quickly to all parts of the world. The development of fertilisers, herbicides and pesticides has made it easier to feed the hungry, while the quality of food has been maintained for longer periods due to preservatives such as anti-oxidants. The use of petrochemicals to produce environmentally friendly fuels, plastics and polymers allows us to do many different things such as heating and insulating our homes or travelling to the moon. Advances in chemistry have purified our most common chemical, water, have allowed us to communicate with each other via computers and fibre optics, have increased our standards of hygiene through the use of disinfectants, detergents, deodorisers and other sanitary products, have clothed us in better materials by mixing natural and syn-

285

thetic fibres, have probably enhanced every aspect of our lives and have enabled us to appreciate the breadth and beauty of our world.

We should remember that the risks associated with the chemical industry are far outweighed by the benefits.

The chemical industry can be roughly divided into two sections:
- **bulk chemicals**, such as sulphuric acid, nitrogen, oxygen and others, which are used as raw materials to make millions of other chemicals
- **fine chemicals** which are used to make speciality products such as pharmaceuticals, dyes, pigments and polymers.

Today, the majority of chemical processes are modern and highly automated. The development of an effective and successful chemical process relies on many factors which include the following:

Raw materials
Energy requirements
Costs
Location
Rate of reaction and yield of product
Co-products and by-products
Safety
Control of waste and effluents
Plant materials
Operation process and quality control.

Raw materials and preparation of feedstock

The reactants in a chemical process are called the **feedstock**. Different processes require different chemical feedstocks or raw materials. The main raw materials include water, air, petrochemicals, natural gas, metallic ores and minerals. Raw materials are converted into feedstocks and treated for further use by many different purification and separation processes.

For example, metal ores are crushed, separated, refined and converted into pure metals before they are used in a particular process.

In the production of ammonia, the main feedstocks, nitrogen and hydrogen, are first extracted from the raw materials: methane must be converted into hydrogen, and air must be distilled to produce nitrogen. Other chemical and physical processes are used to ensure that there are no catalytic poisons, that the water is sufficiently pure and that the reactants are mixed in the correct proportion.

Water is used as a raw material for many chemical reactions. It is also used to heat and cool other chemicals in a reaction and is used as a solvent. Because it is such a good solvent, water contains many impurities. These must be removed as they can cause corrosion, deactivate catalysts or cause a build-up of scale in boilers.

Energy requirements

Electricity generating. Power station and jetty at Moneypoint, Co. Clare.

The optimum use of energy is often the factor which determines whether a production process is profitable or not. Energy sources, such as electricity, oil, natural gas and coal, are all expensive. Therefore, it is important to balance energy-producing processes with energy-demanding processes. Exothermic reactions and expansion of gases leaving pressurised vessels produce heat, while endothermic reactions and compression of gases require heat. Chemical companies reduce heating costs by recycling the energy from exothermic reactions and use it to raise the temperature of endothermic reactions. In large plants, energy (as steam or hot gases) is transferred to different parts of the plant using heat exchangers. Energy costs are usually higher in small plants.

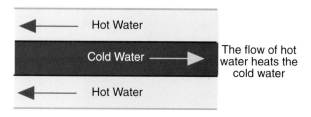

The flow of hot water heats the cold water

Heat exchange

286

Costs

Capital costs are costs involved in setting up the chemical plant. These include research costs, the construction of buildings, reaction vessels, office and services equipment and many other initial costs. Some countries give **grants** to chemical companies to help them to set up. In return, the company helps the local and national economy by employing a range of highly skilled people. When a plant is ready for production, other costs are involved. Raw materials, energy, services (electricity, gas, oil, water), salaries, rates, rental and plant depreciation are all **production costs**. Some production costs may be **fixed**, like salaries, rates and plant depreciation, while other costs are **variable**, such as power and raw materials.

Location

The location of a chemical plant depends on a variety of factors: some are historical, others are practical.

Transporting chemicals by rail, BASF, Ludwigshafen.

Historical factors include the following:
• raw materials available locally
• nearness to ports, road and rail for delivery of raw materials and transport of the final products
• available markets and sales outlets
• skilled and unskilled workers available (employees tend to be more technically skilled than in other manufacturing industries)
• availability of cheap energy and water
• facilities for disposal of effluent. Processes which produce solid waste require land for tip-heads; those producing liquid effluent require tidal rivers or estuaries.

Practical factors include the following:
• environmental and social: the existence of similar industries often helps in establishing new companies even though they may not be economically viable. Sometimes plants are located in regions of high unemployment because of the social benefits.
• government inducements through grants for capital investment, reduction in taxes, and exemption of rates and other charges.

Rate of reaction and yield of product

The speed of a chemical reaction is determined by the concentration of the reactants, temperature, pressure and catalysts, while the yield of product is determined by concentration of reactants, temperature and pressure.

Temperature and pressure are variables which dictate the speed of the reaction and the amount of product formed. It might seem sensible to use high temperatures and pressures to speed up and increase the yield in all chemical reactions. However, it is expensive to use high temperatures and pressures. Sometimes the use of high temperature or high pressure may even cause a decrease in the yield of product.

Consider the conditions used in the main step in the synthesis of ammonia. The reaction is a equilibrium reaction and the conditions which favour the production of ammonia are high pressure and low temperature.

$$N_2(g) + 3H_2(g) \rightleftharpoons 2NH_3(g); \Delta H = -92 \text{ kJ mol}^{-1}$$

The use of high pressure is expensive and the use of low temperature slows down the rate of the reaction too much. A compromise is reached where moderate pressures and temperatures are used to produce a moderate amount of ammonia at a reasonable rate. The use of a catalyst helps to speed up the reaction rate, while the percentage yield is increased by recycling the unreacted nitrogen and hydrogen.

Co-products and by-products

Many chemical reactions give more than one product. The manufacturer may be interested in producing only one product but the particular process may produce co-products or by-products. These may have saleable value and may make a significant contribution to the profit made. At other times, the co-products and by-products may have no commercial

287

value and may need to be separated and disposed of at a high cost.

When hydrogen is produced from natural gas, carbon dioxide is formed as a **co-product**. The CO_2 is often sold for use in carbonated drinks.

$$CH_4(g) + H_2O(g) = 3H_2(g) + CO_2(g)$$

In some reactions, other reactions (side reactions) occur along with the main reaction. Sometimes these side reactions produce useful **by-products**. At other times side reactions may produce by-products which are hazardous and may involve additional costs in separation and disposal.

Epoxyethane is used to make anti-freeze. It is made, in a one step reaction, by reacting ethene with oxygen in the presence of a silver catalyst. Sometimes the ethene is oxidised in a side reaction to form carbon dioxide and water, two by-products.

ethene + oxygen $\rightleftharpoons$ epoxyethane

ethene + oxygen $\rightleftharpoons$ carbon dioxide and water

Safety

Xn -harmful   T -toxic   T+ -very toxic

These substances are dangerous if they enter the body, either through the nose, mouth or the skin.

Xi -irritant

These substances irritate the eyes, throat, nose and skin. They may also be allergens.

C -corrosive

These substances are corrosive and will destroy living tissues (skin, mucous membranes).

F -flammable

The vapours of these substances will catch fire, at room temperature, especially in the presence of a spark or flame.

F+ -extremely flammable

The vapours of these substances are more readily flammable even at temperatures less than 0°C, especially in the presence of a spark or flame.

O -oxidising

These substances, when in contact with some other substances, particularly flammable ones, can start a fire or make a fire worse.

E -explosive

These substances will explode when exposed to heat, shock or friction.

Symbols used on labels and their meaning.

All industries must conform to the laws and guidelines given in the **Health and Safety Act**. The chemical industry in Ireland is also licensed and regulated by the **Environmental Protection Agency**. New chemical industries and extensions or modifications to existing industries are subjected to detailed hazard and environmental analysis. Existing industries must conform to regulations laid down by the local corporations or county councils.

Chemical plants should preferably be located downstream of the prevailing wind: this helps keep foul smells away from the local community. Hazardous gases and effluent must be treated before being released into the atmosphere or into rivers or the sea. Potential hazards must be constantly monitored. Today, most modern chemical plants are computer controlled and can be shut down automatically in the event of the release of hazardous gases or effluent.

Staff must be properly trained in all aspects of safety. They must be able to recognise hazards and know how to react to them. They must be aware of all the safety features in a chemical plant, such as eye-baths, showers, breathing apparatus, emergency plans and many other features.

Control of waste and effluents

Water is vital to the chemical industry. It is used as a coolant, as a reactant, as a solvent and as a cleaning agent. Water may become contaminated during the chemical process and there may be hazardous, harmful or poisonous material dissolved in it. This water has to be treated before it is released into the local rivers or the sea. Most of the largest chemical

Monitoring the operating process and checking quality control at Premier Periclase.

companies, like Bayer and BASF in Germany, have developed sophisticated biological waste water treatment plants. This has dramatically improved the quality of water in the Rhine: many species of fish live alongside the production plants. Waste water must be constantly monitored: pipeline junctions must be inspected, cooling waters must be measured and samples must be analysed using chromatography and biological analysis.

Chemical plants emit many gases, e.g. CO_2, SO_2, CO, NO_x and many organic compounds. Some cause environmental damage such as acid rain or damage to the ozone layer, while others, like dioxins, cause cancer. Waste gases are purified by the installation of gas scrubbers, particle separators, adsorbers, catalytic converters, condensers and incinerators. Emissions are constantly monitored and analysed using instrumental analysis.

Desulphurisation stack to remove SO_2 gas from coal in the central power station at BASF, Ludwigshafen.

Plant materials

Construction materials are chemicals. It is important that these do not react with other chemicals, such as feedstocks, catalysts, solvents and others involved in the chemical process. Poor choice of plant construction materials can lead to increased costs due to lower efficiency, contamination or hazardous reactions. Plant materials should be unreactive with reactants and products, resistant to corrosion, cheap and long lasting and be capable of withstanding high pressures and high temperatures.

Operating processes and quality control

The stages usually involved in a chemical process are shown in the diagram at the bottom of the page. The chemical reaction takes place in a **batch reactor** or a **continuous reactor**.

A **batch reactor** is a scaled-up version of laboratory equipment. The raw materials are put into the reactor and allowed to react. When the reaction is over the product is removed and the process is started over again with new raw materials. Batch reactors are more versatile than continuous reactors. They can be shut down easily and can be used for different chemical reactions. They are useful for preparing small batches of chemicals (pharmaceuticals and fine chemicals) but are expensive to run owing to the large number of staff required to operate, handle and clean out each batch.

Continuous processes require specially designed reactors because the reactants are constantly fed into the reaction vessels and the products are removed as they form. Large

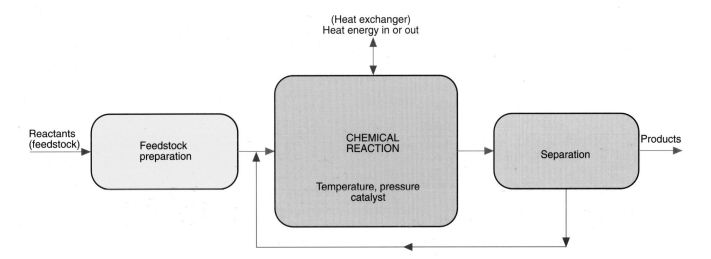

tonnage processes, like the production of ammonia or sulphuric acid, use continuous processes. They are computer controlled and monitored using instrumental analysis and therefore require a small workforce. They are more expensive to set up than batch processes owing to the need for tailor-made reaction vessels (mixers, reactors, separators, heat exchangers etc.) and because of high instrumentation costs. They cannot be shut down easily as closing down a continuous process is very costly and they are uneconomical to run at less than full capacity.

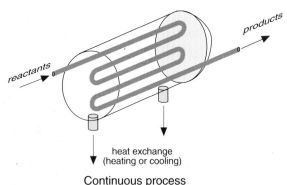

Continuous process

A semi-continuous process is a combination of batch and continuous processes. The main reaction process is usually a batch process, while the separation and purification stages may be continuous.

In each type of process the quality of the reactants and products must be constantly monitored. Chemical companies have staff whose main function is to control quality. Quality control is carried out by taking samples from the various stages of production and analysing them using instrumental analysis.

⬤ ⬤ ⬤ ⬤ ⬤ Example 1A.1 ⬤ ⬤ ⬤ ⬤ ⬤

Calculation of theoretical yield in tonnes

The iron ore, haematite, Fe_2O_3, is reduced to iron by reacting it with carbon monoxide in a blast furnace. How much iron can theoretically be produced from 10 tonnes of Fe_2O_3?

- $Fe_2O_3 + 3CO = 2Fe + 3CO_2$
  1 mol   3 mol   2 mol   3 mol

The balanced chemical equation tells us that 1 mol of Fe_2O_3 produces 2 mol of Fe.

- Number of moles (amount) of Fe_2O_3 used

$$n(Fe_2O_3) = \frac{m}{M} = \frac{1 \times 10^7 \text{ g} \leftarrow 10 \text{ tonnes}}{160 \text{ g mol}^{-1}}$$
$$= 6.25 \times 10^4 \text{ mol } Fe_2O_3$$

- The number of moles (amount) of Fe formed is twice the number of moles (amount) of Fe_2O_3, i.e. $2 \times 6.25 \times 10^4$ mol Fe $= 1.25 \times 10^5$ mol of Fe is produced

- Change this amount (number of moles) to mass using the equation $m = nM$
$= 1.25 \times 10^5$ mol $\times 56$ g mol^{-1} Fe
$= 7 \times 10^6$ g Fe = 7 tonnes Fe

⬤ ⬤ ⬤ ⬤ ⬤ ⬤ ⬤ ⬤ ⬤ ⬤ ⬤ ⬤ ⬤ ⬤ ⬤ ⬤ ⬤ ⬤ ⬤

Visit to a Local Chemical Plant

Many areas in Ireland have obvious chemical industries. Cork has many large chemical and pharmaceutical industries, Dublin has many pharmaceutical and fine chemical industries, while others like Drogheda have very specialised chemical industries. While your own area may not have an obvious chemical industry it should be remembered that chemical process take place in lots of areas such as:
- farming, dairying and horticulture
- food processing, bakeries and confectioners
- brewing and distilling
- painting, dyeing and printing
- energy production
- environmental control.

These and many other industries are very suitable places to visit and study. Your teacher should make a preliminary visit to the company and obtain some background information. A questionnaire (similar to the one below) should be made out before the visit and a report made out on the visit. The company should be thanked for their help and perhaps they may be interested in some feedback from your report.

Company_____
Location of company _____
Products made _____
Type of process _____
Type of people employed _____
Raw materials used_____
Source of raw materials _____
Preparation of feedstock_____
Energy requirements _____
Capital costs _____
Production costs _____
Factors affecting reaction rate and yield _____
Co-products and by-products_____
Safety _____
Waste and pollution control _____
Plant construction_____

1A.2 CASE STUDIES

Irish Chemical Industry

The Irish chemical industry is a relatively new industry. Some of the world's leading chemical companies have established production plants in Ireland since 1970. The chemical industry is the second largest industry (agriculture is the largest) in Ireland. The Irish chemical industry is the fastest growing chemical industry in the western world. Approximately 16 000 people, many of whom are highly skilled graduates, are employed in the chemical and pharmaceutical industry. There is very little production of bulk chemicals (sulphuric acid, nitric acid, petrochemicals etc.) owing to the lack of raw materials in Ireland. The industry is small and concentrates on producing high value chemicals like pharmaceuticals and fine chemicals (dyes, pigments etc.).

The Irish chemical industry can be roughly divided up into the following areas:

• **pharmaceuticals and fine chemicals:** this is the largest sector of the industry and produces a range of products like medicines and health care products.

• **agrochemicals and animal health products:** this sector produces agrochemicals such as fertilisers, fungicides and herbicides and animal health care products to control parasites and other illnesses in animals.

• **cosmetics and household products:** essential oils, perfumes, toiletries, cleansing materials and many other household products are produced.

• **bulk chemicals and plastics:** the only bulk chemicals produced are ammonia and nitric acid. Some plastics and polymers are produced using imported raw materials.

Many other areas, such as the brewing and distilling industries, are not regarded as part of the chemical industry. However, industries like brewing and distilling rely on chemical processes to produce their products. These and other industries in the food and drink area use chemical processes to produce goods which contribute greatly to the wealth of our country.

Crop protection in a rape field.

Plastics in medical technology.

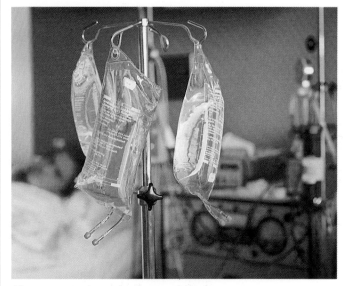

Pharmaceuticals used in haemodialysis.

291

Case Studies

A case study based on the Irish chemical industry must be studied.

Only *one* of the following three industrial processes must be studied.

Case Study 1: Manufacture of Ammonia and its Conversion to Urea

Ammonia, NH_3, is one of the most important chemicals manufactured: without ammonia, artificial fertilisers could not be made easily. In 1908, a German chemist, **Fritz Haber**, discovered the necessary reaction conditions to make nitrogen and hydrogen combine to give a reasonable yield of ammonia. A chemical engineer, **Carl Bosch**, developed the necessary manufacturing plant which enabled ammonia to be made cheaply. The process used by Haber and Bosch has changed very little and is essentially the same today.

Irish Fertilisers Industries (IFI) is the only manufacturer of bulk chemicals in Ireland. The other chemical companies produce pharmaceuticals, fine and speciality chemicals. IFI produces ammonia by a continuous process, using the **Haber process**, from natural gas and from atmospheric nitrogen. Most of the ammonia is used directly in the manufacture of fertilisers, while some of it is shipped by rail to another IFI plant in Arklow and converted into nitric acid which is used to manufacture more fertilisers.

• **Company:** Irish Fertiliser Industries Ltd, Marino Point, Cork

• **Products made:** Ammonia, carbon dioxide and urea

• **Location:** The plant is located in Marino Point, Cork, on the south coast, close to the Kinsale Head gas field. The final products can be distributed by sea from the company jetty or by rail and road. The local workforce is highly skilled in all aspects of chemical production. Cork has many well known chemical companies producing high quality products which require skilled personnel at all levels, from chemists and engineers, technicians and craftsmen, plant operatives who operate complex machines and instruments, to staff with sales and marketing skills.

• **Raw materials:** Natural gas, the source of hydrogen, is obtained from the Kinsale Head gas field, while nitrogen is extracted from the air by distillation. The other raw material, water, is obtained from the river or the sea.

• **Preparation of feedstock**

Production of hydrogen

Sulphur (a catalyst poison) is removed from the natural gas in a desulphurisation reactor.

The natural gas, CH_4, is reacted with water to produce hydrogen and carbon monoxide by steam reforming. The gas is cracked at a pressure of 3.5×10^6 Pa and at a temperature of 750 °C using a nickel catalyst

$$CH_4(g) + H_2O(g) \overset{\text{high temperature}}{\underset{\text{low pressure, Ni}}{\rightleftharpoons}} CO(g) + 3H_2(g);$$

$$\Delta H = +210 \text{ kJ mol}^{-1}$$

Air is then injected into the mixture and reacts with the hydrogen forming steam. The composition of nitrogen gas to hydrogen gas is now in the ratio of $N_2:3H_2$

$$2H_2(g) + O_2(g) \rightleftharpoons 2H_2O(g);$$

$$\Delta H = -482 \text{ kJ mol}^{-1}$$

The heat generated at this stage is used to drive turbines which are used to generate heat for reactions requiring energy. The carbon monoxide (a catalyst poison) is oxidised to carbon dioxide and hydrogen.

$$CO(g) + H_2O(g) \rightleftharpoons CO_2(g) + H_2(g)$$

The product gases now contain a mixture of nitrogen, hydrogen, steam and carbon dioxide.

Removal of carbon dioxide

The carbon dioxide and steam are removed by scrubbing the gas, by passing it through concentrated potassium carbonate.

$$CO_2(g) + K_2CO_3(aq) + H_2O(l) \rightleftharpoons 2KHCO_3(aq)$$

• Reaction system

Synthesis of ammonia

The pressure of the system is raised to over 2×10^7 Pa using a compressor and the temperature is lowered to 400–500 °C by heat exchange. The mixture of 3 volumes of hydrogen to 1 volume of nitrogen is reacted over an iron catalyst. Not all of the mixture forms ammonia at first but the unreacted gases are constantly recycled until they are combined.

$$N_2(g) + 3H_2(g) \rightleftharpoons 2NH_3(g); \Delta H = -92 \text{ kJ mol}^{-1}$$

Ammonia is removed by condensation and the unreacted nitrogen and hydrogen gases are returned to the reaction system. By recycling, yields of up to 98 % ammonia can be obtained.

• Factors affecting rate and yield

The basis of the Haber process is the reversible reaction:

$$N_2(g) + 3H_2(g) \rightleftharpoons 2NH_3(g); \Delta H = -92 \text{ kJ mol}^{-1}$$

In the reaction, the position of equilibrium depends on temperature and pressure. In this case, in accordance with Le Chatelier's principle, the highest yield of ammonia is obtained using a high pressure and a low temperature. However, the use of high pressure is expensive and the reaction rate is slow at low temperatures.

The optimum operating conditions involve the use of a catalyst and are:
Temperature: 380–450 °C
Catalyst: iron
Pressure: 200–300 atm
Promoter: potassium hydroxide

Urea synthesis

Urea, NH_2CONH_2, is used as a fertiliser and animal feed and in the manufacture of pharmaceutical products. Most of the urea produced is used in Ireland as fertilisers, while the rest is exported directly from the Marino Point jetty.

Anhydrous ammonia and pure carbon dioxide are combined at high pressure to produce ammonium carbamate, which is decomposed at low pressure to produce urea.

$$CO_2(g) + 2NH_3(l) \rightleftharpoons NH_2COONH_4(s)$$
ammonium carbamate

$$NH_2COONH_4(s) \rightleftharpoons NH_2CONH_2(s) + H_2O(g)$$
urea

• A yield of approximately 60% is obtained by recycling the unconverted ammonium carbamate. The aqueous urea solution is concentrated by evaporation under vacuum into a 99.7% urea melt. The molten urea is prilled, to produce droplets, by spraying downwards in a prilling tower. As the droplets fall through a rising air current, they cool and solidify into white spherical particles called prills.

The prills are then bagged and stored, ready for distribution by ship, road and rail.

IFI – prilling tower.

• **Co-products:** Carbon dioxide is produced as a co-product. Some is used to manufacture urea, while the rest is sold as CO_2, which is used to carbonate beers and soft drinks. IFI is the largest supplier of CO_2 in Ireland. The company also manufactures carbon black, which is used to provide anti-static characteristics to electrical cables and magnetic tapes.

• **Capital costs and production costs:** The capital costs involved were extremely high owing to the scale of the plant and the extent of the technology involved. The plant was constructed with the aid of Government grants and went into production in 1979. Production costs are also high, these are offset by the utilisation of waste gases, through direct interchange for the various reactions and for driving the steam turbines. Of the total energy demand for turbines, pumps and other electrical equipment, only 20% is derived from electrical power. The recovery of heat is perhaps best illustrated by the fact that temperatures of approximately 1000 °C are required for reforming, while the condensation of ammonia is carried out at –33 °C. Large chemical plants are not very economical and require constant renewal of equipment. Large plants usually have a lifetime of approximately 20 years.

• **Energy requirements:** Large supplies of energy are required for most of the production stages in the synthesis of ammonia. Costs are minimised by the use of heat exchangers.

• **Quality control:** The work of the quality control and research departments ensures that the products produced conform strictly to specification.

• **Safety:** The synthesis of ammonia is not an environmentally friendly process. The chemical processes can release harmful gases and effluents into the atmosphere and sea. The site is enormous, covering many acres and extending high into the sky. Extreme care and vigilance are required at all times to monitor potential hazards. Staff are constantly trained on site in all aspects of safety and hazards are constantly monitored. The entire plant is computer controlled. This allows all processes to be visually monitored and controlled at all times.

• **Waste and pollution control:** Strict environmental controls are enforced at each stage of the production process to ensure that emission of gases and waste water conforms to environmental regulations.

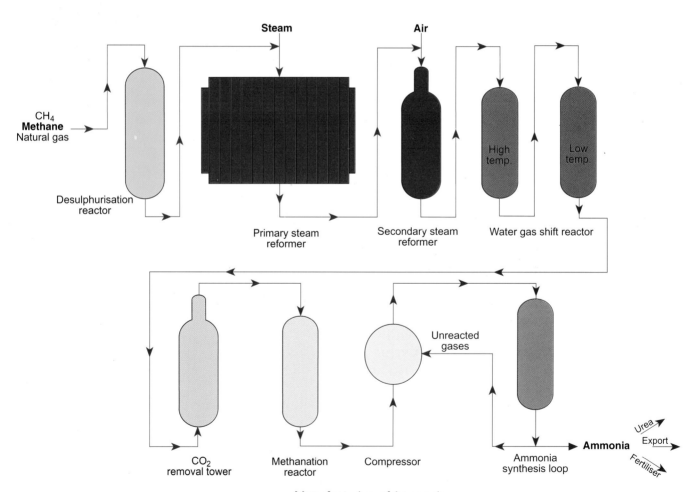

Manufacturing of Ammonia

Case Study 2: Manufacture of Nitric Acid

About 10% of the ammonia produced is oxidised using a catalyst to manufacture nitric acid. The nitric acid manufactured by IFI in Arklow is used to make urea and CAN fertilisers.

- **Company:** Irish Fertiliser Industries, Arklow

- **Products made:** Nitric acid and calcium ammonium nitrate (CAN) and urea fertilisers

- **Location:** The nitric acid plant is situated in Arklow, on the east coast. The original plant started production in 1965. Since 1982, only nitric acid and CAN fertilisers have been produced at the plant. The plant is situated close to a port and is well served by rail and road networks. The main raw material, ammonia, is taken by rail from IFI's Marino Point plant to Arklow. The local area is rich in human resources who are highly skilled in all aspects of chemical production.

IFI Arklow – nitric acid plant.

- **Raw materials:** Ammonia, air, water and a catalyst, composed of platinum and rhodium, react to form nitric acid.

- **Preparation of feedstock**
Pure ammonia is sent to the nitric acid plant from Marino Point and requires no further treatment. Air is used to oxidise the ammonia. A platinum–rhodium gauze catalyst is used to catalyse the reaction.

Reaction System

(1) The ammonia is oxidised to nitric oxide by a continuous process using a platinum-rhodium catalyst at 900 °C and at pressures of approximately 10^6 Pa. The mixture of gases is passed at high speed through a series of catalytic converters to minimise the production of unwanted products such as nitrogen.

$$4NH_3(g) + 5O_2(g) \rightleftharpoons 4NO(g) + 6H_2O(g);$$
$$\Delta H = -909 \text{ kJ mol}^{-1}$$

(2) The gases are cooled to below 600 °C and the nitric oxide (nitrogen(II) oxide) is oxidised to nitrogen dioxide using air.

$$2NO(g) + O_2(g) \rightleftharpoons 2NO_2(g); \Delta H = -115 \text{ kJ mol}^{-1}$$

At least 96% of the ammonia has been converted to nitrogen oxides at this stage.

(3) The nitrogen dioxide (nitrogen(II) oxide) is passed through an absorption tower where it meets a stream of water and forms a mixture of nitric acid and water containing 68.5% nitric acid by mass.

$$3NO_2(g) + H_2O(l) \rightleftharpoons 2HNO_3(aq) + NO(g)$$

The by-product, NO, is recycled. Sulphuric acid is used to reduce the water content and give concentrated nitric acid.

- **Conversion into fertilisers**
Most of the nitric acid produced is reacted with more ammonia to produce ammonium nitrate. Pure ammonium nitrate is not sold in Ireland as a fertiliser because of its explosive properties. It is mixed with calcium carbonate to produce calcium ammonium nitrate (CAN).

Manufacture of ammonium nitrate
Gaseous ammonia and weak (56–60%) nitric acid are reacted in stainless steel vessels:

$$NH_3(g) + HNO_3(aq) \rightleftharpoons NH_4NO_3(aq);$$
$$\Delta H = -92 \text{ kJ mol}^{-1}$$

The reaction is exothermic; the steam produced is used to preheat the ammonia as being fed into the reaction vessel.

Calcium ammonium nitrate (CAN) is produced by mixing the ammonium nitrate with limestone. Calcitic limestone, $CaCO_3$, from Dublin is used to produce granulated CAN, while dolomitic limestone, $MgCO_3.CaCO_3$, from Kilkenny is used to produce prilled CAN. Prilling involves spraying a concentrated solution of the fertiliser through holes at the top of a prilling tower. As the droplets fall they are dried in hot air and form small hard spheres called prills.

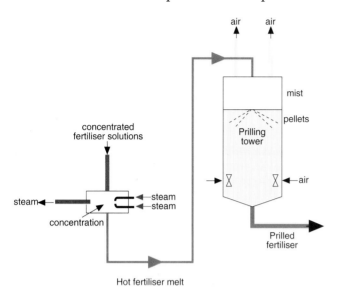

The fertiliser grains are then coated with oil and other materials to make them water resistant and free flowing. The final product is bagged and transported around the country by road and rail.

• Factors affecting rate and yield

The main reaction, like the other reactions, is an equilibrium reaction.

$$4NH_3(g) + 5O_2(g) \rightleftharpoons 4NO(g) + 6H_2O(g);$$

$$\Delta H = -909 \text{ kJ mol}^{-1}$$

The conditions which favour the formation of nitric oxide, NO, are low pressure, excess oxygen and high temperature. In practice, the reaction is carried out at a moderate pressure (10^6 Pa) and at a temperature of 900 °C using a platinum–rhodium catalyst. Under these conditions at least 96% conversion takes place. The Pt–Rh catalyst is replaced every 6–8 weeks as its surface area becomes clogged up and contaminated.

• Co-products and by-products

The co-products produced are all involved in equilibrium reactions and therefore are recycled automatically. Some side reactions occur which produce by-products, like nitrogen.

$$4NH_3(g) + 3O_2(g) \rightleftharpoons 2N_2(g) + 6H_2O(g)$$

These side reactions are favoured by using high pressure and overhot catalysts. Careful control of reaction conditions minimises formation of unwanted by-products like nitrogen.

• Capital costs and production costs:

The capital costs involved were extremely high due to the scale of the plant and the extent of the technology involved. The plant was constructed with the aid of Government grants. The original plant went into production in 1965. Since then many of the production processes have been closed down. Production costs are also high; these are offset by the utilisation of waste gases. The hot gases leaving the converters are used to preheat some reactions and are used to generate electricity which drives the air compressors. Large chemical plants are not very economical and require constant renewal of equipment. Large plants usually have a lifetime of approximately 20 years.

• Energy requirements:
Large supplies of energy are required for most of the production stages in the synthesis of nitric acid. Costs are minimised by using the steam produced to generate power.

• Quality control:
The work of the quality control and research departments ensures that the products produced conform strictly to specification.

• Safety:
The synthesis of nitric acid is not an environmentally friendly process. The process releases large quantities of the oxides of nitrogen into the atmosphere. Nitric acid is an extremely powerful oxidising agent: great care must be taken in handling it. The site is large and contains possible hazardous reaction systems. Extreme care and vigilance are required at all times to monitor potential hazards. Staff are constantly trained on site in all aspects of safety and hazards are constantly monitored.

• Waste and pollution control:
Strict environmental controls are enforced at each stage of the production process to ensure that emissions of oxides of nitrogen are minimised. Waste water is constantly tested and treated to ensure that it contains no pollutants. Emissions of gases and effluent must conform to environmental regulations.

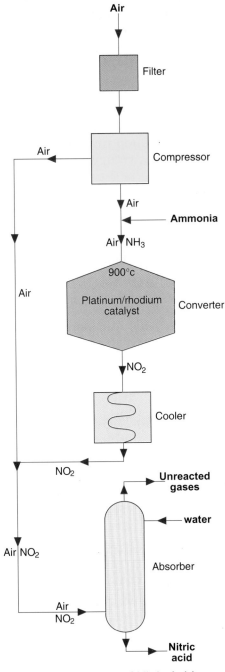

Manufacture of Nitric Acid

Chemistry in Action: Fertilisers

Plants need essential nutrients from the soil to grow well. These nutrients include nitrogen, phosphorus and potassium (NPK). Other elements, such as calcium, magnesium, sodium, copper, iron, manganese, zinc and boron, are also necessary for healthy plant growth but only in very small amounts. The use of manure and compost returns some of the nutrients to the soil, but because of the intensive cultivation of crops and the increasing world population the use of **artificial fertilisers** is necessary.

Improper use of artificial fertilisers often causes environmental problems. For example, when excess nitrates in the soil are washed into rivers or lakes excessive plant growth occurs which disturbs the balance of plant and animal life in the water and results in lack of oxygen. Thus, over a period of time many fish and other animals die.

Fertilisers are bulky and are sold in large bags. It is, therefore, important to have as high a percentage of nitrogen as possible in the fertiliser in order to minimise storage and transportation costs.

Fertiliser	Formula	%N
Ammonia	NH_3	82
Ammonium nitrate	NH_4NO_3	35
Urea	NH_2CONH_2	47

Commercial fertilisers can be single compounds such as ammonium nitrate, but are often mixed compounds, containing nitrogen, phosphorus and potassium (NPK). The ratio of nitrogen to phosphorus to potassium must be shown, by law, on the outside of the fertiliser bag.

The role of NPK fertilisers in plant growth is summarised briefly in the table below:

Nutrient	Role in plant growth	Effect of shortage
N	Essential for synthesis of proteins and chlorophyll	Stunted growth and yellowing of leaves
P	Essential for synthesis of DNA	Slow growth and small fruit
K	Synthesis of carbohydrates and proteins	Yellow curled leaves

Case Study 3: Magnesium Oxide Manufacture from Seawater

Premier Periclase, Drogheda.

Premier Periclase Limited in Drogheda produces high quality **sintermagnesia**, MgO, by reacting lime with the dissolved magnesium salts present in seawater.

• **Company:** Premier Periclase is a subsidiary of CRH plc, one of the largest business corporations in Ireland and a leading international building materials group.

• **Product made:** High quality sintermagnesia, MgO, which is used in the manufacture of refractory linings, particularly for steel making, where chemical attack and high temperatures are commonplace. The pure product, magnesium oxide, MgO, is also called **periclase**. It has a very high melting point and is chemically unreactive. This makes it an ideal material for lining high temperature furnaces.

• **Location:** The plant is located in Drogheda, on the east coast, an area rich in the natural materials involved. The final product, sintermagnesia, is exported to worldwide markets from a private wharf within the port of Drogheda. A distribution centre and warehouse is located within the port of Antwerp, Belgium, which facilitates distribution to all parts of the world. The local workforce is highly skilled in many areas: technical and scientific, sales, marketing and distribution and many others.

• **Raw materials:** Seawater, the source of magnesium ions, is taken from the mouth of river Boyne and is purified before it is reacted with slaked lime produced from limestone taken from local quarries.

• **Preparation of feedstock**
Slaked lime: Limestone, $CaCO_3$, is quarried and then crushed and washed to remove impurities. The washed limestone is then calcined (burnt) in a rotary kiln to produce lime, CaO. The lime is then continuously slaked with decarbonated water to produce a slurry of slaked lime, $Ca(OH)_2$.

$$CaCO_3 + heat \Rightarrow CaO + CO_2$$
limestone   lime   carbon dioxide

$$CaO + H_2O \Rightarrow Ca(OH)_2$$
lime   water   slaked lime

Seawater: The seawater is pumped continuously into the plant where it is degassed to remove carbon dioxide. Suspended sand, silt and clay are removed by passing the seawater through a clarifier.

• **Reaction system:** The clarified and degassed seawater is added in a batch reactor to the slaked lime slurry to produce a precipitate of magnesium hydroxide, $Mg(OH)_2$.

$$Mg^{2+}(aq) + Ca(OH)_2 \Rightarrow Mg(OH)_2(s) + Ca^{2+}(aq)$$

The precipitate is allowed to settle and is thickened before the remaining water is removed by vacuum filtration. The resulting filter cake is then calcined (heated) in furnaces to produce a low density caustic magnesium oxide powder.

$$Mg(OH)_2(s) + heat \Rightarrow MgO(s) + H_2O(g)$$

Mixing of seawater and slaked lime at Premier Periclase.

The caustic magnesium oxide, MgO, is pressed into pellets before it is sintered at a temperature of 2300 °C in kilns. During the final process, crystal growth occurs, the product densifies and boron impurities are volatilised. The high density magnesium oxide formed is called periclase.

• **Energy requirements:** Most of the processes involved rely on large amounts of energy. Quarrying, washing and burning of limestone and piping, pumping, degassing and clarification of seawater from the local estuary all demand energy. After the main reaction extremely high temperatures are involved in calcination and in converting the low density magnesium oxide into high density magnesium oxide, known as periclase. Natural gas and fuel oil are the main sources of energy.

• **Capital costs and production costs:** Costs in setting up such a sophisticated plant are very high, and production costs, because they rely heavily on energy, are also high. The use of natural gas and the nearness of the raw materials lessens production costs to some extent. However, the local quarry is difficult to work, but this is offset by the high purity of the limestone obtained.

• **Quality control:** The work of the quality control and research departments ensures that the products produced conform strictly to specification.

• **Factors affecting rate and yield:** The chemical reaction involved, the reaction of lime with the magnesium ions in seawater, is a very fast precipitation reaction. It is so fast it is known as 'flash precipitation'. The main factors affecting the yield of product are related to product quality and purity.

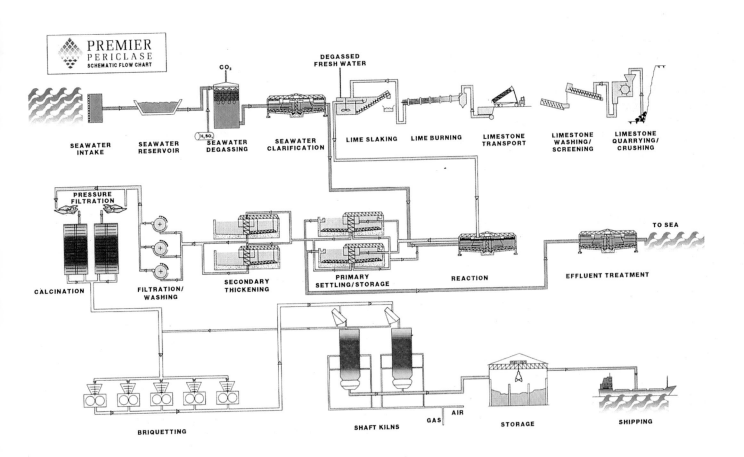

- **Co-products:** The calcium oxide, CaO, formed during the reaction is discharged along with the 'spent' seawater back to the sea.

- **Safety:** The entire process can be regarded as environmentally friendly: of the raw materials, limestone is a naturally occurring rock and seawater is piped from the local estuary. The chemical process releases no harmful gases or effluents. Staff are constantly trained on site in all aspects of safety and hazards are constantly monitored. The filtration, calcination and sintering processes are computer controlled. This allows the processes to be visually monitored and controlled at all times.

- **Waste and pollution control:** Premier Periclase relies on natural resources to produce sintermagnesia and great care is taken to protect the environment. The company has recently reclaimed derelict land beside the plant. Trees and shrubs have been planted to create an area abundant in flora and fauna. Strict environmental controls play an integral part in each stage of the production process.

Exercise 1A

•

(1) (a) Make a list of chemicals used in the following: medicine, agriculture, travel, communications, health care and fuels. In each case name two chemicals.
(b) Mention how each chemical has helped enhance the quality of our lives.

(2) (a) Make a list of some natural resources which are used in the chemical industry. In each case mention the process where the resource is used.
(b) Make a list of the natural resources which are used in the Irish chemical industry.
(c) What does a comparison of the resources in (a) and (b) tell you about the Irish chemical industry?

(3) List some factors which are characteristic of a successful chemical industry.

(4) In terms of raw materials and location, why did each of the following chemical industries set up in Ireland?
(a) pharmaceutical industries (b) ammonia production (c) nitric acid production
(d) production of sintermagnseia.

(5) 'Our standards of living have improved along with the expansion of the chemical industry.' Discuss this statement in terms of the risks and benefits involved.

(6) Discuss, briefly, the factors which must be taken into account when choosing a site for a chemical plant.

(7) (a) Why are the following industries located along the coastline?
(i) magnesium oxide manufacture (ii) ammonia manufacture (iii) nitric oxide manufacture.
(b) In the case of one of the above industries, write an account giving the reasons for and against siting that particular industry in the area.

(8) In the case of the Irish chemical industry, list the main sources of pollution arising from the industry. How does the industry attempt to (a) control emission of waste gases, (b) control spillages, and (c) control emission of solvents into water courses?

(9) Write a brief account of safety in a chemical plant, using the following headings:
(a) location of site, (b) materials used in construction, (c) on-site training, (d) monitoring of hazards and (e) safety features.

(10) (a) Write a brief note on the capital costs and the production costs involved in the setting up and running of a chemical plant.

(11) Chemical plants are usually located in terms of historical factors and practical factors. Discuss some of the historical and practical factors involved.

(12) 'The rate of a chemical reaction and the yield of product are related to several factors.'
Discuss this statement, using as an example a chemical process from the Irish chemical industry.

(13) 'Many chemical reactions give more than one product, while in some reactions, side reactions occur.' Discuss this statement in terms of (a) reduction in costs and (b) increase in costs.

(14) (a) Draw a simple flow chart outlining the steps involved in a chemical process.
(b) Write a brief note on each of the following: (i) batch processes (ii) continuous processes (iii) semi-continuous processes.

(15) (a) Calculate the mass of ethene necessary to manufacture 2 tonnes of ethanol using the process:

$$C_2H_4 + H_2O = C_2H_5OH$$

(b) How much iron can be produced from 20 tonnes of Fe_2O_3?

$$Fe_2O_3 + 3CO = 2Fe + 3CO_2$$

(c) What mass of slag, $CaSiO_3$, can be produced from 10 tonnes of silica, SiO_2, using the process:

$$CaO + SiO_2 = CaSiO_3$$

(d) How many tonnes of ammonia can be produced from 1 tonne of hydrogen in excess nitrogen?

$$N_2 + 3H_2 = 2NH_3$$

(16) Draw up a list of possible questions you might ask on a visit to a local industry which uses a chemical process.

(17) In the case of one of the following chemical processes, answer the questions which follow:
(i) ammonia manufacture and its conversion to urea (ii) nitric acid manufacture and its use to make fertilisers (iii) magnesium oxide manufacture from seawater.
(a) List several factors which made the particular area a favourable location for the particular plant.

(b) Explain how the feedstocks involved are prepared.
(c) Write balanced equations for each of the main stages involved in the particular chemical process.
(d) What type of process is involved: batch, continuous or semi-continuous? Why is the particular type of process suitable for that chemical process?
(e) What are the co-products and by-products (if any) involved in the process? How do they contribute to the production costs?
(f) What are the main capital and production costs involved in setting up and running the plant? Were government grants and other inducements, such as exemption from taxes, used in order to locate the plant in the particular area?
(g) Discuss the effect of changes in concentration, pressure, temperature and the use of catalysts, in relation to speed of the reactions and the yield of product obtained in the particular process.
(h) How modern is the plant? How does a modern plant monitor (i) quality, (ii) emission of waste gases and (iii) emission of liquid effluent?
(i) What particular safety considerations arise in the plant? How are the staff trained to cope with all the safety aspects involved?
(j) Why is it considered safe to discharge liquid effluent into the sea?
(k) How is the emission of waste gases minimised?
(l) How does recycling of reactants contribute to higher yields?
(m) How are harmful solvents disposed of?
(n) Does the company contribute to environmental problems? How does it help to solve those problems?
(o) What type of people are involved in running the plant? What qualifications do they have and what type of work do they do?

1B ATMOSPHERIC CHEMISTRY

Introduction

Smog in Los Angeles.

The atmosphere is made up of a thin layer of gases, consisting mainly of nitrogen (78%) and oxygen (21%). Other gases, such as carbon dioxide, water vapour, the noble gases, methane, oxides of nitrogen, CFCs and ozone, are present in very small concentrations. Some, like carbon dioxide, water vapour, methane, ozone and CFCs, have variable concentrations which depend on human activities.

The lower part of the atmosphere is called the troposphere. This is where most mixing of gases occurs, where the greenhouse effect keeps the Earth warm and where acid rain causes problems. Further out is the strato-sphere: this is where ozone, O_3, acts as a giant sunscreen which protects us from ultraviolet radiation.

The atmosphere around us protects us and supports life. It is important that we take great care in protecting it.

1B.1 OXYGEN

Oxygen is the most abundant element on Earth, making up about 48% of the total Earth. Oxygen is a substance which we often take for granted; it is essential to all forms of life. It is a colourless, odourless gas, slightly soluble in water. Fish and other aquatic creatures rely on this last property in order to breathe. It was identified independently by Joseph Priestley and by Carl Scheele, but was named by the French scientist Antoine Lavoisier. It exists on Earth as oxygen, O_2, and as its allotrope, ozone, O_3.

Extraction of Oxygen from Air

Pure oxygen is an extremely important industrial material. It is prepared directly from the air, which contains about 21% oxygen. The raw material, air, is free. The only major cost involved is electrical energy to drive the air compressors. Air is a very useful resource because it also contains other important gases which can be extracted by fractional distillation of liquid air.

The main stages in the process are:
(1) Compressed air is filtered to remove dust particles.
(2) Carbon dioxide is removed by passing the air through sodium hydroxide.
(3) Water vapour condenses out during compression. The remaining water is removed in drying towers containing silica gel or activated alumina.
(4) The dry carbon-dioxide-free air is cooled and the pressure is reduced in stages. As a result of the cooling and the decrease in pressure, the air liquefies.
(5) The liquefied air is distilled in a fractionating tower. Oxygen-rich air collects at the bottom of the column, while nitrogen (lower b.p.) is removed from the top of the column as a gas.
(6) The nitrogen and oxygen fractions are further purified by removal of the inert gases, neon, krypton and xenon.

Aurora borealis, the northern lights, photographed over spruce trees, Alaska. Auroras are luminous displays that appear in the night sky and are caused by charged sub-atomic particles from the sun interacting with atoms and molecules in the Earth's upper atmosphere.

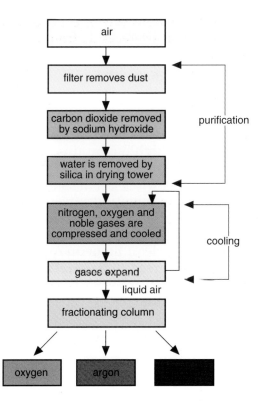

Extraction of gases from air

Some Reactions of Oxygen

Although oxygen, O_2, is a reactive gas, it reacts quite slowly at first because the oxygen–oxygen bond is quite strong. Otherwise, all the organic material on the Earth would quickly burn up and life would not have evolved. However, once combustion starts, most materials burn quickly in oxygen. Most of the oxygen on Earth is in combined form; the majority of the elements form stable oxides with oxygen.

Reactions with Metals

Many metals react readily with oxygen to form oxides.

$$4Li(s) + O_2(g) \Rightarrow 2Li_2O(s)$$
lithium   oxygen   lithium oxide

$$4Fe(s) + 3O_2(g) \Rightarrow 2Fe_2O_3(s)$$
iron   oxygen   iron oxide

Reactions with Non-metals

Non-metals react with oxygen to form covalent oxides.

$$2H_2(g) + O_2(g) \Rightarrow 2H_2O(g)$$
hydrogen   oxygen   water

$$C(s) + O_2(g) \Rightarrow CO_2(g)$$
carbon   oxygen   carbon dioxide

If a limited amount of oxygen is used for the combustion of carbon, carbon monoxide, CO, is the usual product.

$$S(s) + O_2(g) \Rightarrow SO_2(g)$$
sulphur   oxygen   sulphur dioxide

Sulphur also forms another oxide, sulphur trioxide, SO_3, but only small amounts are formed during combustion with oxygen.

Uses of Oxygen

• **Steel manufacture:** Oxygen is used to remove carbon, silicon and phosphorus impurities from cast iron. These are removed by converting them to their gaseous oxides.

• **Medicinal uses:** Patients in hospitals often require pure oxygen because their lungs cannot provide enough oxygen quickly from the air. The concentration of oxygen in the patient's air supply is increased to approximately 32%.

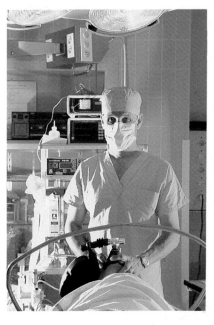

Patient receiving oxygen from an anaesthetist.

• **Welding:** When ethyne (acetylene) reacts with oxygen it produces heat. The heat of the oxyacetylene flame can be used to weld metals together.

• **Manufacture of chemicals:** Many reactions use oxygen as an oxidising agent to produce a variety of chemicals.

• **Purification of rivers:** Sometimes when rivers become heavily polluted oxygen is pumped in to help purify them.

(1) 'Oxygen is often called the reactive gas.'
(a) Why is oxygen so reactive?
(b) Describe, using equations, some reactions of oxygen with metals and with non-metals.
(c) How does oxygen get into the atmosphere?

(2) 'Oxygen reacts with non-metals, such as carbon, sulphur and nitrogen.'
Discuss the effect of the reactions of oxygen with these non-metals on the atmosphere.

(3) (a) Describe the main steps involved in the extraction of oxygen from the air, using fractional distillation.
(b) Describe another method of producing oxygen on a large scale.

(4) Describe some everyday uses of oxygen.

(5) Describe the regions of the Earth's atmosphere. Name the main gases present in each region.

(6) Explain each of the following statements:
(a) Mountain climbers need oxygen supplies when they climb high mountains.
(b) When ethyne burns in oxygen the flame is much hotter than when it burns in air.
(c) Polluted rivers sometimes need more oxygen added to them in summer than in winter.
(d) Ethyne burns with a smoky flame in air, but burns with a clean flame in oxygen.
(e) Anaesthetists need to be careful when using oxygen.

1B.2 NITROGEN

Nitrogen, like carbon, hydrogen and oxygen, is an important component of all living organisms. Without nitrogen plants and animals would have no protein to build up living cells such as muscle and blood cells.

Molecular nitrogen, N_2, is the principal constituent of air (78%), from which it can be obtained by liquefaction and by fractional distillation. Unlike oxygen it is a rather unreactive gas. The nitrogen molecule is unreactive due to the high bond energy of the triple bond in N_2.

$$N \equiv N(g) \Rightarrow 2N(g); \Delta H = +941 \text{ kJ mol}^{-1}$$

The lack of chemical reactivity of nitrogen accounts for the fact that few nitrogen compounds are found in the earth's crust.

Uses of Nitrogen

• **Food freezing:** Liquid nitrogen is very cold and can be used to fast freeze many foods. Fast freezing does less damage to the food than normal freezing.
• **Making flammable chemicals safe:** Many chemicals, such as ethanol, are flammable. Nitrogen is used to reduce the oxygen content in chemical storage tanks, thereby preventing combustion. Sometimes dangerous vapours can lodge in oil tankers. Nitrogen is used to flush vapours out.
• **Food packaging:** Crisps and other fried foods are packed in an atmosphere of nitrogen. This prevents oxidation of the oil which would cause nasty flavours and smells.
• **Shrink fitting of metal components:** When a metal shaft needs to be fitted into a sleeve, the shaft is frozen in liquid nitrogen to reduce its size before fitting. As the shaft returns to room temperature it expands and fits tightly into the sleeve.
• **Ammonia and nitric acid:** Nitrogen is the main raw material used in the manufacture of ammonia, nitric acid and fertilisers.

The Nitrogen Cycle

Special conditions are often necessary to make nitrogen react with other elements. Most plants cannot take in atmospheric nitrogen directly. Some plants, the leguminous plants (such as peas, beans and clover), contain nitrogen-fixing bacteria in their root nodules which can absorb nitrogen from the air and convert it into a form that can be used by the plant to make protein.

During thunderstorms more nitrogen-fixing processes occur. Here, the high energy evolved when lightning flashes forces some nitrogen and oxygen to combine to form nitrogen oxide:

$$N_2(g) + O_2(g) \Rightarrow 2NO(g)$$

The nitrogen oxide then combines with more oxygen to form nitrogen dioxide:

$$2NO(g) + O_2(g) \Rightarrow 2NO_2(g)$$

The nitrogen dioxide then reacts with water to form nitric acid:

$$3NO_2(g) + H_2O(l) \Rightarrow 2HNO_3(l) + NO(g)$$

The nitric acid is washed into the ground where it forms nitrates which are absorbed by the plants.

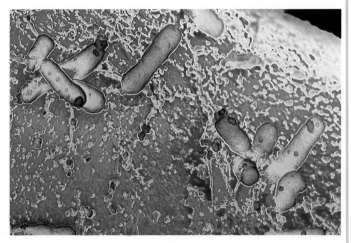

False-colour scanning electron micrograph of nitrogen-fixing bacteria on the root hair of a pea plant. In return for breaking down atmospheric nitrogen and converting it into a form the plant can use, the bacterium gains access to the plant's carbohydrate food stores.

The nitrogen-fixing processes described earlier are the first stages in a natural process called the nitrogen cycle, but are only a minor source of plant nitrogen. The majority of plants obtain their nitrogen from dead and decaying animals and plants. The plant and animal protein is converted into nitrites and nitrates which either are reabsorbed by other plants or are converted by bacteria into nitrogen which is released into the atmosphere.

However, this natural process cannot supply the amounts of nitrogen required to enable us to produce all the food we require today. Scientists have developed many new nitrogen-

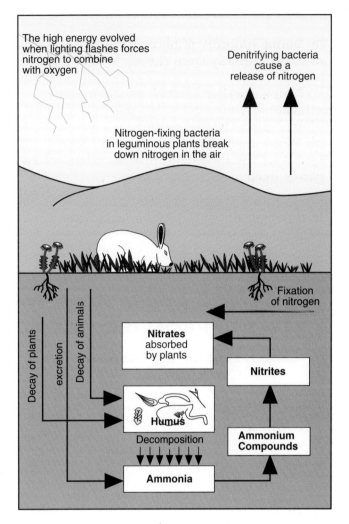

The nitrogen cycle

containing compounds which are now used as artificial fertilisers. One of the most commonly used fertilisers is ammonium nitrate, which is made from two important nitrogen compounds, ammonia and nitric acid.

Exercise 1B.2

•

(1) (a) Draw the electron-dot structure of the nitrogen molecule, N_2.
(b) Explain why nitrogen gas is unreactive.

(2) (a) How abundant is nitrogen, N_2, in the atmosphere?
(b) Why is nitrogen so abundant in the atmosphere?

(3) (a) What is the nitrogen cycle?
(b) During electrical storms, nitrogen combines with oxygen in the atmosphere. Explain how this occurs.
(c) What happens to the oxides of nitrogen which are formed in the atmosphere?
(d) What is natural fixation of nitrogen?
(e) What effect does natural fixation of nitrogen have on the overall supply of plant nitrogen?

(4) Draw a diagram of the nitrogen cycle.

(5) A simplified version of the nitrogen cycle is given below.

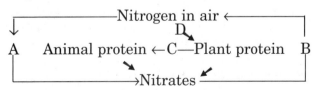

(a) Name the processes A, B, C and D.

305

(b) Name a family of plants that can use nitrogen directly from the air to make protein.

(c) Name another process which enables atmospheric nitrogen to be used directly by plants to synthesise protein.

(d) What is meant by 'nitrogen fixation'?

(e) Name the industrial process where nitrogen is fixed.

(6) 'Liquid nitrogen is used to preserve foods, to flush out dangerous chemicals from tankers and has other uses.'

(a) Why is nitrogen suitable for the specific uses mentioned?

(b) Write a brief account of the main uses of nitrogen.

(7) What is meant by:

(a) natural nitrogen fixation

(b) industrial nitrogen fixation?

1B.3 CARBON DIOXIDE

Occurrence of Carbon in Nature

Industrial diamonds. Diamond is a naturally occurring form of carbon which has crystallised under great pressure.

Carbon in the form of graphite.

Carbon occurs as a free element in the Earth's crust in two different allotropic forms, graphite and diamond. Graphite occurs as the mineral plumbago or 'black lead', while diamond is a very precious stone.

Carbon also occurs in fossil fuels, such as natural gas, petroleum and coal. These fossil fuels were formed thousands of years ago by the decay and compression of living organisms. They consist of carbon atoms covalently bonded mainly to hydrogen atoms and other carbon atoms. Fossil fuels are a major source of energy.

Carbon occurs as carbonates in many rocks, such as chalk, limestone, marble and dolomite. It also occurs to a small (0.035%) but very significant extent in the atmosphere as carbon dioxide.

The Carbon Cycle

The carbon cycle is a complex, though seemingly simple, balance of the distribution of carbon on Earth. The small yet constant amount of carbon dioxide (0.035%) plays a very important role in this equilibrium system, involving the formation and decomposition of carbon dioxide.

The principal steps in the carbon cycle include:

(1) removal of carbon dioxide in the air through formation of carbonates and by photosynthesis

(2) release of carbon dioxide by respiration in plants and animals

(3) death and decay of animals and plants to form fossil fuels and carbonates

(4) combustion of fossil fuels to form carbon dioxide.

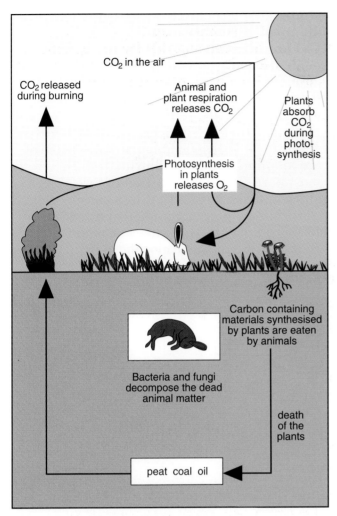

The carbon cycle

Combustion of Natural Substances

The fossil fuels coal, petroleum and natural gas consist largely of compounds of carbon and hydrogen only and are termed hydrocarbons. Coal contains less hydrogen than petroleum or natural gas; the macromolecules of coal frequently consist only of carbon atoms linked together.

Because fossil fuels contain carbon atoms linked together with hydrogen atoms or with other carbon atoms they release large amounts of energy when they are burnt.

The combustion of carbon in fossil fuels yields carbon dioxide gas:

$$C(s) + O_2(g) \Rightarrow CO_2(g)$$

whereas incomplete combustion gives carbon monoxide gas:

$$C(s) + \tfrac{1}{2}O_2(g) \Rightarrow CO(g)$$

The burning of too many fossil fuels, produc-ing large quantities of carbon dioxide, is disturbing the balance of the carbon cycle. This over-production and release of carbon dioxide into the atmosphere is thought to increase the temperature of the Earth. This process, called the greenhouse effect, is believed to be responsible for alterations in weather patterns.

Reaction of Carbon Dioxide with Water

Carbon dioxide is a colourless, odourless gas. It is moderately soluble in water, forming a weak acidic solution: it is an acidic oxide. If carbon dioxide solution is boiled, it loses its acidity completely. This indicates that the reaction of carbon dioxide with water is a reversible reaction.

$$CO_2(g) + H_2O(l) \rightleftharpoons CO_2(aq)$$

Aqueous carbon dioxide is often called carbonic acid: it provides the fizz in fizzy (carbonated) drinks.

Most of the Earth's surface is covered by the oceans, which absorb large amounts of carbon dioxide from the atmosphere. The plankton in the water use up most of the carbon dioxide, while the remainder reacts with the water, forming a weak acid solution. This aqueous

Carbon dioxide is used in fizzy drinks.

carbon dioxide, $CO_2(aq)$, forms an equilibrium mixture containing carbonate and hydrogen-carbonate ions.

$$CO_2(aq) \rightleftharpoons H^+(aq) + HCO_3^-(aq)$$
$$\text{hydrogencarbonate ion}$$
$$\rightleftharpoons 2H^+(aq) + CO_3^{2-}(aq)$$
$$\text{carbonate ion}$$

Here, the 'carbonic acid' solution mainly consists of loosely hydrated CO_2 which ionises into hydrogen ions, hydrogencarbonate ions and carbonate ions. This reaction is responsible for the acidity of carbon dioxide. It should be noted that carbonic acid, H_2CO_3, cannot be isolated as a single substance.

307

Because it is moderately soluble in water and because it is acidic, carbon dioxide is added to carbonated drinks like mineral waters, cola and other soft drinks. It is also present in beers and lagers where it is produced during fermentation.

The Greenhouse Effect

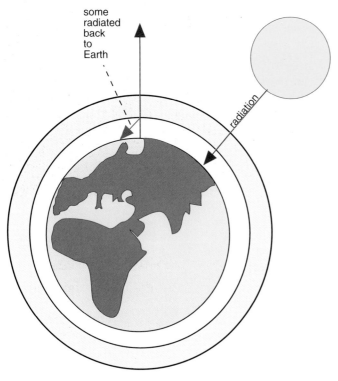

some radiated back to Earth

radiation

The gases in the Earth's atmosphere trap heat and redistribute it around the world; this results in an average global temperature of 15 °C. If the atmosphere did not exist, the temperature would be as low as −18 °C. The gases in the atmosphere which absorb most of the solar energy from the Sun are called 'greenhouse gases', because they absorb infra-red radiation. These gases are only present in small amounts but are powerful absorbers of infra-red radiation. They include carbon dioxide, CO_2, methane, CH_4, water vapour, H_2O, nitrous oxide, N_2O, ozone, O_3, and chlorinated fluorocarbons, CFCs. The main gases in the atmosphere, nitrogen and oxygen, absorb little infra-red radiation. The increase in the quantity of these 'greenhouse gases' is warming the Earth at an increasing rate and may cause major climatic changes. The two main greenhouse gases are carbon dioxide and water vapour, because they are so abundant and can absorb a lot of infra-red radiation. The increased amounts of carbon dioxide, CFCs and methane are mainly responsible for the build up of 'greenhouse gases'.

• **Carbon dioxide:** This is the most common greenhouse gas and is responsible for nearly half of the extra global warming taking place. The percentage of carbon dioxide (0.035%) in the atmosphere has remained almost constant over the centuries. The increase in the burning of fossil fuels is responsible for the release of large amounts of carbon dioxide into the atmosphere. The deforestation of forests in many countries means that less carbon dioxide is absorbed by plants via photosynthesis than before. This means that the concentration of carbon dioxide in the atmosphere is increasing rapidly. The oceans act as a sink and absorb some of the carbon dioxide.

Deforestation

• **Water:** As the Earth gets warmer more water from the oceans evaporates and is changed into water vapour. This may cause an increase in global warming, but the increase in the amount of liquid water in clouds may also block out the Sun's rays, causing a decrease in global warming.

• **Methane:** This is the second most important greenhouse gas. The level of methane in the atmosphere is only 1.7 ppm, but it is 20 times more effective as a greenhouse gas than carbon dioxide. Millions of tonnes of methane are released annually mainly due to:
(1) the increase in the world cattle population. Bacteria in the intestines of cattle produce methane by breaking down organic matter
(2) the increase in urban wastes

Cattle.

(3) emission of methane from swamps, bogs, paddy fields and termites
(4) aerobic fermentation.

Methane also exists trapped in ice crystals. If large-scale global warming resulting from the greenhouse effect started, the ice would melt and the methane locked up in the ice would be released, thereby accelerating the global warming.

• **CFCs:** The level of CFCs in the atmosphere is very low, but CFCs are 40 times more effective as greenhouse gases than carbon dioxide.

Residence Times of Greenhouse Gases

Some greenhouse gases, such as carbon dioxide and CFCs, can reside in the atmosphere for 50–200 years, while others, such as ozone and water vapour, reside for only a few days or weeks. Obviously, the longer a particular gas remains in the atmosphere, the more effect it can have on global warming.

Aerosol spray.

Enhanced Greenhouse Effect

The main implications of the enhanced greenhouse effect include:
• rise in sea levels: levels may rise due to expansion of sea water and melting of ice
• possible extinction of some forests: the increased frequency of drought might cause reproductive failure in vegetation
• local disruptions of agricultural activity: drought may cause changes in food supplies locally
• increase in CO_2 and CH_4 emissions: increase in mean winter temperatures could release carbon dioxide and methane from the ice caps, thus further enhancing the greenhouse effect.

Role of Greenhouse Gases in the Greenhouse Effect

Gas	Warming power	Concentration	Share in increased warming	Residence time
CO_2	1	375 ppm	50%	50–200 years
CH_4	25	2 ppm	15%	10 years
CFCs	20 000	trace	13%	50–100 years
N_2O	250	0.4 ppm	9%	150 years
$H_2O(g)$	Variable	Variable	Variable	Days

Activity 1B.1

To examine the effect of carbon dioxide on universal indicator solution.

CO₂ being bubbled into universal indicator. The solution changes through a range of colours, from navy through green to red.

Procedure

- Bubble carbon dioxide into a beaker containing some universal indicator solution.
- Observe the colour changes.
- Discuss the results.

Aerobic Fermentation

When substances such as sugar, malt, barley and others containing carbohydrates are fermented in the presence of oxygen, carbon dioxide is produced. This aerobic fermentation is used to produce carbon dioxide and alcohol in alcoholic drinks and to produce carbon dioxide in bread, enabling the bread to rise.

Carbon monoxide

Carbon monoxide does not react with water. It is a neutral oxide. Carbon monoxide is a poisonous gas. Once inside the body it occupies the place normally taken by haemoglobin in blood. This deprives the body of oxygen, causing asphyxiation. Carbon monoxide is not present in the atmosphere in large quantities, but is present in large quantities in cigarette smoke and in motor car exhaust fumes. It can also be produced by faulty domestic heating appliances.

Exercise 1B.3

•

(1) (a) In what forms does carbon occur in nature?
(b) Draw a brief outline of the carbon cycle.
(c) How is the balance of the carbon cycle being affected by the burning of large amounts of fossil fuels?
(d) Why does carbon form carbon monoxide and carbon dioxide when it burns in air?

(2) (a) What are fossil fuels?
(b) Describe, using equations, how any three fuels burn in oxygen.

(3) Describe how carbon dioxide gas reacts with water. Why is this an equilibrium reaction?

(4) Write equations showing how carbon burns in air to form (a) carbon dioxide and (b) carbon monoxide.

(5) (a) Explain why carbon dioxide is an acidic oxide.
(b) What happens when carbon dioxide is bubbled into universal indicator solution?
(c) What is the concentration of carbon dioxide in the atmosphere?
(d) 'Large amounts of carbon dioxide are absorbed by the oceans.'
(i) What happens to the carbon dioxide in the oceans?
(ii) What effect does this have on the concentration of carbon dioxide in the atmosphere?

(6) Explain, using equations, what happens to carbon dioxide as it dissolves in water.

(7) (a) Why is the concentration of carbon dioxide in the atmosphere increasing?
(b) How does life in the oceans change the level of carbon dioxide?
(c) What might happen if the temperature of the oceans was increased by 2–3 °C?

(8) 'Carbon dioxide is an acidic oxide.'
(a) Why is it used in soft drinks?
(b) How is it produced in alcoholic drinks and in bread?

(9) (a) What are the main sources of carbon monoxide?

(b) Explain how carbon monoxide is poisonous.

(10) (a) What is the greenhouse effect?
(b) Whereabouts in the atmosphere does the greenhouse effect occur?
(c) What gases cause the greenhouse effect?
(d) How is man increasing the concentration of the greenhouse gases?

(11) (a) How does the greenhouse effect help life on Earth?
(b) Draw a simple diagram illustrating what happens to the radiation from the Sun.
(c) List some possible effects of increased greenhouse warming.

(12) 'Nitrogen and oxygen make up most of the atmosphere.'
(a) Explain why oxygen and nitrogen have no real importance in the greenhouse effect.
(b) Why are carbon dioxide, water vapour, CFCs, methane and the oxides of nitrogen the main contributors to global warming?
(c) What are CFCs?

(d) How do CFCs contribute to the greenhouse effect?
(e) Why do very small concentrations of CFCs play a large role in the greenhouse effect?

(13) (a) Why is methane an important greenhouse gas?
(b) What are the main sources of methane?

(14) (a) What is meant by the residence time of a greenhouse gas?
(b) What are the residence times of the major greenhouse gases?
(c) What are the possible implications of the enhanced greenhouse effect?

(15) (a) Can the problem of the enhanced greenhouse effect be solved by nuclear power?
(b) What role does the EU take in tackling the problem of global warming?
(c) How can individuals play a part?
(d) Why is it difficult to predict greenhouse effects?
(e) Why should we leave predictions about the greenhouse effect to scientists?

1B.4 ATMOSPHERIC POLLUTION

The atmosphere is made up of more than just the gases found in pure air. It also contains many gases and suspended solids which are there mainly as a result of the burning of fuels. When these gases or suspended solids are potentially harmful either to man or to the environment they are known as pollutants. When compared to more industrialised countries, Ireland is relatively free of atmospheric pollution. However, overuse of bituminous coal in urban areas like Dublin and emissions from large power stations have caused problems.

Air pollution is caused either by small suspended solids or liquids in the air or by poisonous or obnoxious gases.

Sources of Pollution

Domestic Fires

Most fuels contain sulphur and carbon. When carbon and sulphur are burnt in air, carbon dioxide and sulphur dioxide are formed.

$$C(s) + O_2(g) \Rightarrow CO_2(g)$$

$$S(s) + O_2(g) \Rightarrow SO_2(g)$$

• Solid fuels, such as coal or coke, contain much more sulphur than oil, natural gas and smokeless fuels, such as anthracite. The burning of large quantities of coal in open fires produces large amounts of suspended solids, sulphur dioxide and carbon dioxide in the air.
• Suspended particles of smoke and dust settle on buildings, making them black and unsightly. If the fine particles pass into lungs they can cause bronchitis and many chest diseases.
• Sulphur dioxide dissolves in water, forming an acidic solution. This can cause respiratory problems in all living things and is the main constituent of acid rain.
• Carbon dioxide, when present in excessive quantities, upsets the balance of the carbon cycle. This causes an increase in the temperature of the Earth, called the greenhouse effect.

Industrial Sources

Industries need fuel to heat factories and to run machinery. Most industries use oil and natural gas for fuel and these release sulphur dioxide and carbon dioxide as pollutants. Some industries, depending on the product being made, produce obnoxious odours or poisonous substances; for example, smells from a meat or fish processor or hydrogen fluoride from a fertiliser plant.

Motor Vehicles

Traffic.

Internal combustion engines, mainly petrol engines, are a major cause of air pollution. Car engines do not burn fuel completely to form carbon dioxide and water. The partly burnt gases from the exhaust consist of unburned hydrocarbons (smoke), carbon monoxide and oxides of nitrogen and lead.

• **Carbon monoxide**, CO, is a colourless odourless gas that has a simple yet devastating effect on the body. It attaches itself, in place of oxygen, to the haemoglobin in the red blood cells. If sufficient carbon monoxide combines with the haemoglobin in the blood, poisoning occurs which may be fatal.

• The **oxides of nitrogen**, NO and NO_2, are produced because the incoming air in the internal combustion engine contains approximately 80% nitrogen gas. This gas mixes with the oxygen in the engine and when heated is converted into nitrogen monoxide, NO. The nitrogen monoxide is then further oxidised to nitrogen dioxide after it is discharged into the atmosphere. The oxides of nitrogen cause damage to eyes and lungs and are a constituent of acid rain. The exhaust gases can also react with sunlight to produce ozone.

• **Lead compounds** are serious pollutants because they affect the brain. Lead is toxic in low concentrations and inhibits IQ. Lead is used in petrol to boost the 'octane' rating of the fuel thereby reducing 'knocking' or 'pinking'. In some countries it is forbidden by law to use

lead additives in petrol; this makes the petrol more expensive. Most countries are gradually turning over to using lead-free petrol. As a result, the concentration of lead in the air is decreasing.

• The use of **catalytic converters** helps convert potential pollutants such as unburned hydrocarbons, NO and CO into harmless substances such as CO_2, N_2 and water. The catalyst will not function if leaded petrol is being used.

Acid Rain

Rainfall is acidic because carbon dioxide is absorbed naturally from the atmosphere to produce a weak acid, carbonic acid. The pH of rain is normally 5–6 as a result.

$$CO_2(g) + H_2O(l) \rightleftharpoons H_2CO_3(aq)$$

When fossil fuels are burned, large quantities of sulphur dioxide and the oxides of nitrogen are released into the atmosphere. These react with water vapour to form sulphuric and nitric acids and this sometimes lowers the pH of rainfall down to pH 3: this is the cause of acid rain.

$$SO_2(g) + H_2O(l) \rightleftharpoons H_2SO_3(aq)$$
$$\text{sulphurous acid}$$

The sulphurous acid is then converted into sulphuric acid, H_2SO_4.

Effects of Acid Rain on the Environment

• **Buildings:** Acid rain corrodes metal and stonework in buildings. This makes maintenance more costly.

Corrosion of stone by acid rain.

- **Crop yields:** Acid soil reduces crop yields and prevents pollination.
- **Damaged trees:** Sulphur dioxide interferes with photosynthesis, causing trees to die.
- **Leaching of metals:** Acid rain causes aluminium and cadmium to be washed out of (leached from) the soil. These toxic metals are taken in by the trees, causing poor growth.
- **Water courses:** In some cases, aluminium and cadmium may leach directly into rivers and lakes. These metals are harmful to humans and animals in very small concentrations.
- **Decline in fish numbers:** When snow melts in spring, there is a quick release of all the pollution accumulated over the winter. This sudden surge in acid rain can cause a decrease in the population of young fish. A normal healthy lake has a pH of about 6.5; a lake becomes dead at pH 4.5.

Prevention of Acid Rain

Acid rain can only be reduced by a reduction in the use of fossil fuels. This process will take a long time. Some action has been taken to reduce the effects of acid rain in the following ways.

1. Use of lime to neutralise lakes: This is an effective but expensive process.

2. Use of catalytic converters: The level of nitrogen oxides released from car exhausts is reduced by using catalytic converters.

3. Use of low sulphur fuels: Natural gas contains very little sulphur. Electricity power stations need to use more natural gas as a fuel.

4. Scrubbing of waste gases: Some electricity supply boards use limestone scrubbers to remove pollutants from waste gases. The sulphuric acid formed reacts with the calcium carbonate (limestone) forming carbonic acid and calcium sulphate.

Some Aspects of Recent Legislation

The Air Pollution Act, which became law in June 1987, provides a legal framework within which relevant EU requirements can be implemented and domestic problems tackled.

Some aspects and consequences of the Act include:

(1) Monitoring of air quality standards: Certain local authorities are required to monitor air quality standards for smoke, sulphur dioxide, lead and nitrogen dioxide in specific locations.

New industrial and other plants must obtain compulsory licensing. Licensing will gradually be extended to existing plants.

(2) Dust monitoring equipment: The most up-to-date automated monitoring equipment is being used for the continuous measurement of total suspended solids (dust).

(3) Unleaded petrol: In Ireland, the level of lead in petrol is now 0.15 g L^{-1}, while the EU limit is 0.40 g L^{-1}. Oil companies have recently extended their network of lead-free petrol outlets.

(4) Protecting the ozone layer: The ozone layer acts as a shield which protects the environment from potentially harmful ultraviolet rays. The use of chlorofluorocarbons (CFCs) is being phased out.

Exercise 1B.4

●

(1) (a) What is atmospheric pollution?
(b) List the main atmospheric pollutants.
(c) What is the source of each pollutant?
(d) What effects do these pollutants have on animal and plant life?
(e) Describe, briefly, the effects they have on the environment.

(2) (a) Write equations showing how carbon, nitrogen and sulphur react with oxygen.
(b) What effect do the oxides of carbon, nitrogen, and sulphur have on the environment?

(3) 'The waste gases from the exhausts of motor cars are a major cause of pollution.'
(a) Name the main pollutants in the exhaust gases.
(b) Describe the effect, if any, each of the pollutants has on humans.

(4) (a) What is meant by natural sources of pollution?
(b) What effect does natural pollution have on the environment?

(5) Carbon dioxide, sulphur dioxide, nitrogen monoxide and nitrogen dioxide are all atmospheric pollutants. Discuss how each of the gases contributes to (a) acid rain and (b) global warming.

(6) (a) Write equations showing the reaction of (i) sulphur dioxide with water and (ii) carbon dioxide with water.
(b) What type of solutions are obtained in (i) and (ii)?
(c) How would you verify this?
(d) What effect would the solutions have on the environment?

(7) (a) Why has acid rain become a problem in the last 20 years?
(b) List the effects of acid rain.
(c) 'If electricity generating stations had very high chimneys, the effects of acid rain would be reduced.' Discuss this statement.
(d) What types of solid fuel could be used in industries in order to minimise acid rain?
(e) How can waste gases be treated to limit emissions of pollutant gases?

(8) (a) Suggest three ways of controlling sulphur dioxide pollution.

(b) What are the main disadvantages of each method?

(9) (a) What is the pH of natural rainwater?
(b) What causes the value to decrease?
(c) What is the pH of acid rain?

(10) Discuss the damage caused by acid rain using the following headings:
(a) the built environment (b) farming
(c) water courses and lakes.

(11) Describe how each of the following helps reduce acid rain:
(a) lime (b) catalytic converters (c) low sulphur fuels (d) limestone scrubbers.

(12) 'Reducing the burning of fossil fuels is the only satisfactory way of reducing acid rain.'
Discuss this statement in relation to each of the following:
(a) legislation (b) technology (c) personal action.

1B.5 THE OZONE LAYER

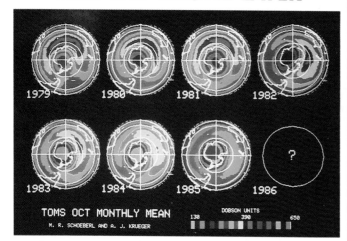

Satellite maps showing the development of a 'hole' in the ozone layer over Antarctica.

Ozone, O_3, and molecular oxygen, O_2, are allotropes – that is, different physical forms of the one element. Ozone is a pale blue gas which is usually present in the air at ground level in low concentrations. In low concentrations ozone is pleasant and refreshing; on sunny days this is noticeable. However, when air is polluted it contains a much higher concentration of ozone; at this level it is toxic and a pollutant to plant life. It reacts with hydrocarbons and oxides of nitrogen to form smog

and possibly causes the death of many trees. Ozone is often used to disinfect water in swimming pools and is used to bleach materials.

Higher up, in the stratosphere, ozone acts as a giant sunscreen, by absorbing dangerous ultraviolet radiation. The amount of ozone present is very small, but a thin layer of it filters out the harmful radiation. Some chemicals, mainly CFCs, are damaging this vital layer.

Chloroalkanes

In Unit 7.1 we saw that methane, CH_4, reacted in the presence of sunlight to form a series of compounds called chloroalkanes. The reactions can be shown as follows:

$$CH_4 + Cl_2 \Rightarrow CH_3Cl + HCl$$
chloromethane

$$CH_3Cl + Cl_2 \Rightarrow CH_2Cl_2 + HCl$$
dichloromethane

Many other chloroalkanes can form, such as trichloromethane, tetrachloromethane and trichloroethane. Chlorine molecules, Cl_2, are broken up into chlorine free radicals, $Cl^{\cdot}$, by the ultraviolet radiation from the Sun. These free radicals are very reactive and can cause many chemical reactions to occur. It is known that chlorine atoms (free radicals) are thousands of times more likely to react with ozone than with anything else.

314

The chlorofluorocarbons (CFCs) and hydrochlorofluorocarbons (HCFCs) are similar to chloroalkanes.

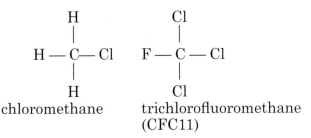

chloromethane   trichlorofluoromethane (CFC11)

CFCs and HCFCs are thought to be mainly responsible for the break-up of the ozone layer.

Formation of Ozone in the Stratosphere

Ozone is formed when ultraviolet radiation breaks up molecular oxygen, O_2, into oxygen atoms, O. Oxygen atoms are free radicals, O˙, because they have an unpaired electron. The oxygen radicals then combine with molecular oxygen to form ozone, O_3.

$$O_2 + uv \Rightarrow 2O˙$$
oxygen   radical

$$O_2 + O˙ \Rightarrow O_3$$
ozone

Once ozone has been produced it protects life on Earth from the ultraviolet radiation which produced the ozone in the first place.

Break-up of Ozone

Ozone, O_3, once it has formed, can also be dissociated again by the energy of the ultraviolet radiation. The photodissociation of ozone can be shown as:

$$O_3 + uv \Rightarrow O_2 + O˙$$

This indicates that the ozone layer is constantly being formed and destroyed owing to the effect of the ultraviolet radiation. The layer is, in fact, a giant equilibrium system, consisting of ozone, oxygen molecules and oxygen atoms. The concentration of ozone is said to be at a steady state when the rate of formation of ozone equals the rate of destruction of ozone. When this steady state is altered, damage to the ozone layer can occur, causing a 'hole' in the ozone layer.

Removal of Ozone and CFCs

Chlorofluorocarbons (CFCs) contain chlorine, fluorine and carbon. They were developed in the 1930s by Thomas Midgely. They are very useful compounds because they are very unreactive, are non-toxic, have low flammability and are cheap.

Uses of CFCs

• **Aerosol propellants:** CFCs are used to propel perfumes, deodorants, insecticides and other compounds into the atmosphere.
• **Refrigerants:** Refrigerators and air conditioning units use CFCs as coolants. When leakage occurs or when the refrigerators are scrapped, the CFCs escape into the atmosphere.
• **Solvents:** CFCs are used in dry cleaning and for cleaning electronic circuits.
• **Blowing agents:** Expanded plastics such as polystyrene are made into foams using volatile CFCs.

Residence Times of CFCs

CFCs are very stable compounds. They are so stable that they can remain unreacted in the atmosphere for hundreds of years. This means that the aerosol sprays and refrigerants which we use today may still cause damage to the ozone layer in ten, twenty or even in a hundred years time.

Breakdown of CFCs in the Stratosphere

Higher up in the atmosphere, the stratosphere, CFCs are broken down by the intense ultraviolet radiation present.

$$F-CCl_3 + uv \Rightarrow F-C˙Cl_2 + Cl˙$$

trichlorofluoromethane (CFC11)

The breakdown releases chlorine atoms, which are very reactive free radicals. These chlorine atoms break ozone up into molecular oxygen and chlorine monoxide, ClO.

$$Cl˙ + O_3 \Rightarrow ClO + O_2$$

$$ClO + O˙ \Rightarrow Cl˙ + O_2$$

$$O_3 + O˙ \Rightarrow 2O_2$$

The chlorine acts as a catalyst and is not consumed during the reactions. It is constantly regenerated in this catalytic cycle. While the concentration of chlorine atoms is small compared to the concentration of the oxygen atoms, a single chlorine atom can remove about 1 million ozone atoms.

Other free radicals are also destroying ozone. Oxygen atoms and the oxides of nitrogen, NO and NO_2, also take part in catalytic cycles which break up the ozone layer. While Cl, O and NO_x free radicals are thought to be mainly responsible for the break-up of ozone, many other complex reactions affect the gases present in the atmosphere.

Methane and CFCs

Chlorine free radicals are so reactive that they should have destroyed most of the ozone in the atmosphere. However, other gases react with the chlorine and remove it.

For example, methane reacts with chlorine atoms and forms chloromethane and hydrogen chloride.

$$CH_4 + Cl \Rightarrow CH_3Cl + HCl$$

Some of the HCl can fall down to the Earth in rain, thereby removing some of the chlorine from the atmosphere.

CFCs and the Future

• In 1974, Mario Molina and Sherry Rowland predicted that CFCs were destroying the ozone layer more quickly than it could be produced.
• In 1985, Joe Farman reported that the level of ozone over the Antarctic had been declining for some years. The researchers also noticed that when ozone is depleted, chlorine monoxide, ClO, is present. CFCs are suspected as the source of chlorine atoms.
• Nimbus satellites have measured the thickness of the ozone layer using spectroscopy.
• In 1990, the environment ministers of 59 countries and the EU signed a declaration to phase out the use of CFCs by 1999.
• Safer replacements for CFCs are being investigated. Hydrochlorofluorocarbons (HCFCs) are thought to be safer because they have lower ozone depletion potentials.

Exercise 1B.5

(1) (a) What is a chloroalkane?
(b) How are chloroalkanes produced?
(c) Draw the structure of chloromethane.
(d) Show, using structural formulae, how CFCs are similar to chloromethane.

(2) (a) What is ozone?
(b) List some common properties of ozone.
(c) How is ozone used commercially?
(d) 'Ozone causes problems in the atmosphere and also protects life on earth.' Discuss this statement, briefly.

(3) (a) What is the stratosphere?
(b) How is ozone formed in the stratosphere?
(c) Write equations for the formation of ozone in the stratosphere.
(d) Why is the presence of ozone of vital importance to us?

(4) (a) What causes ozone to break down?
(b) How is ozone reformed?
(c) Is the concentration of ozone constant in the stratosphere?
(d) What is changing the concentration of ozone from its steady state?

(5) (a) What are CFCs and HCFCs?
(b) Why are CFCs useful?

(c) List some uses of CFCs.
(d) What is meant by the residence time of a CFC?
(e) Why does the residence time of a CFC cause a problem?

(6) (a) How are CFCs broken down in the stratosphere?
(b) What is released into the stratosphere when CFCs break down?
(c) What happens to the ozone molecules when CFCs break down?
(d) Write equations showing how ozone is removed by Cl atoms.
(e) What role does chlorine monoxide, ClO, play in the process?

(7) (a) Why have the chlorine atoms not destroyed all of the ozone in the atmosphere?
(b) What role does methane play in preserving the ozone layer?
(c) Why is there a relatively high concentration of methane in the atmosphere?

(8) (a) What compound did scientists first notice when they reported the 'hole' in the ozone layer?
(b) What steps have been taken to limit the use of CFCs?
(c) What compounds are being used as replacements for CFCs?

Some of the more important key terms are listed below. Other terms not listed may be located by means of the index.

1. Characteristics of an effective chemical industry: The development of an effective and successful chemical process relies on many factors which include the following:

Raw materials
Energy requirements
Costs
Location
Rate of reaction and yield of product
Co-products and by-products
Safety
Control of waste and effluents
Plant materials
Operation process and quality control

2. Irish chemical industry: The Irish chemical industry can be roughly divided up into the following areas:
Pharmaceuticals and fine chemicals
Agrochemicals and animal health products
Cosmetics and household products
Bulk chemicals and plastics

3. Oxygen: Oxygen is the most abundant element on Earth, making up about 48% of the total Earth. Pure oxygen is an extremely important industrial material. It is prepared directly from the air which contains about 21% oxygen. Oxygen, O_2, is a reactive gas; it reacts quite slowly at first because the oxygen–oxygen bond is quite strong. Otherwise, all the organic material on the Earth would quickly burn up and life would not have evolved. However, once combustion starts, most materials burn quickly in oxygen.

4. Nitrogen: Molecular nitrogen, N_2, is the principal constituent of air (78%), from which it can be obtained by liquefaction and by fractional distillation. The nitrogen molecule is unreactive due to the high bond energy of the triple bond in N_2.

5. Carbon dioxide: Carbon dioxide is a colourless, odourless gas. It is moderately soluble in water, forming a weak acidic solution: it is an acidic oxide. Aqueous carbon dioxide is often called carbonic acid: this provides the fizz in fizzy (carbonated) drinks.

The combustion of carbon in fossil fuels yields carbon dioxide gas, whereas incomplete combustion gives carbon monoxide gas.

6. Atmospheric pollution: Air pollution is caused either by small suspended solids or liquids in the air or by poisonous or obnoxious gases.

7. Sources of pollution: The main sources include: domestic fires, industrial sources and motor vehicles.

8. Acid rain: When the water vapour in the atmosphere absorbs large quantities of sulphur dioxide and the oxides of nitrogen, the pH of rain can become as low as pH 3.

9. Effects of acid rain on the environment: Acid rain affects buildings, crops, trees, water courses and fish.

10. Prevention of acid rain: Acid rain can only be reduced by a reduction in the use of fossil fuels. The effects of acid rain can be reduced in following ways:
(a) use of lime to neutralise lakes
(b) use of catalytic converters
(c) use of low sulphur fuels
(d) scrubbing of waste gases.

11. Ozone: Ozone, O_3, and molecular oxygen, O_2, are allotropes. When air is polluted it contains high concentration of ozone; at this level it is toxic and a pollutant to plant life. It reacts with hydrocarbons and oxides of nitrogen to form smog and possibly causes the death of many trees. In the stratosphere, ozone acts as a giant sunscreen, by absorbing dangerous ultraviolet radiation. CFCs and HCFCs are thought to be mainly responsible for the break-up of the ozone layer. However, HCFCs have a lower ozone depletion potential than CFCs.

2A.1 CRYSTALS

Vitamin A crystals.

Diamond crystals.

Crystals are easily recognisable by their beautiful forms. The crystalline structure may vary from the wonderful patterns in snowflakes to the stunning simplicity of a diamond.

The study of crystals is of interest to a variety of people: the geologist interested in mineral identification, the engineer developing a new computer micro-chip or the scientist attempting to make a better superconductor.

For the student of chemistry, the crystalline state provides a means of studying the properties of many substances and their structures by examining the relationship between crystal type and its binding forces. The property which differentiates a crystal from all other substances is its precisely ordered structure.

Crystals are formed by atoms, ions or molecules arranged in a regular geometric arrangement called a lattice.

In 1912, a German scientist, **von Laue**, discovered that crystals consisted of an orderly arrangement of particles by passing a beam of X-rays through a crystal and observing the effect on a photographic plate on the far side of the crystal. **Lawrence Bragg** and his father, **William**, studied the interference patterns produced in this way and calculated the distances between the layers of particles, which enabled them to work out the different types of crystalline structures. The Braggs were jointly awarded the Nobel Prize for Physics in 1915 for their work on crystal structure using X-ray diffraction.

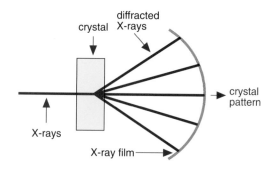

Crystal structures of many complex organic molecules were later determined using X-ray diffraction techniques. In 1957, **Dorothy Hodgkin** worked out the precise structure of vitamin B_{12}, a molecule consisting of 180 atoms. She used an X-ray diffractometer and a computer to produce a model of vitamin B_{12}.

318

She was awarded the 1964 Nobel Prize for Chemistry for her years of careful research on the structures of biological molecules. Many other complex structures, such as penicillin and DNA, were determined using X-ray diffraction techniques.

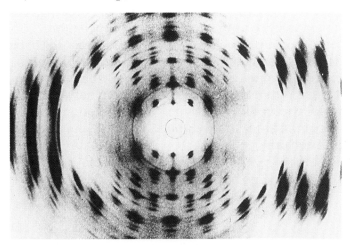

X-ray diffraction image of the DNA molecule. An image such as this played the major role in the unravelling of the double helix structure of the DNA molecule by Crick and Watson in 1953.

Sir William Henry Bragg 1862–1942 and
Sir William Lawrence Bragg 1890–1971

William Bragg and his son Lawrence shared the 1915 Nobel Prize for Physics for their study of crystal structures by means of X-rays. William Bragg was born in Cumberland, England. He was described as generous in spirit, warm and free from pretence and as a lover of tradition and craftsmanship. In 1885, J. J. Thomson helped him to obtain an appointment as Professor of Mathematics and Physics in Adelaide, Australia. He was so busy building up the resources of his laboratory and helping to develop the new university that he never thought of doing research for nearly 20 years. In his forties he was invited to give a lecture on radioactivity and decided to do some experiments relating to the subject. Thereafter he became world-famous as a pioneer of radioactivity. In 1912, a German scientist, von Laue, published a paper showing beautiful photographs which were obtained by passing a beam of X-rays through a crystal. Lawrence Bragg interpreted the patterns obtained by von Laue as a reflection of X-rays from atom-rich planes in a crystal. His father, William, designed an X-ray spectrometer which enabled Lawrence to examine many crystals. Together the Braggs discovered the structure of many substances and started the science of X-ray crystallography.

In 1985, Robert Curl, Harold Kroto and Richard Smalley discovered that the element carbon can also exist in the form of spheres. They called the new carbon ball, C_{60}, buckminsterfullene, after the architect, R. Buckminster Fuller, who used hexagons and a small number of pentagons to create the "curved" surface on the Montreal World Exhibition building in 1967. C_{60} consists of 12 pentagons and 20 hexagons with carbon atoms at each corner — the same as a soccer ball. Hence the name "bucky balls". Bucky balls can enclose metals in them, can be used as superconductors and may be used as catalysts.

Panels on the surface of the ball are made up of hexagons and pentagons.

Crystalline solids which show regular, orderly structure are called lattices. A lattice is a three-dimensional array of points in a regular repeating pattern and is characterised by the distances between each repeating point and by the angles between the axes. The smallest repeating array of points which will make up an entire crystal is called the 'unit cell'.

A simple lattice and unit cell are shown below.

unit cell is cubic

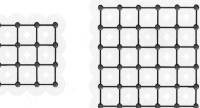

Final crystal is cubic
(For clarity only the front face is shown)

Classification of Crystals

Crystals are classified into four types according to the kind of particles that make up the crystal lattice.

(1) Ionic crystals
(2) Molecular crystals
(3) Covalent macromolecular crystals
(4) Metallic crystals

(1) Ionic Crystals

Ionic crystals are formed when positive and negative ions bind together by electrostatic forces.

The electrostatic force extends in all directions around each ion. Each positive ion is surrounded by negative ions and each negative ion is surrounded by positive ions in a stable regular geometric arrangement. The ions pack together as closely as possible in accordance with their size and in such a way that ions of like charge can avoid close contact.

There are several types of ionic lattices. Sodium chloride, common salt, is perhaps the best known.

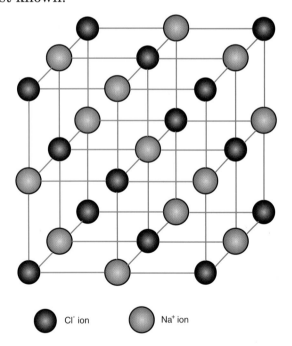

● Cl⁻ ion　　● Na⁺ ion

In sodium chloride each positive sodium ion is surrounded by six negative chloride ions and each negative chloride ion is surrounded by six positive sodium ions. In this way a giant cubic ionic lattice is built up. All other ionic compounds consist of orderly arrangements of ions, but the lattice need not necessarily be cubic in shape.

Physical Properties and Binding Forces

As the forces which bind ionic crystals together are relatively strong, such crystals are hard and have high melting points and high boiling points. While ionic crystals are hard, they are also very brittle, i.e. they cannot be easily deformed by application of a force without breaking up. The application of a force disturbs the regular geometric arrangement of the ions, bringing ions of the same charge close to each other. When this occurs, the forces between the ions are no longer attractive but repulsive and the crystal layers fly apart, causing the crystal to fracture.

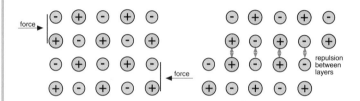

Effect of a deforming force on a crystal.

The same rigid geometric arrangement of fixed ions is responsible for the fact that ionic compounds do not conduct electricity. In order to conduct electricity an ion must move under the influence of an electrical field. This mobility is not possible in an ionic lattice where the ions occupy fixed positions. When ionic compounds are melted or in solution, the ions are freed from their fixed positions and are then excellent conductors of electricity.

The binding force is an ionic crystal is high (1000 kJ mol⁻¹) for most typical ionic crystals. This large binding energy is overcome when the crystals are dissolved in polar solvents because of the interaction between the ions and the polar solvent molecules. Consequently, ionic compounds like sodium chloride dissolve in polar solvents like water. Ion–dipole interaction pulls the crystal apart because it is stronger than the ion–ion interaction and the hydrogen bonds in the solvent.

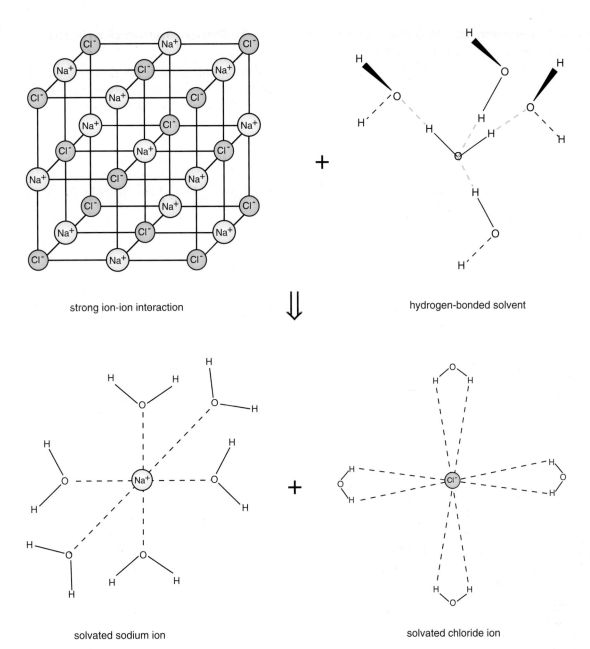

strong ion-ion interaction

hydrogen-bonded solvent

solvated sodium ion

solvated chloride ion

Dissolving sodium chloride in water.

(2) Molecular Crystals

Molecular crystals consist of two types: (i) non-polar molecular crystals and (ii) polar molecular crystals.

Non-polar Molecular Crystals

Non-polar molecular crystals are formed when non-polar molecules bind together by weak van der Waal's forces.

These may be simple molecules containing only non-polar covalent bonds, such as H_2, O_2, I_2 or S_8, or they may be molecules whose individual bonds are polar, such as CO_2 or CH_4, but due to their symmetric arrangement the polarities cancel each other.

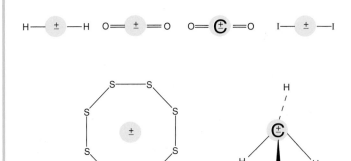

Iodine, I_2, is perhaps the most common simple non-polar molecular lattice. It consists of a regular arrangement of diatomic molecules held together in a cubic lattice by weak van der Waal's forces.

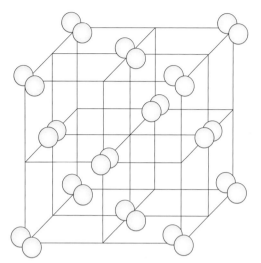

The crystal structure of iodine

Physical Properties and Binding Forces

Non-polar substances tend to be soft and volatile and have low boiling points due to the weak intermolecular forces which hold them together. They do not conduct electricity because they contain no mobile ions or mobile electrons. Non-polar molecular liquids are good solvents for other non-polar substances. Iodine is soluble in tetrachloromethane (non-polar) but not in water (polar).

The binding forces in iodine and tetrachloromethane are both similar in size (non-polar). This allows the two substances to intermingle with each other and form a solution.

The interaction between the non-polar iodine and the polar water is not strong enough to disturb the strong hydrogen bonds between the water molecules. As a result no solution forms.

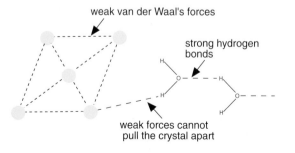

weak van der Waal's forces

strong hydrogen bonds

weak forces cannot pull the crystal apart

Polar Molecular Crystals

Polar molecular crystals are formed when polar molecules bind together by dipole–dipole interactions.

Common examples of molecules with dipoles include hydrogen chloride, HCl, iodine chloride, ICl, monochloromethane, CH_3Cl, and water, H_2O, the most important dipole of all.

$$H^{\delta+}\!\!-Cl^{\delta-}$$
$$I^{\delta+}\!\!-Cl^{\delta-}$$

Physical Properties and Binding Forces

Polar molecules (dipoles) can form relatively strong bonds with each other arising from the unequal distribution of charge.

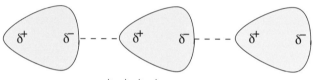

dipole-dipoleinteraction

When the interaction is between hydrogen and a very electronegative element such as fluorine, chlorine, oxygen or nitrogen, the interaction is even stronger and is called hydrogen bonding.

$$H^{\delta+}\!\!-Cl^{\delta-}----H^{\delta+}\!\!-Cl^{\delta-}----H^{\delta+}\!\!-Cl^{\delta-}----H^{\delta+}\!\!-Cl^{\delta-}$$

hydrogen bonding

Because the dipole–dipole interaction is strong, polar molecules have higher boiling points, are less volatile and are harder than non-polar substances. Polar liquids are excellent solvents for ionic solutes and for other polar substances because they can form ion–dipole and dipole–dipole interactions, respectively. Although polar liquids contain some free ions, there are usually not enough ions to conduct an electric current.

(3) Covalent Macromolecular Crystals

Covalent macromolecular crystals are crystals in which all the atoms are linked together in a vast three-dimensional network by covalent bonds.

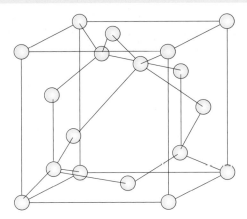

The structure of diamond

The most familiar examples include diamond, a vast network of tetrahedral carbon atoms, and quartz, SiO_2, which is also tetrahedral in shape.

Properties and Binding Forces

The covalent bonds which hold the lattice together are strong (1000 kJ mol^{-1}) and make the crystalline structure very rigid. The entire crystal can be considered as a giant covalent molecule. Covalent network solids are extremely hard and have high melting points and high boiling points, as large numbers of covalent bonds must be broken to destroy the crystal lattice. Because the breaking of this arrangement requires such high energy, covalent network crystals are insoluble in most polar and non-polar solvents. Covalent crystals do not conduct electricity because all the available electrons are used in bonding and are not free to move throughout the crystal structure and conduct an electrical current.

Graphite is Soft and Conducts Electricity

Graphite is not a pure covalent crystal. In graphite, each carbon atom is covalently bonded to three other carbon atoms, arranged in flat hexagonal sheets. The fourth outer electron of the carbon is not required in this bonding arrangement and is free to move throughout the crystal. The hexagonal layers are held together by weak van der Waal's forces. As the

binding forces between the hexagonal layers are weak, the carbon layers are free to move easily over one another just like a pack of cards. For this reason, graphite is an excellent lubricant. Graphite, unlike diamond, is extremely soft. The 'free' electrons in graphite form an electron cloud between the layers and consequently graphite can conduct electricity.

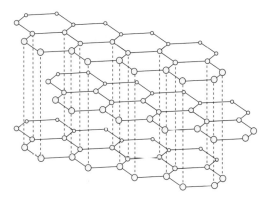

The structure of graphite

(4) Metallic Crystals

Metallic crystals may be considered as a regular geometric arrangement of positive ions surrounded by a sea of electrons.

The crystals are held together by strong metallic bonds, formed by the interaction of the freely moving electrons and the fixed positive metal ions.

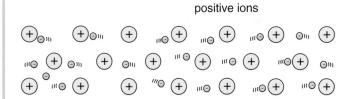

positive ions

'sea' of electrons

The structure of a metallic crystal

Atoms in metals have several unoccupied orbitals in their outer shells. Overlap of these orbitals enables the valence electrons of the metal to move from one valence shell to another. The valence electrons are said to be 'delocalised', that is, they do not belong to one particular atom but rather to the entire metallic crystal.

The crystal is like an orderly array of oranges packed in a box, where the oranges represent the positive metal ions and the air spaces in the box represent the sea of electrons. X-ray diffraction shows that metals have a closely packed structure.

Physical Properties and Binding Forces

• Most metals have high melting points because the metallic bond is strong.
• Metals generally have high densities because the positive metal ions are closely packed together.
• Metals are good electrical conductors because electrons can easily move throughout the metal. The incoming electrons (from the applied field) can easily replace electrons in the electron sea and move throughout the system.
• Metals are good conductors of heat because the mobile electrons acquire high kinetic energy and move rapidly throughout the metallic crystal transporting heat.

Electrons moving through a metal

• Metals are easily distorted. Since the electrons are delocalised throughout the crystal, the positive ions can be moved about with relative ease. Metals can, therefore, be distorted without disrupting the crystal structure, enabling them to be stretched (ductile), hammered (malleable) or bent (elastic) without fracturing the crystal.

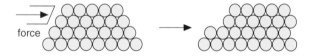

The positions of the positive ions in a metal before and after a displacing force

Chemistry in Action: Semiconductors

Pure silicon is an insulator. The addition of trace amounts of certain elements to the pure covalent structures of silicon or germanium increases their ability to conduct an electric current.

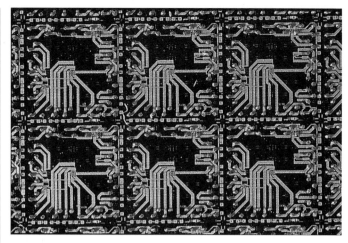

Light micrograph showing a section of a semiconductor wafer featuring several chips. Each chip comprises a complex pattern of semiconductor materials.

Each silicon atom has four valence electrons, involved in bonding in a crystalline arrangement similar to that of diamond. If trace amounts of boron, or any other group III element, are added to pure silicon the conductivity increases dramatically. The boron atom, unlike silicon, has only three valence electrons and consequently creates a 'hole' in the crystal lattice. This hole allows other electrons to 'hop' through the crystal system.

Semiconductors of this type are called p-type semiconductors (p for positive). If phosphorus or any other group V element is added to pure silicon another type of semiconductor is produced, called an n-type semiconductor (n for negative). n-type semiconductors have an excess of electrons which carry the current throughout the crystal.

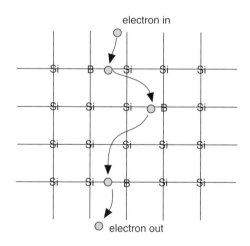

p-type semiconductor

In Unit 5.2, the use of space filling and ball and stick models was introduced to illustrate the structure of organic molecules. Similar models can be used to demonstrate crystalline structures.

Try to build some of the following using a set of molecular models:

Diamond: Select some carbon atoms (these are black with four holes). Try and fit as many as possible together. What do you observe?

Graphite: Select the carbon atoms again. Using only three holes, build as large a structure as possible. What is the difference between this structure and the diamond structure?

Metal: Select as many atoms of the same size as you can (the colour does not matter) to represent metal atoms. Build them up into a pile in a small rectangular box. Are the metal atoms closely packed?

Classification of Crystal Structures

Type	Structural unit	Bonding	Properties	Examples
Ionic	Positive and negative ions	Strong electrostatic attraction	Hard, brittle, high melting points, conduct when molten or in solution	NaCl MgO $NaNO_3$
Non-polar molecular	Non-polar molecules	Weak van der Waal's forces	Soft, low melting points, non-conductors	I_2 CO_2
Polar molecular	Polar molecules	Dipole–dipole interactions or hydrogen bonding	Soft, low melting points, poor conductors, solvents for ionic and polar substances	H_2O NH_3
Covalent macromolecular	Atoms	Strong covalent bonds	Very hard, high melting points, non-conductors	Diamond SiO_2
Metallic	Metallic ions	Metallic	High melting points, easily distorted, good conductors	Cu Fe Na

(1) (a) What is a crystal?
(b) List the different types of crystal structures.
(c) Name the binding force in each crystal type.

(2) (a) How did William and Lawrence Bragg study crystal structures?
(b) What is a unit cell?
(c) What contribution did Dorothy Hodgkin make to the development of crystal structures?

(3) Compare the bonding forces in ionic, non-polar molecular, polar molecular, covalent macromolecular and metallic crystals with each other.

(4) Some crystals are extremely brittle. Explain why by referring to the bonding forces between the crystal units.

(5) Ionic crystals contain ions. Explain why they are poor electrical conductors.

(6) The binding energy in an ionic crystal is high ($100–1000$ kJ mol^{-1}). Explain how this high energy is overcome when an ionic compound dissolves in water.

(7) Compare the properties of polar molecular and non-polar molecular compounds with reference to the following:
(a) binding forces (b) crystal unit (c) hardness
(d) solubility (e) electrical conduction.

(8) Sodium chloride, $NaCl$, is soluble in water; ethanol, C_2H_5OH, is also soluble in water; while iodine, I_2, is practically insoluble in water. Discuss the reasons for this trend.

(9) Iodine is soluble in tetrachloromethane and practically insoluble in water. Give reasons for this.

(10) Explain the meaning of the following terms. In each case give an example of each force.
(a) Van der Waal's forces (b) dipole–dipole interaction (c) ion–dipole interaction.

(11) What type of forces must be overcome in order to melt crystals of the following:
(a) KCl (b) SiO_2 (c) C (diamond) (d) H_2O
(e) BaO (f) Ag (g) Br_2?

(12) The chemical bonds in diamond are covalent. The chemical bond in iodine is also covalent, yet iodine melts at a much lower temperature than iodine. Explain this difference in properties.

(13) Diamond is a hard crystalline solid which does not conduct electricity, while graphite is a relatively soft crystalline solid which does conduct electricity. Why are their properties so different?

(14) Compare the crystal structures in
(a) CO_2, (b) SiO_2 and (c) MgO, by referring to the following:
(i) crystal unit (ii) binding forces (iii) hardness (iv) solubility (v) electrical conduction.

(15) Classify each of the following solid elements as ionic, metallic, molecular or covalent macromolecular:
(a) tin, Sn (b) sulphur, S_8 (c) iodine, I_2.

(16) On the basis of the description given classify each of the following solids as polar molecular, non-polar molecular, metallic, ionic or covalent macromolecular.
(a) A hard white solid with a high melting point. It is a non-conductor of electricity but when it dissolves in water it does conduct electricity.
(b) A hard black solid which does not conduct electricity. It dissolves in hexane but not in water.
(c) A lustrous yellow solid that conducts electricity.
(d) A black solid, which is insoluble in water and which conducts an electrical current.

(17) Use your knowledge of crystal structures to explain each of the following:
(a) sodium chloride does not conduct electricity when solid but does when it is molten
(b) copper is a hard dense metal
(c) sodium chloride crystals grow in a regular cubic shape
(d copper conducts electricity
(e) iron is malleable.

2A.2 ADDITION POLYMERS

What are Polymers?

Polymers are large chain-like molecules which are built up from smaller molecules called monomers. The word polymer comes from the Greek 'poly', meaning many and 'meros' meaning part.

For example, poly(ethene) is made up from many ethene molecules. The ethene molecules are joined together something like a chain of paper clips.

$$n \left(\begin{array}{c} H \quad\quad H \\ \diagdown \quad\quad \diagup \\ C = C \\ \diagup \quad\quad \diagdown \\ H \quad\quad H \end{array} \right) \longrightarrow \left[\begin{array}{c} H \quad H \\ | \quad | \\ C - C \\ | \quad | \\ H \quad H \end{array} \right]_n$$

ethene   poly(ethene)

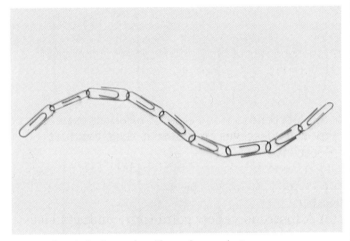

Paper clips linked together like polymer chains.

All living things are made from natural polymers like protein, starch and cellulose. The first synthetic polymers were made less than a hundred years ago. The first plastic was, perhaps, discovered in 1846 when **Christian Schoenbein** spilled some nitric and sulphuric acids and wiped up the mess with his wife's cotton apron. This resulted in the formation of cellulose nitrate which formed the basis of many industries: gun powder, artificial fibres and photographic films. The first real synthetic polymers were developed by chemists like **Baekeland**, who discovered bakelite in 1909 and **Carothers**, who discovered nylon in 1937.

Today we take polymers for granted, whether they are natural polymers, such as rubber, silk and cotton, or synthetic polymers, such as elastomers, plastics or fibres.

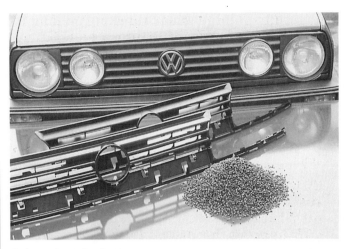

Polymers used as car components.

Polymerisation of Alkenes: Addition Polymerisation

Ethene is used to make poly(ethene) (polythene) by making the ethene monomers squeeze closer together. This is done using an extremely high pressure and a catalyst. The double bond in ethene is quite reactive and breaks open, allowing the ethene molecules to react with each other. This is called addition polymerisation.

Ethene + Ethene + Ethene + - - - - - $\longrightarrow$ Polythene chain

$$\begin{array}{c} H \quad H \quad H \quad H \quad H \quad H \\ \diagdown\diagup \quad\quad \diagdown\diagup \quad\quad \diagdown\diagup \\ C=C \; + \; C=C \; + \; C=C \longrightarrow \\ \diagup\diagdown \quad\quad \diagup\diagdown \quad\quad \diagup\diagdown \\ H \quad H \quad H \quad H \quad H \quad H \end{array} \quad \begin{array}{c} H \; H \; H \; H \; H \; H \\ | \; | \; | \; | \; | \; | \\ -C-C-C-C-C-C- \\ | \; | \; | \; | \; | \; | \\ H \; H \; H \; H \; H \; H \end{array}$$

Low-Density and High-Density Poly(ethene)

Poly(ethene) can be made in two ways: with branches (low density) and in straight chains (high density).

Low-density and high-density poly(ethene) differ in the amount of branching which occurs during polymerisation. Side chains may form during the reaction; the more side chains formed, the less dense the polymer. The side chains prevent other side chains from packing closely together, which results in a low-density polymer. The reaction conditions and the particular catalyst used determine whether the poly(ethene) produced is low-density poly(ethene) (LDPE) or high-density poly(ethene) (HDPE).

Low-density poly(ethene) is produced when free radicals are used under high pressure.

This results in large scale branching making it impossible for the chains to fit neatly together.

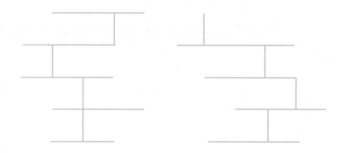

Branched - cannot fit the two together tightly

High-density poly(ethene) is produced when ionic catalysts, called Ziegler–Natta catalysts, are used at low pressure. The poly(ethene) formed has no branching along the polymer chain. This allows the chains to be packed neatly as in a crystal. This makes the high-density poly(ethene) more crystalline than the low-density poly(ethene).

individual strands

strands packed close together

high density

Chains can fit together tightly

More Addition Polymers

Monomer	*Polymer*

$$n\left(\begin{array}{c}\text{H}\quad\ \ \text{H}\\ \diagdown\ \ \diagup\\ \text{C}=\text{C}\\ \diagup\ \ \diagdown\\ \text{H}\quad\ \ \text{Cl}\end{array}\right) \longrightarrow \left[\begin{array}{c}\text{H}\ \ \ \text{H}\\ |\ \ \ \ |\\ \text{C}-\text{C}\\ |\ \ \ \ |\\ \text{H}\ \ \ \text{Cl}\end{array}\right]_n$$

Vinylchloride — Polyvinylchloride

$$n\left(\begin{array}{c}\text{F}\quad\ \ \text{F}\\ \diagdown\ \ \diagup\\ \text{C}=\text{C}\\ \diagup\ \ \diagdown\\ \text{F}\quad\ \ \text{F}\end{array}\right) \longrightarrow \left[\begin{array}{c}\text{F}\ \ \ \text{F}\\ |\ \ \ |\\ \text{C}-\text{C}\\ |\ \ \ |\\ \text{F}\ \ \ \text{F}\end{array}\right]_n$$

Tetrafluoroethene — Polytetrafluoroethene

$$n\left(\begin{array}{c}\text{H}\quad\ \ \text{H}\\ \diagdown\ \ \diagup\\ \text{C}=\text{C}\\ \diagup\ \ \diagdown\\ \text{H}\quad\ \ \text{C}_6\text{H}_5\end{array}\right) \longrightarrow \left[\begin{array}{c}\text{H}\ \ \ \text{H}\\ |\ \ \ |\\ \text{C}-\text{C}\\ |\ \ \ |\\ \text{H}\ \ \ \text{C}_6\text{H}_5\end{array}\right]_n$$

Styrene — Polystyrene

$$n\left(\begin{array}{c}\text{H}\quad\ \ \text{H}\\ \diagdown\ \ \diagup\\ \text{C}=\text{C}\\ \diagup\ \ \diagdown\\ \text{H}\quad\ \ \text{CN}\end{array}\right) \longrightarrow \left[\begin{array}{c}\text{H}\ \ \ \text{H}\\ |\ \ \ |\\ \text{C}-\text{C}\\ |\ \ \ |\\ \text{H}\ \ \ \text{CN}\end{array}\right]_n$$

Acrylonitrile — Polyacrylonitrile

Properties and Uses of LDPE and HDPE

	Density/g cm^{-3}	Crystallinity/%	Tensile strength / atm	Softening point/°C	Uses
LDPE	0.92	55	90–135	633	Plastic film for packaging food
HDPE	0.96	80	200–310	673	Injection moulded products such as food containers and buckets

328

Industrial and Domestic Importance of Polymers

Polymers for sport.

Polymers have widespread uses because they are strong, flexible, easy to shape and can often be tailor-made to suit almost any purpose. They consist of long thin chains, like strands of spaghetti. If the strands are linked strongly together this produces a solid mass, e.g. a polypropene milk crate. If they are not strongly linked together this produces a soft solid which can be bent, e.g. 'polythene'.

Teflon, poly(tetrafluoroethene), is almost inert and extremely resistant to wear. It is used as a replacement for hip and knee joints and as surfaces for frying pans and skis.

HDPE is strong and resistant to heat. It can be blow moulded into any shape and can be sterilised by heating.

Many polymers can be drawn through a small hole, making the polymer molecules line up alongside one another to form fibres as in poly(propene) which is used to make carpets. Many other polymer fibres are used to make fabrics and clothes.

Discoveries of Polymers
LDPE

In 1933, **Gibson** and **Fawcett**, two chemists in ICI, were working on the reaction between ethene and benzaldehyde under high pressure. They had hoped to form a ketone but some of the reaction mixture leaked out. When they added more ethene they found a white waxy solid: this was poly(ethene). They repeated the reaction many times, sometimes with success, but at other times the reaction mixture blew up. They later identified oxygen as a source of free radicals, which started the polymerisation of the ethene. The use of free radicals caused large-scale branching of the polymer chains, forming what is now called low-density poly(ethene), LDPE.

HDPE

In 1953, **Karl Ziegler** passed ethene gas at atmospheric pressure through a small amount of an organometallic compound, $(C_2H_5)_3Al$, and $TiCl_4$. This caused the production of poly(ethene) with no branching along the polymer chain. These polymer chains can pack much more easily than those made by the original process using high pressure. This polymer is high-density poly(ethene), HDPE. This catalyst is extremely efficient but catches fire quite easily. Recently a catalytic system was developed in the USA which uses a chromium(III) oxide, Cr_2O_3, aluminosilicate system which is much safer to use.

Some Common Addition Polymers and Their Uses

Common name	Polymer	Monomer	Uses
Polythene	Poly(ethene)	Ethene	Plastic packaging, squeezy bottles
Polypropylene	Poly(propene)	Propene	Kitchenware, carpets
Polystyrene	Poly(phenylethene)	Phenylethene	Plastic cups, insulation
Polytetrafluoroethylene (PTFE)	Poly(tetrafluoroethene)	Tetrafluoroethene	Knee and hip joints, ski surfaces
Polyvinylchloride (PVC)	Poly(chloroethene)	Chloroethene	Pipes and gutters, floor covering

Teflon

In 1938, **Roy Plunkett**, a chemist working for Du Pont, discovered Teflon quite by chance. He was studying gaseous tetrafluoroethene which was stored in steel cylinders. He opened the valve on one of the cylinders which he assumed was full but found that no gas came out. He was sure the cylinder was full and decided to examine it by cutting it in half. When the cylinder was cut open it contained a white powder, which was poly(tetrafluoroethene), now commonly called Teflon.

Activity 2A.2

Demonstration of the Physical Properties of Some Addition Polymers

Introduction

The purpose of the demonstration is to compare the density, flexibility, softening point and hardness of poly(ethene), poly(chloroethene) and poly(phenylethene).

Requirements

Safety glasses, equal-sized strips of LDPE, HDPE, poly(chloroethene) and poly(phenylethene), balance, graduated cylinder, water, two metal stands, weights and metre stick, Bunsen burner, tripod and metal plate, hammer and nail.

Procedure

[1] Density
(1) Measure the density of each of the polymer samples by finding the mass and the volume of each in the usual way.
(2) Record your results in your table .

[2] Flexibility
(1) Pull each sample to see if it can be stretched.
(2) Try to break each sample with your hand.
(3) Measure how much each polymer can bend before breaking, as shown.

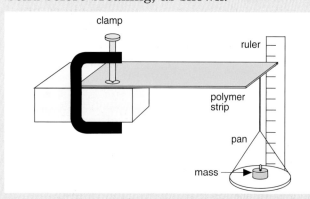

(4) Record your observations in your table.

[3] Softening point
(1) Place a strip of each polymer on the metal sheet.
(2) Place the metal sheet on the tripod and heat.

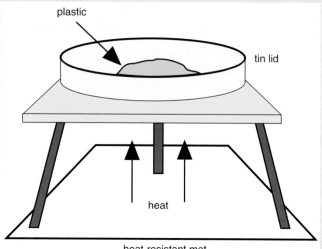

(3) List the polymers in order as they soften.
(4) Record your observations in your table.

[4] Hardness

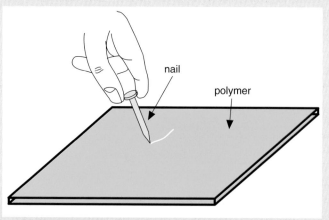

(1) Scratch each polymer with a sharp nail.
(2) Try to break each polymer with a hammer.
(3) Record your observations in your table.

Plastic Recycling

Plastics have become almost indispensable. Cars, trains and aeroplanes would not function without plastics. Tennis players on the professional circuit would not play as well without plastics. People rarely dress without putting on a garment made of some polymer or other. Many engineering processes and some life-saving medical techniques depend on plastic. Plastics have proved to be exceptionally versatile and convenient materials and yet are seen as a major problem in terms of disposal. Plastics are not the main culprits in the volumes of waste which need to be recycled or treated. Other materials, such as paper, metals, glass and textiles make up far more of our waste. It should also be remembered that the use of plastics has a far less harmful effect on the environment than many alternative materials.

Management of Plastics

• **Landfill:** Plastic used as landfill is wasteful in terms of energy. Most plastics have a calorific value close to heating oil. Waste plastics can be used as an energy source.
• **Incineration:** Using proper technology energy can be recovered from plastics which reduces the need for landfill.
• **Recycling:** Many plastics can be converted back into pellets. However, it is difficult even for experts to identify all plastics during sorting. Recycling of mixed plastics lowers the value of the plastic but mixed plastics have been used to make seats, fences, flooring and many other products.
• **Degradeable plastics:** Plastics have been developed which are biodegradeable, photodegradeable and which are degraded by the environment (air, heat, sun, water, wind and rain). Biodegradation of plastics, like the biodegradation of vegetable and animal wastes, relies on

Recyclable plastics developed by Bayer.

the presence of oxygen to degrade the material by bacterial action.
• **Reduction at source:** The most important step in the management of plastics is the need to reduce the amount of plastics used. New technology is being used to make plastics thinner and lighter. Recently the thickness of polythene cling film has been halved and the weight of soft drink bottles has been cut by a third.

Recycling of Polystyrene

Poly(phenylethene) or polystyrene has widespread applications. It is used as plastic cups, as ceiling tiles and as an insulator in cavity walls and freezers. Recycling of the large amount of polystyrene used helps solve the problem of plastic waste, in terms of the environment and financially.

The main stages involved include:
• **Collection:** Collection of plastics for recycling is a not as simple as it seems. It is labour intensive and costly.
• **Sorting:** Each type of plastic has its own particular properties. Polystyrene must be carefully sorted out from other plastics. Sometimes even experts have difficulty knowing one plastic from another.

331

• **Shredding:** The polystyrene is ground up and shredded into granules.
• **Washing and drying:** The polystyrene pellets are washed to remove any labels or other materials present and then dried to remove any excess moisture present.
• **Re-extrusion:** The polystyrene granules are melted down and reformed by extrusion. The re-extruded polystyrene can then be used to form new products.
• **Recycled products:** The products formed lose some of the properties of the original polymer. Recycled polymers are not as high in quality as the original material.

Exercise 2A.2

•

(1) (a) What are monomers and polymers?
(b) How do monomers join together to form an addition polymer?
(c) Name three natural polymers.
(d) Name three synthetic polymers.

(2) Name the monomer in each of the following polymers.
(a) polythene (b) polvinylchloride (c) polystyrene (Teflon).

(3) (a) What is the difference between high-density and low-density poly(ethene)?
(b) Draw diagrams to illustrate the difference.
(c) Give two uses for each type of polymer.

(4) Describe, using diagrams, how you would demonstrate some physical properties of the following polymers:
(a) poly(ethene) (b) poly(chloroethene)
(c) poly(phenylethene).

(5) (a) Why are some plastics flexible?
(b) Give some examples of flexible plastics.
(c) What is a fibre?
(d) Mention two domestic uses of fibres.

(6) 'Man is becoming more and more dependent on plastics.'
Discuss this statement in relation to the following:
(a) sports equipment (b) leisure wear
(c) medicine (d) motor cars (e) kitchenware
(f) buildings.

(7) 'The discovery of low-density poly-(ethene), LDPE, revolutionised materials.'
(a) Write a brief note on the discovery of LDPE.
(b) Comment on the statement above.

(8) Write a brief note on the discovery of HDPE and of Teflon.

(9) 'Teflon is an extremely versatile material.' Give reasons which support this statement.

(10) (a) Why does plastic have a poor public image?
(b) How is plastic of benefit to us?
(c) Why does plastic waste cause disposal problems?

(11) (a) What is a polymer? Describe how ethene can be converted into poly(ethene).
(b) Explain why polythene is referred to as an addition polymer.
(c) Name some other addition polymers. Draw the structure of the monomer and polymer of each named polymer. Give the uses of each polymer.
(d) On the one hand, poly(ethene) can be used as a plastic film for packaging and on the other hand it can be used for making hard plastic containers. Explain why it can be used to make such different products.

(12) Alkenes such as ethene and propene have often been described as the building blocks of the plastics industry. Discuss the properties of alkenes which enable them to be used in so many ways.

(13) 'Plastics have widespread industrial and domestic uses.'
Discuss this statement in relation to the poor public image that the word plastic creates.

(14) 'Effective management of solid waste should use a balance of the following methods:
(a) waste prevention and reduction at source
(b) material recycling and re-use
(c) energy recycling
(d) safe disposal of residual waste.'
Discuss each method in terms of plastic waste in the environment.

(15) Polystyrene is often recycled.
(a) Write a brief note on the stages involved.
(b) Comment on the economic aspects of recycling.

2A.3 METALS

Metals used in construction of steam crackers.

More than three-quarters of all the known elements are metals. Until the development of plastics during the twentieth century, metals were the most important material known to man. Metals have some superior properties which make them very useful: high electrical and thermal conductivities, high melting points, characteristic lustre and the ability to be bent or stretched without cleaving. The extent of each of these properties varies from metal to metal. The properties of each metal depend on the arrangement of the atoms within the crystalline state and on the size of the metallic crystal.

The periodic table is used to divide the elements into families or groups of elements in accordance with their properties. Sometimes, we do not need to know all about the properties of an element. Often we only need to know whether an element is a metal or a non-metal. Metals such as copper have very distinct and useful properties. Copper is ductile, which means it can be drawn into wire form for use in electrical wires. The non-metals oxygen, nitrogen, carbon, sulphur and chlorine have very different properties to metals. Carbon is a non-metal. Carbon in the form of graphite is used as a lubricant because of its chemical structure. It is important to know the different properties of metals and non-metals so that we can use these properties for our benefit.

Comparison of Metals and Non-metals

Metals are on the left-hand side and the middle of the periodic table, while non-metals are on the right-hand side of the periodic table.

Most metals are hard lustrous solids. Some metals, like the alkali metals, are soft, while others, like mercury, are liquids. Most non-metals are liquids or gases. Others, like sulphur and carbon in the form of graphite, are soft solids, while carbon as diamond is an extremely hard solid.

Hardness

Most metals are hard. Since time began metals have been used as weapons and as tools because of their hardness.

Most non-metals are liquids or gases. Therefore, they are generally soft. Diamond is an exception as it is a non-metal which is very hard. Diamond is used to cut many materials because it is so hard. Graphite is a soft material which is used as a lubricant.

Melting Points and Boiling Points

Metals generally have high melting and boiling points. For this reason they are used whenever high temperatures are involved, such as in car engines, machinery and cooking utensils.

The majority of non-metals have very low melting points and boiling points and exist as liquids or gases.

Lustre

Metals are shiny (lustrous) until they corrode. Some, like silver and gold, maintain their lustre and are used as jewellery.

All non-metals, except diamonds, crystalline sulphur and iodine, are dull and non-lustrous.

Silver – a lustrous metal.

333

Malleability and Ductility

When a force is applied to a metal crystal, the layers of metallic ions can slide and slip over each other. This allows metals to be stretched into wire form (ductility) and to be hammered and moulded into shape (malleability) without breaking. Non-metals, such as graphite and sulphur, are brittle and shatter when they are stretched, bent or hammered.

Heat Conductivity

Metals are good conductors of heat. When a metal is heated at one end, the thermal energy can be transferred from one metal ion to the next because they are so closely packed together.

The particles in non-metals are relatively far apart from each other. For this reason, heat cannot be transfered easily from particle to particle. Therefore, non-metals are poor conductors of heat.

Electrical Conductivity

Metals are good electrical conductors because each metal ion is surrounded by a sea of electrons. These electrons are free to move without disturbing the metal structure.

Non-metals are poor conductors of electricity because their electrons are fixed. Graphite is an exception and has mobile electrons. Graphite is a good electrical conductor.

Alloys

An alloy is a mixture of a metal with one or more other elements.

From early times our ancestors mixed metals together in furnaces to make alloys. The first alloy made was probably bronze, which was a mixture of tin and copper. Alloys are used instead of pure metals because they have better properties. For example, stainless steel is better than iron because it is harder and does not rust.

The addition of small amounts of other elements to a metal breaks up the regular lattice structure of the metal and prevents 'slip' which makes the metal harder, tougher and less malleable.

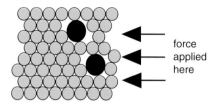

Slip cannot occur in the alloy because the atoms of different size cannot slide over each other

Carbon in Steel

There are many different types of steel depending on which elements have been added. Some steels are extremely hard. In hard steels, carbon is added to the molten iron. The small carbon atoms fit into the holes in the lattice structure making slip between the iron layers difficult. This formation of iron carbide makes the steel very hard. Iron carbide can be used to cut concrete and other materials.

Exercise 2A.3
•

(1) (a) Explain why metals are hard, while non-metals are normally liquids or gases or soft solids.
(b) Explain why diamond is an extremely hard non-metal.

(2) (a) Explain why metals are good electrical conductors and the majority of non-metals are very poor conductors.
(b) Why is graphite, a non-metal, a good conductor of electricity?

(3) 'Metals are malleable and ductile, while non-metals are brittle.'
Use some examples to explain and confirm this statement.

Some Alloys

Alloy	Composition	Properties
Brass	Copper, zinc	Harder than copper
Bronze	Copper, tin	Harder than copper
Stainless steel	Iron, chromium, nickel	Does not rust; harder than iron
Mild steel	Iron, carbon	Harder and stronger than iron
Duralumin	Aluminium, copper	Stronger than aluminium

(4) Explain each of the following:
(a) Kitchen sinks are usually made of stainless steel
(b) Letterboxes and doorknockers are made of brass and no copper.
(c) Aeroplanes are made from Duralumin and not from pure aluminium.
(d) Discs for cutting concrete contain carbon.

(5) (a) What is an alloy?
(b) Name four alloys. State the use of each alloy.
(c) How does carbon strengthen steel?

(6) Name alloys which would be suitable to make the following:
(a) washing machine drums
(b) reinforcement for concrete
(c) metal statues.
In each case, state why the particular alloy is suitable.

(7) Explain why metals:
(a) have high melting points (b) are good conductors of electricity (c) are malleable
(d) have high density.

(8) (a) Explain how alloying alters the properties of a metal.
(b) Name an alloy of iron, an alloy of copper and an alloy of aluminium. In each case name the metals in the alloy.
(c) Give one use for each alloy.

2B ADDITIONAL ELECTROCHEMISTRY AND THE EXTRACTION OF METALS

2B.1 THE ELECTROCHEMICAL SERIES

The discovery of electricity is generally credited to an Italian scientist, **Luigi Galvani**. Galvani and his wife, **Lucia**, did many experiments on the bodies of dead animals and in particular on frog's legs. They concluded from their experiments that animals must contain electricity.

Alessandro Volta repeated many of Galvani's experiments and invented the first battery around 1800. He assembled a pile of zinc and silver discs which were separated by paper discs that had been dipped in salt water. He received a small electric shock when he touched the two ends of the pile. He discovered that an electrical voltage is produced when two different metals are placed in contact by means of a conducting solution. Volta discovered that the size of the electrical current produced depended on the particular pair of metals used. He arranged the metals in a series and formed the first electrochemical series of metals.

Volta's invention of the galvanic pile inspired **Humphrey Davy** to conduct many experiments on electricity. In 1807, he passed electricity through molten potassium hydroxide. He danced with joy when he noticed small globules of a silvery metal floating to the surface. The metal burst into lilac flames: Davy had discovered potassium. He developed many theories on electrolysis and as a result isolated six new elements: sodium, potassium, magnesium, calcium, strontium and barium.

Michael Faraday attended one of Davy's lectures and presented him with a bound version of the lecture. Davy was so impressed with Faraday that he gave him a position as a laboratory assistant. From that position, Faraday became the most famous experimental scientist of his time. He invented the electric motor and generator and developed the principles of electrolysis.

Today, we are familiar with more sophisticated batteries and cells such as long-life cells,

Sir Humphry Davy 1778–1829
Davy was born in Cornwall, the eldest son of a woodcarver. He had to leave school at the age of sixteen to support his widowed mother. He was widely read and largely self-taught – he read Lavoisier's Traité Élémentaire in the original French. He discovered the anaesthetic properties of nitrous oxide, 'laughing gas', and before his twenty-second birthday he virtually created the new science of electro-chemistry.

At the Royal Institution, he proposed theories of electrolysis which enabled him to isolate sodium, potassium, magnesium, calcium, strontium and barium. His last and most popular discovery was the Davy lamp which helped prevent explosions caused by methane igniting in coal mines. He was a romantic – he wrote poetry all his life. Science for him was a way of 'interpreting the works of nature so as to unfold the wisdom and glory of the Creator'.

Perhaps his greatest gift to science was his own assistant and colleague – Michael Faraday.

used in heart pacemakers, and rechargeable batteries. The earlier Voltaic pile and today's modern cells each depend on a spontaneous chemical reaction which produces an electric current. The chemical reaction and the flow of the electric current occur when electrons move; it is an oxidation–reduction reaction.

Electrochemical Cells

An electrochemical cell is a system consisting of electrodes that dip into an electrolyte which causes a chemical reaction. The chemical reaction can either use or generate an electric current. If an electrochemical cell generates an electric current, it is called a voltaic or galvanic cell; if it uses an electric current to drive a chemical reaction, it is called an electrolytic cell and the process is called electrolysis.

Production of Voltage

If two different metals are dipped into an electrolyte and are brought into contact, an electric current flows. The presence of the current can be demonstrated by placing a suitable bulb or voltmeter in the circuit. By experimenting with different pairs of metals it is possible to increase or decrease the voltage in the circuit. The production of electricity seems to depend on one metal being more reactive than the other.

For example, zinc is a more reactive metal than copper. In a simple zinc–copper cell, with copper sulphate as electrolyte, the zinc electrode is negative with respect to the copper electrode. This indicates that the zinc is supplying electrons to the external circuit. This oxidation–reduction process can be represented by two half-reactions:

$$Zn(s) \rightarrow Zn^{2+}(aq) + 2e^-$$

$$Cu^{2+}(aq) + 2e^- \rightarrow Cu(s)$$

Here, the zinc metal is oxidised to zinc ions in solution, while the copper ions in solution pick up the electrons released from the zinc and change into copper atoms, which are deposited on the copper electrode.

The overall reaction can be written as:

$$Zn(s) + Cu^{2+}(aq) \rightarrow Zn^{2+}(aq) + Cu(s)$$

Zinc, being the more reactive metal, displaces copper from the solution. Electrons pass from the more reactive metal into the circuit and are then removed from the circuit by the less

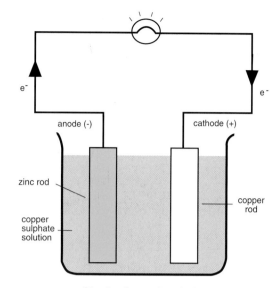

Simple electrochemical cell

reactive metal; thus the cell produces electricity.

The simple cell just described is not an effective way of producing electricity. Changes at the surfaces of the electrodes tend to cause a rapid fall in the voltage produced. For example, the zinc electrode may become coated in copper which stops the cell reaction. In other cells, a gas such as hydrogen may be produced which gathers at an electrode and changes the nature of the electrode. These changes at the electrode are called polarisation and must be minimised in order to increase the efficiency of a cell. One method of doing this is to separate the zinc–copper cell into two halves and join the two halves together by means of a salt bridge. The salt bridge is a tube of electrolyte which allows the flow of ions but prevents mixing of the two different half-cell solutions.

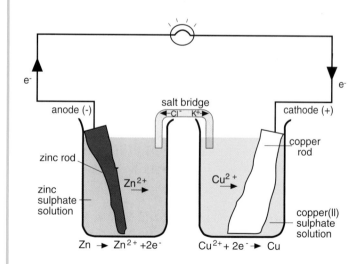

This zinc–copper arrangement produces a fairly steady voltage of 1.1 volts. If the copper and zinc electrodes are replaced by two other metals a different voltage (or potential difference) is obtained. The maximum voltage obtained in a cell depends on the difference in potential of the two electrodes. It is not possible to measure the potential of each electrode directly; it is only possible to measure the potential difference. A standard or reference electrode is used, which all other electrode potentials are related to.

The standard or reference electrode is the hydrogen electrode, which consists of hydrogen gas bubbling over an inert platinum electrode dipped into a 1.0 mol L^{-1} H^+ ion solution at STP. This standard hydrogen electrode is arbitrarily assigned a value of 0 volts.

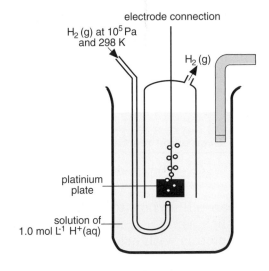

Other electrode potentials are obtained by comparison with the standard hydrogen electrode. If the electrode potential of, say, zinc is required, we set up two electrodes, one being the standard hydrogen electrode, the other the zinc electrode consisting of zinc dipping into a 1.0 mol L^{-1} solution of zinc ions. The voltage or potential difference measured on a voltmeter should read 0.76 volts, the deflection of the needle indicating that zinc has a greater tendency to give off electrons than hydrogen.

Values of other electrode potentials are obtained in a similar way; in this way we can calculate the potential differences of most cells. The value of the standard electrode potential is a measure of the reactivity not only of metals but also of non-metals and of many ions.

The list of standard electrode potentials is called the electrochemical series and is similar to the simpler activity series of metals. It should be remembered that each cell reaction is an oxidation–reduction reaction; thus, the electrode potential is often called the redox potential.

Some standard electrode potentials are given below.

Electrode reaction	Standard electrode potential/volts
$Li \rightleftharpoons Li^+ + e^-$	+ 3.04
$K \rightleftharpoons K^+ + e^-$	+ 2.92
$Mg \rightleftharpoons Mg^{2+} + 2e^-$	+ 2.38
$Zn \rightleftharpoons Zn^{2+} + 2e^-$	+ 0.76
$H_2 \rightleftharpoons 2H^+ + 2e^-$	**0.00**
$Cu \rightleftharpoons Cu^{2+} + 2e^-$	–0.34
$Ag \rightleftharpoons Ag^+ + e^-$	–0.80
$Hg \rightleftharpoons Hg^{2+} + 2e^-$	–0.80

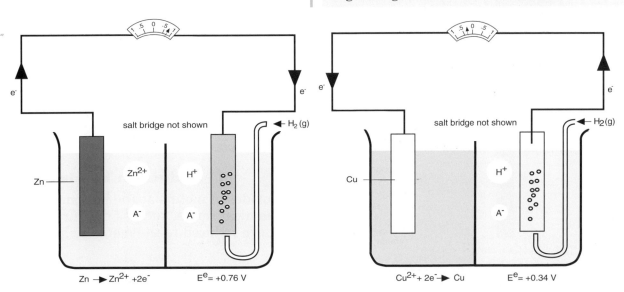

Measuring standard electrode potentials.

337

The larger positive value of potassium (2.92 volts) tells us that it has a greater tendency to lose electrons than zinc (0.76 volts).

For example, when potassium is placed in a solution of zinc ions the potassium will donate electrons to the zinc ions, changing them into zinc metal, while the potassium metal changes into potassium ions.

$$K(s) + ZnCl_2(aq) \rightarrow 2KCl(aq) + Zn(s)$$

However, silver does not donate electrons to zinc ions because silver is lower in the electrochemical series than zinc. When silver is placed in aqueous zinc chloride no reaction occurs.

$$Ag(s) + ZnCl_2(aq) \rightarrow \text{No reaction}$$

Dry Cells

The dry cell is the cell which is used in radios, torches and toys. It was developed in the nineteenth century by a French chemist called **Leclanche** and produces a maximum voltage of 1.5 V. It is called the dry cell because the electrolyte is a thick paste made of ammonium chloride. Modern dry cells are nearly leak-proof. The anode is the zinc case which surrounds the cell, while the cathode is a graphite rod surrounded by a paste of ammonium chloride, NH_4Cl, zinc chloride, $ZnCl_2$, manganese dioxide, MnO_2, and carbon black. The carbon black is used to increase the surface area of the cathode.

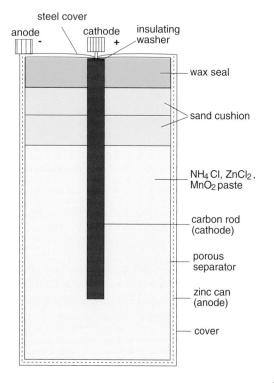

The reactions occurring at the electrodes can be represented simply as:

Anode reaction: The zinc case acts as the anode donating electrons which form the cell current.

$$Zn(s) \rightarrow Zn^{2+}(aq) + 2e^-$$

Cathode reaction: The cathode reaction is complex and is represented as

$$2NH_4^+(aq) + 2e^- \rightarrow 2NH_3(g) + H_2(g)$$

$$H_2(g) + 2MnO_2(s) \rightarrow MnO_3(s) + H_2O(l)$$

Dry cells are fairly inexpensive but the voltage drops quite quickly and they cannot be recharged. Alkaline batteries are more advanced forms of the dry cell. They are more expensive but produce a voltage slightly higher than 1.5 V.

The Electrochemical Series

The electrochemical series was introduced in Unit 1.5. The electrochemical series tells us how reactive a metal is and also tells us how a compound of a metal reacts with another substance.

Decreasing reactivity
→
Li—K—Ca—Na—Mg—Al—Zn—Fe—Sn—Pb—H—Cu—Hg—Ag—Au

The series can tell us how to predict chemical reactions such as:
• reactions of metals with acids, water and oxygen
• extraction of metals from ores
• effect of heat on hydroxides, carbonates and nitrates
• reduction of metal oxides.

Reduction of Metal Oxides

Except for the very unreactive metals (those metals low in the electrochemical series, such as silver and gold), most metals occur in nature combined with other elements. Therefore, most metals must be extracted from their compounds by reduction.

(1) Reduction by Heat

The oxides of metals at the bottom of the electrochemical series, such as mercury, silver and gold, are reduced to the metal by heat alone.

$$HgO(s) \xrightarrow{\text{heat}} Hg(l) + \tfrac{1}{2}O_2(g)$$

338

(2) Chemical Reduction with Hydrogen or Carbon

The oxides of metals in the middle of the electrochemical series are reduced on heating with carbon or hydrogen.

Copper can be extracted from copper oxide by heating with hydrogen.

$$CuO(s) + H_2(g) \rightarrow Cu(s) + H_2O(g)$$

The use of carbon as a reducing agent is important in the manufacture of a number of metals. It is used in the extraction of zinc, lead and iron from their oxides.

In the extraction of iron from its ores, carbon is used as a reducing agent.

$$\overset{\text{heat}}{Fe_2O_3(s) + 3C(s) \rightarrow 2Fe(l) + 3CO(g)}$$

(3) Reduction by electrolysis

The ores of metals higher up in the series are more stable and are usually reduced by electrolysis. The method is expensive, due to the cost of electricity and the high temperatures involved. It is usually carried out in countries where electricity is cheap. Metals such as sodium and aluminium are extracted from their ores using electrolysis. Electricity is passed through the heated salt of the metal: the metal ions are reduced at the cathode.

$$Na^+(l) + e^- \rightarrow Na(l)$$

$$Al^{3+}(l) + 3e^- \rightarrow Al(l)$$

Exercise 2B.1

•

(1) Write a brief account of the contributions each of the following made to the development of electrochemistry:
(a) Galvani (b) Volta (c) Davy (d) Faraday.

(2) (a) Define oxidation and reduction in terms of electron transfer.
(b) Describe the reaction of zinc with dilute hydrochloric acid, in terms of electron transfer.

(3) Explain why some metals are found free in nature, some can be extracted quite easily from their ores, while others are difficult to extract from their ores. In your answer refer to at least three different metals.

(4) Define, giving relevant examples, each of the following terms:
(a) electrolyte (b) electrode (c) electrochemical cell.

(5) When two different metals, such as zinc and nickel, are placed in an electrolyte an electric current flows.
(a) Draw a labelled diagram of the cell.
(b) Why does an electric current flow?
(c) Show the direction of the electron flow.
(d) Write equations for the two half reactions.

(6) When different metals are used as electrodes in a cell, different voltages are obtained.
(a) Suggest why different voltages are obtained.
(b) What standards are used to compare cell voltages?

(7) (a) What is a dry cell?
(b) What are the main uses of dry cells?
(c) Draw a labelled diagram of a typical dry cell.
(d) How is the surface area of the anode increased?
(e) Write half equations for the reactions which occur at the electrodes.

(8) 'Metal oxides can be reduced by heat alone, with hydrogen or with carbon, and by electrolysis.'
Describe, using equations, each method of reducing a metal oxide.

(9) What is the electrochemical series? Using the electrochemical series, list four metals in order of their reactivity.

(10) (a) Describe what happens when potassium is placed in a solution of zinc ions.
(b) Describe what happens when silver is placed in a solution of zinc ions.

(11) (a) Describe, using a cell diagram, the difference between a voltaic cell and an electrolytic cell.
(b) Write a half equation for the reaction occurring at the anode in a voltaic cell.

(12) 'Potassium was discovered by Humphrey Davy when he electrolysed potassium hydroxide.'
Write a half equation for the reaction occurring at the anode.

2B.2 ELECTROLYSIS OF MOLTEN SALTS

Electrolysis

Electrolysis is the process where an electric current is used to drive a non-spontaneous chemical reaction.

When two electrodes are immersed in an electrolytic solution and connected to the opposite poles of a battery or power supply an electrolytic cell is formed. Chemical changes take place at the electrodes as soon as the current begins to flow. A redox reaction causes the chemical changes: oxidation occurs at the anode and reduction occurs at the cathode. The current is transferred through the electrolytic solution (electrolyte) by means of mobile ions. The surface of the electrodes is changed during the chemical reaction by decomposition, by deposition or by evolution of a gas.

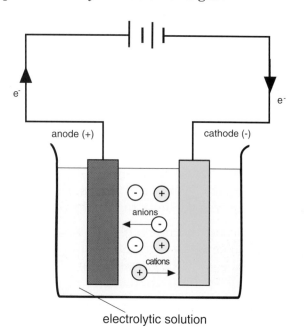

electrolytic solution

Electrolysis of molten salts

Many salts are used as a source of chemicals: many of them are made using electrolysis. For example, sodium and chlorine can be extracted from sodium chloride. Sodium chloride, like other ionic salts, does not conduct electricity. However, sodium chloride can conduct electricity when molten or in aqueous solution. Sodium chloride melts at 800 °C, which makes it difficult to electrolyse in the school laboratory. Sodium and chlorine are extracted from sodium chloride using a Down's cell (see Unit 2B.4).

Electrolysis of molten lead bromide

Lead bromide, $PbBr_2$, is an ionic solid. The lead and bromide ions are fixed in the ionic lattice. When it is melted (at 373 °C) it breaks up into mobile ions: positive lead ions, Pb^{2+}, and negative bromide ions, Br^-. In an electrolytic cell, the positive lead ions move towards the negative electrode, while the negative bromide ions move towards the positive electrode. The electrodes are made of an inert material, such as platinum or carbon, so that they do not become involved in the cell reaction.

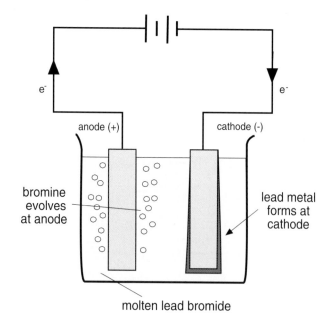

molten lead bromide

Anode reaction (+): oxidation
Bromide ions, Br^-, are attracted to the anode and give up their electrons. Bromine atoms are formed first and these then combine with one another forming bromine gas which is liberated at the anode. The gas is easily identified by its red-brown colour.

$$2Br^-(l) \rightarrow Br_2(g) + 2e^-$$

Cathode reaction (–): reduction

Lead ions are attracted to the cathode, where they find a supply of electrons and are changed into lead metal.

$$Pb^{2+}(l) + 2e^- \rightarrow Pb(s)$$

Overall reaction:

Bromine gas is liberated at the anode, while lead metal is deposited at the cathode.

$$Pb^{2+}(l) + 2Br^-(l) \rightarrow Pb(s) + Br_2(g)$$

Other Molten Salts

Many other salts can be melted and then broken up into their respective elements. For example, when sodium chloride (NaCl, melting point 801 °C) is electrolysed, sodium metal is produced at the cathode, while chlorine gas, Cl_2, forms at the anode.

Anode reaction (+): oxidation

$$2Cl^-(l) \rightarrow Cl_2(g) + 2e^-$$

Cathode reaction (–): reduction

$$2Na^+(l) + 2e^- \rightarrow 2Na(s)$$

Exercise 2B.2

•

(1) (a) What is meant by electrolysis?
(b) Draw a labelled diagram, illustrating how electrolysis occurs.

(2) What use can be made of electrolysis of molten salts?

(3) 'Lead bromide is often electrolysed in the school laboratory.'
(a) Why is lead bromide easy to electrolyse?
(b) Write half equations for the reactions which occur at the electrodes during the electrolysis.

(4) (a) Why is it difficult to electrolyse most salts?
(b) Where do the positive ions migrate to during electrolysis?
(c) What happens to the negative ions during electrolysis?
(d) Draw a diagram illustrating the movement of ions during electrolysis.

(5) (a) Mention any two metals which can be extracted from their ores by electrolysis of their molten salts.
(b) Why would a simple electrolytic cell need to be adapted for commercial use?

(6) 'Sodium chloride can be broken up into sodium metal and chlorine gas in a similar way to lead bromide.'
(a) Draw a diagram of the electrolytic cell.
(b) Write half reactions for the reactions occurring at the electrodes.

(7) (a) Explain why solid sodium chloride is a poor conductor of electricity, yet when melted, it readily conducts electricity.
(b) State the products formed when molten sodium chloride is electrolysed.
(c) Give ionic equations to account for the products.

(8) In the laboratory, it is relatively easy to demonstrate the electrolysis of molten lead bromide.
(a) Why is it more difficult to electrolyse solid sodium chloride?
(b) Describe, using a labelled diagram and giving the cell half reactions, how you would electrolyse molten sodium chloride.

2B.3 CORROSION

Corrosion is the name given to the processes which take place when a metal reacts slowly with air, water or any other substance in the environment.

The most common example of corrosion is the rusting of iron. Iron, if left in the air for some time, becomes coated with an orange powder called rust. Rusting is a wasteful process and has serious economic consequences. The corrosion of iron, rusting, occurs in the presence of oxygen and water and is an electrochemical or redox reaction.

Corrosion of steel plates.

Two simultaneous reactions occur on the surface of the iron and cause rust.

$$Fe(s) \xrightarrow{\text{oxidised}} Fe^{2+}(aq) + 2e^-$$

$$O_2(g) + 2H_2O + 4e^- \xrightarrow{\text{reduced}} 4OH^-(aq)$$

The end product of this redox reaction, iron(II) hydroxide, $Fe(OH)_2$, is then oxidised by atmospheric oxygen to form hydrated iron(III) oxide, $Fe_2O_3.H_2O$, which we call rust. Rusting of iron is very common in areas near the sea, as the sodium chloride in salt water increases the electrical conductivity of the water and accelerates the redox reaction.

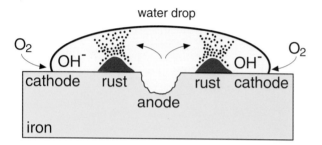

Rusting of iron is oxidised to iron (II) ion at the anode, and oxygen is reduced to hydroxide ion at the cathode

Corrosion of iron returns the pure iron to its state as an iron ore. Other metals, like aluminium, are also returned back to their natural ore state when they corrode.

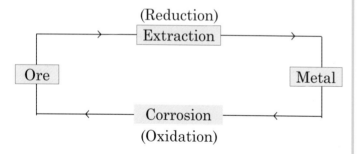

Relative corrodability

Metals high up in the electrochemical series are the most reactive metals: they are easily corroded. Often metals, like zinc, which are high up in the electrochemical series are used as a sacrifice to protect more valuable metals, like iron and steel. Zinc is used to protect oil rigs in the North Sea: corrosion occurs on the zinc and not on the steel structure of the oil rigs.

Prevention of Corrosion

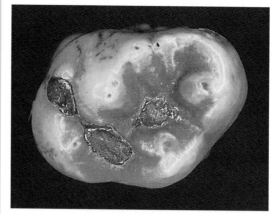

Tooth filling is a powdered alloy consisting of silver, tin, copper and zinc mixed with mercury.

Protective coating being applied to a metal tank.

(1) Application of Protective Layers

• **Galvanising:** Iron is often galvanised in order to protect it from corrosion. The iron is dipped into a bath of molten zinc giving it a fine surface coat of zinc. If the surface coat of the zinc is damaged or scratched the iron still does not corrode because the zinc is oxidised in preference to the iron as zinc, being higher in the electrochemical series, is a more reactive metal than iron. This is an example of sacrificial corrosion.

• **Electroplating:** Often iron is protected by electroplating another metal onto its surface. Tin is electroplated onto iron because it is non-toxic (unlike zinc) and can then be used to can foods – 'tin cans'. Many other metals are often electroplated. For example, chromium is used to protect iron in motor cars and bicycles, while nickel is electroplated with silver (EPNS, electroplated nickel silver) in the making of jewellery.

- **Surface coating:** Air can be excluded from the surface of a metal by painting, by greasing or by coating it in a layer of plastic. These processes place a barrier between the metal and the air and prevent corrosion. However, constant attention is required to ensure that the surface coating is not broken, because once the surface is broken or damaged corrosion takes place quite quickly.

(2) Anodising

Sometimes corrosion is useful. For example, a freshly cut piece of aluminium corrodes to form a layer of aluminium oxide on its surface. The surface coating of aluminium oxide prevents further oxidation from taking place. The process of applying this protective oxide coating is known as anodising. Anodising of aluminium is usually carried out by electrolysis.

(3) Alloys

It is possible to change the properties of a metal by mixing it with other metals or with non-metals. Steel is an alloy of iron and other elements, such as carbon, chromium and nickel, mixed together in varying amounts to give different qualities of steel. 'Stainless steel' has a protective coating of chromium oxide, Cr_2O_3, on its surface. This oxide is not hydrated like rust and is not affected by air or water. It is used in kitchen sinks, washing machines and cutlery.

(4) Sacrificial Protection

Iron corrodes more rapidly in the presence of copper and corrodes more slowly in the presence of magnesium. If we look at the electrochemical series we see that magnesium is higher up the series and more reactive than iron, while copper is lower down the series and less reactive than iron. When two pairs of metals such as iron and copper or iron and magnesium come into contact with water an electrochemical cell is set up. In the case of magnesium and iron in contact with water, the magnesium is oxidised and not the iron. Iron-containing steel pipes buried under ground are protected from corrosion by connecting the steel to a highly reactive metal such as magnesium; an electrochemical cell is then set up in the wet soil. The reaction

$$Mg(s) \xrightarrow{\text{oxidised}} Mg^{2+}(aq) + 2e^- \text{ occurs}$$

and not

$$Fe(s) \xrightarrow{\text{oxidised}} Fe^{2+}(aq) + 2e^-$$

Here, the magnesium is acting as an anode and is oxidised instead of the iron. In this process an active metal, magnesium, is used as a sacrificial anode to protect the iron.

Exercise 2B.3

•

(1) (a) What is corrosion?
(b) What is rusting?
(c) What chemical process is involved in rusting?

(2) (a) Which metals corrode most easily?
(b) How is corrosion prevented?
(c) Name some of the methods used to prevent corrosion.
(d) What is the end product of corrosion?

(3) 'Extraction of metals from their ores and corrosion are part of a redox cycle.'
Explain and comment on the implications of this statement.

(4) 'When iron rusts it forms an oxide, which does not protect it from further corrosion, while aluminium and chromium form oxide layers which protect the metal from further corrosion.'
Explain this statement.

(5) 'The corrosion of iron has serious economic consequences.'
(a) Define corrosion.
(b) Describe, briefly, how iron corrodes.
(c) What type of reaction occurs when iron corrodes?
(d) Why do motor cars rust more quickly if they are left near the sea?

(6) Corrosion can be prevented in many ways, using the following methods:
(i) sacrificial protection (ii) galvanising (iii) electroplating (iv) anodising (v) use of surface coatings (vi) alloying.
(a) Describe each process briefly, commenting on its advantages and disadvantages.
(b) Write cell reactions for each of the electrolytic methods used.

(7) 'Sacrificial anodes are used to prevent corrosion.'
Describe the process involved and name two industries which use this process.

(8) Explain each of the following statements:
(a) Iron rusts more quickly in sea water than in distilled water.
(b) Iron rusts faster than steel.
(c) Iron in shipwrecks in deep sea water rusts slowly.

(9) (a) What is an alloy?
(b) Name an alloy that is resistant to corrosion.
(c) Describe how the metal in the particular alloy was made corrosion resistant.

(10) Write a brief note on the advantages and disadvantages of the use of surface coatings in the prevention of corrosion.

2B.4 STRONGLY ELECTROPOSITIVE METALS

Extraction of metals from their ores

Metal ores being extracted from the ground.

Most metals occur naturally as inorganic compounds in the Earth's crust as ores; these ores are usually carbonates, oxides or sulphides. The method used to extract a metal from its ore depends on the position of the metal in the electrochemical series. For example, iron can be extracted quite easily from its principal ore, Fe_2O_3, by heating, whereas aluminium is extracted from alumina, Al_2O_3, by electrolysis, a process requiring a large amount of energy.

Reactive metals such as sodium, calcium, magnesium and aluminium are difficult to extract from their ores. The reactive metal tends to hold strongly onto the other element or elements to which it is bonded. Metals high in the electrochemical series are strongly electropositive, because they tend to donate electrons and form positive ions. It is not possible to extract a reactive metal from its ore by heating the ore with a reducing agent such as carbon or carbon monoxide. The metal must be extracted using an electrochemical process called electrolysis.

Rubies. Ruby is a red transparent crystal of aluminium oxide. Small amounts of chromium oxide give it its red colour. Natural rubies are highly prized as gemstones, while synthetic rubies are used in lasers and watches.

Extraction of Sodium

Sodium, like the other alkali metals, lithium and potassium, is a very reactive metal. Because they are so reactive, the alkali metals are not found free in nature. Sodium is found as rock salt, $NaCl$, and as Chile saltpetre, $NaNO_3$.

Sodium is extracted from purified molten rock salt by electrolysis in a Down's cell. The rock salt ($NaCl$) is broken down into sodium metal and chlorine gas. The cell reaction is similar to the electrolysis of molten lead bromide (see Unit 2B.2). Sodium chloride is an ionic compound and when it is molten, the sodium and chloride ions are free to move. The positive sodium ions migrate to the steel cathode and are deposited as sodium metal, while the chloride ions migrate to the graphite anode and are liberated as chlorine gas. The molten sodium floats to the top and is drawn off, while the chlorine gas is collected using a hood. The molten sodium is kept apart from the chlorine gas by a steel gauze.

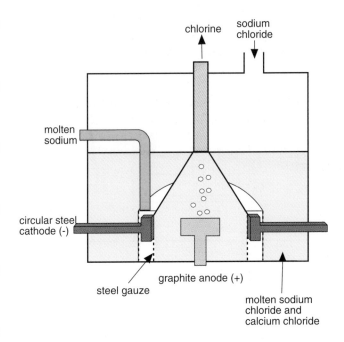

The Down's cell for the production of sodium

Cathode reaction: reduction

$$Na^+(l) + e^- \rightarrow Na(s)$$

Anode reaction: oxidation

$$2Cl^- \rightarrow Cl_2(g) + 2e^-$$

The electrolyte is a 40%:60% mixture by mass of sodium chloride and calcium chloride. A mixture is used rather than pure sodium chloride because the mixture melts at a lower temperature (600 °C) than sodium chloride (800 °C). This reduces the energy required to operate the cell. A small quantity of calcium is formed at this operating temperature; this is returned to the cell where it reforms calcium chloride or is removed by filtration.

Uses of Sodium and Chlorine

The sodium metal is used mainly in the production of anti-knocking agents (tetraethyl lead and tetramethyl lead), in the production of titanium and as a coolant in nuclear reactors. The by-product of the process, chlorine, is used in the manufacture of plastics and in water treatment.

Extraction of Aluminium

Aluminium is the most abundant metal in the Earth's crust. Aluminium only occurs combined with other elements. The only important ore of aluminium is bauxite, which is purified to form alumina. Bauxite is mined in north Africa, Jamaica, Brazil and Australia. Bauxite is purified to make pure aluminium oxide in Aughinish in Ireland and then exported to Anglesea in the UK. The purified alumina is then electrolysed to produce aluminium. The electrolytic process uses large amounts of electrical energy, firstly for electrolysis and secondly to keep the electrolyte in a molten state.

Aluminium can be fashioned and shaped by most metal-working processes. It is light and yet can form alloys which are stronger than steel. It is corrosion resistant and a good conductor of both heat and electricity. As a result of these properties aluminium has a wide range of uses; it is used in buildings, as kitchen foil, as long range electrical lines and in the manufacture of aeroplanes and ships.

There are two main steps in the extraction of aluminium from bauxite: the first stage involves the purification of the ore to form aluminium oxide, while the second involves the electrolysis of the aluminium oxide to form aluminium metal.

(1) Purification of the Ore

The ore is crushed and then heated under pressure with hot sodium hydroxide solution. Aluminium oxide, being amphoteric, dissolves in the hot alkali and forms sodium aluminate.

$$Al_2O_3.3H_2O(s) + 2NaOH(l) \rightarrow 2NaAlO_2(aq) + 4H_2O(l)$$
aluminium oxide   sodium aluminate

The alumina dissolves and the solution is filtered leaving behind the main impurities, the oxides of silicon, iron and titanium. The impurities form an unsightly red mud which is pumped into large lagoons and allowed to settle. At Aughinish in Ireland the lagoons have been successfully converted into a nature reserve containing many species of birds.

The sodium aluminate solution is then 'seeded' with pure crystals of aluminium hydroxide and agitated with air. On cooling pure aluminium oxide trihydrate precipitates.

$$2NaAlO_2(aq) + 4H_2O(l) \rightarrow Al_2O_3.3H_2O(s)$$

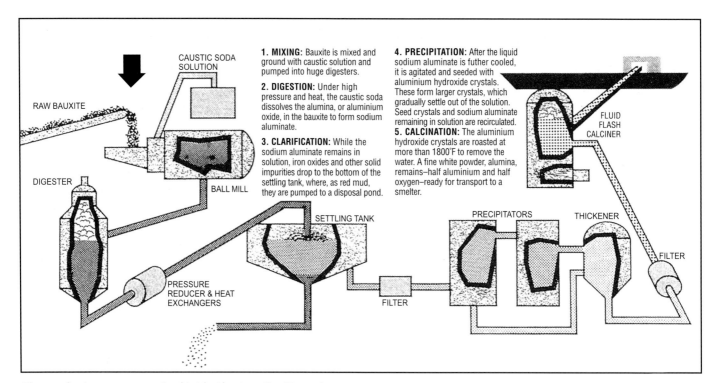

The production process at Aughinish Alumina, Co. Limerick.

Aluminium hydroxide crystals are roasted at more than 1800°C to remove the water.

$$Al_2O_3.3H_2O(s) \xrightarrow{\text{heat}} Al_2O_3(s) + 3H_2O(g)$$

(2) Electrolytic Reduction of Aluminium Oxide to Aluminium

Aluminium forms strong bonds with oxygen. It requires a lot of energy to break these bonds. Aluminium oxide cannot be reduced by heating alone; it must be reduced by electrolysis.

The electrolytic cell is a shallow steel container lined with carbon which acts as the cathode. The anode consists of blocks of carbon which are suspended in the cell. The reaction

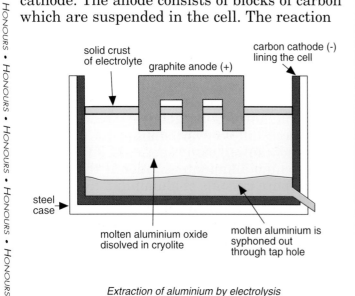

Extraction of aluminium by electrolysis

temperature is lowered by dissolving the alumina in molten cryolite, Na_3AlF_6. The pure alumina melts at 2020 °C, but dissolves in molten cryolite at 900 °C.

Aluminium is produced at the cathode: this is reduction.

$$Al^{3+}(l) + 3e^- \rightarrow Al(l)$$

It collects as a liquid at the bottom of the tank and is siphoned off and allowed to solidify.

Oxygen gas is produced at the anode: this is oxidation.

$$2O^{2-}(l) \rightarrow O_2(g) + 4e^-$$

The production of oxygen gas at the carbon anode presents a problem because it combines with the carbon and produces carbon dioxide. Because of this the carbon anodes have to replaced about every three weeks. The pure aluminium is transferred to holding furnaces where it is blended with other metals such as copper, manganese, magnesium and iron to form high strength alloys.

Anodised Aluminium

Aluminium can be made corrosion resistant by anodising it by electrolysis. The aluminium acts as an anode in an electrolytic cell containing dilute sulphuric acid. Oxygen gas which is produced during the cell reaction converts the surface of the aluminium to aluminium oxide.

The aluminium oxide forms a tough even coating on the aluminium which makes it resistant to corrosion. The surface coating of aluminium oxide is hydrated and porous: this allows the anodised aluminium to be dyed.

Economics of the Process

The electrolysis of aluminium oxide is costly. It is usually carried out in countries which have cheap sources of electricity, usually hydro-electric plants. Large currents (3×10^7 amps for every 100 tonnes of aluminium) are required to heat the molten electrolyte. Electricity costs make up approximately 25–30% of the total cost involved in making aluminium.

Recycling of aluminium

Many consumer goods are made from aluminium. Many soft drinks and alcoholic drinks are sold in aluminium cans which often end up on the rubbish tip. Aluminium from these cans may be recycled.
• By using recycled aluminium, approximately 95% of the energy needed to produce aluminium cans from raw materials can be saved.
• Recycling aluminium reduces litter and reduces the volume of waste collected by local authorities.
• Aluminium is worth approximately £800 per tonne. Collection of aluminium cans can be a source of income for interested groups, like schools and community groups.

Exercise 2B.4

•

(1) (a) Name and give the formula of a common ore of each of the following metals: sodium, aluminium and iron.
(b) Explain why some metals are difficult to extract from their ores.

(2) 'Most metals occur in nature as <u>ores</u>. The metal can be extracted from the ore by <u>chemical reduction</u> or by <u>electrolysis.</u>'
Explain, using an example in each case, the meaning of the underlined terms.

(3) 'Sodium is extracted from rock salt using a Down's cell.'
(a) Draw a labelled diagram of the cell, indicating clearly the anode and the cathode.
(b) The working temperature of the cell is lower than the melting point of rock salt. Explain how the cell operates at the lower temperature.
(c) Write equations for the anode and cathode reactions.
(d) Give two uses of each product of the cell reaction.

(4) 'Aluminium is extracted from bauxite, an ore containing aluminium oxide.'
(a) Explain why aluminium is extracted from aluminium oxide by electrolysis rather than by chemical reduction.
(b) Describe, using a labelled diagram, the electrolytic cell used.
(c) Explain why the anode must be replaced frequently during the process.

(d) Why is molten cryolite used as the electrolyte?
(e) What are the main impurities remaining after the impure aluminium oxide is dissolved in molten cryolite? How are they disposed of?
(f) Aluminium can be made corrosion resistant by anodising. How is this done?
(g) Give two different uses of aluminium.

(5) 'The production of aluminium is costly.' Discuss this statement. In your answer refer to the location of the electrolysis plant.

(6) Outline, using relevant equations, the main stages in the extraction of aluminium from bauxite.

(7) (a) Make a list of some consumer goods which are made from aluminium.
(b) Name some which are difficult to recycle and explain why.
(c) Which goods are easily recycled?
(d) What are the reasons for recycling aluminium?

(8) Write a brief note on the environmental aspects of aluminium production. In your answer refer to the following:
(a) type of impurities found during extraction
(b) removal of impurities (c) effects of aluminium on the environment.

(9) (a) What is anodised aluminium?
(b) Describe two properties of anodised aluminium which make it a useful product.

(10) Metallic sodium is obtained by electrolysis in the Down's cell shown in the diagram.

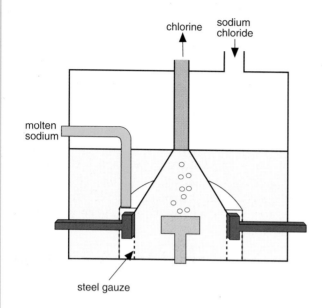

(a) Make a copy of the diagram and indicate the anode and the cathode.
(b) What are the anode and cathode made from?
(c) How is the chlorine removed from the cell?
(d) How is sodium removed from the cell?
(e) The molten electrolyte usually consists of sodium chloride mixed with another salt. What is the salt and why is it used?

(11) There are normally two stages in the production of aluminium from its ore. In the first stage pure alumina is obtained, and in the second stage aluminium is extracted from the alumina by electrolysis.
(a) Name the main ore of aluminium.

(b) Outline the main stages in the prodution of aluminium from that ore.
(c) Why is it difficult to electrolyse alumina?
(d) How is this difficulty overcome?
(e) During electrolysis the anodes have to be replaced. What chemical reaction is responsible for this?
(f) How is aluminium anodised?
(g What are the benefits of anodised aluminium?

(12) Aluminum is obtained from alumina by electrolysis in the cell shown in the diagram.

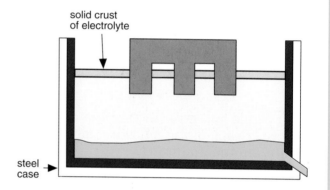

(a) Indicate, on a copy of the diagram, the anode and the cathode.
(b) What are the anode and cathode made from?
(c) How is the oxygen removed from the cell?
(d) How is aluminum removed from the cell?
(e) The molten electrolyte usually consists of sodium chloride mixed with another substance. What is the substance and why is it used?

2B.5 D-BLOCK METALS

Transition Metals

The metals in the middle of the periodic table are known as the transition metals. They have a number of special properties which set them apart from the main group elements.

The transition metals are strictly defined as those elements having a partially filled d or f subshell in any common oxidation state.

The d-block transition elements are those transition elements with an unfilled d subshell in common oxidation states. Although zinc, cadmium and mercury are not d-block transition metals by strict definition, they are often included with them because they have many similar properties.

Chemical Properties

The transition metals have a number of properties that set them apart from the main group elements.

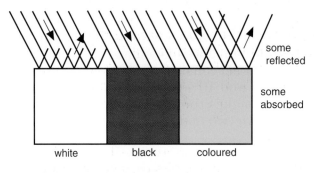

Colour depends on the amount of
light absorbed and reflected.

$MnO_4^-(aq)$ is purple
$Cr_2O_7^{2-}(aq)$ is orange
$Co^{2+}(aq)$ is pink

(1) All the transition elements are metals and, with the exception of zinc, cadmium and mercury, are hard solids with high melting points and boiling points.

(2) They have a number of oxidation states (variable valency). Iron forms Fe^{2+} ions and Fe^{3+} ions. Copper forms Cu^+ ions and Cu^{2+} ions. Vanadium exists in all the oxidation states from +2 to +5. Because they have many oxidation states, transition metals are often involved in redox reactions.

(3) Transition metal compounds are usually coloured, in contrast to the main group elements which are usually white with colourless solutions.

(4) Transition metal compounds can act as catalysts. For example, iron is used as a catalyst in the Haber process to manufacture ammonia, while nickel is used in many hydrogenation reactions.

Uses of Transition Metals

• The transition metals have many familiar commercial applications: iron tools, copper wires, silver jewellery, chromium and titanium alloys.
• Because they have a wide range of oxidation states, they are used as catalysts in many important industrial processes.
• In addition to their commercial applications, they have many have extremely important biological functions. For example, iron is present in haemoglobin in blood and myoglobin in muscle.

Extraction of Iron

Iron is the fourth most abundant element in the Earth's crust. It is found in many parts of the world, mainly as the ores magnetite, Fe_3O_4, haematite, Fe_2O_3, and as iron pyrites, FeS.

Iron and its alloys, which are called 'steels', are of enormous commercial importance; iron and steel products are used in transport, in building and in agriculture.

Iron ore is smelted in a furnace to reduce impurities in much the same way as in early Roman times. The iron ore is reduced to iron using coke. The prefered ore used today is haematite, Fe_2O_3, which contains 70% iron. In practice a number of ores from different sources are blended together to give a uniform composition.

Manufacture of Iron and Steel

Iron is manufactured in a blast furnace, a steel cylindrical vessel usually about 70 m high. The furnace is lined with heat resistant fire bricks made from a material which refracts the heat. Iron ore, limestone and coke (carbon) are fed continuously into the top of the furnace and at the same time blasts of hot air are blown in through nozzles called tuyeres. The hot waste gases are recycled and used again to preheat the hot air.

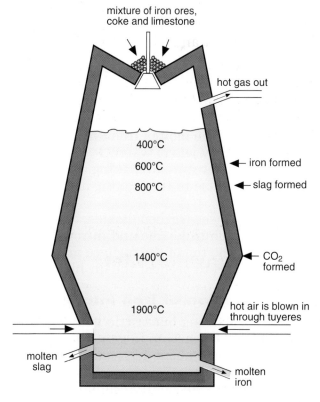

A blast furnace

349

Chemistry of the Process

(1) Coal is first converted to coke, by removal of oxygen by heat treatment, and then sintered. The porous form of the coke enables oxygen to circulate through it in the furnace converting it first to carbon dioxide and then to carbon monoxide.

$$C(s) + O_2(g) \rightarrow CO_2(g)$$

$$C(s) + CO_2(g) \rightarrow 2CO(g)$$

The first reaction is exothermic, while the second is endothermic and lowers the temperature to about 1100 °C.

(2) The iron ore is reduced either by the carbon monoxide or by the coke to iron and carbon dioxide.

$$Fe_2O_3(s) + 3CO(g) \rightarrow 2Fe(s) + 3CO_2(g)$$

$$2Fe_2O_3(s) + 3C(s) \rightarrow 4Fe(s) + 3CO_2(g)$$

The temperature falls to about 600 °C. The molten iron trickles off to the hearth at the bottom of the furnace where it is tapped off and run off into moulds to form cast iron (pig iron) or taken in giant ladles called torpedoes directly to the steel making part of the plant.

(3) The silica impurities are removed by the limestone.

$$\overset{heat}{CaCO_3(s) \rightarrow CaO(s) + CO_2(g)}$$

$$CaO(s) + SiO_2(s) \rightarrow CaSiO_3(s)$$
$$\text{calcium} \quad \text{silicon} \quad \text{calcium}$$
$$\text{oxide} \quad \text{dioxide} \quad \text{silicate (slag)}$$

The calcium silicate in the form of liquid slag falls to the bottom of the hearth, where it floats on top of the molten iron and is skimmed off. It is cooled by jets of water. It can be used directly as a filler in road making or converted into pellets for use in the making of heat insulators.

The waste gases (carbon dioxide, carbon monoxide, sulphur dioxide and nitrogen) are drawn off at the top of the furnace and are reused as a fuel to preheat the air blast.

Conversion of Iron Into Steel

Cast iron is a strong but brittle material, containing about 4% carbon. In general, if iron contains less than about 1.7% carbon it is classified as a steel.

All steels are alloys of iron containing other metals and non-metals. The composition of the steel depends on its end use. For instance, molybdenum is added to increase malleability, tungsten is added to provide a sharp cutting edge and manganese is added for hardness. Stainless steel consists of 70% iron, 20% chromium and 10% nickel: chromium in stainless steel makes the steel rust proof.

There are two main stages in the production of steel.

Removal of impurities from the iron (basic oxygen process)

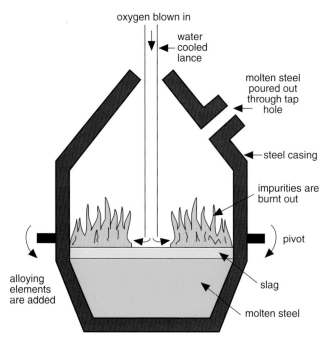

Basic oxygen steelmaking

The main impurities include carbon, sulphur, silicon and phosphorus. The carbon, silicon and phosphorus are removed during the basic oxygen process, where the impurities are burned out of the molten iron by a blast of oxygen which is blown onto the surface of the iron through a water-cooled lance. The carbon is converted to carbon dioxide, while the silicon and phosphorus are converted to their oxides, SiO_2 and P_4O_{10}. The silicon and phosphorus oxides then combine with limestone to form a slag. The slag floats on top and can be easily separated. The sulphur is removed by reacting it with a magnesium lance, which is lowered into the molten iron. Magnesium sulphide is formed and floats to the top and is then withdrawn.

350

Flowsheet of iron and steel production

Raw materials   *Process*

Iron ore $\longrightarrow$ Blast furnace $\longrightarrow$ or direct reduction $\longrightarrow$ Hot metal $\longrightarrow$ Basic oxygen furnace $\longrightarrow$ Liquid steel $\downarrow$ Ladle refining $\downarrow$ Continuous casting $\downarrow$ Rolling mill $\downarrow$ Steel products

Steel scrap $\longrightarrow$ Electric arc furnace

Addition of alloying elements

When the impurities are removed the required quantities of the other elements, such as carbon, molybdenum, tungsten, nickel and chromium manganese, are added. Often specially selected scrap metal is added and reused during the steel making process. Argon, being unreactive, is used to stir the mixture of elements. The molten steel is shaped by squeezing it through a mould, like toothpaste through a tube. This process of continuous casting enables the precast steel to be cut into any length directly.

The Electric Arc Process

The electric arc process, which was mainly used for small batches of special steels, is now being used more and more in the production of steel. In this process an electric arc is generated between two carbon electrodes. The electric arc melts the charge, consisting of iron and steel scrap. The liquid steel is then refined, cast, rolled and made into the required steel products. The process produces steel with fewer impurities than steel made using the basic oxygen process as no fuel (coke) is used and the higher operating temperature removes sulphur and phosphorus more effectively.

The electric arc process is used in the steel industry to make high quality specialty steels. The main stages in the process are as follows:

• **Charging:** Scrap steel is added into the furnace in stages and melted. Lime (CaO) is added to remove the sulphur and phosphorus impurities.

• **Melting:** The furnace lid is closed and the electric arc produces a temperature of 3500 °C which melts the scrap.

• **Refining:** When the scrap is melted, the carbon, phosphorus, sulphur and silicon impurities are removed.

• **Tapping:** The furnace is tilted backwards to remove the liquid slag. The temperature is raised and the molten steel is transferred to a ladle. Often extra alloying materials are added to make the required steel.

• **Continuous casting:** The molten steel is transferred from the ladle into a continuous caster. The steel is passed through a water cooled mould making a continuous slab of very hot steel. The slab is straightened and cut into lengths.

• **Rolling:** The hot steel is reheated and rolled into the finished products such as round bars, steel plates and other shapes.

The electric arc process

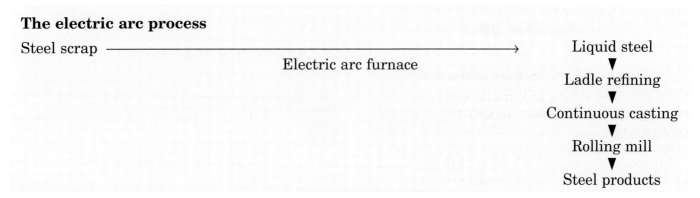

Steel scrap $\longrightarrow$ Electric arc furnace $\longrightarrow$ Liquid steel $\downarrow$ Ladle refining $\downarrow$ Continuous casting $\downarrow$ Rolling mill $\downarrow$ Steel products

Environmental considerations

The iron and steel making industry is a major cause of pollution. Quarrying of the ore causes severe devastation to the landscape, sulphur dioxide causes acid rain, while smoke and other suspended particles cause many respiratory diseases. The need for great environmental control is obvious.

The electric arc process is a cleaner process than the basic oxygen process as it uses electrical energy and not solid fuel. Emissions of gases and water are monitored in order to meet the required statutory limits.

Chemistry in Action: Biological Importance of Some Metals

Anaemic child.

Many metals play crucial roles in the structure of all living things.

Sodium, potassium, calcium and magnesium ions are found in body fluids. Sodium ions and chloride ions are found in blood plasma (controls pressure) and in sweat (controls temperature). Sodium and potassium ions are essential in the transmission of electrical impulses in our nervous system.

The average adult human body contains about 1 kg of calcium, the majority of which occurs in bones and teeth. Normal growth is impaired if calcium is deficient. Calcium is essential for blood clotting and muscular contractions and is used in our nervous system. The level of calcium in the blood is controlled by hormones.

Magnesium is necessary in the functioning of muscles and the nervous system. Magnesium is obtained from green plants. All green plants contain chlorophyll, a giant molecule containing magnesium bonded to four nitrogen atoms. Chlorophyll is essential for photosynthesis, without which plants could not manufacture food from carbon dioxide and water.

Iron is found in blood haemoglobin and is essential in the transport of oxygen throughout the body. Iron is also found in protein myoglobin and is stored in some organs such as the liver. A healthy adult contains approximately 4 g of iron.

Small amounts of chromium, manganese, cobalt, nickel, copper and zinc are found in biological systems. These elements are often called trace elements. Chromium is involved in the use of glucose as an energy source in the body, while manganese is involved in many enzyme systems. Cobalt is found in vitamin B_{12}, which is necessary for the formation of DNA and RNA. Copper is necessary for the synthesis of haemoglobin and is necessary for the prevention of many diseases in animals. Deficiency of copper in animals often results in paralysis. Zinc has several roles; it is a constituent of many enzymes and is involved in the production of the hormone insulin. Zinc is used to treat severe cases of malnutrition and inherited diseases such as sickle cell anaemia. Zinc is also used to treat the modern slimming disease anorexia nervosa.

Exercise 2B.5

•

(1) Compare the properties of the main group metals with those of the transition metals.

(2) (a) Define (i) transition metal and (ii) d-block element.
(b) Describe the main properties of the transition metals.

(3) Iron is a transition metal. Describe, using iron and its compounds as examples, any four properties of transition metals.

(4) (a) How do the electronic configurations of the transition metals differ from those of the elements in the main groups of the periodic table?
(b) Using copper and its compounds as an example, describe three properties of transition compounds.

(c) Write the electronic configurations for the copper(I) ion and the copper(II) ion.

(d) Copper(I) compounds are often white and have little ability as catalysts, while copper(II) compounds are usually coloured and often act as catalysts. Explain why this is so.

(5) Iron is extracted from its principal ores haematite and magnetite by heating them with a mixture of coke and limestone in a blast furnace.

(a) Describe briefly, using equations, the main stages involved in the process.

(b) Draw a labelled diagram of the blast furnace. Why is it called a blast furnace?

(c) Explain the purpose of adding limestone to the furnace.

(d) How is sulphur removed from the furnace? What was the source of the sulphur?

(e) What happens to the waste gases?

(f) How is steel made by the basic oxygen process?

(g) Cast iron is a strong and brittle material. How are its characteristics changed when it is converted to steel?

(h) The iron and steel industry is a major cause of pollution. Name the main pollutants and describe how they may be kept to a minimum.

(6) Aluminium is extracted from its ore by electrolysis, while iron is extracted from its ore by heating with coke. Explain why both processes are reduction reactions.

(7) (a) Draw a labelled diagram of the main equipment used in the basic oxygen process.

(b) How does the process work?

(8) The method used to extract a metal from its ore depends on the position of the metal in the electrochemical series. Suggest a method for extracting each of the following from their ores:

(a) silver from silver sulphide, Ag_2S

(b) zinc from zinc sulphide, ZnS

(c) calcium from calcium chloride, $CaCl_2$

(d) iron from iron(III) oxide, Fe_2O_3

(e) sodium from rock salt, $NaCl$.

(9) The following metals are in the order in which they appear in the electrochemical series:

Na   Al   Zn   Fe   Sn   Cu

(a) Which metal can occur free in nature? Explain why this is possible.

(b) Two of the metals are used to prevent corrosion of iron. Identify the two metals. By which method do they protect the iron?

(c) Show, using an equation, the displacement of one metal from a salt solution of another metal in the series. Show that this is a redox reaction.

(d) Which of these metals are usually extracted from their ores by an electrolytic method? In each case give a brief description of the electrolytic process involved in the extraction.

(10) Iron is extracted from its ore, haematite, in the blast furnace as shown.

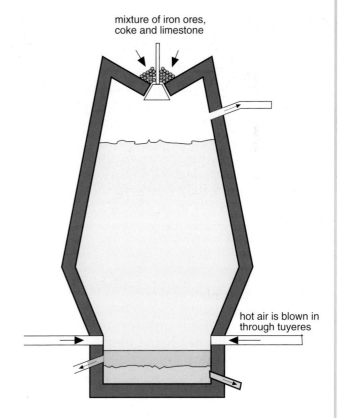

mixture of iron ores, coke and limestone

hot air is blown in through tuyeres

(a) What is the formula of the ore haematite?

(b) Why is haematite used as pellets and not in powder form?

(c) What are the main impurities in the process?

(d) How is each impurity separated from the ore and removed from the furnace?

(e) Write balanced equations for the reduction of the iron(II) oxide by (i) carbon monoxide and (ii) carbon.

(f) Describe, briefly, how pig iron is converted to steel.

(11) (a) How is cast iron converted into steel? In your answer refer to different types of steels.
(b) Mention two ways in which steel is more useful than iron.

(12) 'Iron is extracted from iron ore in a blast furnace.'
(a) Name the usual ore used and the names of the two other substances which make up the charge which is fed into the furnace.
(b) Give the function of each of the three materials.
(c) A substance floats on the top of the molten iron. Give the name and any common use of this substance.
(d) Iron rusts easily. Write equations for the reactions involved in rusting.
(e) How does the addition of carbon prevent rusting in stainless steel?

(13) The following is an account of the extraction of iron from a blast furnace. 'Iron ore and substances A and B are fed into the top of the furnace. A burns to form carbon dioxide which reacts, higher up in the furnace, with more of A to form C. C is the main reducing agent which converts the oxide ore to D. B decomposes in the blast furnace to form E and carbon dioxide. E reacts with impurities to form F.'
(a) Name the substances represented by A, B, C, D, E and F.
(b) How is iron converted to steel?
(c) State three conditions which promote rusting.
(d) How is rusting prevented?

(14) (a) What is the electric arc process?
(b) Why is this process used in preference to the basic oxygen process?
(c) Why is the electric arc process suitable for a small country like Ireland?
(d) How is energy saved using the electric arc process?

(15) (a) Draw a labelled diagram of the electric arc furnace.
(b) What is the main raw material used?
(c) Why is this material graded and sorted before being used in the furnace?
(d) Why is lime added to the furnace?
(e) Why does the steel made by this process contain fewer impurities than steel made using the basic oxygen process?

(16) Outline the main stages used in converting scrap steel into structural steel.

▼▼▼▼ Key Terms ▼▼▼▼

Some of the more important terms are listed below. Other terms not listed may be located by means of the index.

1. **Crystals:** Crystals are formed by atoms, ions or molecules arranged in a regular geometric arrangement called a lattice.

2. **Ionic crystals:** Ionic crystals are formed when positive and negative ions bind together by electrostatic forces.

3. **Non-polar molecular crystals:** Non-polar molecular crystals are formed when non-polar molecules bind together by weak van der Waal's forces.

4. **Polar molecular crystals:** Polar molecular crystals are formed when polar molecules bind together by dipole–dipole interactions.

5. **Covalent macromolecular crystals:** Covalent macromolecular crystals are crystals in which all the atoms are linked together in a vast three-dimensional network by covalent bonds.

6. **Metallic crystals:** A metallic crystal may be considered as a regular geometric array of positive ions surrounded by a 'sea' of electrons.

7. **Polymers:** Polymers are large chain-like molecules which are built up from smaller molecules called monomers.

8. **Low-density and high-density poly(ethene):** Poly(ethene) can be made in two ways: with branches (low density) or in straight chains (high density).

9. **Management of plastics:** Landfill, incineration, recycling, use of degradeable plastics and reduction at source are the main methods used to control the build-up of plastics.

10. **Recycling of polystyrene:** The main stages involved include: collection, sorting, shredding, washing, drying and re-extrusion.

11. Metals: Metals have some superior properties which make them very useful: high electrical and thermal conductivites, high melting points, characteristic lustre and the ability to be bent or stretched without cleaving. The extent of each of these properties varies from metal to metal.

12. Electrochemical cells: An electrochemical cell is a system consisting of electrodes that dip into an electrolyte which causes a chemical reaction. The chemical reaction can either use or generate an electric current.

13. Dry cells: The dry cell was developed in the nineteenth century by a French chemist called Leclanche and produces a maximum voltage of 1.5 V. It is called the dry cell because the electrolyte is a thick paste made of ammonium chloride. Alkaline batteries are more advanced forms of the dry cell. They are more expensive but produce a voltage slightly higher than 1.5 V.

14. The electrochemical series: The electrochemical series tells us how reactive a metal is and also tells us how a compound of a metal will react with another substance.

15. The position of a metal in the electrochemical series: The series can tell us how to predict chemical reactions such as:
reactions of metals with acids, water and oxygen
extraction of metals from ores
effect of heat on hydroxides, carbonates and nitrates
reduction of metal oxides.

16. Electrolysis: Electrolysis is the process in which an electric current is used to drive a non-spontaneous chemical reaction.

17. Corrosion: Corrosion is the name given to the processes which take place when a metal reacts slowly with air, water or any other substance in the environment.

18. Extraction of metals from their ores: The method used to extract a metal from its ore depends on the position of the metal in the electrochemical series. Reactive metals, such as sodium, calcium, magnesium and aluminium, are difficult to extract from their ores. The metal must be extracted using an electrochemical process called electrolysis.

19. Transition metals: The transition metals are strictly defined as those elements having a partially filled d or f subshell in any common oxidation state.

20. Extraction of iron: Iron ore is smelted in a furnace to reduce impurities in much the same way today as in early Roman times. The iron ore is reduced to iron using coke.

APPENDIX 1. PHYSICAL QUANTITIES

The Seven SI Base Units

There are seven base SI units from which units may be derived for all other physical quantities.

Table A The seven SI base units

Physical quantity	Symbol for quantity	Base unit	Symbol for base unit
length	l	metre	m
mass	m	kilogram	kg
time	t	second	s
electric current	I	ampere	A
thermodynamic temperature	T	kelvin	K
amount of substance	n	mole	mol
luminous intensity	l_v	candela	cd

Derived SI Units

The recommended names and symbols of some SI derived units are given in Table B. It should be noted that other SI derived units are often recommended for general use. For example cm^3, dm^3 or L(litre) is used for volume and $mol\ L^{-1}$ is used for concentration.

Table B Some SI derived units

Physical quantity		Derived unit		
Name	*Symbol*	*Name*	*Symbol*	*Other Units*
volume	V	cubic metre	m^3	cm^3, ml, L
density	p	kilograms per cubic metre	$kg\ m^{-3}$	$g\ cm^{-3}$
speed of electromagnetic waves	c	metres per second	$m\ s^{-1}$	
frequency	f	hertz	$Hz(s^{-1})$	
force	F	newton	$N(kg\ m\ s^{-2})$	
pressure	p	pascal	$Pa(kg\ m^{-1}\ s^{-2} = N\ m^{-2})$	
energy	E			
enthalpy	H	joule	J	$kg\ m^2\ s^{-2}$
concentration of solute B	$[B]$	moles per cubic metre	$mol\ m^{-3}$	$mol\ L^{-1}$
elementary charge (of a proton)	e	coulomb	$C(As)$	

Physicochemical Constants

Table C Physicochemical constants

Quantity	Symbol	Value
speed of light in a vacuum	c	2.998×10^8 m s^{-1}
elementary charge (of proton)	e	1.602×10^{-19} C
mass of electron	m_e	9.110×10^{-31} kg
mass of proton	m_p	1.673×10^{-27} kg
mass of neutron	m_n	1.675×10^{-27} kg
atomic mass unit	m_{14}	1.661×10^{-27} kg
Avogadro constant	N_A	6.022×10^{23} mol^{-1}
gas constant	R	8.314 J K^{-1} mol^{-1}
Boltzmann constant	$k = R/L$	1.381×10^{-23} J K^{-1}
Faraday constant	$F = N_A e$	9.648×10^4 C mol^{-1}
Planck constant	h	6.626×10^{-34} J s
molar volume of a gas at s.t.p.	V_m	2.24×10^{-2} m^3 mol^{-1}

APPENDIX 2. RELATIVE ATOMIC MASSES

The relative atomic masses of elements are given correct to the nearest unit, with the exception of chlorine.

Element	Symbol	A_r	Element	Symbol	A_r
Aluminium	Al	27	Molybdenum	Mo	96
Antimony	Sb	122	Neodymium	Nd	144
Argon	Ar	40	Neon	Ne	20
Arsenic	As	75	Nickel	Ni	59
Barium	Ba	137	Niobium	Nb	93
Beryllium	Be	9	Nitrogen	N	14
Bismuth	Bi	209	Osmium	Os	190
Boron	B	11	Oxygen	O	16
Bromine	Br	80	Palladium	Pd	106
Cadmium	Cd	112	Phosphorus	P	31
Caesium	Cs	133	Platinum	Pt	195
Calcium	Ca	40	Potassium	K	39
Carbon	C	12	Praseodumium	Pr	141
Cerium	Ce	140	Rhenium	Re	186
Chlorine	Cl	35.5	Rhodium	Rh	103
Chromium	Cr	52	Rubidium	Rb	86
Cobalt	Co	59	Ruthenium	Ru	101
Copper	Cu	64	Samarium	Sm	150
Dysprosium	Dy	163	Scandium	Sc	45
Erbium	Er	167	Selenium	Se	79
Europium	Eu	152	Silicon	Si	28
Fluorine	F	19	Silver	Ag	108
Gadolinium	Gd	157	Sodium	Na	23
Gallium	Ga	70	Strontium	Sr	88
Germanium	Ge	73	Sulphur	S	32
Gold	Au	197	Tantalum	Ta	181
Hafnium	Hf	179	Tellurium	Te	128
Helium	He	4	Terbium	Tb	160
Holmium	Ho	165	Thallium	Tl	204
Hydrogen	H	1	Thorium	Th	232
Indium	In	115	Thulium	Tm	170
Iodine	I	127	Tin	Sn	119
Iridium	Ir	192	Titanium	Ti	48
Iron	Fe	56	Tungsten	W	184
Krypton	Kr	84	Uranium	U	238
Lanthanum	La	139	Vanadium	V	51
Lead	Pb	207	Xenon	Xe	131
Lithium	Li	7	Ytterbium	Yb	173
Lutetium	Lu	175	Yttrium	Y	89
Magnesium	Mg	24	Zinc	Zn	65
Manganese	Mn	55	Zirconium	Zr	91
Mercury	Hg	201			

APPENDIX 3. IDENTIFICATION OF SOME INORGANIC COMPOUNDS

Flame Tests

Metal ion	Flame colour
Sodium	golden yellow
Potassium	violet
Copper	sharp green
Barium	dull green
Lithium	bright red
Calcium	brick red
Strontium	crimson
Lead	bluish

The following metal ions give no visible colour to the flame:
aluminium, iron, magnesium, nickel, silver, tin, zinc

Tests for Anions

Anion	Test	Observation	Confirmatory test
Carbonate CO_3^{2-}	Add dilute HCl	CO_2 evolved turns limewater milky	Add $MgSO_4 \rightarrow$ white precipitate
Hydrogen carbonate HCO_3^-	Add dilute HCl	CO_2 evolved turns limewater milky	Add $MgSO_4 \rightarrow$ no precipitate
Nitrate NO_3^-	Add $FeSO_4$ + concentrated H_2SO_4	Brown ring	
Chloride Cl^-	Add dilute HNO_3 + $AgNO_3$	White precipitate	Soluble in NH_3 solution
Sulphate SO_4^{2-}	Add $BaCl_2$ or $Ba(NO_3)_2$ solution	White precipitate	Precipitate is insoluble in HCl or HNO_3
Sulphite SO_3^{2-}	Add $BaCl_2$ or $Ba(NO_3)_2$ solution	White precipitate	Precipitate is soluble in HCl or HNO_3
Phosphate PO_4^{3-}	Add NaOH Add $Ba(NO_3)_2$ solution	White precipitate	

APPENDIX 4. SOLUBILITY OF SOME COMMON SUBSTANCES IN COLD WATER

Anion Cation:	carbonate	chloride	hydroxide	nitrate	sulphate
calcium	I	sss	s	sss	s
copper	I	ss	I	sss	ss
iron(II)	I	sss	I	–	ss
iron(III)	I	sss	–	sss	ss
lead	I	I	I	ss	I
magnesium	I	ss	I	ss	ss
potassium	sss	ss	sss	ss	ss
silver	–	I	–	ss	s
sodium	ss	ss	sss	sss	ss
zinc	I	sss	I	sss	ss

Key:
I = insoluble or very slightly soluble, will form a precipitate
s = only a little will dissolve
ss = soluble
sss = very soluble
– = unstable compound

APPENDIX 5. OCTANE NUMBERS

Compound	Octane numbers
heptane	0
butane	94
pentane	62
hexane	25
cyclohexane	83
benzene	>100
methyl benzene	>100
2 - methylpropane	>100
2 - methylbutane	93
2, 3 - dimethylbutane	103
2 - methylpentane	73
3 - methylpentane	75
2, 3 - dimethylpentane	91

APPENDIX 6. ORGANIC SYNTHESIS ROUTE MAP (Example)

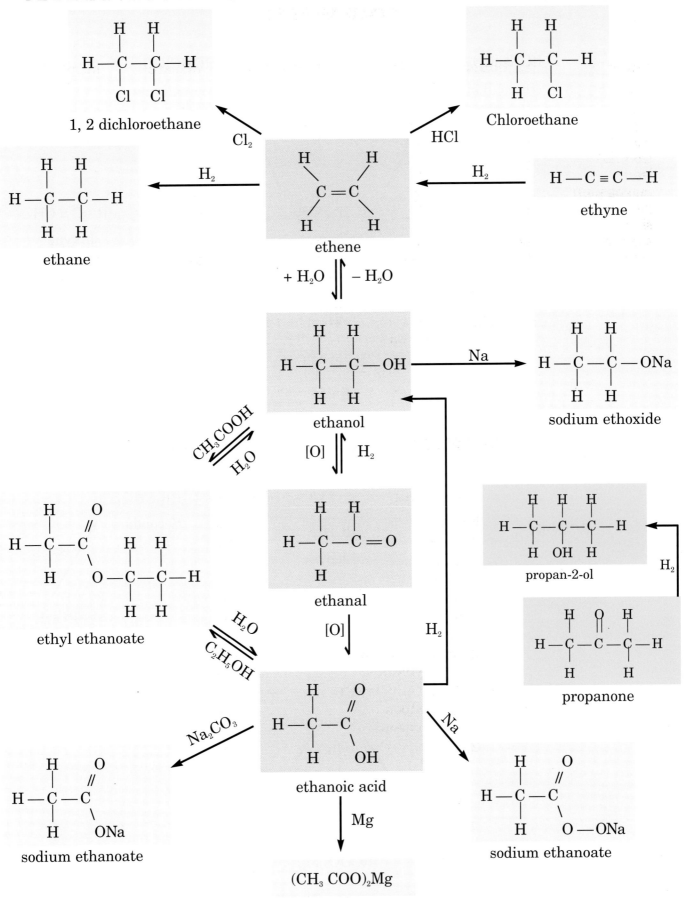

APPENDIX 7. ORGANIC SYNTHESIS ROUTE MAP (GENERAL)

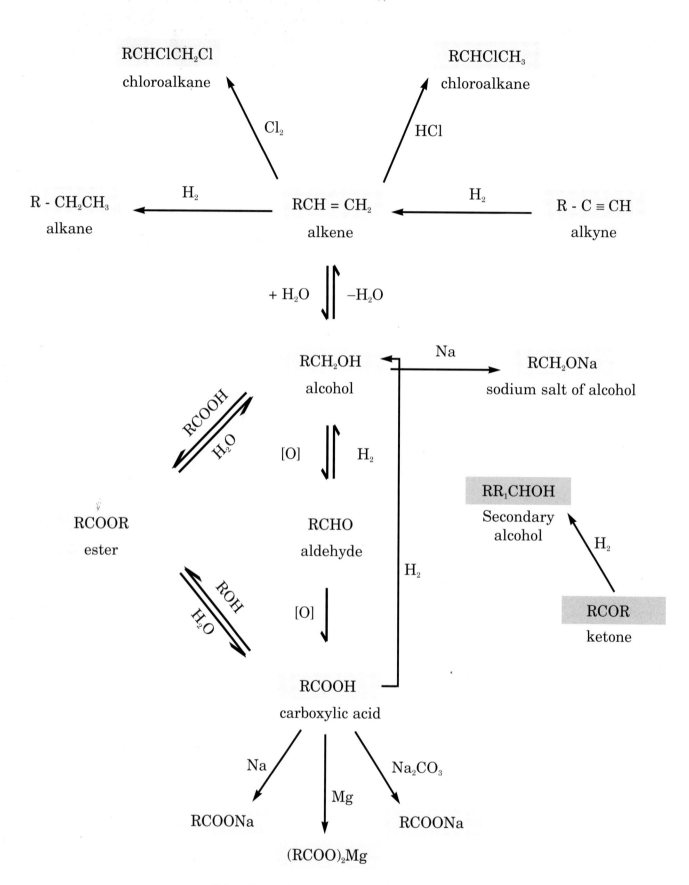

ANSWERS TO PROBLEMS
Unit 1

1.2 (14) (c) 69.79; Ar(Ga) = 69.7.

(15) Ar(O) = 16.0; Ar(Si) = 28.1;
Ar(Mg) = 24.3.

(16) Ar(Ir) = 192.2; Ar(Pb) = 207.2.

1.3 (8) $^{241}_{95}Am \rightarrow {}^{237}_{93}Np + \alpha$

(a) 237 (b) 93 (c) Np

(9) $^{238}_{92}U \rightarrow {}^{234}_{90}Th + \alpha$

(a) 234 (b) 90 (c) 90

(10) $^{14}_{6}C \rightarrow C + \beta$

(b) 6 (c) C

1.4 (34) 114 pm.

(35) 232 pm.

Unit 2

2.4 (1)(b) (i) 1.9 (ii) 0.4 (iii) 0.9 (iv) 1.0, (v) 0 (vi) 0 (vii) 2.1 (viii) 2.3 (ix) 3.2.

2.6 (1)(a) (i) 0 (ii) −1 (iii) +1 (iv) +2 (v) −2 (vi) +1.

(1)(b) (a) +7 (b) +7 (c) +6 (d) +2 (e) +2 and +3 (f) +2 (g) +5 (h) +5, (i) +3 (j) −3.

(3)(a) +3 (b) +5.

(5)(a) (i) +7 (ii) +6 (iii) +6 (d) +2.

(6)(a) +5 (b) +3 (c) +5.

Unit 3

3.2 2(i) 28 (ii) 34 (iii) 28 (iv) 44 (v) 63 (vi) 158 (vii) 74.5 (viii) 162.

(4)(a) 119.5 g mol^{-1} (b) 74.5 g mol^{-1} (c) 23 g mol^{-1} (d) 61 g mol^{-1} (e) 74 g mol^{-1} (f) 1 g mol^{-1} (g) 34 g mol^{-1}.

(5)(a) 180 g mol^{-1} (b) 60 g mol^{-1} (c) 246 g mol^{-1} (d) 60 g mol^{-1} (e) 158 g mol^{-1} (f) 148 g mol^{-1} (g) 160 g mol^{-1}.

(6)(a) 0.05 mol (b) 0.025 mol (c) 0.25 mol (d) 1.0 mol (e) 0.5 mol (f) 0.5 mol (g) 0.0125 mol.

(7)(a) 0.25 mol (b) 1.31 mol (c) 0.15 mol(d) 0.2 mol.

(8)(a) 0.5 mol H_2O = 9 g; 2 mol He = 8 g.

(b) 2 mol C_2H_5OH = 92 g; 5 mol H_2O = 90 g.

(c) 2.5 mol NaCl = 146.25 g; 1.25 mol $CaCO_3$ = 125 g.

(9)(a) 6.35 g, 115 g, 54 g, 192 g.

(b) 46 g, 80 g, 24 g, 152.5 g.

(c) 61.6 g, 8.5 g, 1024 g, 69 g.

(d) 400 g, 408 g, 441 g.

(10)(a) 1 mol N and 3 mol H (b) 2 mol N, 4 mol H, 1 mol C and 1 mol O (c) 1 mol Na, 1 mol N and 3 mol O (d) 2 mol N (3) 2 mol N and 4 mol H.

(11)(a) 1 mol Co, 2 mol Cl, 12 mol H and 6 mol O.

(b) 1 mol Cu, 1 mol S, 12 mol H and 10 mol O.

(c) 1 mol Mg, 1 mol S, 14 mol H and 11 mol O.

(12)(a) 180 g mol^{-1} (b) 0.0028 mol.

(13)(a) 2.5 mol (b) 0.217 mol (c) 0.1 mol (d) 0.83 mol.

(14)(a)(i) 1.2×10^{24} (ii) 6×10^{22} (iii) 9×10^{23} (iv) 6×10^{23} (v) 3×10^{23} (vi) 1.5×10^{22} (b) (i) 4.2×10^{24} (ii) 2.4×10^{24} (iii) 8.1×10^{24} (iv) 6×10^{24} (v) 1.2×10^{24} (vi) 2×10^{23}.

(15)(a) 2.73×10^{22} (b) 8.6×10^{22} (c) 5.2×10^{22} (d) 4×10^{22} (e) 2.38×10^{22} (f) 7.5×10^{22}.

(16) 2.4×10^{23} mol H, 1.2×10^{23} mol O.

(17) 3.07×10^{20} molecules.

(18)(a) 3×10^{23}, (b) 6.66×10^{22} and (c) 3×10^{22}.

(19)(a) 0.033 mol (b) 1666.7 mol (c) 2.5×10^{-12} mol (d) 2.92×10^{-9} mol (e) 5×10^{-6} mol (f) 0.066 mol (g) 2.92×10^{-9} mol (h) 3.33×10^{-4} mol.

(20)(c) 0.15 mol (d) 2.3×10^{23} atoms.

(21)(a) 0.896 L (b) 0.056 L (c) 2.24 L (d) 6×10^{22} atoms (e) 30 g mol^{-1}.

3.3 (2)(b) 186.5 cm^3 (c) 465 cm^3.

(4) 2.38×10^{4} Pa.

(5) 3.63×10^{6} Pa.

(6)(a) 3.68 cm^3 (b) 3.41 cm^3.

(7)(a) 40.2 cm^3 (b) 112 kPa.

(8) 34.94 L.

(9) 16.6 L.

(10) 7.63 L.

(11) 8605%.

(12)(b) 0.009 mol (c) 0.5 mol.

(13)(d) 2.39 L (e) 1.2 g.

(14) 47 g mol^{-1}: ethanol

(15) 49.9; 58

(16) 8.036 J K^{-1} mol^{-1}(8.314 J K^{-1} mol^{-1}).

(17)(c) 58.2.

3.4 (1) C_3H_6O.

(2)(a) CH_2O (b) CH_4 (c) $C_3H_4O_3$ (d) CH_2O (e) CH (f) CH.

(3) Na_2O.

(4) $C_3H_{10}N_2$.

(5) C_6H_8.

(6) N_2H_4CO.

(7)(a) 82.35% (b) 22.22% (c) 35.0% (d) 30.43% (e) 30.43% (f) 21.21% (g) 46.66%.

(8)(a) 39.3% Na, 60.7% Cl (b) 69.6% K, 28.6% O, 1.8% H (c) 2% H, 32.7% S, 65.3% O (d) 54.1% Ca, 43.2% O, 2.7% H (e) 40% C, 6.67% H, 53.33% O (f) 40% Ca, 12% C, 48% O (g) 17.6% Na, 39.7% Ca, 42.7% O.

(9)(a) 62.9% (b) 36.1% (c) 51.1% (d) 45.5%.

(10) Not quite; $C_{17.4}H_{26}N$.

(11) CH_2.

(12)(a) $C_7H_6O_2$ (b) $C_9H_8O_4$ (c) CF_2.

(13)(a) C_6H_6 (b) $C_6H_{12}O_6$ (c) $C_2H_4O_2$ (d) $C_{10}H_{14}N_2$.

(14)(a) $C_{18}H_{22}O_2$ (b) $C_{20}H_{25}ON_3$ (c) CHO_2.

3.5 (7) 71.43 g.

(8) 26.4 g CO_2, 9.9 g H_2O.

(9) 9200 g.

(10)(a) 550 g (b) 10.8 kg (c) 225 g.

(11) 0.125 mol Fe_3O_4.

(12)(a) 20.41 mol (b) 653 g.

(13)(a) 141.1 g (b) 8 g.

(14) 1.933 kg.

(15)(a) 56 L (b) 33.6 L (c) 134.4 L.

(16) (a) 5.2 g (b) 4.48 L.

(17) 33.6 L.

(18) 156.8 L.

(19) (a) 0.02 mol (b) 0.01 mol (c) CaO (d) CaO.

(20) (a) 0.8 kg O_2 (b) 144000 g S.

(21) (a) 1 mol H_2 (b) 0.5 mol H_2 (c) 1 g H_2
(d) 100 g O_2 (e) 4 kg O_2.

(22) 280 mol O_2 is the limiting reactant.

(23) (a) 2 kg N_2 is the limiting reactant,
(b) 571.4 g H_2.

(24) (a) 130 g Zn is the limiting reactant (b) 8 g S.

(25) (a) 5 mol O_2 (b) 52 g C_2H_2 is the limiting
reactant.

(27) 76.36%.

(28) 69.77%.

(29) 0.15 mol Zn is the limiting reactant;
0.15 mol H_2.

(30) 0.70 mol HCl is the limiting reactant;
0.233 mol $AlCl_3$.

(31) (a) 1020 g Al_2O_3 (b) 264.7 g Al (c) 432 kg Al.

(32) (a)(i) 2 mol Cl_2 (ii) 146 g HCl (b) 79.5%.

Unit 4

4.1(2) (a) 0.5 mol L^{-1} NaOH (b) 0.4 mol L^{-1} Na_2CO_3
(c) 0.2 mol L^{-1} HCl (d) 0.05 mol L^{-1} H_2SO_4
(e) 0.5 mol L^{-1} $NaHCO_3$.

(3) 2 mol L^{-1} NaOH (b) 0.2 mol L^{-1} $Na_2Cr_2O_7$
(c) 0.1 mol L^{-1} HCl (d) 0.1 mol L^{-1} H_2SO_4
(e) 0.25 mol L^{-1} $NaHCO_3$.

(4) (a) Dissolve 21.2 g Na_2CO_3 in 1 L solution.
(b) Dilute 0.04 L (40 cm^3) of 2.5 mol L^{-1} Na_2CO_3
solution into 1 L solution.
(c) Dilute 0.2 L (200 cm^3) of 1.5 mol L^{-1} $NaHCO_3$
solution into 1 L solution.
(d) Dilute 0.4 L (400 cm^3) of 0.5 mol L^{-1} HCl
solution into 1 L solution.
(e) Dilute 0.1 L (100 cm^3) of 1.0 mol L^{-1} H_2SO_4
solution into 1 L solution.

(5) (a) (i) Dilute 8.33 cm^3 of conc. HCl into 1 L
solution.
(ii) Dilute 5.55 cm^3 of conc. H_2SO_4 into 1 L
solution.
(iii) Dilute 6.25 cm^3 of conc. HNO_3 into 1 L
solution.

(b) (i) Dilute 33.3 cm^3 of conc. HCl into 2 L
solution.
(ii) Dilute 22.2 cm^3 of conc. H_2SO_4 into 2 L
solution.
(iii) Dilute 25 cm^3 of conc. HNO_3 into 2 L
solution.

(6) (a) 0.2 L (200 cm^3) of 0.1 mol^{-1} NaCl solution.
(b) 0.375 L (375 cm^3) of 1.5 mol L^{-1} NaOH
solution.
(c) 0.025 L (25 cm^3) of 0.5 mol L^{-1} H_2SO_4
solution.

(7) 0.05 L (50 cm^3) of stock solution.

(8) 0.08 L (80 cm^3) of stock HCl solution in 2 L
solution.

4.1(9) (a) 5 mol NaOH (b) 0.5 mol HCl
(c) 0.1 mol H_2SO_4.

(10) (a) 2.5 g NaOH dissolved in 200 cm^3 solution.
(b) 4.2 cm^3 of 12 mol L^{-1} HCl dissolved in
100 cm^3 solution.
(c) 2.12 g anhydrous Na_2CO_3 dissolved in
200 cm^3 solution.

(11) 50 cm^3 ethanol in 200 cm^3 solution.

(12) Add 50 g NaCl to 150 g H_2O.

(13) (a) Dilute 20 cm^3 of 100% w/v into 100 cm^3
solution.
(b) Add 10 g sucrose to 100 cm^3 solution.
(c) Add 25 cm^3 methanol to 75 cm^3 ethanol.
(d) Add 25 mg $CaCO_3$ to 1 L solution.

(14) 1290 ppm.

(15) Yes, 0.0042 ppm < 0.01 ppm.

(16) No, 160 ppm < 200 ppm.

(17)(a) 5.43 mol L^{-1} (b) 8.7 mol L^{-1}.

(19) 3.48 mol L^{-1}.

4.3(a)(5)(i) 12.5 cm^3.

(6)(a) 0.086 mol L^{-1} (b) 3.44 g L^{-1}.

(7) 0.25 mol L^{-1}.

(8)(a) 0.048 mol L^{-1} (b) 1.74 g L^{-1}.

(9)(a) 0.022 mol L^{-1} (b) 0.08% w/w.

(10) 6.3 g $NaHCO_3$.

(11)(a) 50 cm^3 (b) 600 cm^3 (c) 20 L (d) 312.5 cm^3
(e) 250 cm^3 (f) 125 cm^3.

(12) 0.179 mol L^{-1} HCl.

(13) 0.304 mol L^{-1} CH_3COOH.

(14) (c) 0.4125 mol L^{-1} (d) 2.475 w/w %.

(15) (a) 63% (b) 10

4.3(b)(3) (e) 0.0196 mol L^{-1}.

(4) x = 7 moles.

(5)(a) 0.0992 mol L^{-1} (b) 95.8%.

(6) 0.0984 mol L^{-1} Fe^{2+} solution.

(7) (d) 0.1282 mol L^{-1}.

(8) (a) 0.0533 mol L^{-1} (b) 13.54 g L^{-1}.

(9) (d) 0.0905 mol L^{-1} (e) 45% H_2O (f) x = 7.

(10) 44.7%.

(12) (a) 0.0294 mol L^{-1} (b) 2.1903 g L^{-1}.
(13) (g)(i) 0.0488 mol^{-1} (ii) 0.488 mol L^{-1} (iii) 3.64%.
(14) 3.4% Cl$_2$.

Unit 5

5.4 (9) (c) –3800 kJ mol^{-1}.
(15) –1124 kJ mol^{-1}.
(16) –233 kJ mol^{-1}.
(17) 88 kJ mol^{-1}.
(18) –276 kJ mol^{-1}.
(19) (a) 220 kJ mol^{-1} (b) –731 kJ mol^{-1}
    (c) –286 kJ mol^{-1} (d) –196 kJ mol^{-1}
    (e) –92 kJ mol^{-1} (f) –104 kJ mol^{-1}
(20) –56.7 kJ mol^{-1}.

Unit 6

6.1 (3) B
(5) (a) After 7 minutes (f) 1.0 minute
    (g) –0.02 g min^{-1}.
(6) (b) 2.5 g (e) at 16 minutes (f) after 16 minutes
    (g) 0.0112 mol O$_2$ (h) 0.4464 mol dm^{-3}
    (i) 10.2 cm^3 min^{-1}; initial rate was 80 cm^3 min^{-1}.
(7) (f) 0 –10 sec. (g) at 80 sec. (h) 126 cm^3
    (i) 1.575 cm^3 s^{-1} (j) 0.52 cm^3 s^{-1}.
(8) (a) 0.64 cm^3 s^{-1} (b) 0.2286 cm^3 s^{-1} (c) 0.178 cm^3 s^{-1}
6.2 (10) (d) 600 cm^3 s^{-1}.
(11) (d) 66.6 sec.
(12) (b) (i) 37 s^{-1} (ii) 4 s^{-1} (c) 0.046 g S.
(13) (d) 100 sec.
(14) (d) 64 cm^3; 75 cm^3 (e) 8.6 cm^3 min^{-1}
    (f) 8.25 cm^e min^{-1}.
(15) (b) 170 sec (c) 33°C (e) approx. 100 sec.

Unit 7

7.3d (10) (g) 10.4 g (h) 48%.
(11) (e) (i) Na$_2$Cr$_2$O$_7$.2H$_2$O (ii) 50%.
(12) (f) C$_2$H$_5$OH (g) 88%.
(13) (a) CH$_3$COOH (b) 70%.

Unit 8

8.1 (4) 33.33 mol^{-1} L.
(5) 0.33 mol C$_2$H$_5$OH.
(6) 0.0253 mol L^{-1} HCOOH.
(8) Approx. 0.2 mol L^{-1} CO, 0.2 mol L^{-1} and
    O mol L^{-1} HCOOH; reaction is nearly 100% to
    R.H.S.
(9) 0.0817 mol L^{-1}.
(10) 1.2 mol L^{-1}NO.
(11) 0.0218 mil L^{-1}.
(12) 10.56 mol^{-2} L^2.
(13) 0.243 mol L^{-1} H$_2$, 0.043 mol L^{-1}I$_2$,
    0.714 mol L^{-1} HI.
(14) 0.068 mol L^{-1} BrCl.

(15) 0.035 mol L^{-1} Br$_2$, 0.025 mol L^{-1}Cl$_2$,
    0.03 mol L^{-1} BrCl.
(17) 1.5625 x 10^{-3}.
(18) 7.62 x 10^{-3} mol L^{-1}.
(19) 0.248.
(20) 0.003 mol L^{-1} CO, 0.023 mol L^{-1} H$_2$O,
    0.017 mol L^{-1} CO$_2$, 0.017 mol L^{-1} H$_2$.
(21) 0.51 mol L^{-1} H$_2$.
(22) (a) 43.2 (b)(i) 1.07 mol H$_2$, 1.07 mol I$_2$,
    0.86 mol HI (ii) 2.14 g H$_2$, 135.89 g I$_2$,
    110.08 g HI.
(23) (a) 4 (b) 5.16 g CH$_3$COOH, 17.76 g C$_2$H$_5$
    32.03 g CH$_3$COOC$_2$H$_5$, 6.55 g H$_2$O.
(24) 0.0675 mol^2 L^{-2}.
(25) (a) 2.5 mol N$_2$, 7.5 mol H$_2$, 25 mol gas (b)

Unit 9

(5) (d) 240 mg L^{-1}, 240 ppm.
(6) (a) (i) 200 mg L^{-1} (ii) 1600 mg L^{-1}.
(7) (d) 250 ppm.
9.2 (13) 7.2 ppm.
(14) (f) 6.4 x 10^{-3} g L^{-1}; 6.4 ppm.
(16) (i) 10 mg L^{-1} (ii) 60 kg (iii) 6 x 10^{6}L.
(20) (e) 128 ppm.
(21) (d) 184 ppm.
9.3 (3) (a) 2 (b) 1.699 (c) 12.
(4) (a) 12.6021 (b) 0.699.
(6) >1.
(7) 1.8 x 10^{-9}.
(8) 6.631.
(9) [OH$^-$] > [H$^+$] => solution is basic.
(10)(a) 1.0 x 10^{-10} mol L^{-1} (b) 3.9811 x 10^{-10} mol L^{-1}
    (c) 0.0398 mol L^{-1} (d) 1.5136 x 10 L^{-1}
    (e) 0.8913 mol L^{-1} (f) 1.9952 x 10^{-8} mol L^{-1}.
(11) 2.9348.
(12) 2.7218.
(13) 1.3848.
(14) 2.5134.
(15) 11.4225.
(16) 10.1021.
(17) 1.71 x 10^{-5} mol L^{-1}.
(18) 1.58 x 10^{-6} mol L^{-1}.
(19) 5.7143 mol L^{-1}.
(20) 1.2649 x 10^{-4} mol L^{-1}.
(23) 3.9811 x 10^{-8} mol L^{-1}.
(24)(b) 3 x 10^{-7} mol L^{-1}.
    (d) 2.9604.

Option 1

(15) (a) 1.2174 tonnes
    (b) 14 tonnes.
    (c) 16.66 tonnes
    (d) 5.66 tonnes.